Lecture Notes in Physics

For information about Vols. 1–131, please contact your bookseller or Springer-Verlag.

Vol. 132: Systems Far from Equilibrium. Proceedings, 1980. Edited by L. Garrido. XV, 403 pages. 1980.

Vol. 133: Narrow Gap Semiconductors Physics and Applications. Proceedings, 1979. Edited by W. Zawadzki. X, 572 pages. 1980.

Vol. 134: γγ Collisions. Proceedings, 1980. Edited by G. Cochard and P. Kessler. XIII, 400 pages. 1980.

Vol. 135: Group Theoretical Methods in Physics. Proceedings, 1980. Edited by K. B. Wolf. XXVI, 629 pages. 1980.

Vol. 136: The Role of Coherent Structures in Modelling Turbulence and Mixing. Proceedings 1980. Edited by J. Jimenez. XIII, 393 pages. 1981.

Vol. 137: From Collective States to Quarks in Nuclei. Edited by H. Arenhövel and A. M. Saruis. VII, 414 pages. 1981.

Vol. 138: The Many-Body Problem. Proceedings 1980. Edited by R. Guardiola and J. Ros. V, 374 pages. 1981.

Vol. 139: H. D. Doebner, Differential Geometric Methods in Mathematical Physics. Proceedings 1981. VII, 329 pages. 1981.

Vol. 140: P. Kramer, M. Saraceno, Geometry of the Time-Dependent Variational Principle in Quantum Mechanics. IV, 98 pages. 1981.

Vol. 141: Seventh International Conference on Numerical Methods in Fluid Dynamics. Proceedings. Edited by W. C. Reynolds and R. W. MacCormack. VIII, 485 pages. 1981.

Vol. 142: Recent Progress in Many-Body Theories. Proceedings. Edited by J. G. Zabolitzky, M. de Llano, M. Fortes and J. W. Clark. VIII, 479 pages. 1981.

Vol. 143: Present Status and Aims of Quantum Electrodynamics. Proceedings, 1980. Edited by G. Gräff, E. Klempt and G. Werth. VI, 302 pages. 1981.

Vol. 144: Topics in Nuclear Physics I. A Comprehensive Review of Recent Developments. Edited by T.T.S. Kuo and S.S.M. Wong. XX, 567 pages. 1981.

Vol. 145: Topics in Nuclear Physics II. A Comprehensive Review of Recent Developments. Proceedings 1980/81. Edited by T. T. S. Kuo and S. S. M. Wong. VIII, 571-1.082 pages. 1981.

Vol. 146: B. J. West, On the Simpler Aspects of Nonlinear Fluctuating. Deep Gravity Waves. VI, 341 pages. 1981.

Vol. 147: J. Messer, Temperature Dependent Thomas-Fermi Theory. IX, 131 pages. 1981.

Vol. 148: Advances in Fluid Mechanics. Proceedings, 1980. Edited by E. Krause. VII, 361 pages. 1981.

Vol. 149: Disordered Systems and Localization. Proceedings, 1981. Edited by C. Castellani, C. Castro, and L. Peliti. XII, 308 pages. 1981.

Vol. 150: N. Straumann, Allgemeine Relativitätstheorie und relativistische Astrophysik. VII, 418 Seiten. 1981.

Vol. 151: Integrable Quantum Field Theory. Proceedings, 1981. Edited by J. Hietarinta and C. Montonen. V, 251 pages. 1982.

Vol. 152: Physics of Narrow Gap Semiconductors. Proceedings, 1981. Edited by E. Gornik, H. Heinrich and L. Palmetshofer. XIII, 485 pages. 1982.

Vol. 153: Mathematical Problems in Theoretical Physics. Proceedings, 1981. Edited by R. Schrader, R. Seiler, and D.A. Uhlenbrock. XII, 429 pages. 1982.

Vol. 154: Macroscopic Properties of Disordered Media. Proceedings, 1981. Edited by R. Burridge, S. Childress, and G. Papanicolaou. VII, 307 pages. 1982.

Vol. 155: Quantum Optics. Proceedings, 1981. Edited by C.A. Engelbrecht. VIII, 329 pages. 1982.

Vol. 156: Resonances in Heavy Ion Reactions. Proceedings, 1981. Edited by K.A. Eberhard. XII, 448 pages. 1982.

Vol. 157: P. Niyogi, Integral Equation Method in Transonic Flow. XI, 189 pages. 1982.

Vol. 158: Dynamics of Nuclear Fission and Related Collective Phenomena. Proceedings, 1981. Edited by P. David, T. Mayer-Kuckuk, and A. van der Woude. X, 462 pages. 1982.

Vol. 159: E. Seiler, Gauge Theories as a Problem of Constructive Quantum Field Theory and Statistical Mechanics. V, 192 pages. 1982.

Vol. 160: Unified Theories of Elementary Particles. Critical Assessment and Prospects. Proceedings, 1981. Edited by P. Breitenlohner and H.P. Dürr. VI, 217 pages. 1982.

Vol. 161: Interacting Bosons in Nuclei. Proceedings, 1981. Edited by J.S. Dehesa, J.M.G. Gomez, and J. Ros. V, 209 pages. 1982.

Vol. 162: Relativistic Action at a Distance: Classical and Quantum Aspects. Proceedings, 1981. Edited by J. Llosa. X, 263 pages. 1982.

Vol. 163: J. S. Darrozes, C. Francois, Mécanique des Fluides Incompressibles. XIX, 459 pages. 1982.

Vol. 164: Stability of Thermodynamic Systems. Proceedings, 1981. Edited by J. Casas-Vázquez and G. Lebon. VII, 321 pages. 1982.

Vol. 165: N. Mukunda, H. van Dam, L.C. Biedenharn, Relativistic Models of Extended Hadrons Obeying a Mass-Spin Trajectory Constraint. Edited by A. Böhm and J.D. Dollard. VI, 163 pages. 1982.

Vol. 166: Computer Simulation of Solids. Edited by C.R.A. Catlow and W.C. Mackrodt. XII, 320 pages. 1982.

Vol. 167: G. Fieck, Symmetry of Polycentric Systems. VI, 137 pages, 1982.

Vol. 168: Heavy-Ion Collisions. Proceedings, 1982. Edited by G. Madurga and M. Lozano. VI, 429 pages. 1982.

Vol. 169: K. Sundermeyer, Constrained Dynamics. IV, 318 pages. 1982.

Vol. 170: Eighth International Conference on Numerical Methods in Fluid Dynamics. Proceedings, 1982. Edited by E. Krause. X, 569 pages. 1982.

Vol. 171: Time-Dependent Hartree-Fock and Beyond. Proceedings, 1982. Edited by K. Goeke and P.-G. Reinhard. VIII, 426 pages. 1982.

Vol. 172: Ionic Liquids, Molten Salts and Polyelectrolytes. Proceedings, 1982. Edited by K.-H. Bennemann, F. Brouers, and D. Quitmann. VII, 253 pages. 1982.

Lecture Notes in Physics

198

Recent Progress in Many-Body Theories

Proceedings of the Third International Conference
on Recent Progress in Many-Body Theories
Held at Odenthal-Altenberg, Germany
August 29 – September 3, 1983

Edited by H. Kümmel and M. L. Ristig

Springer-Verlag
Berlin Heidelberg GmbH 1984

Editors

H. G. Kümmel
Institut für Theoretische Physik, Ruhr-Universität Bochum
Postfach 10 21 48, D-4630 Bochum 1

M. L. Ristig
Institut für Theoretische Physik, Universität zu Köln
Zülpicher Straße 77, D-5000 Köln 41

ISBN 978-3-540-12924-0

Library of Congress Cataloging in Publication Data. International Conference on Recent Progress in Many-Body Theories (3rd: 1983: Odenthal, Germany) Recent progress in many-body theories. (Lecture notes in physics; 198) 1. Many-body problem–Congresses. I. Kümmel, H. (Hermann), 1922-. II. Ristig, M. L. (Manfred L.) III. Title. IV. Series.
QC174.17.P7I57 1983 530.1'44 84-5395
ISBN 978-3-540-12924-0 ISBN 978-3-540-38808-1 (eBook)
DOI 10.1007/978-3-540-38808-1

Originally published by Springer-Verlag Berlin Heidelberg New York Tokyo in 1984

2153/3140-543210

ORGANIZING COMMITTEE

C.E. Campbell (Minneapolis)

H. Kümmel (Bochum)

M.L. Ristig (Köln)

S. Rosati (Pisa)

SESSION CHAIRMEN

N.W. Ashcroft (Cornell)

L.W. Bruch (Madison)

C.E. Campbell (Minneapolis)

J.W. Clark (Washington U.)

I.M. Khalatnikov (Moscow)

H. Kümmel (Bochum)

C. Mahaux (Liege)

S.A. Moszkowski (UCLA)

J.K. Percus (Courant)

S. Rosati (Pisa)

K.S. Singwi(Evanston)

LIST OF SPEAKERS

J. Arponen (Helsinki)

N.W. Ashcroft (Cornell)

Y. Avishai (Beer-Sheva)

K.S. Bedell (Stony Brook)

O. Benhar (Rome)

R.F. Bishop (Manchester)

L.W. Bruch (Madison)

T. Chakraborty (Minneapolis)

J.W. Clark (Washington U.)

F. Coester (Argonne)

K. Emrich (Bochum)

A. Fabrocini (Pisa)

S. Fantoni (Pisa)

B. Friman (Urbana)

V. Goldman (Amsterdam)

F. Green (New South Wales)

R. Guardiola (Granada)

P. Haensel (Warsaw)

A. Kallio (Oulu)

R. Kishore (S.P. Brazil)

E. Krotscheck (Heidelberg)

K.E. Kürten (Courant)

W. Kutzelnigg (Bochum)

A. Lande (Groningen)

L. Lantto (Oulu)

M.A. Lee (Kent State)

A. Lejeune (Liege)

C. Lhuillier (Paris)

M. de Llano (Mexico)

E.V. Ludena (Caracas)

C. Mahaux (Liege)

M. Miller (Washington State)

H.J. Monkhorst (Gainsville)

H. Müther (Tübingen)

J.W. Negele (MIT)

L. Nosanow (NSF)

C.J. Pethick (Urbana, Nordita)

W. Piechocki (Warsaw)

S.C. Pieper (Argonne)

D. Pines (Urbana)

D. Rainer (Bayreuth)

L. Reatto (Parma)

D. Sandler (Urbana)

K.S. Singwi (Evanston)

R.A. Smith (Texas A & M)

S. Szprynger (Wroclaw)

T. Takatsuka (Morioka)

J.P. Vary (Iowa)

P. Vashishta (Argonne)

R. Wiringa (Urbana)

Chia-Wei Woo (San Diego and San Francisco)

Lu Yu (Beijing)

J.G. Zabolitzky (Köln)

Table of Contents

PREFACE

The Third International Conference on Recent Progress in Many-Body Theories was held at Odenthal-Altenberg, Germany, August 29 to September 3, 1983. About one hundred and thirty scientists from many countries met in the pleasant and stimulating environment and beautiful surroundings of Haus Altenberg with its excellent lecturing facilities. This meeting, like the previous ones held at Trieste, Italy, in 1978 and Oaxtepec, Mexico, in 1981, brought together an exciting group of physicists working in various fields ranging from elementary particle physics, nuclear physics, quantum fluids, solid-state and statistical physics to quantum chemistry. The broad spectrum of interests documented the fact that one of the major aims of this series of conferences has been successfully achieved: to strengthen the exchange and collaboration among physicists doing research in what sometimes seem to be disparate fields. Talks and discussions focused on Monte-Carlo, perturbative and variational approaches to nuclear and elementary particle systems, quantum liquids, electronic systems and the fast developing new area of polarized quantum fluids.

This volume presents the lecture notes on these topics delivered by the invited speakers and a summary by L.H. Nosanow. To facilitate rapid publication we have kept the editing of the papers to a minimum.

The success of the meeting could not have been possible without the great help of many interested participants, the advisory committee, coworkers, secretaries, and the staff of Haus Altenberg. We are very grateful to the Deutsche Forschungsgemeinschaft and the Ministerium für Wissenschaft und Forschung des Landes Nordrhein-Westfalen for financial support.

Bochum and Köln, Germany
December 1983

H. Kümmel   M.L. Ristig

QUARK CLUSTERS IN NUCLEI

James P. Vary†
Arizona Research Laboratories, University of Arizona, Tucson, Arizona
and Physics Department, Iowa State University, Ames, Iowa

and

Hans J. Pirner
Institut für Theoretische Physik, Universität Heidelberg, and
Max Planck Institut für Kernphysik, Heidelberg, FRG

Deep inelastic lepton scattering has proven to be a powerful tool for discovering the substucture of the hadrons. In the same way, we expect that deep inelastic scattering (DIS) from nuclei reveals how that substructure is modified by the nuclear medium. We have presented a model[1] which successfully explained DIS on ^{3}He[2] in terms of 3, 6, and 9-quark clusters. We have recently improved the model by incorporating more realistic nuclear wave functions to determine the effects of Fermi motion and to evaluate the geometrical overlap probabilities of cluster formation.[3] Furthermore, we have included the contributions of sea quarks[4] which are important for understanding the new data with iron targets.[5,6] We have also extended the model to obtain a description of the elastic charge form factor of ^{3}He[7].

Here, we summarize the principal ingredients of the model and present illustrative results for DIS. For a lepton of incident lab energy E, final energy E' at lab scattering angle θ, we define the lab energy loss $\nu = E-E'$ and the negative of the invariant four-momentum transfer squared, $Q^2 = 4EE'\sin^2\theta/2$. The scattering occurs from a target nucleus of A nucleons, N neutrons, and Z protons whose mass M is approximated as Am, where m is the nucleon mass. We employ the Bjorken variable $x \equiv Q^2/2m\nu$ which has values $0<x<A$ and the Nachtmann variable $\xi \equiv 2x[1+(1+Q^2/\nu^2)^{1/2}]^{-1}$ which goes over to x as $Q^2\to\infty$, $\nu\to\infty$ but x fixed. The variable ξ accounts[8] for finite mass corrections to scaling which are important over the range of the ^{3}He data[2] and the low Q^2 Fe data[6], but which are not important for the high Q^2 Fe data.[5]

We write the total invariant structure function for the nucleus as

$$\frac{\nu}{\sigma_M}\frac{d^2\sigma}{d\Omega dE'} = \nu W_2^{tot} = \nu W_2^{in} + \nu W_2^{q-el} , \tag{1}$$

where the Mott cross section $\sigma_M = 4\alpha^2(E')^2Q^{-4}\cos^2\theta/2$; α is the fine structure constant; and the inelastic structure function

$$\nu W_2^{in} = \sum_{\substack{\text{quarks}\\ j}} e_j^2 \frac{\xi}{A} \mathcal{P}(\xi) . \tag{2}$$

The function $\mathcal{P}(\xi)$ is the distribution of quarks in the nucleus with momentum fraction ξ/A of the total nuclear light-cone momentum $P^+ = E+P_z$. In our quark cluster model[1] we assume the quarks are found in an i-quark (i-q) cluster (i=3,6,9,...,3A) within the nucleus with probability $\tilde{p}_i$ so that

$$\mathcal{P}(\xi) = \sum_{\substack{\text{clusters}\\ i}} \tilde{p}_i \bar{P}_i(\xi) , \tag{3}$$

where $\bar{P}_i(\xi)$ is the ξ-distribution of quarks within the cluster. We then write

$$\bar{P}_i(\xi) = \int_0^{\xi^{th}_{i/A}} dy \int_0^{\xi^{th}_{q/i}} du\, \bar{n}_{q/i}(u)\, N_{i/A}(y)\, \delta\!\left(\frac{u}{i/3} \cdot \frac{y}{A} - \frac{\xi}{A}\right) , \tag{4}$$

where $\bar{n}_{q/i}(u)$ is the distribution of quarks in the i-q cluster with momentum fraction $u(i/3)^{-1}$, $N_{i/A}(y)$ is the distribution of i-q clusters in the nucleus with momentum fraction y/A, and the δ function guarantees momentum conservation. We take cluster masses $m_i = mi/3$ and define the thresholds by

$$\xi^{th}_{i/A} = A \bullet \frac{\left(1 + \frac{m_i^2}{M^2}\frac{Q^2}{\nu^2}\right)^{1/2} + 1}{\left(1 + \frac{Q^2}{\nu^2}\right)^{1/2} + 1} \tag{5}$$

$$\xi^{th}_{q/i} = \frac{i}{3} \bullet \frac{2}{\left(1 + \frac{4m_i^2}{Q^2}\right)^{1/2} + 1} . \tag{6}$$

Equation (4) may be simplified to read

$$\bar{P}_i(\xi) = A \int_0^{\xi^{th}_{i/A}} dy\, \frac{1}{y}\, n_{q/i}(\xi/y) N_{i/A}(y)\, \theta\left(\frac{i\xi}{3y} - \xi^{th}_{q/i}\right) , \tag{7}$$

where $n_{q/i}(3u/i) = (i/3)\bar{n}_{q/i}(u)$ and θ represents the usual step function. For 3-q clusters we take $n_{q/3}$ distributions determined by QCD and best fits to data.[9] We terminate the cluster sum in Eq. (3) at 9-q clusters, and we take $n_{q/i}$ for i = 6,9 from counting rules and Regge behavior to be:

$$n_{q/6}(z) = 1.762(z)^{-1/2}(1-z)^9 \tag{8}$$

$$n_{q/9}(z) = 2.239(z)^{-1/2}(1-z)^{15} \ . \tag{9}$$

For the Fermi motion of the clusters we take, at present, a simple Gaussian distribution

$$N_{i/A}(y) = \frac{1}{(2\pi)^{1/2}\sigma} \exp\left[- \frac{(y-1/3)^2}{2\sigma^2} \right], \tag{10}$$

with $\sigma = 0.1152$ based upon a simplified analysis of ^{3}He wave functions.[1] Results with more realistic treatment of the Fermi motion in Eq. (7) will be presented elsewhere.[3]

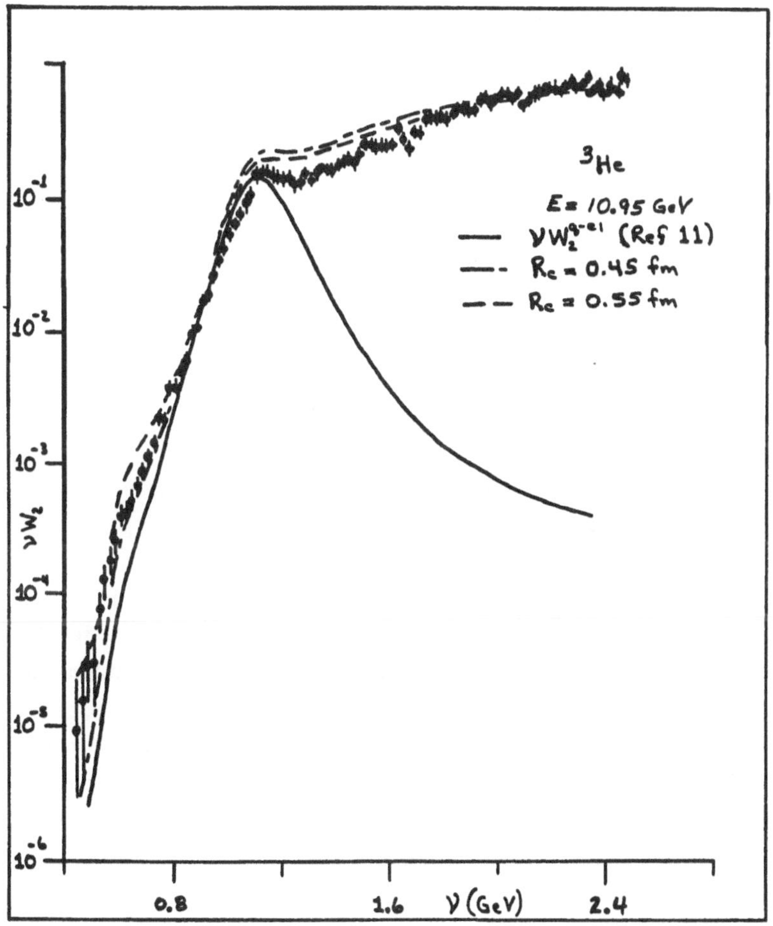

Figure 1

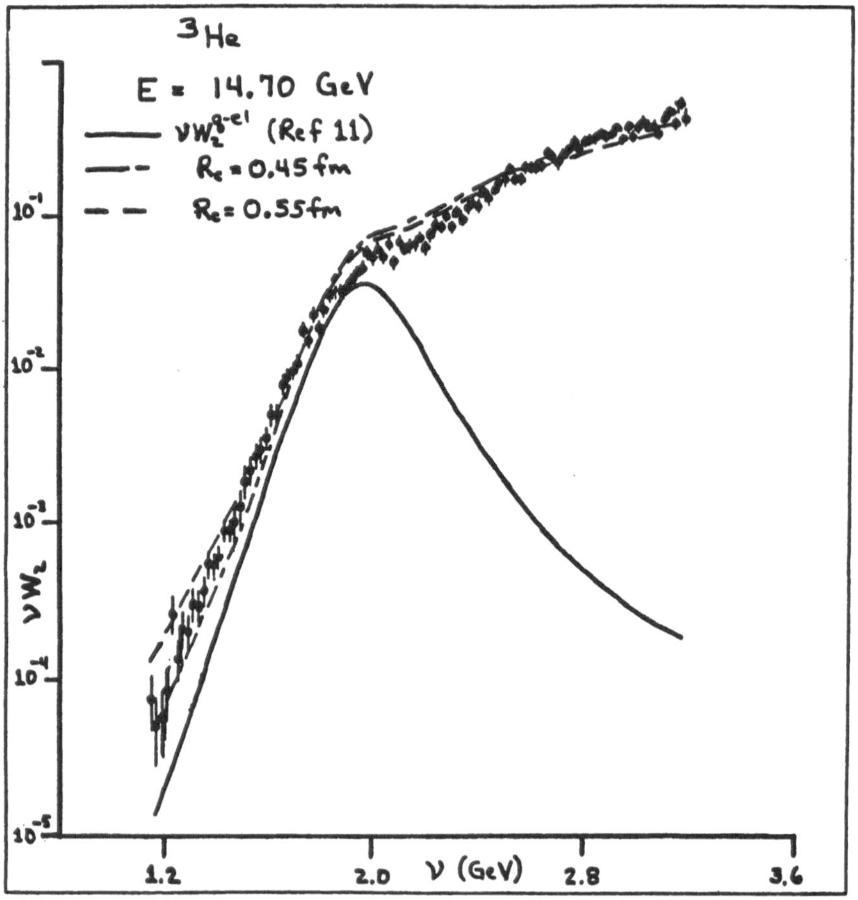

Figure 2

For the ^{3}He data, we may readily use realistic nuclear wave functions to determine the quasi-elastic contribution νW_2^{q-el} in Eq. (1) and to determine the cluster probabilities $\tilde{p}_i$. For these purposes we employ semi-realistic[10] and realistic[11] three-body wave functions for ^{3}He obtained by solutions of the Faddeev equations. For the present analysis of ^{3}He data we will take νW_2^{q-el} to be $\tilde{p}_3$ times the results obtained by the Hannover group.[11] For the probability that a quark is found in an i-q cluster, we compute geometrical overlaps as a function of a critical bag radius R_c[1] using the semi-realistic results of the Los Alamos group.[10] A visual fit to the highest energy ^{3}He data sets is used to determine an optimum R_c value and, hence, the cluster probabilities $\tilde{p}_i$.

We show in Figs. 1 and 2 an envelope of results for DIS on ^3He corresponding to the choices $R_c \approx 0.45$ fm ($\tilde{p}_3 = 0.93$, $\tilde{p}_6 = 0.07$, $\tilde{p}_9 = 0.00$) and $R_c = 0.55$ fm ($\tilde{p}_3 = 0.83$, $\tilde{p}_6 = 0.15$, $\tilde{p}_9 = 0.02$) along with the pure quasi-elastic results of the Hannover group[11] for comparison. In order to fit the low ν (high x or ξ) data, we require a substantial contribution for the 6-q cluster. Based upon the results shown here, we estimate a "best fit" to the data is obtained with this improved version of our model using $R_c = 0.50$ fm and the resulting $\tilde{p}_3 \approx 0.88$, $\tilde{p}_6 \approx 0.11$, and $p_9 \approx 0.01$. Analyses with different choices of semi-realistic nuclear wave functions suggest an uncertainty of ±0.05 fm in R_c and ±0.02 on each

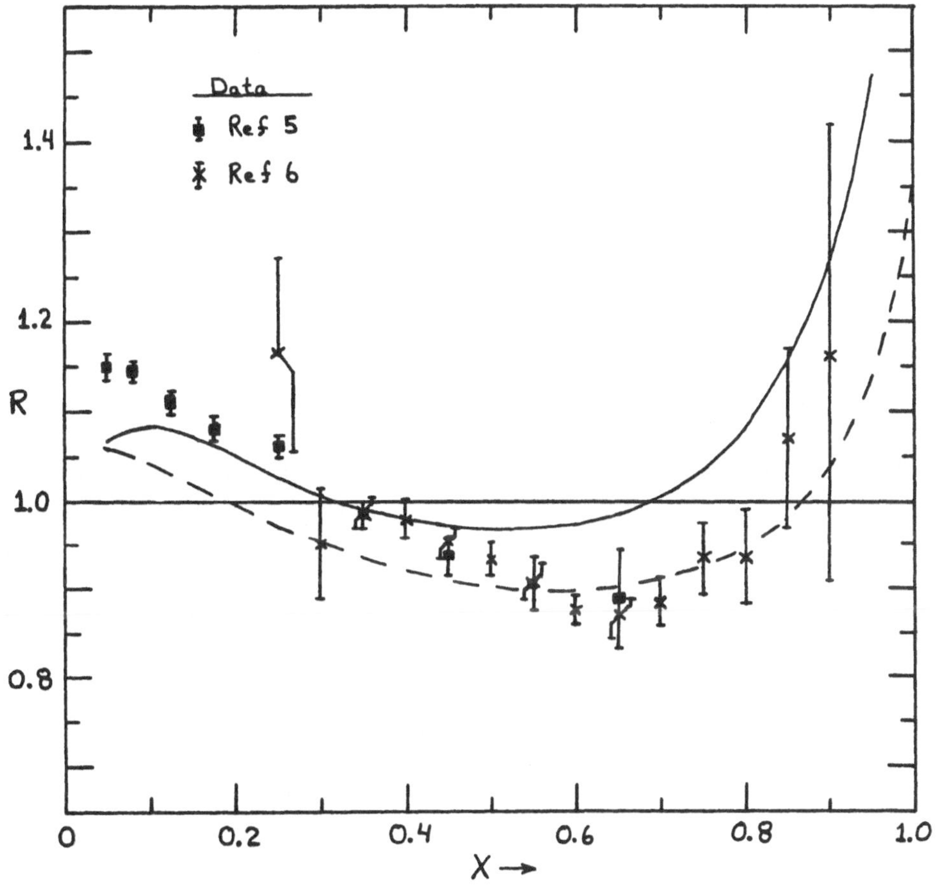

Figure 3

of the $\tilde{p}_i$ values. In particular, the 9-q cluster probability is determined almost entirely by conservation of probability since the 9-q cluster has only small contributions in this range of data. These results differ only slightly in R_c from the earlier fit[1] with the more substantial changes occuring in the $\tilde{p}_i$ and in νW_2^{q-el}. The nearly unchanged R_c reflects an offsetting influence of short-range correlations between $\tilde{p}_i$ and νW_2^{q-el}.

Let us fix the ingredients determined to fit the ^3He data and now address the DIS of leptons from Fe.[4] Here the data[5,6] are presented as a ratio R of cross sections per nucleon of iron to deuterium (D). We assume that R_c is the same in all nuclei and hold it fixed at 0.50 fm but realize that D is less dense than ^3He while Fe is more dense. We estimate the effects of these density differences[3] to yield $(\tilde{p}_3, \tilde{p}_6, \tilde{p}_9) = (0.964, 0.036, 0)$ and $(0.777, 0.180, 0.043)$, respectively for D and Fe. Then if we evaluate R for lepton scattering at E = 100 GeV, $\theta = 8^o$ and include the strange and charm sea contributions[4] appropriate to $Q^2 = 20(\mathrm{GeV/c})^2$ we obtain the solid curve in Fig. 3. The low x region is primarily sensitive to the sea quark contributions. The dip in R below unity is sensitive to the ratio of 6-q cluster contributions in Fe and D. To fit this dip we need to decrease the 6-q cluster probability in D and/or enhance the 6-q cluster probability in Fe. The rise near x = 1 is sensitive to the treatment of Fermi motion which is approximated here as the same in Fe and D. Compare the dashed curve for E = 20 GeV, the sea at $Q^2 = 1.8(\mathrm{GeV/c})^2$ and an arbitarily increased density effect in ^{56}Fe wherein $(\tilde{p}_3, \tilde{p}_6, \tilde{p}_9) = (0.70, 0.22, 0.08)$.

While the quark cluster model is successful in describing the ^3He data, there is a need to improve the description of Fe/D data. Further work on this is in progress.[12]

REFERENCES

†Supported by USDOE contract no. DE-AC02-82ER40068, HENP Division.

[1]H. J. Pirner and J. P. Vary, Phys. Rev. Lett. 46, 1376 (1981); H. J. Pirner and J. P. Vary, Nucl. Phys. A 358, 413C (1981).

[2]D. Day et al., Phys. Rev. Lett. 43, 1143 (1979).

[3]H. J. Pirner and J. P. Vary, to be published.

[4]H. J. Pirner and J. P. Vary, University of Heidelberg preprint UNI-HD-83-02; and to be published.

[5]J. J. Aubert et al., CERN-EP/83-14 (unpublished); Phys. Lett. 105B, 322 (1981).

[6]A. Bodek et al., Phys. Rev. Lett. 50, 1431 (1983).

[7]J. P. Vary, S. A. Coon, and H. J. Pirner, Proceedings of the Tenth International Conference on Few Body Problems in Physics, Karlsruhe, 21-27 August 1983; Proceedings of the International Conference on Nuclear Physics, Florence, 29 August-3 September 1983; and to be published.

[8]O. Nachtmann, Nucl. Phys. B 63, 237 (1973).

[9]A. J. Buras and K. J. F. Gaemers, Nucl. Phys. B 132, 249 (1978).

[10]J. L. Friar, B. F. Gibson, E. L. Tomusiak, and G. L. Payne, Phys. Rev. C 24, 665 (1981); and private communication.

[11]H. Meier-Hajduk, Ch. Hajduk, P. U. Sauer, and W. Theis, Nucl. Phys. A 395, 332 (1983); and private communication.

[12]O. Nachtmann and H. J. Pirner, University of Heidelberg preprint THEP-83-08.

STOCHASTIC SOLUTION OF NUCLEAR MODELS HAVING SUB-NUCLEAR DEGREES OF FREEDOM

J.W. Negele

Massachusetts Institute of Technology, Cambridge, MA 02139 U.S.A.

I. Introduction

Whereas the traditional description of nuclei in terms of static potential interactions between nucleons with perturbative corrections for meson exchange currents can account for observed nuclear structure at low energy, there is presently no fundamental understanding of how such an approximation arises from the underlying QCD picture of a nucleus as a dense fluctuating melange of extended color-singlet clusters of quarks and glue. The goal of this present work, therefore, is to begin to learn how the interactions between sub-nuclear degrees of freedom give rise to simple effective interactions involving only nucleon coordinates. Traditional approximate solutions to the nuclear many-body problem are subject to such uncontrolled errors that in any comparison between solutions of sub- nuclear and nuclear models of the same system, physical effects would be unavoidably entangled with errors arising from the many-body approximations. Therefore, in this work, stochastic methods are used which can solve the many-body problems to any required degree of accuracy.

This paper summarizes the results of an initial sequence of small, controlled steps. Since the stochastic method described in Section II is especially easy to use for many-fermion problems in one spatial dimension, only one-dimensional Fermion models have been considered. After solving a nuclear model with strongly repulsive potential interactions in Section III, we show in Section IV how an adiabatic approximation to a meson-nuclear field theory yields an excellent approximation to the full theory. The application of our stochastic method in Section V to a lattice gauge theory lays the foundation for future treatment of QCD. In Section VI, a confining quark model is solved exactly, and compared with a nuclear model based on a potential obtained from the phase shifts derived from the quark model. This work reported here represents only the modest beginning of a research effort, and the prospects for future extension are assessed in the final section.

II. Stochastic Evaluation of Path Integrals Using Guided Random Walks

Stochastic methods for calculating ground state or thermal properties of many-body systems may be thought of as Monte Carlo evaluations of exact functional integrals for observables, and the different variants correspond to alternative choices for the observable, functional integral, and stochastic sampling techniques.

The properties of physical interest to us may be expressed in terms of specific matrix elements of $e^{-\beta H}$, the imaginary time (Euclidean) evolution operator.[1,2] For example, in the limit of large β, the ground state energy may be written $E = \frac{\langle\phi|He^{-\beta H}|\phi\rangle}{\langle\phi|e^{-\beta H}|\phi\rangle}$ where $|\phi\rangle$ is an approximate wave function. The expectation value of an operator O may be written exactly as $\langle O\rangle = \frac{\langle\phi|e^{-\beta H}Oe^{-\beta H}|\phi\rangle}{\langle\phi|e^{-2\beta H}|\phi\rangle}$ or approximately as $\langle O\rangle = 2\frac{\langle\phi|Oe^{-\beta H}|\phi\rangle}{\langle\phi|e^{-\beta H}|\phi\rangle} - \frac{\langle\phi|O|\phi\rangle}{\langle\phi|\phi\rangle}$. The phase shift for the scattering of two composite particles at energies below the excitation threshold for the first excited state is obtained by imposing the boundary condition that the wave functions vanish at a specified fragment separation distance and evaluating the energy E of the total system as a function of this distance.[3] The fission lifetime may be calculated exactly below the threshold for fragment excitation by calculating the scattering phase shift and approximately for any high barrier by calculating the overlap matrix element V between interior and exterior states[2]

$$\frac{\langle\phi_{int}|e^{\beta H}|\phi_{ext}\rangle}{[\langle\phi_{int}|e^{-\beta H}|\phi_{int}\rangle\langle\phi_{ext}|e^{-\beta H}|\phi_{ext}\rangle]^{\frac{1}{2}}} = \tanh(V\beta).$$

Alernative functional integrals for the evolution operator may be obtained by integrating over an auxiliary field, or inserting at infinitesimal time increments a resolution of unity expressed as an integral over boson coherent states, Grassman coherent states, Slater Determinants, or coordinate eigenstates. Of these alternatives, coordinate eigenstates are appropriate for our purposes.

The essence of the method is illustrated by the calculation of the matrix element $\langle\phi_a|e^{-\beta H}|\phi_b\rangle$ for a single particle in a potential $V(x)$. For an infinitesimal step

$$\begin{aligned}\langle x_n|e^{-\epsilon(\hat{H}-E)}|x_{n-1}\rangle &\approx \int dp\langle x_n|e^{-\epsilon\frac{p^2}{2m}}|p\rangle\langle p|e^{-\epsilon V(\hat{x})-E}|x_{n-1}\rangle \\ &= \sqrt{\frac{m}{2\pi\epsilon}}e^{-\frac{m}{2\epsilon}(x_n-x_{n-1})^2}e^{-\epsilon(V(x_{n-1})-E)} \\ &\equiv P(x_n,x_{n-1})W(x_{n-1})\end{aligned} \tag{1}$$

so that

$$\begin{aligned}\langle\phi_a|e^{-\beta(H-E)}|\phi_b\rangle = &\int dx_1\ldots dx_n\langle\phi_a|x_n\rangle\langle x_n|e^{-\epsilon(H-E)}|x_{n-1}\rangle\langle x_{n-1}|\ldots \\ &|x_3\rangle\langle x_3|e^{-\epsilon(H-E)}\ldots|x_2\rangle\langle x_2|e^{-\epsilon(H-E)}|x_1\rangle\langle x_1|\phi_b\rangle \\ = &\int dx_1\ldots dx_1\phi_a(x_n)P(x_n,x_{n-1})W(x_{n-1})\ldots P(x_3,x_2)W(x_2)P(x_2,x_1)W(x_1)\phi_b(x_1)\end{aligned} \tag{2}$$

The simplest way to evaluate this multiple integral using the Monte Carlo method is as follows. First x_1 is randomly selected according to the distribution function $\phi_b(x_1)$, which may be chosen to be positive, and the temporary value of the score is defined to be $W(x_1)$. Given x_1, x_2 is chosen to be Gaussian distributed about x_1 according to the probability $P(x_2, x_1)$ and the score is multiplied by $W(x_2)$. This procedure is repeated for all n and finally x_n is chosen to be Gaussian distributed around x_{n-1} and the score is multiplied by $\phi_a(x_n)$. For an ensemble of such calculations, each score $\phi_a(x_n)\prod_{i=1}^{n-1} W(x_i)$ is obtained with probability $\left(\prod_{i=1}^{n-1} P(x_{i+1}, x_i)\right)\phi_b(x_1)$ so that the average value of the score for a large ensemble of random samples approaches $\langle\phi_a|e^{-\beta(H-E)}|\phi_b\rangle$.

Two significant improvements may be made to this simple method. First, the statistics may be improved greatly by replicating points at each step with probability proportional to $W(x_n)$ instead of accumulating the weight $W(x_n)$ in the score. The calculation may then be viewed as a diffusion process with a source-sink term $W(x)$ and diffusion term $P(x_m, x_{m-1})$. An initial ensemble of points distributed according to $\phi_b(x)$ is first diffused by the Gaussian $P(x_2, x_1)$. In regions where $V(x) > E$, $W(x^i) < 1$ and the point x^i is deleted from the ensemble with probability $1 - W(x^i)$. When $W(x^i) > 1$ the point x^i is always replicated $[W(x^i)]$ times (where $[W]$ denotes the greatest integer in W) and with probability $W(x^i) - [W(x^i)]$ it is replicated one additional time. In each successive step, points diffuse according to $P(x^i_m, x^i_{m-1})$ and are replicated according to $W(x^i_m)$. Thus elements of the ensemble are created in regions of attractive $V - E$ and deleted in regions of repulsive $V - E$ such that the final ensemble of points $\{x_n\}$ is distributed according to $\prod_{i=1}^{n-1} P(x_i, x_i)W(x_i)\phi(x_i)$ and $\langle\phi_a|e^{-\beta(H-E)}|\phi_a\rangle$ is given by the average value of $\phi_a(x)$ evaluated with the ensemble $\{x_n\}$. Whereas the first method retains ensembles having products of weights $\prod_{i=1}^{n} W(x_i)$ which may vary over many orders of magnitude, with corresponding loss of statistical accuracy, in the replication method each member of the final ensemble $\{x_n\}$ contributes with the same weight. In practice, the value of E is selected to maintain a constant average ensemble size and yields an independent evaluation of the ground state energy.

The second improvement is to guide the random walk using an approximate trial wave function $\phi(x)$ which contains as much of the essential physics as may be understood at the outset. The infinitesimal evolution operator for the product $\phi(x)\psi(x)$ is $\phi(\hat{x})e^{-\epsilon H(\hat{x})}\frac{1}{\phi(\hat{x})}$ which to leading order in ϵ has the matrix element

$$\langle x_n|\phi(\hat{x})e^{-\epsilon(\hat{H}-E)}\frac{1}{\phi(\hat{x})}|x_{n-1}\rangle = \sqrt{\frac{m}{2\phi\hat{\epsilon}(x_{n-1})}}\, e^{\frac{m}{2\hat{\epsilon}(x_{n-1})}\left(x_n - x_{n-1} - \frac{\epsilon}{m}\frac{\phi'(x_{n-1})}{\phi(x_{n-1})}\right)^2} \times e^{-\epsilon\left[\frac{-\frac{1}{2m}\phi''(x_{n-1}) + V(x_{n-1})\phi(x_{n-1})}{\phi(x_{n-1})}\right]} \tag{3}$$

This evolution operator differs from Eq. (1) in three respects. The Gaussian diffusion term is shifted by a drift term proportional to $\frac{\phi'}{\phi}$ which guides members of the ensemble away from regions where the wave function is small so that the points sample most densely the region in which the contribution is the largest. The source- sink term is now $\frac{(H-E)\phi}{\phi}$ instead of $V - E$. In the limit in which ϕ is exact, this term is just a constant, there are no fluctuations in the population, all the physics is contained in the drift term, and the statistical variance is a minimum. To the extent to which ϕ incorporates much of the essential physics, the evolution is guided by that portion of the physics through the drift term, and the stochastic treatment of the source term is only required to treat the remnant of the physics which is left out of the trial function. Finally in principle the size of the Gaussian step $\hat{\epsilon}(x_{n-1})$ depends upon x_{n-1}, $\hat{\epsilon}(x) = \epsilon\left[1 - \frac{\epsilon}{m}\frac{d^2}{dx^2}ln\phi(x)\right]^{-1}$, although in the special case of an oscillator this yields an irrelevant scale factor and in most practical applications the effect is negligible.[6] Trial functions have been utilized extensively in the closely related Greens Function Monte Carlo Method, and in a variety of physical applications are absolutely necessary to render a calculation practical.[6]

The many-fermion problem may be treated formally in the same way as the one-body problem. The matrix element of the infinitesimal evolution operator between antisymmetrized states is

$$\langle x_1^n \ldots x_A^n|\phi e^{-\epsilon(\hat{H}-E)}\frac{1}{\phi}|x_1^{n-1}\ldots x_A^{n-1}\rangle = \sum_P (-1)^P \left[\frac{m}{2\pi\hat{\epsilon}}\right]^{\frac{A}{2}} e^{-\frac{m}{2\hat{\epsilon}}\left(x^n_{P_i} - x_i^{n-1} - D_i(x^{n-1})\right)^2 - \epsilon S(x^{n-1})} \tag{4}$$

where the drift term is $D_i(x^{n-1}) = \frac{\epsilon}{m}\frac{d}{dx_i^{n-1}} ln\phi(x_1^{n-1}\ldots x_A^{n-1})$ and the source term is $S(x^{n-1}) = \frac{(\hat{H}-E)\phi(x^{n-1})}{\phi(x^{n-1})}$. In more than one spatial dimension, interference between positive and negative contributions to the functional integral degrades the statistical accuracy such that very good trial functions and exceedingly large ensembles are required to obtain useful results.[8] In one dimension, however, antisymmetry completely specifies the nodal points and a positive definite result may be obtained by simply evolving the wavefunction in the subspace $x_1 < x_2 < x_3 \ldots < x_A$. As described in Ref. 1, it is useful to approximate the determinant by

$$\begin{aligned}
&\sum_P (-1)^P e^{-\frac{m}{2\epsilon}(x^n_{P_i} - x_i^{n-1} - D_i)^2} \\
&\quad = e^{-\frac{m}{2\epsilon}\sum_i (x_i^n - x_i^{n-1} - D_i)^2} det\left|e^{-\frac{m}{2\epsilon}\left[(x_i^n - x_j^{n-1} - D_j)^2 - (x_i^n - x_i^{n-1} - D_i)^2\right]}\right| \\
&\quad \approx e^{-\frac{m}{2\epsilon}\sum_i (x_i^n - x_i^{n-1} - D_i)^2} \prod_{i=2}^{A}\left(1 - e^{-\frac{m}{\epsilon}(x_i^n - x_{i-1}^n)(x_i^{n-1} - D_i - x_{i-1}^{n-1} + D_{i-1})}\right)
\end{aligned} \tag{5}$$

which includes the effect of the first images surrounding each node. Because of the computational simplicity, all the fermion models treated in this work are restricted to one dimension and are solved using Eq. (4).

III. Many-Nucleon Problem with a Static Potential

As a prelude to the study of models involving sub-nuclear degrees of freedom, it is first necessary to solve a many-nucleon problem with a two-body potential having strong short range repulsion analogous to that arising in phenomenological nucleon-nucleon potentials. The nuclear model discussed in detail in Ref. 2 uses a potential that has been defined to reproduce approximately the binding energy per particle, Fermi gas kinetic energy per particle, core radius, and maximum attraction observed in realistic three-dimensional potential theories.

The quality of solutions obtained with the stochastic techniques of Section II is indicated by the nuclear matter saturation curve and ground state density distributions shown in Figures 1 and 2. As shown in Ref. 2, the extrapolation of observable quantities in the limit as the time step ϵ approaches zero is straightforward, yielding sufficient precision in the binding energy and density distributions for subsequent study of the role of sub-nuclear degrees of freedom.

The static potential model is also useful in its own right for the exploration of a number of approximations commonly used in nuclear systems. Obvious topics include the use of effective interactions of the Skyrme form, the validity of the stationary-phase theory of fission, Hartree-Fock, Brueckner- Hartree-Fock, and coupled cluster approximations, and quantitative study of quantum oscillations in the nuclear density distribution.

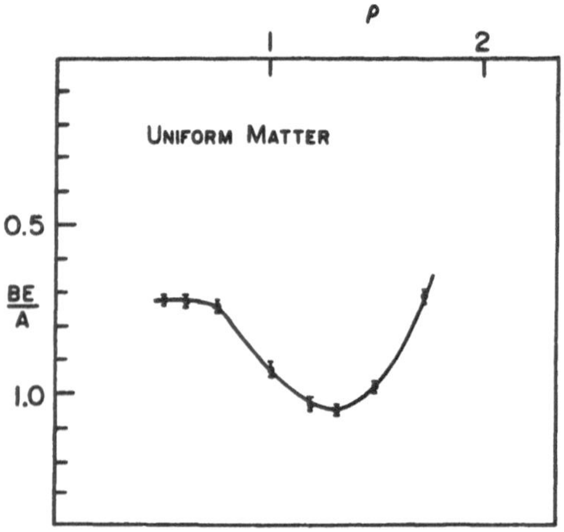

Fig. 1 Nuclear matter saturation curve.

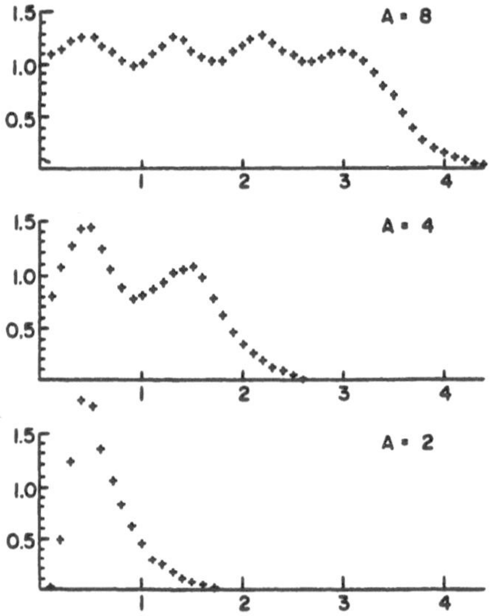

Fig. 2 Ground state density distributions for finite nuclei.

IV. Meson Nucleon Field Theory

The simplest nuclear theory containing sub-nuclear degrees of freedom is a theory of the Walecka form containing vector and scalar mesons coupled to nucleons[9] and a one-dimensional version of this theory has been studied in Ref. 10. Eliminating the time component of the vector field, which is a static mode, and defining E to be the momentum conjugate to the space component of the vector field, V_1 the Hamiltonian may be written

$$
\begin{aligned}
\mathcal{H} = &\psi^\dagger\left(-i\alpha\frac{\partial}{\partial x} + \beta M\right)\psi + \frac{1}{2}\left(\pi_s^2 + (\phi')^2 + m_s^2\phi^2\right) + \frac{1}{2}\left(E^2 + \frac{(E')^2}{m_v^2} + m_v V_1^2\right) \\
&+ \psi^\dagger\left[-g_v\alpha V_1 - \frac{g_v}{m_v^2}E' - g_s\beta\phi\right]\psi + \frac{1}{2}\frac{g_v^2}{m_v^2}(\psi^\dagger\psi)^2
\end{aligned}
\tag{6}
$$

where ϕ and π_s are the scalar field and momentum respectively and $m_{v,s}$ and $g_{v,s}$ denote masses and coupling constants for the vector and scalar fields. Taking the non-relativistic limit and fourier transforming to discrete modes in a domain with periodic boundary conditions yields

$$
\begin{aligned}
\mathcal{H} = &-\frac{1}{2M}\sum_{i=1}^{A}\frac{\partial^2}{\partial x_i^2} + \sum_{k=1}^{\infty}\left\{-\frac{1}{2}\left[\frac{\partial^2}{\partial\phi_k^2} - w_k^2\phi_k^2\right] - \frac{1}{2}\left[\frac{\partial^2}{\partial\hat{E}_k^2} - \Omega_k^2\hat{E}_k^2\right]\right. \\
&\left.-g_s\phi_k\rho_k + \frac{1}{2}\frac{g_v^2}{m_v^2}\rho_k^2 - \sum_{i=1}^{A}\frac{g_v}{m_v}\hat{E}_k\frac{\partial\rho_k}{\partial x_i}\right\} + H_{zero}
\end{aligned}
\tag{7}
$$

where $\hat{E} = \frac{E}{m_v}$, $w_k^2 = m_s^2 + k^2$, $\Omega_k^2 = m_v^2 + k^2$ and H_s includes the static contribution of all $k = 0$ modes (which is identical to the mean field approximation) as well as the zero-point contribution $-\frac{1}{2}\sum_{k=1}^{\infty}(w_k + \Omega_k)$.

The appropriate variables for the stochastic evolution described in Section II. are the A fermion coordinates $\{r_i\}$ and the Fourier components ϕ_k and E_k of the scalar field and momentum conjugate to the vector field. For static nucleons, the meson modes are just decoupled displaced harmonic oscillators

$$\left(-\frac{1}{2}\frac{\partial^2}{\partial\phi_k^2} + \frac{1}{2}w_k^2\phi_k^2 - g_s\rho_k\phi_k\right)\Psi_s(\phi_k) = \epsilon_k^s\Psi_s(\phi_k)$$

$$\left(-\frac{1}{2}\frac{\partial^2}{\partial E_k^2} + \frac{1}{2}\Omega_k^2 E_k^2 - \frac{g_v}{m_v}E_k\frac{\partial p_k}{\partial x_i} + \frac{1}{2}\left(\frac{g_v}{m_v}\right)^2\rho_k^2\right)\Psi_v(E_k) = \epsilon^v\Psi_v(E_k) \tag{8}$$

with solutions $\Phi_s(\phi_k) = e^{-\frac{1}{2}w_k[\phi_k - \chi_k]^2}$ and $\Phi_v(E_k) = e^{-\frac{1}{2}\Omega_k[E_k - \varsigma_k]^2}$, where $\chi_k = \frac{g_s\rho_k}{w_k^2}$, $\varsigma_k = \frac{g_v}{\Omega_k^2 m_v}\sum_i \frac{\partial\rho_k}{\partial x_i}$, $\epsilon_k^s = -\frac{1}{2}g_s^2\frac{\rho_k^2}{w_k^2}$, and $\epsilon_k^v = \frac{1}{2}g_v^2\frac{\rho_k^2}{\Omega_k^2}$.

For the static two-body problem, the interaction energy yields the familiar static potential with scalar attraction and vector repulsion:

$$\sum_k\left(\frac{-g_s^2}{k^2+m_s^2} + \frac{g_v^2}{k^2+m_v^2}\right)\rho_k^2 \to U(x) = \left[\frac{-g_s^2}{m_s^2}e^{-m_s|x|} + \frac{g_v^2}{m_v^2}e^{-m_v|x|}\right] + \text{periodic images.} \tag{9}$$

The physical difference between the full meson field theory and a static potential theory based on the two-body potential $U(x)$ in Eq. 9 is the non-adiabatic coupling between nucleons and mesons. Physically, for sufficiently high frequencies, the meson motion is sufficiently fast relative to the nucleon motion that the static solutions Eq. (8) become exact. The sum over dynamics meson modes may then be truncated by summing the static contributions for all modes above a cutoff k_{max} as follows

$$\mathcal{H} \Rightarrow \sum_{k=1}^{k_{max}}\left[\mathcal{H}_k - \left(-\frac{g_s^2}{w_k^2} + \frac{g_v^2}{\Omega_k^2}\right)\right] + \sum_{i<j}U(x_i - x_j)$$

The obvious trial wave function to use for the nuclear coordinates and dynamic meson modes is the product of a nucleon Slater determinant and displaced meson oscillator functions

$$\Psi_{\text{Trial}} = \psi_{SD}(x_1\ldots x_n)\prod_k e^{-\frac{1}{2}w_k[\phi_k - \chi_k(x_1\ldots x_n)]^2 - \frac{1}{2}\Omega_k[E_k - \varsigma_k(x_1\ldots x_n)]^2} \tag{11}$$

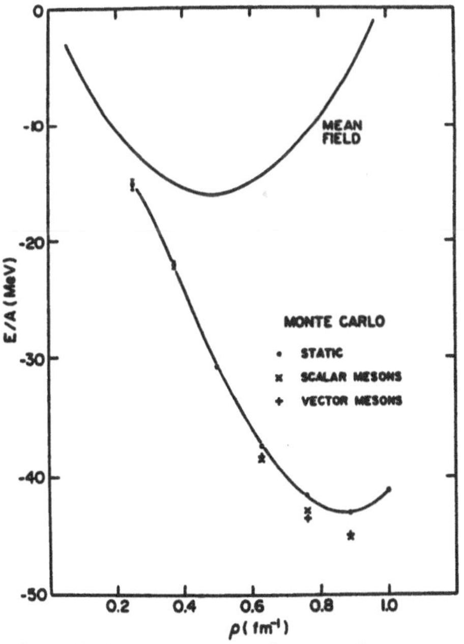

Fig. 3 Nuclear matter saturation curve for meson-nucleon field theory.

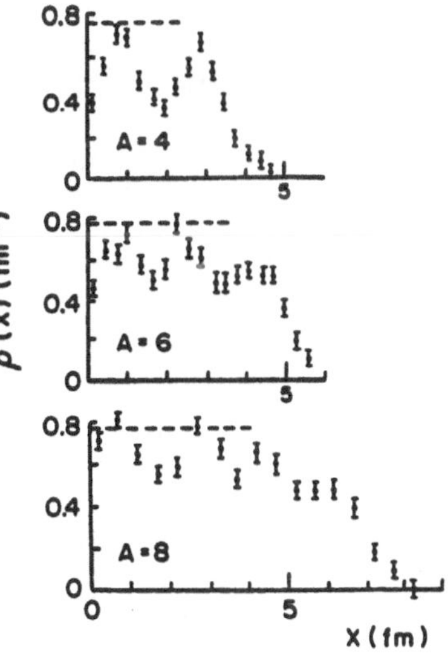

Fig. 4 Ground state nuclear density distributions for finite nuclei.

To the extent to which the low-lying meson modes are adiabatic, all the meson physics is incorporated in the displaced oscillator functions, and the stochastic evolution is only required to deal with the small non-adiabatic effects. The trial function Eq. (11) constitutes a form of Born Oppenheimer approximation in which the dynamic part of the meson wave function is determined by the instantaneous nucleon coordinates and provides a convenient physical description of the role of mesons in nuclei.

The principal result of stochastic solutions of the meson-nucleus field theory is that dynamic contributions of mesons are very small. Dynamic modes were cut off above 3 fm^{-1}, justified *a posteriori* by the fact that lower modes were essentially adiabatic as well. In the adiabatic limit, the meson contribution to the nuclear self-energy is 188.4 MeV and this was shifted only by 3.9 MeV and 2.7 MeV from the effect of dynamic scalar and vector mesons respectively.

The nuclear matter saturation curve for this theory is shown in Figure 3. The stochastic solution of the many-fermion problem with the static potential of Eq. (9) is shown by the solid dots, and the effect of adding dynamic scalar or vector modes, shown by the crosses and pluses respectively, is less than 5%. The parameters of the model were defined such that the mean field approximation, shown by the solid curve, saturated at the usual energy and density, so the vector potential is evidently much weaker than desired for a more realistic model.

Density distributions for finite nuclei are shown in Figure 4, and at the level of precision of the present statistics, no effects of meson dynamics were observable. Binding energies of finite nuclei displayed roughly the same sensitivity to meson dynamics as nuclear matter, with the total effect of dynamic scalar and vector mesons on the binding energy of the A=12 system being 5%. The nucleon-nucleon correlation function showed no effect of meson dynamics at the one percent level, and the meson correlation function $< \phi_k^2 >$ differed by less than one percent from the adiabatic value $\chi_k^2 + \frac{1}{2\omega_k}$ for all modes considered.

In the context of meson theory, these results suggest that the appropriate way to treat subnuclear degrees of freedom is to first solve the nuclear many-body problem with the static interaction derived in the adiabatic limit, and then treat the small dynamic meson effects perturbatively. The physical Born-Oppenheimer argument is independent of dimension and should apply to more realistic models in three dimensions. Extension of the present method to relativistic fermions in one dimension is practical, and for very light nuclei, stochastic calculations in three dimensions may be possible.

V. Lattice Gauge Theory

Ultimately, any treatment of subnuclear degrees of freedom will require the solution of QCD. Although a variety of stochastic approaches are presently utilized to solve lattice gauge theories, the practical and physical appeal of using a trial wave function to guide the stochastic evolution motivated us to try the method described in Section II. As a first application, we have studied $U(1)$ lattice gauge theory in 3 spatial dimensions.[11] This case is both simple enough to display the essence of the method and rich enough, because of the presence of a phase transition, to pose a non-trivial test of the use of a trial function.

The dynamical variables are periodic variables $A_\mu(\vec{n})$ defined on the links of a three dimensional lattice with μ denoting the direction of the link and $\vec{n}$ denoting the position of the end of the link. The Hamiltonian may be written

$$H = \sum_{\mu,\vec{n}} \frac{1}{2} \frac{\partial^2}{\partial A_\mu^2(\vec{n})} - \lambda \sum_{\mu,\vec{n}} (1 - \cos B_\mu(\vec{n})) \tag{12}$$

where $B_\mu(\vec{n})$ is the lattice curl

$$B_\mu(\vec{n}) = \epsilon_{\mu\nu\gamma}[A_\gamma(\vec{n} + \vec{e}_\nu) - A_\gamma(\vec{n})] \tag{13}$$

which is just the sum over all plaquettes of directed links around the plaquette. In terms of the coupling constant g of the lattice theory, $\lambda = \frac{1}{g^4}$. In the weak coupling limit, the first term becomes E^2 and the second term becomes B^2 reproducing continuum QED.

For a periodic angular variable θ, the matrix element of the kinetic evolution operator may be expanded in periodic eigenstates

$$\begin{aligned} \langle\theta'|e^{-\frac{\epsilon}{2}\frac{\partial^2}{\partial\theta^2}}|\theta\rangle &= \sum_n \langle\theta'|n\rangle\langle n|e^{-\frac{\epsilon}{2}\frac{\partial^2}{\partial\theta^2}}|\theta\rangle = \sum_n e^{in(\theta'-\theta)}e^{-\epsilon\frac{n^2}{2}} \\ &= \sum_m e^{-\frac{1}{2\epsilon}(\theta'-\theta+2\pi m)^2} \rightarrow_{\epsilon\rightarrow 0} e^{-\frac{1}{2\epsilon}(\theta'-\theta)^2} \end{aligned} \tag{14}$$

In the limit of infinitesimal time step ϵ, the sum of the Gaussian propagators over an infinite set of images obtained in the last line using the Poisson sum formula may be replaced on the interval $|\theta' - \theta| < \pi$ by a single Gaussian. Thus, the evolution of a periodic angular variable becomes identical to that of a coordinate in quantum mechanics, and the lattice gauge theory is equivalent to $3L^3$ variables $A_\mu(n)$ evolving with a potential

$$V(A) = -\lambda \sum_{\text{plaquettes}} (1 - \cos(B[A])).$$

A convenient gauge-invariant trial function is the product over all plaquettes P

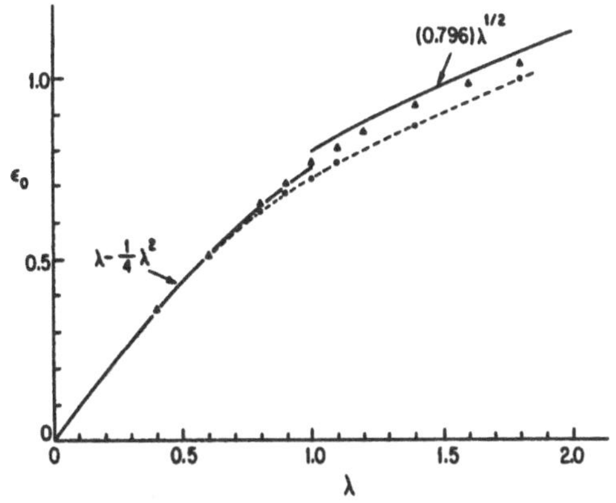

Fig. 5 Ground state energy per plaquette as a function of λ. Solid dots and triangles denote exact Monte Carlo and variational bounds respectively and strong and weak coupling expansions are indicated by solid lines.

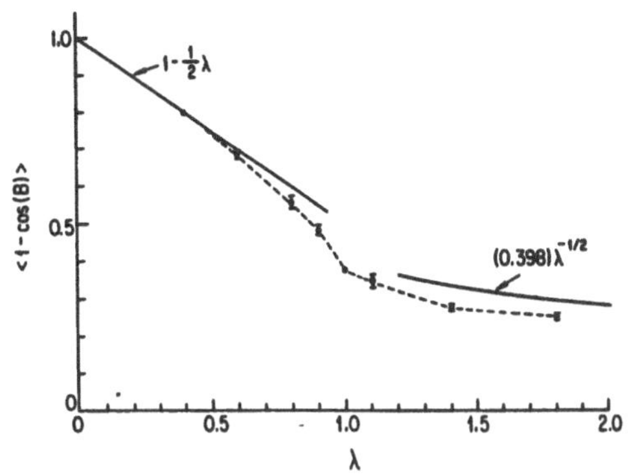

Fig. 6 The plaquette energy as a function of λ. Error bars denote Monte Carlo results and the solid lines indicate strong and weak coupling expansions.

$$\Phi(A) = \prod_P f(B_p[A]) \tag{15}$$

where

$$f(B) = e^{-b(\lambda)(1-\cos B)-\frac{1}{3}(a(\lambda)-b(\lambda))\sin^2 B} \tag{16}$$

B_P denotes the lattice curl Eq. (13) for the plaquette P and $a(\lambda)$ and $b(\lambda)$ are variational parameters to optimize the wave function. In the strong coupling limit, that is $\lambda \to 0$, the ground state is dominated by single plaquette contributions $\psi_0 = \prod_P e^{-\frac{1}{2}\lambda[1-\cos B_P]} + O(\lambda^2)$ so that Eq. (15) becomes exact for $a(\lambda) = b(\lambda) = \frac{1}{2}$. In the weak coupling limit, the ground state is just an ensemble of independent harmonic oscillations $\psi_0 = e^{-\frac{\sqrt{\lambda}}{2}\sum_{PP'} B_P \Gamma_{PP'} B_{P'}}$. It is shown in Ref. 11 that the diagonal terms strongly dominate the off-diagonal terms and are reproduced by $a(\lambda) = 0.428\lambda^{\frac{1}{2}}$ and $b(\lambda) = \frac{4}{3}a(\lambda)$. The trial function smoothly interpolates between these two limits for intermediate λ and gives an economical representation of much of the local structure of the ground state wave function.

The primary result of this calculation is that the stochastic method of Section II is well suited to lattice gauge theory calculations. Results by S. Chin for the binding energy per plaquette and plaquette energy $\langle -1\cos(B)\rangle$ are shown in Figures 5 and 6. The use of a trial function yields high statistical accuracy throughout the region of the phase transition, and the results are consistent with strong and weak coupling expansions in the appropriate domains and with other Monte Carlo results.

Whereas the results thus far are not directly relevant to the study of sub-nuclear degrees of freedom, the method may be extended straightforwardly to non-abelian lattice gauge theories. Just as a physically motivated trial function for the meson fields was extremely fruitful in the case of a meson field theory, it is hoped that an appropriate trial function will be useful in the subsequent treatment of non-abelian gauge theories and quarks.

VI. Confining Quark Model

The confining quark model of Ref. 12 provides a convenient system in which to study how nuclear structure arises from underlying quark degrees of freedom. Physically, the model may be thought of as an adiabatic limit in which for any configuration of $2N$ spinless quarks, the color fields instantaneously adjust themselves to form the lowest energy configuration in which N distinct pairs of quarks are connected by flux tubes. Mathematically, the model is defined by specifying a potential energy $v(r)$ which is an increasing function of r associated with a flux tube of length r and defining the total potential energy of the quarks to be

$$V = \min_P \left\{ v(x_{P1} - x_{P2}) + v(x_{P3} - x_{P4}) + \ldots + v\left(x_{P(N-1)} - x_{PN}\right)\right\} \tag{17}$$

where the minimum over all permutations selects the lowest energy assignment of pairs. The theory exhibits the desired separability, confinement, and exchange symmetry and is free of Van der Waals interactions. Although finding the optimal pairing in three dimensions entails solution of a non-trivial assignment problem, in one dimension with periodic boundary conditions, the optimal pairing of consecutively labeled quarks either pairs q_{2m} with q_{2m+1} or q_{2m} with q_{2m-1} for all m. Nucleons are composed of two quarks in the simplest version of the model used here and are thus Bosons. The potential $v(x)$ is taken to be quadratic, yielding a quark density in a free nucleon proportional to $x^2 e^{-\frac{x^2}{\sqrt{3}}}$. Preliminary results by C. Horowitz comparing this model with a nuclear potential model obtained from the exact phase shifts are presented below.[3]

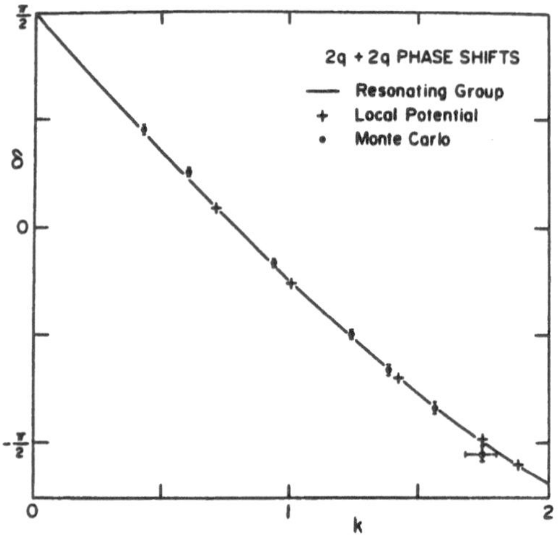

Fig. 7 Phase shifts for the scattering of a two-quark nucleon from a two quark nucleon.

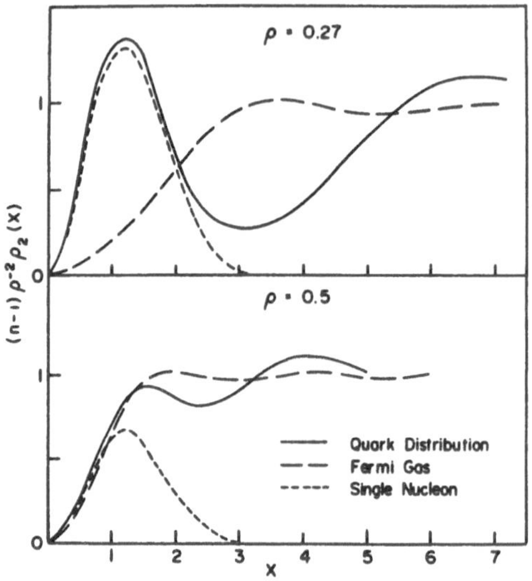

Fig. 9 Quark-quark correlation functions at two densities.

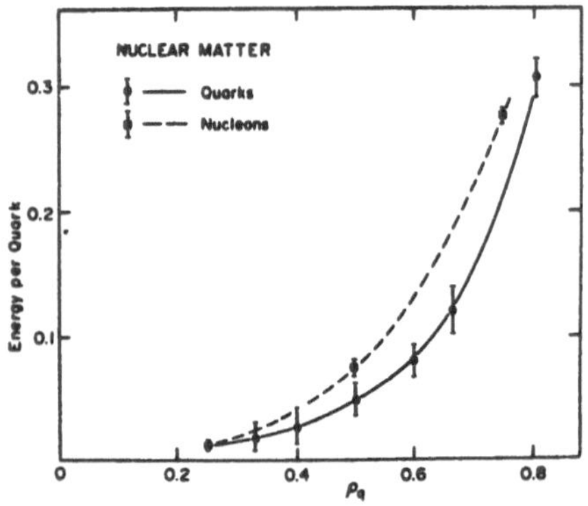

Fig. 8 Binding energy per quark of nuclear matter.

Figure 7 shows the phase shifts for the scattering of a two-quark nucleon from a two-quark nucleon. As noted in Section II this scattering problem may be solved stochastically below threshold by calculating the ground state of the four quark systems with the appropriate boundary condition on the distance between the cm of the left two quarks and the right two quarks, and the Monte Carlo results are shown by the error bars in Fig. 7. An accurate approximation is also provided by the single channel resonating group method, with the result shown by the solid line. Since the phase shifts in this energy range are evidently characterized by a scattering length and an effective range, they are easy to reproduce with a simple local potential. The fit obtained using $V_N(x) = \frac{52}{\sqrt{2\pi}} e^{-\frac{x^2}{3}}$ is shown by the crosses. Of the two effects which occur for overlapping nucleons, diminuation of the potential interaction energy and increased kinetic energy due to the Pauli principle, the later effect dominates and the interaction is purely repulsive.

The binding energy per quark of uniform nuclear matter are shown in Figure 8. Stochastic solution of the many-quark problem yields the solid dots connected by the solid line. In contrast, the many-nucleon problem defined by bosons interacting via the phenomenological potential $V_N(x)$ yields the square points connected by the dashed line. Although the model is exceedingly simple, it is tantalizing that the discrepancy between the phenomenological and exact descriptions is qualitatively similar to the discrepancy encountered in calculating three dimensional nuclear matter with realistic phenomenological nuclear forces.

A more microscopic view of the quark behavior is provided by the quark-quark correlation functions shown in Fig. 9 which specify the probability of finding another quark a distance x from a given quark. The normalization is defined such that at density ρ corresponding to n particles in a periodic box of length L, the integral of the correlation function from 0 to L is $L(n-1)$, thus counting the (n-1) remaining particles. The fermi gas correlation function then approaches 1 in the interior of the box and approaches zero within range $\frac{1}{k_F}$ of 0 and L as shown by the long dashed lines. At very low density, one would expect the correlation functions to look like the undistorted ground state nucleon density at short distances and approach a constant corresponding to the low-density nucleon gas at large distances. Hence the appropriately normalized density distributions for a single nucleon are also shown in Fig. 9 by the short dashed curves. As observed in the top portion of the figure, at $\rho = 0.27$ the quark distribution exhibits the low density behavior of undistorted nucleons. At $\rho = 0.5$, however, the nucleon correlations have nearly disappeared and the correlation function is close to that of a Fermi gas. An analogous transition between the nucleon momentum distribution and a Fermi gas distribution is observed in the quark momentum distribution.

A particularly important result of this calculation is the fact that whereas only minor quantitative differences between the nucleon and quark descriptions of nuclear matter arise in the transition region, major qualitative differences are observed in the quark distributions. In the light of recent measurements of quark structure functions in nuclei using deep inelastic lepton scattering, it will be valuable to explore the observable consequences of this quark behavior and whether comparable effects arise in more realistic models in three dimensions.

VII. Summary and Conclusions

In conclusion, stochastic calculation of ground state properties using random walks guided by physically motivated trial functions is a useful tool for studying the role of sub-nuclear degrees of freedom in nuclear models. The initial investigations summarized here indicate that the present formulation is useful for the many fermion problem, non-relativistic meson-nucleon field theory, and quark models in one spatial dimensions as well as for bosons and $U(1)$ gauge theory in three dimensions.

The present work suggests a number of possible extensions. It is straightforward to treat one-dimensional relativistic field theory by putting the fermions on a lattice and occupying half the states to fill the Fermi sea. This will facilitate the study of the role of antinucleon contributions and provide a test of current prescriptions for the relativistic generalization of Brueckner theory. The generalization of the $U(1)$ calculation to $SU(2)$ and $SU(3)$ is straightforward, and it would be instructive to begin by exploring trial functions for $SU(2)$ in two spatial dimensions.

The biggest practical problem is the stochastic treatment of Fermions with spin or isospin or in more than one dimension. The present technique is feasible for light nuclei with spin-independent forces in three dimensions. With sufficiently good trial functions and very large statistical populations transient estimates can provide exact results for more complicated systems. In the absence of exact results, variational calculations can be quite valuable. Constraining the nodes to be those of a trial function and otherwise stochastically evolving the shape of the wave function can yield an excellent variational bound. In the case of the confining quark model, the trial function $e^{-\lambda V}\Phi_{SD}$ where V is the potential in Eq. (17) and Φ_{SD} is a Slater determinant, is a good approximation throughout the transition region, and this variational ansatz is being studied in 3 dimensions.[12] (Note that it is essentially the same as Eq. (16) for the lattice gauge theory when $a(\lambda) = b(\lambda)$.)

Thus, a number of promising possibilities exist for more realistic studies, and I am confident that the approach outlined in this work will play a continuing role in the quest to understand the physics of sub-nuclear degrees of freedom.

Acknowledgements

The stochastic methods in this work are based on the approach to Monte Carlo developed by Steve Koonin, and have benefitted from discussions with David Ceperley, Malvin Kalos, and Henri Orland. Fellowship support by the John Simon Guggenheim Memorial Foundation and support by the NSF Institute for Theoretical Physics and Department of Energy contract DE-AC02-76ER03069 are gratefully acknowledged.

References

1. S.E. Koonin, Nuclear Theory 1981, ed. G.F. Bertsch (World Scientific, (1981).
2. J.W. Negele, Proc. Int. Symposium on Time-Dependent Hartree-Fock and Beyond, Lecture Notes in Physics, Vol. 171, ed. K. Goeke and P.G. Reinhardt (Springer-Verlag, NY, 1982).
3. C. Horowitz, E. Moniz, and J.W. Negele, to be published.
4. J.W. Negele, Rev. Mod. Phys. 54, 913, 1982.
5. B. Blankenbecler and R.L. Sugar, Institute for Theoretical Physics preprint NSF-ITP-82-95 (1982).
6. J.W. Moskowitz, K.E. Schmidt, M.A. Lee, and M.H. Kalos, J. Chem. Phys. 77, 349 (1982).
7. M.H. Kalos, Phys. Rev. 128, 1791 (1962); J. comp. Phys. 1, 257 (1966); Phys. Rev. A2, 250 (1970); D.M. Ceperley and M.H. Kalos, Monte Carlo Methods in Statistical Mechanics, ed. K. Binder (Springer-Verlag, NY, 1979).
8. D.M. Ceperley and B.J. Alder, Phys. Rev. Lett. 45, 566 (1980).
9. J.D. Walecka, Ann. Phys. 83, 491 (1971); S.A. Chin and J.D. Walecka, Phys. Lett. 52B, 24 (1974); S.A. Chin, Ann. Phys. 108, 403 (1977).
10. B.D. Serot, S.E. Koonin, and J.W. Negele, Phys. Rev. C in press.
11. S.A. Chin, S.E. Koonin, and J.W. Negele, to be published.
12. M.H. Kalos and K.M. Panoff, private communication.

COUPLED-CLUSTER THEORY OF PIONS IN NUCLEAR MATTER AND THE EMC EFFECT

F. Coester

Argonne National Laboratory*, Argonne, Illinois 60439

The conventional theory of nuclei assumes that the interactions of nucleons are due to the exchange of mesons. The usual treatment eliminates the mesons at the outset in favor of two- and three-body potentials and two-body current densities ("exchange current" operators). There is no difficulty in formulating the many-body theory including mesons on the formal level, but the complexity of the formalism easily frustrates efforts to obtain reliable approximations of sufficient accuracy. High accuracy in the potential energies is essential for a significant calculation of binding energies since the latter are much smaller than the former.

Recent measurements of deep-inelastic lepton scattering show a marked difference of the nuclear structure functions $F_2(x,Q^2)$ observed in iron and deuterium[1,2] - the so-called EMC (European Muon Collaboration) effect. This discovery raises the possibility that pion densities in nuclei may be observed in these experiments[3,4,5] and brings a new focus to the theory of pions in nuclei. It was generally assumed that deep-inelastic lepton scattering from a nucleus occurs off the constituents of nucleons whose structure is not affected by their environment. It is a conservative extension of this view to assume that the leptons may scatter either off the constituents of the nucleons or off the constituents of the mesons. Such a model requires nuclear wave functions including mesons and the connection between these wave functions and the structure function of the nucleus.

The many-body Hamiltonian underlying this discussion ~~is~~ of the form

$$H = H^o + H' , \tag{1}$$

where H^o includes the kinetic energy of the mesons, H^o_{mes}, and the nucleons, H^o_N, as well as a self-energy counter term, H_{self}. We have

$$H^o_{mes} = \sum_\mu \int d^3k\, c^\dagger_\mu(\vec{k}) c_\mu(\vec{k}) \omega_\mu(\vec{k}) , \tag{2}$$

where μ labels different mesons and $\omega_\mu(\vec{k}) = (\vec{k}^2 + m_\mu^2)^{1/2}$, and

$$H^o_N = \int d^3p\, c^\dagger_N(\vec{p}) \frac{\vec{p}^2}{2m_N} c_N(\vec{p}) . \tag{3}$$

The operators $c_\mu(\vec{k})$, $c_N(\vec{p})$, $c^\dagger_\mu(\vec{k})$, $c^\dagger_N(\vec{p})$ are annihilation and creation operators

*Work supported by the U. S. Dept. of Energy under contract number W-31-109-ENG-38.

of mesons and nucleons. The interaction Hamiltonian H' is

$$H' = \sum_{\mu} \int d^3p' \int d^3k \int d^3p \; \delta(\vec{p}'+\vec{k}-\vec{p}) \; C_N^{\dagger}(\vec{p}')C_N(\vec{p})$$
$$\times \; \{C_{\mu}^{\dagger}(\vec{k})(\vec{k},\vec{p}'|v_{\mu}^{\dagger}|\vec{p}) + (\vec{p}'|v_{\mu}|\vec{p},-\vec{k})C_{\mu}(-\vec{k})\} \; , \qquad (4)$$

where v_{μ} is the vertex appropriate for the meson μ including a form factor. The pion vertex will serve as an illustration:

$$(\vec{k},\vec{p}'|v_{\pi}^{\dagger}|\vec{p}) = (2\pi)^{-3/2} \frac{f}{m_{\pi}} \left(\frac{\Lambda^2-m_{\pi}^2}{\vec{k}^2+\Lambda^2}\right)^2 \frac{i\,\vec{\sigma}\cdot\vec{k}\;\vec{\tau}}{(2\omega_{\pi}(\vec{k}))^{1/2}} \; . \qquad (5)$$

The pion number density per nucleon is given by the expectation value of the number operator:

$$\rho_{\pi}(\vec{k}) = \frac{1}{A} \langle C_{\pi}^{\dagger}(\vec{k}) \; C_{\pi}(\vec{k})\rangle \; . \qquad (6)$$

It is therefore also the linear response of the energy to a change $\omega_{\pi}(\vec{k}) \to \omega_{\pi}(\vec{k})+\eta(\vec{k})$ in H^{o}_{mes},

$$\rho_{\pi}(\vec{k}) = \frac{1}{A} \left\langle \frac{\delta H(\eta)}{\delta\eta(\vec{k})} \right\rangle_{\eta=0} \; . \qquad (7)$$

If we modify also the self-energy counter term H_{self} to take the change $\omega\to\omega+\eta$ into account then we obtain the excess density of pions according to

$$\rho_{\pi,ex} = \frac{1}{A} \left\langle \frac{\delta E_{pot}}{\delta\eta(k)} \right\rangle_{\eta=0} \; , \qquad (8)$$

where E_{pot} is the potential energy of the nucleus. Thus any formulation of the many-body theory which allows a calculation of the potential energy will also yield a corresponding excess density of pions.[6]

The coupled-cluster form of the nuclear wave function is

$$|\Psi\rangle = e^{(S+S')}|\Phi\rangle \; , \qquad (9)$$

where $|\Phi\rangle$ is the non-interacting Fermi gas containing no mesons, and

$$S = \sum_{n} S_n \quad ; \quad S' = \sum_{m,n} S'_{m,n} \; . \qquad (10)$$

The operator S_n creates n linked particle-hole pairs and no mesons, while $S'_{m,n}$ creates m mesons and n particle-hole pairs.

Only $S'_{1,1}$ enters into the exact expression for the potential energy,

$$E_{pot} = \langle\Phi|H'S'_{1,1}|\Phi\rangle - A\,E_{self}\ , \tag{11}$$

where E_{self} is the self energy of an isolated nucleon. Eq. (11) may be compared to the analogous formula

$$E_{pot} = \langle\Phi|V(1+S_2)|\Phi\rangle \tag{12}$$

of the usual potential theory. It follows from Eq. (8) that

$$\rho_{\pi,ex}(\vec{k}) = \left\{ \frac{1}{A} \langle\Phi|H' \frac{\delta S'_{1,1}}{\delta\eta(k)} |\Phi\rangle - \frac{\delta E_{self}}{\delta\eta(k)} \right\}_{\eta=0} . \tag{13}$$

The coupled-cluster equations for S and S' have the familiar form

$$\left[H^o,(S+S')\right] + \left(e^{-(S+S')}\ H'\ e^{(S+S')}\right)_c = 0\ , \tag{14}$$

where the subscript indicates the creation part of the operator which after normal ordering contains at least one creation operator and no annihilation operators. To the extent that a model based on one-meson-exchange potentials is reasonable we should get a reasonable zero-order approximation by neglecting in the coupled-cluster equations (14) all linked expressions with more than one pion. This means in particular that $S' \to \sum_n S'_{1,n}$ and higher powers of S' are neglected. Then the approximate coupled-cluster equations have the form

$$[H^o,S'] + \left(e^{-S}\ H'\ e^{S}\right)_c = 0\ , \tag{15}$$

$$[H^o,S] + \left(e^{-S}[H',S']e^{S}\right)_c = 0\ . \tag{16}$$

The operator S' creating the mesons can be eliminated from (15) and (16) yielding coupled cluster equations for S only. If we further replace H^o by H^o_{mes} in (15) the result is the usual potential theory with static one-meson exchange potentials. The associated meson amplitudes S' can then be readily obtained from (15).

We now come to the problem of relating the structure functions to the nuclear wave functions. The structure function $F_2(x,Q^2)$ of any target is defined as a quadratic function of the matrix elements of the current densities. The arguments Q^2 and x are the square of the four-momentum transfer, Q, and the Bjorken scaling variable x defined by

$$x = Q^2/(-2p\cdot Q)\ , \tag{17}$$

where p is the initial momentum of the target. The question is whether the structure function of a nucleus can be expressed in terms of the structure functions of constituent nucleons and pions and the nuclear wave function. The essential physical assumption, which allows an affirmative answer to this question, is that the currents in the nucleus are sums of one nucleon and one-pion operators which contribute incoherently to the structure functions.

A consistent calculation of the relevant matrix elements of the currents requires nuclear wave functions that are covariant under Lorentz boosts in the direction of the momentum transfer, $\vec{Q} = \{0,0,|\vec{Q}|\}$, evaluated in the rest frame of the target. The required boosts are included in the subgroups of the Poincaré group that leaves the light fronts perpendicular to $\vec{Q}$ invariant. It is possible to impose this "front-form"[7] symmetry on ordinary nuclear wave functions with all constituents on their mass shells.[5] Appropriate independent variables are the transverse momenta and light-front momentum fractions which are functions of the momenta $\vec{q}_i$ and $\vec{k}_j$ of nucleons and pions in the nucleus at rest $(\sum_i \vec{q}_i + \sum_j \vec{k}_j = 0)$. For the ith nucleon and the jth pion we have respectively

$$z_i = \frac{(q_{i3} + \omega_N(\vec{q}_i))A}{\sum_i \omega_N(\vec{q}_i) + \sum_j \omega_\pi(\vec{k}_j)} , \tag{18}$$

$$y_j = \frac{(k_{j3} + \omega_\pi(\vec{k}_j))A}{\sum_i \omega_N(\vec{q}_i) + \sum_j \omega_\pi(\vec{k}_j)} . \tag{19}$$

Manifestly these variables are constrained by the identity

$$\sum_i z_i + \sum_j y_j = A . \tag{20}$$

The desired expression for the structure function is for $R^2 \gg m_N^2$

$$F_2^A(x) = \int dz\, F_2^N\left(\frac{x}{z}\right) f_N(z) + \int dy\, F_2^\pi\left(\frac{x}{y}\right) f_\pi(y) . \tag{21}$$

The densities $f_N(z)$ and $f_\pi(y)$ are obtained from the square of the wave function by integrating over all transverse momenta and all but one light-front momentum fraction. The normalization conditions

$$\int dz\, f_N(z) = 1 \quad ; \quad \int dy\, f_\pi(y) = \langle n_\pi \rangle , \tag{22}$$

where $\langle n_\pi \rangle$ is the average number of pions per nucleon, follow from the definitions. The momentum balance relation

$$\int dz\, z\, f_N(z) + \int dy\, y\, f_\pi(y) = 1 \tag{23}$$

follows from Eq. (20).

It is clear from Eqs. (18) and (19) that the densities $f_N(z)$ and $f_\pi(y)$ cannot be obtained exactly from the densities $\rho_N(\vec{q})$ and $\rho_\pi(\vec{k})$. An approximate relation results if we replace the denominator in (18) and (19) by its expectation value. The approximation has the virtue that the integral relations (22) and (23) are preserved.

Alternatively one might assume that the wave function of a physical nucleon has two components: a core and a core + one pion component. Eq. (18) and (19) give for A = 1 and one pion

$$y = \frac{k_3 + \omega_\pi(\vec{k})}{\omega_N(\vec{k}) + \omega_\pi(\vec{k})} \quad , \tag{24}$$

$$z = 1 - y \quad , \tag{25}$$

since $\vec{q} = -\vec{k}$ in this case. If, in addition, the Fermi motion of these composite nucleons is neglected one finds $f_N(z)$ related to $f_\pi(y)$ by

$$f_N(z) = (1-\langle n_\pi\rangle)\, \delta(1-z) + f_\pi(1-z) \quad . \tag{26}$$

Since Eq. (24) gives y as a function of $\vec{k}$ it implies a definite relation between $\rho_\pi(\vec{k})$ and $f_\pi(y)$. Clearly the shape of $f_\pi(y)$ obtained from a given $\rho_\pi(\vec{k})$ will be different for the two approximations.

Both approximations have been used.[5] The result is that the experimental x-dependence of the ratio $F_2^A(x)/F_2^N(x)$ can be roughly reproduced but not the magnitude. This is illustrated in Fig. 1, where the calculated ratio R(x) of the structure functions for iron and deuterium (using the first approximation) is plotted versus x and compared to the EMC data.[1] The second approximation gives similar results. There is an obvious need for better calculations of $f_N(z)$ and $f_\pi(y)$ using more information about the cluster structure of the wave functions. However studies of the effect of ad hoc modifications of both $\langle n_\pi\rangle$ and the shape of $f_\pi(y)$ make it very unlikely that improved calculations of $f_\pi(y)$ and a better treatment of the momentum balance will achieve agreement with the data without modifications of the nuclear structure.[5]

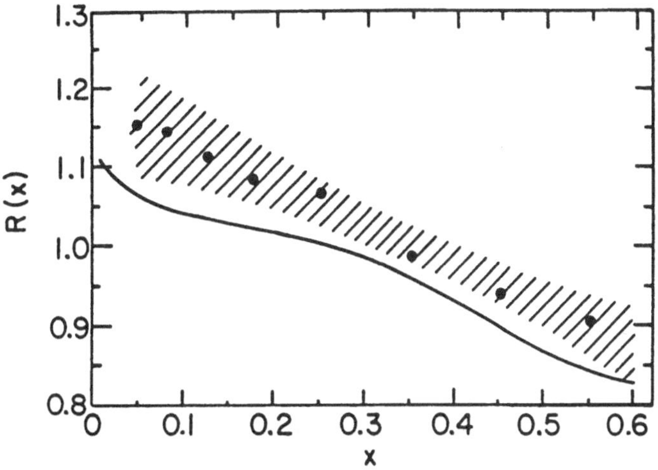

Fig. 1

References

1. European Muon Collaboration, J. J. Hubert et al., Phys. Lett. 123B, 275 (1983).
2. Rochester-SLAC-MIT Collaboration, A. Bodek et al., Phys. Rev. Lett. 50, 1431 (1983).
3. C. H. Llewellyn Smith, Oxford report Ref. 18/83.
4. M. Ericson and A. W. Thomas, CERN report, Ref. TH. 3533; A. W. Thomas, Phys. Lett. 126B, 97 (1983).
5. E. L. Berger, F. Coester and R. B. Wiringa, Argonne National Laboratory report ANL-HEP-PR-83-24 submitted to Phys. Rev. D.
6. B. L. Friman, V. R. Pandharipande and R. B. Wiringa, Phys. Rev. Lett. 51, 763 (1983).
7. P. A. M. Dirac, Rev. Mod. Phys. 21, 392 (1949).

Δ-EXCITATIONS AND MANY-BODY THEORY OF NUCLEAR MATTER

H. Müther
Institut für Theoretische Physik
Universität Tübingen
Auf der Morgenstelle 14

1. Introduction

The classical model of a nuclear many-body system is to consider the nucleus as a system of inert nucleons which interact by static two-body interactions. This means, that one treats the nucleons as elementary particles, which do not have any internal degrees of freedom. Since one typically derives the nucleon-nucleon (NN) interaction from the NN scattering data, one furthermore assumes, that the interaction of two nucleons in the nuclear medium is equal to the interaction of two nucleons in the vacuum. During the last years a lot of effort has been made to get reliable solutions for this nuclear many-body problem with realistic interactions. Now it seems, that the classical many-body perturbation expansions like e.g. the Brueckner theory[1)] the "Exponential S" method[2)] and the Jastrow-variational method[3)] yield results, which are in good agreement, if the effects of 3-body correlations are taken into account in the Brueckner calculation[4)]. This might be interpreted as a good indication for the convergence of the Brueckner expansion after inclusion of 3-body correlations. This indication is supported by the evaluation of m-body ring diagrams[5)]. In figure 1 we display results for the energy of nuclear matter after inclusion of three- and four-body ring diagrams. From the ratio of the four-versus three-body contributions one may hope that the higher order ring diagrams give negligible contributions. Therefor one may conclude, that the technique to perform classical many-body calculations of nuclear matter is essentially solved. If one still finds discrepancies between the calculated result and the empirical values, one may doubt the basic assumptions of the underlined model.

A possible extension of the classical nuclear many-body model is to consider excitations of the nucleons e.g. to the Δ(3,3) resonance. To demonstrate that the inclusions of such isobar terms is not extremely exotic, we recall the fact, that already for the evaluation of the effective NN interaction terms with intermediate Δ-excitation play a very important role. Such isobar terms give a considerable contribution to a medium range attraction of the NN interaction[6)]. In a more phenomenological one-boson-exchange (OBE) model of the NN interaction, these contributions are simulated by the exchange of an effective σ-meson. In the nuclear medium, however, the attractive isobar terms in the effective NN interaction are reduced due to

Pauli-and dispersive effects.

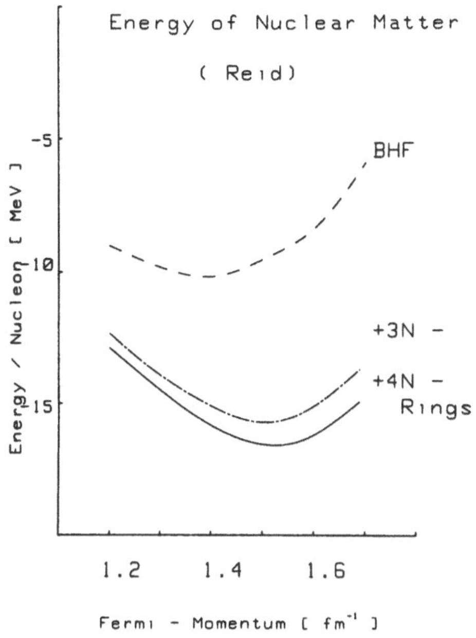

Figure 1:

Contribution of ring diagrams to the binding energy of nuclear matter of various density. While the dashed line represents the result of a Brueckner-Hartree-Fock calculation, the dashed-dotted curve contains in addition the contribution of 3-body rings whereas the solid curve is obtained if also the 4-body rings are included. The Reid soft core potential has been used for the NN interaction.

These quenching effects are absent if the medium range attraction is treated phenomenologically. Therefore a many-body calculation which treats the Δ-excitations in the NN interaction explicitly, typically yields a smaller binding energy for a nuclear system than a more phenomenological calculation[7].

In a nuclear many-body system, however, one can also consider terms with intermediate Δ-excitations, which involve three or more nucleons. In a phenomenological calculation such many-body terms would be taken into account assuming effective many-nucleon forces. Microscopic calculations for such many-body terms have been performed for 3-body systems[8] and light nuclei[9]. Such investigations show, that 3-body terms with intermediate Δ-excitations yield contributions to the binding energy, which are of similar importance as the effect of 3-body correlations. In this contribution we would like to discuss the effects of n-body ring diagrams with intermediate Δ-excitations on the binding energy of nuclear matter. We will see, that the series of such n-body ring diagrams diverges at a density which is about twice the normal nuclear matter density. We will demonstrate, that by a self-consistent renormalization of the residual interaction these divergencies can be counterbalanced. Preliminary calculations show, that the reordered perturbation expansion leads to finite, attractive contributions from many-body terms with Δ-excitations.

2. Ring diagrams and effective meson exchange in nuclear matter

On of the major problems for calculating higher order terms in a many-body theory of nuclear matter, is to find a suitable representation of the two-body interaction. The effective interaction of two nucleons, approximated by the Brueckner G-matrix, or also the transition potentials NN↔NΔ etc. in nuclear matter depend in general on three momenta. For the calculation of ring diagrams, it seems to be most efficient to represent the interaction in terms of particle-hole (ph) momenta[10]. These are the total ph momentum k, which is conserved by the interaction and to relative ph momenta for the initial and final state. In addition one has to consider spin and isospin quantum numbers which are conveniently chosen to be the spin and isospin of the ph states with $\vec{k}$ being the symmetry axis. The ph momentum $\vec{k}$ and the spin-isospin quantum numbers S, M, T are conserved throughout the whole ring diagram. This means, that the contribution to the binding energy per nucleon can be calculated as

$$\Delta E = \frac{3}{4k_F^3} \sum_{STM} (2T+1) \int_0^\infty k^2 dk \; F_{SMT}(k) \tag{1}$$

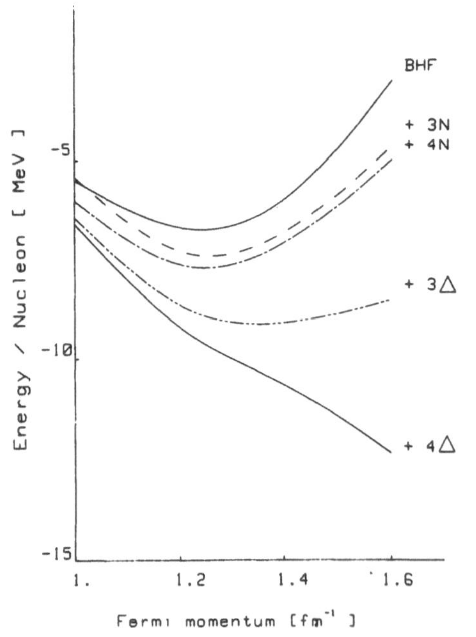

Figure 2:

Contribution of ring diagrams to the binding energy of nuclear matter. The upper solid line displays the result of a BHF calculation. The dashed and dashed-dotted curves are obtained if pure nucleonic ring diagrams are considered with 3 and with up to 4 nucleons respectively. The dahsed-dot-dot curve shows the energies of calculations which include in addition the 3-body ring diagrams with Δ excitations and the lower solid curve is obtained when all 3- and 4-body ring diagrams are considered with inclusion of Δ-terms. The potential MDFPΔ1 of ref. 7 has been used for the NN interaction.

The amplitude $F_{SMT}(k)$ has to be calculated for each ring diagram individually integrating over all relative ph momenta. Using this technique, one can evaluate the three- and four-body ring diagrams without and with intermediate Δ excitations. Results of such calculations[5] are displayed in Fig. 2. One sees immediately that the three-body terms with intermediate Δ excitations are of the same importance as the 3N-body terms. The contribution of four-body ring diagrams with Δ excitations, however, gets increasingly larger with increasing density. Therefore one may ask, if the series of such n-body diagrams converges. To allow a calculation of ring diagrams to any order n with inclusion of Δ excitations, one would like to simplify the expression for the ph interaction by averaging out the dependence on the relative momenta. For that purpose we first calculate the structure function P(q) in the different spin-isospin channels. The structure function or polarization function describes the response of the nuclear system for an external particle-hole probe with spin-isospin quantum number SMT, a momentum $\vec{q}$ and an energy ω. If we restrict ourselves for the moment to the ω=0 limit, the structure function can be calculated as

$$P^{SMT}(q) = \int_R d^3p\,\Pi(q,\vec{p}) + \int_R d^3p \int_R dp'\ \Pi(q,\vec{p})\ \tilde{g}^{SMT}(\vec{p},\vec{p}\,',q)\ \Pi(q,\vec{p}\,') \qquad (2)$$

where $\Pi(q,\vec{p})$ is the unperturbed particle-hole propagator

$$\Pi(q,\vec{p}) = \frac{1}{(2\pi)^3}\ \frac{1}{\varepsilon_p(\frac{1}{2}q+\vec{p})-\varepsilon_n(-\frac{1}{2}q+\vec{p})} \qquad (3)$$

The interaction $\tilde{g}^{SMT}$ is the reducible particle-hole interaction which can be calculated from the irreducible ph-interaction g^{SMT} by iteration to all orders.

$$\tilde{g}^{SMT}(\vec{p},\vec{p},',q)=g^{SMT}(\vec{p},\vec{p}\,',q) + \int_R d^3p''g^{SMT}(\vec{p},\vec{p}\,'',q)\ \Pi(q,\vec{p}\,'')\tilde{g}^{SMT}(\vec{p}\,'',\vec{p}\,',q) \qquad (4)$$

If we now approximate the irreducible ph interaction by the Brueckner G-matrix, the integral equation (4) can be solved and the structure function can be calculated. If on the other hand, we assume an irreducible ph interaction, which does not depend on the relative ph momenta, also the reducible ph interaction $\tilde{g}$ depends on the momentum q only and the equation (4) can be solved immediately

$$\tilde{g}^{SMT}(q) = \frac{g^{SMT}(q)}{1-g^{SMT}(q)\Pi_0(q)} \qquad (5)$$

where $\Pi_0(q)$ is the integrated ph propagator, which is often referred to as the Lindhard function. Now we can also define an irreducible ph interaction g which does not depend on relative momenta by requesting, that the structure functions P(q) calculated for this simple interaction g is the same as the structure function calculated for the complicated G-Matrix. This prescription yields a very reasonable averaging procedure for the relative momenta. Now for such a simple interaction we can calculate the results for the ring diagrams to all order, using the standard method of reference 11.

$$E/A = -\frac{1}{A}\frac{1}{2}\frac{i\hbar}{2(\pi)^4}\int_{-\infty}^{\infty} d\omega \int d^3q \sum_{SMT}\int_0^1 d\lambda\, g^{SMT}(q)\Pi^0(q,\omega)^2\tilde{g}_\lambda^{SMT}(q,\omega) \quad (6)$$

where $\Pi^0(q,\omega)$ is the Lindhard function for energy transfer ω and

$$\tilde{g}_\lambda^{SMT}(q,\omega) = \frac{\lambda g^{SMT}(q)}{1-\lambda g^{SMT}(q)\Pi^0(q,\omega)} \quad (7)$$

which is an extension of eq. (5) for an energy transfer ω and a reduction factor λ for the interaction strength. To take out the terms which are already contained in the BHF approximation, one should subtract from eq. (6) the terms which are of second order of g^{SMT}. It is straight forward to extend the formalism described so far to include also all possible ring diagrams with intermediate Δ excitations. The simplyfied transition potentials for nucleon particle-hole to Δ-hole etc., which depend on momentum transfer only, can be derived from NΔ structure functions in the same way as it has been discussed before for the NN interaction. Also the equations 6 and 7 are easily extended to include Δ states. The energy of nuclear matter with inclusion of isobar ring diagrams is represented by the dashed-dot-dot curve in Fig. 3.

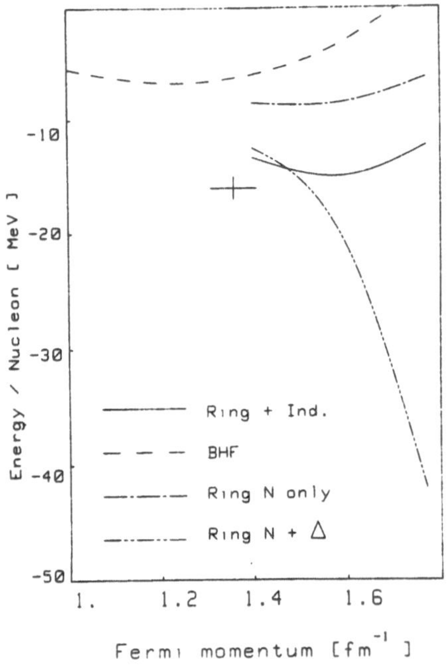

Figure 3:

Contribution of ring diagrams to the binding energy of nuclear matter. While the dashed curve shows the results of the BHF approximation, the dash-dot and dash-dot-dot curves are obtained if the contribution of ring diagrams without and with Δ excitations are added. The solid curve shows the result for the extended ring diagrams using the induced interaction. The potential MDFPΔ2 of ref. 7 is used in the BHF approximation whereas the OBEP of ref. 15 has been used to get the preliminary results for the ring diagrams.

This result confirms very much the suspicion, that the ring diagrams diverge at higher densities. Notice the change of the energy scale. To understand this critical behaviour of the ring diagrams, we analyzed the contribution of the single terms to the expression (6) in detail and find that almost the whole contribution comes from the term with S=T=1, M=0 since the reducible ph interaction $\tilde{g}$ in this channel becomes very big.

For these spin-isospin quantum numbers, the interaction is dominated by a direct one-pion exchange. Keeping this in mind, one would interprete the reducible ph interaction, which contains the coupling of the pion to the ph and Δh-states of the medium which have the same quantum numbers as the pion, as the exchange of a pion which polarizes the surrounding nuclear medium. In this sense the ratio of the reducible ph interaction $\tilde{g}$ versus the irreducible one g can be interpreted as an enhancement factor of the one-pion exchange in the nuclear medium.

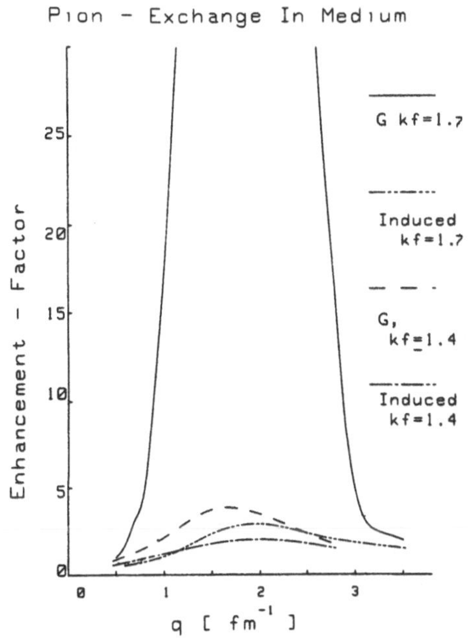

Figure 4:
Enhancement factor for the effective one-pion exchange in the nuclear medium $\tilde{g}/g$ in the limit ω=0. Results are presented using the G-matrix approximation by the dashed and the solid line for a Fermi momentum of 1.4 and 1.77 fm-1, respectively. With inclusion of the induced interaction, the dashed-dot and dashed-dot-dot lines are obtained for these densities

Such enhancement factors for the limit ω=0 are displayed in Fig. 4. One can see, that this enhancement factor diverges at twice the normal nuclear matter density. This behaviour is directly connected to the phenomenon of pion condensation, which has extensively been discussed in the literature[13]. At the critical point for the phase transition, the self-energy of the pion $\Pi_{self}(q)$ becomes so large and attractive, that the pion propagator in the medium

$$\frac{1}{k^2+m_\pi^2+\Pi_{self}(q)-\omega^2} \rightarrow \infty \qquad (8)$$

diverges for ω=0. Now, if the pion propagator diverges, the range of the effective one-pion exchange in the medium gets infinitly large and the normal many-body theory does not converge any more. Therefore also the contribution of the ring diagrams tends to diverge.

Now, in order to improve the standard many-body theory, let us consider the calculation of the pion self-energy. In a model, which considers nucleon ph states only and assumes a simple ph interaction, the pion self-energy can be calculated as

$$\Pi_{self}(q) \propto \frac{\Pi_0(q)}{1-\gamma(q)\Pi_0(q)} \qquad (9)$$

where Π_0 is the Lindhard function and the constant $\gamma(q)$ characterizes the irreducible ph interaction in the pion channel minus the direct bare one-pion exchange contribution. The constant γ represents the Pauli exchange term of the one-pion exchange and other mesons plus correlations effects. Now we have learned, that the one-pion exchange gets very large in the nuclear medium. This means that also the exchange contribution represented by the constant γ should be modified. This renormalization makes γ more repulsive and therefore the pion self-energy less attractive. This means, that the same mechanism, which leads to pion condensation, makes γ more repulsive and thereby prevents this phase transition[14)]. This renormalization of the ph interaction has been discussed under the name of induced interaction[15)]. To calculate this induced interaction, self-consistently, one has to solve a set of nonlinear integral equations[16)]. This can be done by iteration, if one can choose a reasonable starting point for the iteration scheme.

If now the induced interaction tends to prevent the phase transition to pion condensation, one can also expect, that it will lead to a reduction of the enhancement factors for the one-pion exchange in the nuclear medium. This can be seen from Figure 4. With such a reduced one-pion exchange in the nuclear medium, we also expect smaller values for the ring diagrams. Indeed, preliminary calculations show (see solid curve of Figure 3) that the energy contribution from these extended ring diagrams does not increase that dramatically any more with increasing density. This means, that the divergent behaviour of the ring diagrams with Δ excitations at higher densities is compensated and saturation is obtained at a reasonable density.

References

1) B.D. Day, Rev. Mod. Phys., 39 (1967) 771.
2) H. Kümmel, K.H. Lührmann, J.G. Zabolitzky, Phys. Rep. 36.
3) V.R. Pandharipande, Lecture Notes in Physics, 142 (Springer, Berlin, 1981).
4) B.D. Day, Phys. Rev. C24 (1981) 1203.
5) W.H. Dickhoff, A. Faessler, H. Müther, Nucl. Phys. A389 (1982) 492.
6) K. Holinde, R. Machleidt, M.R. Anastasio, A. Faessler, H. Müther, Phys. Rev. C18 (1978) 870.
7) M.R. Anastasio, A. Faessler, H. Müther, K. Holinde, R. Machleidt, Phys. Rev. C18 (1978) 2416.
8) C. Hajduk, P.U. Sauer, Nucl. Phys. A322 (1979) 329.
9) A. Faessler, H. Müther, K. Shimizu, W. Wadia, Nucl. Phys. A333 (1980) 428.
10) W.H. Dickhoff, A. Faessler, J. Meyer-ter-Vehn, H. Müther, Phys. Rev. C23 (1981) 1154.
11) A.L. Fetter, J.D. Walecka, "Quantum Theory of Many-Particle Systems", McGraw Hill (1971).
12) K. Holinde, K. Erkelenz, P. Alzetta, Nucl. Phys. A198 (1972) 598.
13) G.E. Brown, W. Weise, Phys. Rep. 27 (1976) 1.
14) W.H. Dickhoff, A. Faessler, J. Meyer-ter-Vehn, H. Müther, Nucl. Phys. A368 (1981) 445.
15) S. Babu, G.E. Brown, Ann. of Phys. 78 (1973) 1.
16) W.H. Dickhoff, A. Faessler, H. Müther, Shi-Shu Wu, Nucl. Phys. (1982) in press.

RANDOM WALK IN FOCK SPACE *

L. Szybisz and John G. Zabolitzky
Institut für theoretische Physik, Universität zu Köln
Zülpicher Straße 77, 5ooo Köln 41

Abstract

We describe a Monte-Carlo algorithm to solve exactly the ground-state problem for a system of up to four nucleons interacting via a scalar neutral meson field. The mesonic degrees of freedom are treated exactly without recourse to the potential approximation.

It has become clear since several years /1/ that atomic nuclei cannot be described as systems of nucleons interacting via two-body forces to arbitrary accuracy. One possible way to overcome this potential approximation is the inclusion of nucleon interactions via explicit meson fields instead of potentials. This problem can be solved exactly by means of Monte-Carlo methods in the special case of a scalar, neutral meson field providing a test case for other, more approximate and more generally applicable treatments.

The Hamiltonian we wish to discuss and which defines our model is given by

$$H = T - V = \sum \omega(k)\, a_k^+ a_k + \sum t(p)\, a_p^+ a_p - \sum g\, \rho(k\, p\, p')\, (2\omega(k))^{-1/2} \left(a_k^+ a_p^+ a_{p'} + a_k\, a_{p'}^+ a_p\right) \tag{1}$$

where momenta p refer to nucleons, momenta k to mesons, and kinetic energies are given by

$$\omega(k) = \sqrt{\mu^2 + k^2}$$
$$t(p) = \sqrt{m_B^2 + p^2}\,. \tag{2}$$

In order to take care of neglected other meson degrees of freedom and simultaneously avoid any possible divergencies we from the very beginning introduce a nucleon formfactor in the center-of-mass rest frame

$$\rho(q) = e^{-s q^2}. \tag{3}$$

In order to completely specify the model it is then only required to state numerical values for the meson mass μ , the bare nucleon mass m_B, the geometrical size of the nucleon as meson source/sink $r_N=\sqrt{6s}$ and the coupling constant g. The physical nucleon mass will then be obtained from the Schrödinger equation for the Hamiltonian (1)

$$H \Psi = m_{Phys} \Psi \tag{4}$$

within the space of wavefunctions of one nucleon plus 0,1,2,..., ∞ mesons. The binding energy of the A nucleon system, E_A, is then obtained from eq. (4) in the space of A nucleons plus 0,1,2,..., ∞ mesons,

$$H \Psi = (A m_{Phys} - E_A) \Psi . \tag{5}$$

The wavefunction is written as a superposition of states with a definite number of mesons, N=0,1,2,..., ∞

$$\begin{aligned} \Psi &= \sum c_{p_j}^{N\, k_i} \, | N, k_1 \dots k_N ; p_1 \dots p_A \rangle \\ &= \sum_{N=0}^{\infty} \sum_{\substack{k_1 \dots k_N \\ p_1 \dots p_A}} c_{p_1 \dots p_A}^{N, k_1 \dots k_N} \prod_{i=1}^{N} a_{k_i}^{+} \prod_{j=1}^{A} a_{p_j}^{+} \, | 0 \rangle . \end{aligned} \tag{6}$$

Because of our choice (3) we have $\rho(q) \geqslant 0$. Besides this property the functional form of ρ really does not matter for the present calculation and eq. (3) was adopted just as a matter of convenience. With this provision, we have $c \geqslant 0$ for all components in the ground-state wavefunction. The latter statement holds true since we are concerned only with the spatial part of the wavefunction. Limiting our attention to systems with four nucleons maximum the antisymmetry of the nucleon part of the wavefunction is maintained via the spin/ isospin degrees of freedom of the nucleons. Eq. (6) therefore describes a completely symmetric state with respect to nucleon interchange or meson interchange. Thereby we avoid the well-known Fermion disease present in Monte-Carlo methods in general /2/.

The Monte-Carlo algorithm is now spelled out rather easily. Eqs. (4) or (5) are rewritten, denoting the eigenvalue by $E_0 < 0$,

$$\psi = \frac{1}{T-E_0}\int V\psi = \int K\psi . \tag{7}$$

With above provisions, any matrix element of the integral Kernel K between states from eq. (6) is positive. The same holds true for the coefficients describing the wavefunction. Therefore, eq. (7) may be translated to a random walk process with transition probability K and terminal probability distribution $\psi = \psi^{[\infty]}$

$$\psi^{[\nu+1]} = \int K \psi^{[\nu]} \tag{8}$$

since it may be shown that for all eigenvalues λ_i of K

$$-1 \leq \lambda_i \leq +1 . \tag{9}$$

In other words, eq. (7) may be solved by iteration. In order to have a stable iteration E_o is required to be the correct eigenvalue of eqs. (4), (5). By varying some trial eigenvalue used in place of the yet unknown E_o one may stabilize the iteration and thereby determine E_o.

A small technical problem arises because of the eigenvalue -1, eq. (9). Iteration (8) will converge to the eigenstate corresponding to the eigenvalue of largest modulus. Obviously, we are interested in the eigenstate corresponding to $\lambda = +1$. Therefore, iteration sequence (8) is modified to

$$\begin{aligned} \psi^{[\nu+1]} &= \frac{1}{2}\int K\psi^{[\nu]} + \frac{1}{2}\iint KK\psi^{[\nu]} \\ c_\eta^{[\nu+1]} &= \frac{1}{2}\left(\lambda_\eta + \lambda_\eta^2\right) c_\eta^{[\nu]} . \end{aligned} \tag{1o}$$

It is seen that components corresponding to negative eigenvalues are damped out. The random walk (1o), creating and annihilating mesons in its process, converges fastest to the desired probability density $\psi^{[\infty]}$ (in the Monte-Carlo sense; these are <u>not</u> physical probabilities, of course) if the eigenvalues (9) other than +1 are far away from +1. With increasing strength of the interaction, i.e. with increasing nucleon number, increasing coupling constant, or decreasing nucleon size, the convergence gets slower since the eigenvalues approach +1 more closely. Nevertheless, several hundred iterations are sufficient in most cases for three digit accuracy in the energy /3/.

Calculations have been performed using for the meson mass μ = 139 MeV. In the static limit /4/, i.e. $m_B = \infty$, one may fix the nucleons some distance r apart and calculate a static nucleon-nucleon potential as function of r (Born-Oppenheimer approximation). We determine the coupling constant g by requiring that this potential bind the deuteron with E_d=2.22 MeV. As a function of the free parameter r_N, the nucleon radius, the resulting potentials are shown in fig. 1. The limiting case $r_N \rightarrow 0$ will yield the Yukawa potential.

We limit our attention now to the case r_N=1 fm. Having determined numerical values for the meson mass, source radius, and - by above scheme - coupling constant g=2.568 we determine the bare nucleon mass m_B from eq. (4) with the requirement m_{Phys} = 938 MeV. Solving eq. (4) in the static limit which may be done analytically we obtain m_B = 965.3 MeV. Exact solution of eq. (4) my means of the proposed Fock-space Monte-Carlo method yields m_B = 962.5 MeV, i.e. a 1o % correction.

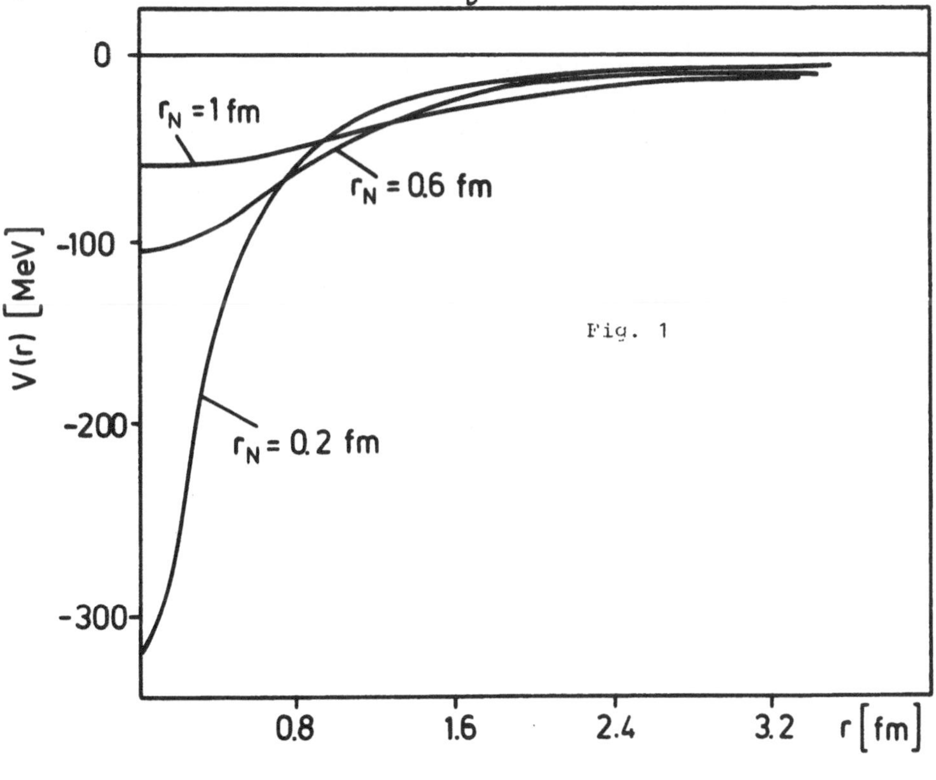

Fig. 1

Employing the potential model for the A = 2,3,4 nuclei we obtain E(A=2) = 2.22 MeV by construction, E(A=3) = 17 MeV, E(A=4) = 5o MeV. The standard Green's-Function Monte-Carlo (GFMC) procedure /5/ has been used to obtain these results free of any approximation but the potential description. It is seen that the binding increases drastically with the number of nucleons leading to severe overbinding. This is of course due to the purely attractive nature of the potentials fig. 1 or equivalently the omission of vector mesons in the Hamiltonian (1). In particular the present model would lead to collapse of nuclear matter. Unfortunately at the present point in time we do not yet have the exact eigenvalues from eq. (5) to compare with above numbers.

Many thanks are due to F. Coester for numerous discussions.

References

1. H. Kümmel, K.H. Lührmann, J.G. Zabolitzky, Phys.Rep.36C(1978)1
2. D. Arnow, M.H. Kalos, M.A. Lee, K.E. Schmidt, J.Chem.Phys.77(1982)1
   D.M. Ceperley, Proceedings of the NATO ARW on Monte-Carlo Methods in Quantum Problems (1982), to be published
   K.E. Schmidt, M.H. Kalos, to be published
3. L. Szybisz, J.G. Zabolitzky, Proceedings of the NATO ARW on Monte-Carlo Methods in Quantum Problems (1982), to be published
4. E.M. Henley, W. Thirring, Elementary Quantum Field Theory, Mc-Graw-Hill, New York 1962
5. M.H. Kalos, Phys.Rev.128(1962)1891

* Supported by the Deutsche Forschungsgemeinschaft and the Alexander-von-Humboldt-Foundation.

NUCLEAR MATTER PROPERTIES IN THE BHF APPROXIMATION WITH THE PARIS N-N POTENTIAL AND MODELS OF 3N INTERACTIONS[*)]

A. LEJEUNE,

Université de Liège,
Physique Nucléaire Théorique
Institut de Physique au Sart Tilman
Bâtiment B.5
B-4000 LIEGE 1, Belgique

and

M. MARTZOLFF and P. GRANGE

Physique Nucléaire Théorique, C.R.N., B.P. 20,
F-67037 STRASBOURG CEDEX, France

*) Talk presented by A. Lejeune at the "Third International Conference on Recent Progress in Many-Body Theories", Altenberg, 1983.

1. Introduction. The theoretical derivation of the free N-N interaction given by the Paris group[1)] takes into account data from the deuteron and from revised analysis of NN , πN and ππ scatterings. It provides a fairly reliable description of the long and medium range of the free N-N interaction. A strong momentum dependence mockes up a non locality at short distance and as most realistic N-N interactions the Paris potential cannot be directly used for calculating nuclear properties. An effective interaction must first be constructed which is in general identified with the Brueckner reaction matrix.

Our main goals concern : i) besides the binding energy, certain aspects of the Paris potential in nuclear matter which to our knowledge have not yet been investigated i.e. single particle properties at the Fermi surface ; ii) the renormalization via a 3-body force of the effective interaction to account for empirical saturation properties of the infinite medium and its effects on single particle properties ; iii) the construction of an effective interaction for nucleon-nucleus scattering at low and intermediate energy which takes into account the results of studies i) and ii) above. This program has already been investigated to a certain extent[2,13)] with however no attention to the last two points. We report here and discuss binding energy calculations and preliminary results concerning point ii).

A necessary step to study the effect of the 3N-force is a reliable and still tractable treatment of the nuclear many body problem with the 2N-force alone. Such a possibility has been repeatedly confirmed by the Liège group[3)]. Their approach is essentially a modified version of the Bethe-Brueckner expansion for the binding energy. Our approach to point ii) is based on an effective 2N-force V_3 which would then be added to the free 2N-force. This V_3 is derived in the spirit of Ref. [4)] either from a 3N-force including ππ , πρ and ρρ exchange or taken from the study of Coon et al.[5)]. Its effects on single particles properties mentioned above as well as on saturation are envisaged.

We give now some specific aspects of our formal approach to the treatment of 2 and 3 body forces and then report on preliminaries results.

2. Theoretical background. The nuclear matter 2-body correlated wave function is solution of the integral equation

$$\Psi(k\underline{r}) = \chi(k\underline{r}) + \frac{m}{\hbar^2}\int d\underline{r}' G^k(\underline{r},\underline{r}')\{V(\underline{r}',\nabla r')\psi(k\underline{r}')\} \quad , \tag{1}$$

where $\chi(k\underline{r})$ stands for a plane wave and the 2N-force V contains an explicit momentum dependence. $G^k(\underline{r},\underline{r}')$ is the nuclear matter Green's function which in the ℓ-partial wave is

$$G_\ell^{k_0}(r,r') = \frac{1}{2\pi^2}\{PP\int_0^\infty \frac{k'^2dk'j_\ell(k'r)j_\ell(k'r')\bar{f}(k')}{\tilde{D}(k')}$$

$$- i\pi k_0^2 j_\ell(k_0r)j_\ell(k_0r')\bar{f}(k_0)|D'(k_0)|^{-1}\} \quad , \tag{2}$$

$j_\ell(x)$ is the spherical Bessel function, $\bar{f}(k)$ the angle averaged Pauli operator and $\tilde{D}(k')$ the angle averaged energy denominator, both specified in Eqs. (22) to (28) of Ref. [6]. In Eq. (2) k_0 is the real root of the denominator $D(k)$.

It is found that $G_\ell^{k_0}(r,r')$ obeys the following differential equation

$$[\frac{1}{r'^2}\frac{\partial}{\partial r'}r'^2\frac{\partial}{\partial r'} + k_0^2 - \frac{\ell(\ell+1)}{r'^2}]\; G_\ell^{k_0}(r,r') = \frac{k_0}{2\pi}\cdot|D'(k_0)|^{-1}\cdot\; \frac{\delta(r-r')}{r^2}$$

$$+ H_\ell(r,r') \quad , \tag{3}$$

with

$$H_\ell(r,r') = \frac{-1}{2\pi^2}\int_0^\infty [\bar{f}(k)-1]j_\ell(kr)j_\ell(kr')k^2\frac{(k^2-k_0^2)}{D(k)}dk \quad . \tag{4}$$

Combining a partial wave reduction of equation (1) and the use of Eq. (3) leads to coupled integral equations for the radial part $u_{\ell\ell'}^{JST}(r)$ of the correlated wave function. Upon specifying the Green's function to the scattering case this equation embodies the proper asymptotic condition and therefore allows for an easy check of the numerical codes in terms of the known values of the N-N phase shifts of the Paris potential.

In the BHF approximation the mass operator is given by[3]

$$M_\rho(k,\varepsilon) = \sum_{j\leq k_F} \langle \underline{j},\underline{k}|G(\varepsilon+e_\rho(j))|\underline{j}\underline{k}\rangle_A \quad , \tag{5}$$

with the following self consistent choice for the energy $e_\rho(k)$

$$e_\rho(k) = \frac{\hbar^2k^2}{2m} + \mathrm{Re}\; M_\rho(k,e_\rho(k)) \quad \text{for all } k \quad . \tag{6}$$

The partial wave matrix elements of G up to $\ell=5$ are obtained as in Ref. [7] with however a proper account of the momentum dependent terms. For partial waves with $\ell>5$ the summation in Eq. (5) is performed, when indicated, analytically in the Born approximation with G replaced by V the free Paris

2N-force. The effect of including higher order partial waves (HOPW) with $\ell>5$ on M_ρ and on the binding energy has already been investigated for model interaction[8] and is discussed below for the specific case of the Paris potential.

In term of the mass operator and Eq. (6), the effective mass m^* is given by[3]

$$\frac{m^*}{m} = 1 - \frac{d\ \mathrm{Re}\ M_\rho(k,e_\rho(k))}{d\ e_\rho(k)} \quad , \tag{7}$$

and characterizes the energy dependence of the potential and the density of single particle levels. A distinctive feature of the self consistent continuous choice of Eq. (6) is that it induces a local enhancement of m^* at the Fermi surface. This enhancement has been identified as due to the excitation of low-lying intermediate states and is the subject of other reports in this conference. It is an open question whether or not the specific nature of the Paris potential affects these excitations as compared to other type of N-N interactions.

We describe now briefly our approach to the renormalization of the Brueckner reaction matrix.

The underlying idea is that if long and medium range N-N potentials are mediated by meson exchanges then 3N-forces due to similar exchanges should exist. Even if they were to account for all the defect in saturation obtained from the N-N potential alone, only a 10 to 15 % change in the potential energy is required to bring agreement with the empirical saturation properties of nuclear matter. Thus a perturbative treatment of the 3N-force may be justified as discussed in Refs.[4,10].

Let $W(\underline{r}_1,\underline{r}_2,\underline{r}_3)$ be the 3-body potential to be specified later. From this W an effective density dependent 2 body force $V_3(\underline{r}_{12})$ is introduced[10] using the $\ell=0$ on-shell defect function obtained from the Paris potential. The effect of $V_3(\underline{r}_{12})$ in nuclear matter is evaluated perturbatively[9,10]. A nuclear matter calculation is performed with the Paris potential plus the effective force $V_3(\underline{r}_{12})$. The results are compared with those of a similar calculation with the Paris potential alone. We stress that this procedure involves many successive approximations and conclusions on the three-body force itself must be drawn with cautions.

The 3-body forces we shall envisage arises from the exchange of 2π , $\pi\rho$ and 2ρ between three nucleons as derived in Ref.[16] and from the exchange of 2π only as derived by Coon et al.[5] using current algebra and PCAC constraints.

3. Numerical procedure and checks. First to be calculated are the Green's function $G_\ell(r,r')$. Different numerical treatments are retained for s-waves (Filon integration) and waves with $\ell>0$ (Chebytchev integration). In both cases, proper care is taken of the principal value integral present in the real part of $G_\ell(r,r')$. The integral equation for the radial part of the correlated wave function is then solved using a combination of matrix inversion and iterative schemes starting from an initial guess for $u^{JST}_{\ell\ell'}(r)$, either the spherical Bessel function $j_\ell(kr)$ if $\ell=\ell'$ or a very small constant if $\ell \neq \ell'$. Through the iterative procedure a better precision in the small r region is reached more efficiently than from a simple matrix inversion technic alone. The continuous choice retained for the auxiliary single particle field induces a scattering-type behaviour of the real part of the correlation function $u^{JST}_{\ell\ell'}(r)$. To check our numerical procedure in this respect, we specialize Green's propagator to the free scattering case and look for the smallest size of the grid (r,r') which leads to N-N phase shifts in agreement with known results[1)] to within 0.1 degree on the average. Specific to the Paris potential we achieve this goal only with a dense scanning of the small (r,r') region. Typically the r integration of the integral equations is performed using Chebytchev summations in the intervals 0.-0.133, 0.133-0.321, 0.321-0.686, 0.686-2.05 fm with respectively 7, 4, 7 and 5 points. A modified Gauss integration[11)] with 7 points is retained for $2.05<r\leq 12$ fm . Our numerical results indicate that an uncertainty $|\Delta\delta_\ell| \approx 0.2$ degree on the average within the scattering energy range 150-350 MeV leads —through imprecisions in the single particle field $U(k)$ for $k\geq 2.5\ fm^{-1}$— to an uncertainty in the binding energy B/A which we estimate as $|\Delta B/A| \approx 1$ MeV/A .

4. Preliminary results and discussion. Our preliminary results concern the binding energy of nuclear matter only and were checked by independent calculations in momentum space[11)]. However we have not yet available definite results about the detailed behaviour of single particle quantities at the Fermi surface such as the effective mass. These particular calculations are lengthly and need special care.

In table 1, we show the various self consistent partial waves contribution to the potential energy of nuclear matter at four different Fermi momenta, coupled states $^3G_5$-$^3I_5$ and $^3H_6$-$^3J_6$ are not included. The real and imaginary parts of the G-matrix have been kept coupled during the iterative procedure to self-consistency. Saturation is found at $k_F = 1.62\ fm^{-1}$ with $B/A = -16.1$ MeV/A . However, the uncertainty is estimated to be of the

Table I : Potential energies in MeV per nucleon in partial waves states for different Fermi momenta in fm^{-1} . The 2N-force is the Paris interaction used in the continuous version of the BHF approximation.

States	$k_F = 1.10\ fm^{-1}$	$k_F = 1.36\ fm^{-1}$	$k_F = 1.6\ fm^{-1}$	$k_F = 1.7\ fm^{-1}$
$^3S_1$	-12.44	-18.24	-23.55	-24.67
$^3D_1$	0.59	1.45	2.76	3.52
$^3D_2$	- 1.73	- 4.16	- 7.89	- 9.98
^{3}even ($\ell \leq 4$)	- 0.15	- .47	- 1.17	- 1.48
T.E.	-13.73	-21.42	-29.85	-32.61
$^1S_0$	- 9.61	-15.63	-22.10	-24.74
$^1D_2$	- 1.13	- 2.83	- 5.60	- 7.21
^{1}even ($\ell \leq 4$)	- 0.13	- .45	- 1.14	- 1.51
S.E.	-10.87	-18.91	-28.84	-33.46
$^3P_0$	- 2.00	- 3.73	- 5.62	- 6.36
$^3P_1$	4.73	10.34	18.76	23.44
$^3P_2$	- 3.17	- 7.36	-13.57	-16.77
$^3F_2$	- 0.20	- .61	- 1.26	- 1.61
^{3}odd ($\ell \leq 5$)	0.51	1.40	2.75	3.50
T.O.	- 0.13	0.04	1.06	2.20
$^1P_1$	2.12	4.41	7.64	9.31
^{1}odd ($\ell \leq 5$)	0.36	1.03	2.14	2.80
S.O.	2.48	5.44	9.78	12.11
$\frac{1}{2}\bar{U}$	-22.25	-34.85	-47.85	-51.76
B/A	- 7.20	-11.84	-16.00	-15.80

order of ± 2.0 MeV/A . Of this, ± 0.5 MeV/A is coming from the numerical procedure, i.e. small variations of the N-N phase shifts with different sets of grid points in (r,r') . To reduce five times this uncertainty would require grids in (r,r') of too large size with our available computer capacity. The remaining ± 1.5 MeV/A is due to the use of the angle averaged Q/e operator[3,7] and the neglect of the HOPW with $\ell>5$. If included using the Paris potential in the Born approximation, this particular contribution is found to be 0.22 MeV/A.

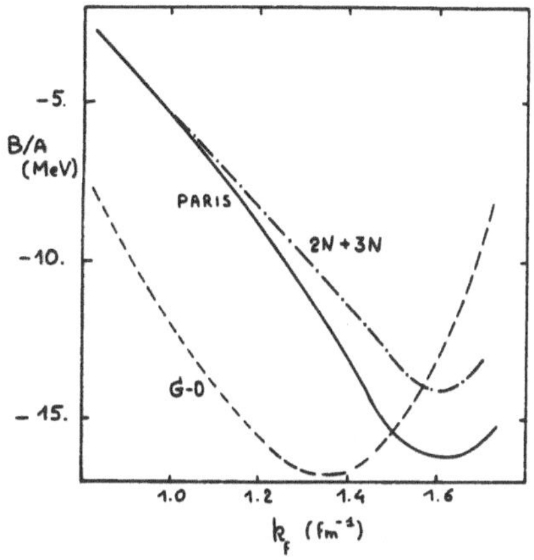

Binding energy of nuclear matter versus k_F . Our resul (full curve). Evaluation based on G-0 interaction[12] (dashed curve). Renormalization with a 3N-force (dash-dotted curve).

In Fig. 1 the continuous curve shows the binding energy of nuclear matter as a function of k_F . Also shown —dashed curve— is the renormalized saturation curve obtained with the effective force G-0 of Ref. [12].
At first sight important changes in the G-matrix elements would be required to reach the empirical saturation point. Recently, Matin and Dey[13] have studied the saturation properties of the Paris potential in the framework of the BHF approximation. For the standard choice of the single particle field their result for the binding energy is -13.8 MeV/A at variance with the value quoted in Ref. [1] : -11.2 MeV/A . An error in $^3D_2$ channel — -1.5 MeV/A in excess— and the neglect of HOPW $^3F_3$ and higher — +1 MeV/A— accounts for the difference. For the continuous version of the BHF approximation systematic deviations at all densities also occur in Ref. [13] with respect to the results reported here. It is therefore important to identify the origins of these differences.

In the continuous version of the BHF approximation the G-matrix is complex. Following the procedure of Ref. [6], we have kept the real and imaginary part coupled during the self-consistent calculations. To our understanding it seems that a decoupling procedure was adopted in Ref. [13] as only the real part of G is calculated. It is a simple matter to evaluate the consequences of this approximation. At $k_F = 1.6\ fm^{-1}$ we have performed a complete self-consistent calculation ignoring the imaginary part of the Green's function. An additional binding energy of 2.5 MeV/A is obtained with respect to the complete calculation of Table 1. The HOPW neglected in Ref. [13] give a repulsive contribution of 0.5 MeV/A. As we have not yet investigated in detail the behaviour of the single particle spectrum and of the effective mass at $k \approx k_F$, it is still not known how sensitive is the binding energy to variation around k_F of the single particle potential U. Hence following our analysis, the above approximations lead to an additional attractive contribution to B/A of the order of 3 MeV/A i.e. —within the accuracy of the calculation— essentially the difference with the value of -21.3 MeV/A reported by Matin and Dey[13)] and our result of -16.1 MeV/A. The latter is consistent with the value reported by Day[14)].

In Fig. 1 the dash-dotted curve shows the result of our first attempt to renormalization via a 3N-force in a manner explained above. The 3-body potential retained in this case includes contributions from 2π, $\pi\rho$ and $\rho\rho$ exchanges as derived in Ref. [16)] with form factors IV of masses $\eta_\pi = 7.5\ \mu_\pi$ and $\eta_\rho = 9.5\ \mu_\pi$ and a weak ρ coupling constant. A similar calculation with the Tucson model[5)] of the 3N-force is in progress. Although all possible systematic studies have not yet been carried out in terms of form factors and with the presently available 3N-forces, earlier results[15)] and the dash-dotted curve of Fig. 1 indicate that an extra component of the 3N-force is required to achieve an adequate renormalization. The perturbative approach used here to evaluate the contribution of the 3N-force makes it difficult to justify the addition of a hard-core type of component to this force as in Ref. [15)]. On the other hand the $2\pi - \pi\rho - \rho\rho$ 3N-force mentioned above has been obtained in a way which keeps in the effective 2N-force V_3 invariants build out of spin and isospin only. It is an open question whether or not a weak enough density dependent additive terms to V_3 of the L.S and $(L.S)^2$ type could be found phenomenologically and possibly justified formally which would achieve the desired renormalization.

In summary, we have presented and discussed some results obtained with the Paris potential in the framework of the continuous version of the BHF approximation. These results concern only the binding energy and are not yet complete as the behaviour of single particle properties near the Fermi surface remains to be investigated.

A possible approach to the renormalization of the Brueckner reaction matrix via 3-body forces is emphasized. Its plausibility is discussed through a preliminary treatment with a simple 3N-force. Systematic studies of the saturation effect of various 3-body forces within different treatment of the many body system will hopefully shed some light on the puzzling problem of the saturation of nuclear forces.

We would like to thank the NATO Scientific Affairs Division for the research grant n° 025.81.

References

[1] M. Lacombe, B. Loiseau, J.M. Richard, R. Vinh Mau, J. Côté, P. Pirès and R. de Tourreil, Phys.Rev. C21(1980)861

[2] H.V. von Geramb, in "The Interaction Between Medium Energy Nucleons in Nuclei", 1982, edited by H.O. Meyer, (APS 1983), p. 44

[3] J.-P. Jeukenne, A. Lejeune and C. Mahaux, Phys.Rep. 25C(1976)85

[4] B.A. Loiseau, Y. Nogami and C.K. Ross, Nucl.Phys. A165(1975)601, erratum A176(1971)665

[5] S.A. Coon, M.D. Scadron, P.C. McNamee, B.R. Barrett, D.W.E. Blatt and B.H.J. McKellar, Nucl.Phys. A317(1979)242

[6] J.-P. Jeukenne, A. Lejeune and C. Mahaux, Phys.Rev. C10(1974)1391

[7] K.A. Brueckner and J.L. Gammel, Phys.Rev. 109(1958)1023

[8] P. Grangé, A. Lejeune and C. Mahaux, Nucl.Phys. A319(1979)50

[9] P. Grangé, M. Martzolff, Y. Nogami, D.W.L. Sprung, C.K. Ross, Phys.Lett. 60B(1976)237 ; P. Grangé and M. Martzolff, Lett.Nuovo Cim. 16(1976)156

[10] B.H.J. McKellar, R. Rajaraman, in "Mesons in Nuclei", North-Holland Publ. Comp. (1979) (M. Rho and D.H. Wilkinson, editors)

[11] M.I. Haftel and F. Tabakin, Nucl.Phys. A158(1970)1

[12] X. Campi and D.W.L. Sprung, Nucl.Phys. A194(1972)401

[13] M.A. Matin and M. Dey, Phys.Rev. C27(1983)2356

[14] B.D. Day, Phys.Rev.Lett. 47(1981)226 ; Comm.Nucl.Part.Phys. 11(1983)115

[15] J. Carlson, V.R. Pandharipande and R.B. Wiringa, Nucl.Phys. A401(1983)59

[16] M. Martzolff, B. Loiseau and P. Grangé, Phys.Lett. 92B(1980)46.

THREE-BODY FORCES, RELATIVISTIC EFFECTS, ISOBARS AND PIONS IN NUCLEAR SYSTEMS

R. B. Wiringa

Physics Division, Argonne National Laboratory, Argonne IL 60439, USA

Conventional microscopic calculations in nuclear physics start from a nonrelativistic Hamiltonian of the form

$$H = \sum_i T_i + \sum_{i<j} V_{ij} \tag{1}$$

where T_i is the one-body kinetic energy and V_{ij} is a two-body potential fit to deuteron properties and low-energy scattering data. Realistic V_{ij} that give good data fits can often be written in a convenient operator notation:

$$V_{ij} = \sum_p v^p(r_{ij})O^p_{ij} \tag{2}$$

where the O^p_{ij} are operators such as 1, $\tau_i\cdot\tau_j$, $\sigma_i\cdot\sigma_j$, S_{ij}, $\vec{L}\cdot\vec{S}$, $\vec{L}^2$, etc. (Many recent V_{ij} use p=1,14; we refer to these as v14 models.) The many-body Schrödinger equation is then solved to obtain the ground state energy, wave function, and expectation values of other quantities of interest. Such a procedure gives a qualitative description of nuclear saturation properties, but it is now well established that the simple H of eq.(1) is quantitatively inadequate. For example, the light nuclei are underbound with too large a charge radius,[1] while nuclear matter is overbound at far too high a density.[2] This note reviews recent studies that go beyond the simple H of eq.(1). These include 1) the introduction of three-nucleon potentials, 2) estimates of relativistic effects, 3) the introduction of isobar degrees of freedom in the two-body potential, and 4) probing the influence of pion degrees of freedom on nuclear systems.

The many-body technique used is the variational method.[3] A variational wave function, ψ_v, is constructed as a symmetrized product of two-body correlation operators, F_{ij}, multiplying an unperturbed ground state ϕ:

$$\psi_v = (S \prod_{i<j} F_{ij})\phi. \tag{3}$$

The F_{ij} should reflect the correlations induced by V_{ij}, so an operator expansion similar to eq.(2) is used:

$$F_{ij} = \sum_q f^q(r_{ij};\alpha)O^q_{ij} \tag{4}$$

where α denotes a set of variational parameters and the O^q_{ij} are some subset (such as q=1,8) of the operators used in eq.(2). Variational upper bounds E_v to the ground state energy E_o

$$E_v = \langle\psi_v|H|\psi_v\rangle/\langle\psi_v|\psi_v\rangle \geq E_o \tag{5}$$

are calculated for various parameter sets α, and the lowest energy obtained selects the best ψ_v. Expectation values in eq.(5) are evaluated using Monte Carlo integration in the light nuclei,[4] and a diagrammatic cluster expansion with Fermi-hypernetted-chain and single-operator-chain (FHNC/SOC) summations in infinite matter.[3]

The effect of adding a three-nucleon potential, V_{ijk}, to H has been studied in the three- and four-body nuclei, and in infinite nuclear and neutron matter.[5,6] The models studied include 1) the Tucson[7] two-pion-exchange (TPE) force, derived from πN scattering data and constrained by PCAC and low-energy theorems, and 2) various intermediate-isobar-state TPE potentials combined with phenomenological three-nucleon-repulsion (TNR) - in particular the Urbana model V force.[5] The Tucson model can be visualized as arising from both S- and P-wave processes, as shown in Fig.1a-b, while the Urbana model V simulates processes such as Fig.1b-c.

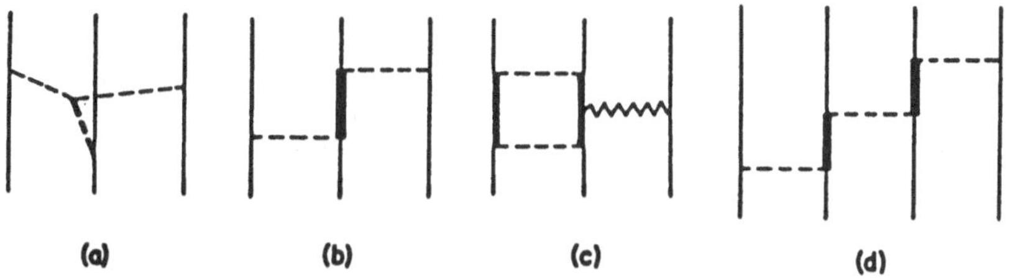

Fig. 1

The results of binding energy caculations for ^3H and ^4He (coulomb force included) for two different V_{ij} - the Argonne v14[8] and Urbana v14[9] models - are shown in Table I, along with the results when either the Tucson or Urbana model V potentials are added. Both V_{ijk} give significant additional attraction to ^3H and ^4He, leaving ^3H slightly underbound and ^4He slightly overbound.

Table I. Binding energy of light nuclei with Argonne v14 (Urbana v14) in MeV

	^3H	^4He
$\langle T_i+V_{ij}\rangle$	-7.0±.1 (-7.2±.1)	-22.1±.3 (-23.8±.2)
$\langle$+Tucson $V_{ijk}\rangle$	-8.1±.1 (-8.1±.1)	-29.8±.5 (-29.3±.5)
$\langle$+Urbana $V_{ijk}\rangle$	-8.2±.1 (-8.1±.1)	-29.6±.4 (-29.1±.4)
Experiment	-8.48	-28.3

In nuclear matter, however, the Tucson model gives additional attraction at all densities, as shown in Fig. 2, thus aggravating the discrepancy with the empirical saturation curve derived from semiempirical mass formulas. The Urbana model V potential saturates matter at a much lower density with less binding, although still some distance from the empirical curve. This improved saturation is a strong argument for including TNR terms in V_{ijk}, in addition to the TPE processes. The results also indicate that the addition of a specific V_{ijk} to different V_{ij} models decreases the differences between the total Hamiltonians in both matter and the light nuclei.[6]

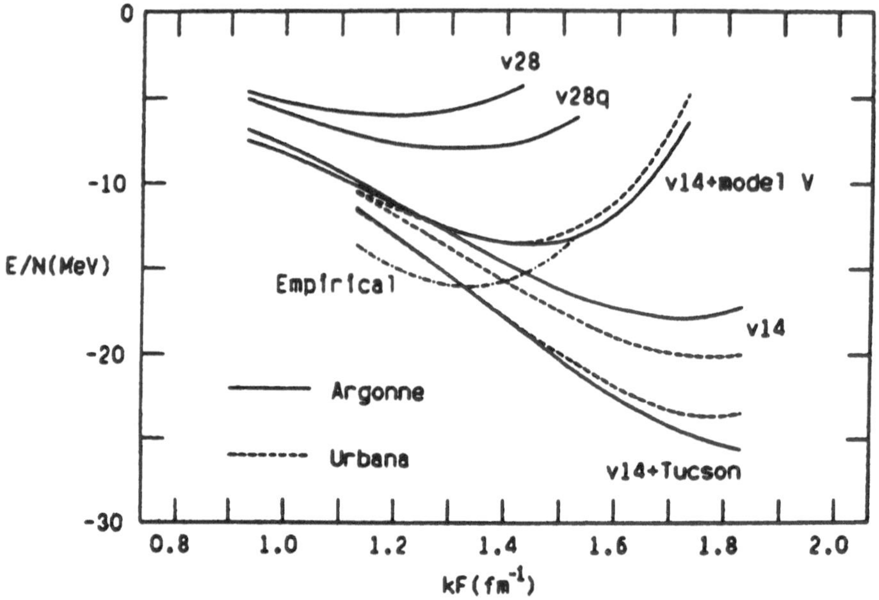

Fig. 2

Calculations of neutron matter can be combined with nuclear matter results to yield the symmetry energy of matter, β.[10] The addition of V_{ijk} to H increases β from 26 to 29 MeV for Argonne v14 + Urbana model V; the empirical value is 30-36 MeV. Further, the TNR term in Urbana model V significantly increases the energy of neutron matter at high density, dominating the energy beyond $\rho \sim 0.3$ fm^{-3}. This leads to a significant stiffening of the equation of state, and a larger maximum mass for neutron stars.

The ^3H results for the Tucson V_{ijk} shown here are in disagreement with some recent momentum-space Faddeev calculations, which find little change in the binding.[11,12] To investigate this discrepancy, the Monte Carlo integration has also been performed with configuration-space Faddeev wave functions of the Los Alamos-Iowa group[13] substituted for the ψ_v of eq.(3). In these calculations the Tucson V_{ijk} adds -1.15 MeV to the Reid soft core and -.94 MeV to the Argonne v14 potentials, in reasonable agreement with the variational results of Table I. Also it appears that the latest momentum-space Faddeev calculations have not converged.[12] The Faddeev wave functions give lower energies than the variational wave functions used to date, however, which indicates a need for improvement in the detailed structure of ψ_V.

Another possible way of going beyond the H of eq.(1) is to include relativistic effects. This can be done by constructing a consistent relativistic multi-body dynamics[14] that has the nonrelativistic limit of eq.(1), and then calculating corrections to this limit in powers of m^{-1}. The relativistic Schrödinger equation for the deuteron,

$$(2(p^2 + m^2)^{1/2} + v)\,\chi = (2\omega + v)\,\chi = M_d\chi \tag{6}$$

where v is some direct two-body interaction, can be manipulated into a more conventional form:

$$E\,\hat{\chi} = (p^2/2m + [1 - Vm/\omega(M_d + 2\omega)]^{-1}V)\,\hat{\chi} \tag{7}$$

where $V=(\omega/m)^{1/2}v(\omega/m)^{1/2}$ and $E=(M_d^2-4m^2)/4m$. This leads to a binding energy for the deuteron:

$$E \simeq E_{NR} + \langle\hat{\chi}|V^2|\hat{\chi}\rangle/4m \tag{8}$$

where E_{NR} is the normal nonrelativistic energy. For the Reid soft core potential, E_{NR}=-2.2 MeV and $\langle V^2\rangle/4m$=1.6 MeV, compared to $\langle V\rangle$=-24.3 MeV. To make E fit the deuteron binding energy, V must be increased by about 8%. In the three- and four-body systems the approximate mass operator is

$$M_n \simeq nm + H_{NR} - \sum_i p_i^4/8m^3 - \sum_{i<j} \{(p_i^2 + p_j^2), V_{ij}\}/8m^2 \tag{9}$$

where H_{NR} is given by eq.(1); in ^{4}He an extra three-body term also appears, but its contribution is completely negligible. The last two terms in eq.(9) plus the rescaling of V constitute the relativistic corrections. They have been calculated[15] in ^{3}H and ^{4}He for the Reid v8 potential using the variational Monte Carlo method described above. The results shown in Table II indicate the relativistic effects may be comparable in magnitude to the three-body potentials.

Table II. Binding energy in light nuclei for Reid v8 in MeV

	$\langle T_i+V_{ij}\rangle$	$.08\langle V_{ij}\rangle$	$-\langle p_i^4/8m^3\rangle$	$-\langle\{(p_i^2+p_j^2),V_{ij}\}/8m^2\rangle$	E
^{3}H	-6.9±.1	-4.1	-2.4	4.8	-8.6±.5
^{4}He	-22.4±.4	-10.7	-6.0	12.4	-26.7±1.4

A proper treatment of the relativistic effects would of course involve refitting V_{ij} to two-nucleon scattering data as well, but the simple scaling used here should provide a reasonable first estimate. Also the m^{-1} expansion of the kinetic energy term probably overestimates that correction; the full relativistic kinetic energy is more difficult to evaluate, however, and is left to future work.

The combination of three-nucleon potential and relativistic correction terms may be able to fix the binding energies of both ^{3}H and ^{4}He, but neither modification alone is likely to do the trick. From Table I it can be seen that the needed correction to the two-body energy is ~ 4 times as large in ^{4}He as in ^{3}H. The V_{ijk} give a correction ~ 6 times as large, since ^{4}He has four triples and is more tightly bound. The relativistic corrections are predominantly one- and two-body effects and give only ~ 2.5 times as much in ^{4}He as in ^{3}H. A combination of the two effects may thus be able to get both binding energies simultaneously.

An alternative to the use of phenomenological V_{ijk} to improve on the H of eq.(1) is to add an additional degree of freedom to the NN potential: the isobar or Δ(1232) resonance.[8] The isobar plays an important role in the TPE processes which generate the intermediate-range attraction of NN interaction and also in the TPE and TNR processes of the three-nucleon interaction. Isobar components can be generated by a potential containing generalized one-pion-exchange (OPE) transition operators with $\pi N\Delta$ couplings. When iterated, such a potential will simulate the contribution of TPE box diagrams in NN scattering, and in three-body clusters will generate contributions corresponding to the TPE and TNR processes of Fig.1b-c. Indeed, many higher-order terms like the four-body process of Fig.1d will also be generated. The transition potential approach does not give as accurate a theoretical picture of the NN interaction as the dispersion theoretic models like Paris,[16] where isobars and higher resonances are implicitly included. However a model with explicit isobars can fit NN data just as well, and may result in a better many-body H because of its ability to generate many-body forces.

Many workers have studied the transition potential approach, both in constructing V_{ij} and using them in many-body calculations. (See ref. 8 for a short review.) Early work frequently considered only a few isobar components in specific channels. Later all possible $\pi N\Delta$ and $\pi\Delta\Delta$ couplings were used, but frequently the important physical feature of a repulsive core interaction in NΔ and $\Delta\Delta$ channels was omitted. Little emphasis was placed on getting high-quality fits to NN data. Many-body calculations using these models were made in lowest-order only, so the benefits of having a better many-body H were largely lost.

Recently we have constructed three new NN potentials: a v14 model that is a conventional NN potential, and two v28 models with explicit isobars[8]. The conventional V_{ij}, intended to serve as a standard for comparison, is the Argonne v14 model, used in the calculations discussed above. The v28 models have 28 operators, i.e. p=1,28 in eq.(2), including an NN part with the same structure as the v14 model, plus 12 transition operators for all possible $\pi N\Delta$ and $\pi\Delta\Delta$ couplings and two central operators for short-range repulsion in NΔ and $\Delta\Delta$ channels. The potential structure is shown schematically in Fig.3. The two v28 models differ in the strength of the $\pi N\Delta$ coupling: one has the "Chew-Low" value $(f^2_{\pi N\Delta}/4\pi)=4(f^2_{\pi NN}/4\pi)$ and is designated Argonne v28, while the other has the value $(f^2_{\pi N\Delta}/4\pi)=2.88(f^2_{\pi NN}/4\pi)$ predicted by the quark model, which we will call Argonne v28q.

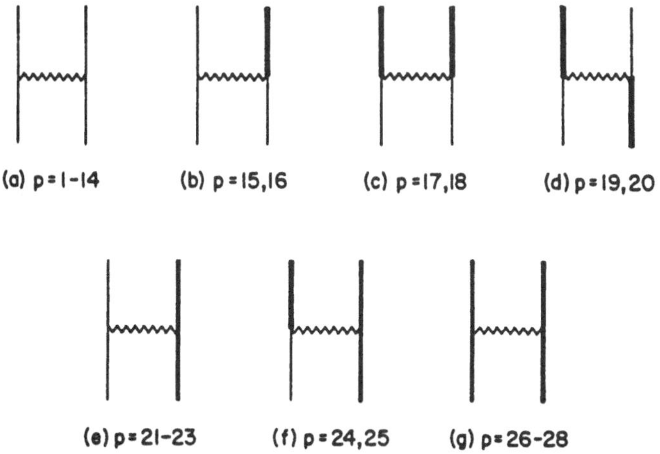

Fig. 3

Particular attention was paid to achieving a good description of NN data. All models were fit to a recent phase shift analysis, WI81,[17] for np data in the range 25-400 MeV, plus deuteron properties and scattering length parameters. A direct comparison to data was also made for some 1762 np points, including total and differential cross sections, polarizations, etc., in the range from 5 to 330 MeV, with a χ^2/point of ~ 1.7 for all three models. The models are essentially phase equivalent, so differences in the results of nuclear structure calculations should not be attributable to differences in phase shifts.

The operator form for the v28 models is convenient for nuclear structure calculations, and they are currently being used to study the light nuclei and nuclear matter. Preliminary variational results for nuclear matter are shown in Fig.2. Isobar components in ψ_v are generated by including transition operator components in the F_{ij} of eq.(4) corresponding to the processes of Fig.3b-d. The energy is evaluated with generalized FHNC/SOC methods which incorporate many-body clusters, and thus the effect of many-body forces. The calculations show a strong saturating effect in matter, even stronger than that induced by the addition of the Urbana V_{ijk} to the v14 models, resulting in a reasonable density but insufficient binding. There is room for improvement in the many-body results, due to the considerable flexibility in the structure of the v28 models, particularly in the intermediate- and short-range parts of the transition, NΔ and ΔΔ potentials which are not well determined by fitting low-energy NN data.

Another degree of freedom of interest is that of the pion. In nuclear physics the pion degrees of freedom are frequently suppressed in favor of two- and three-body potentials and two-body currents. One could hardly do otherwise without a detailed model for nucleon structure, since a free nucleon is surrounded by a pion field with a nonzero expectation value $\langle n^\pi\rangle_N$ for the pion number operator:

$$n^\pi = \int d^3k\, a_\pi^\dagger(k)\, a_\pi(k)/(2\pi)^3. \qquad (10)$$

When A nucleons are brought together in a nucleus, their interaction by OPE and TPE processes increases the number of pions to $\langle n^\pi\rangle_A$, creating a pion "excess" $\langle\delta n^\pi\rangle_A$:

$$\langle\delta n^\pi\rangle_A = \langle n^\pi\rangle_A - A\langle n^\pi\rangle_N. \qquad (11)$$

The pion excess is intimately related to exchange currents in nuclear electromagnetic processes, and gives a partial explanation for recently observed differences in the deep-inelastic lepton scattering from ^{56}Fe and ^{2}H targets.[18]

The pion excess can be estimated using the same static potential approximation that leads to the OPE potential, V_{ij}^π. This approximation yields a number operator[19]

$$\delta n_{ij}^\pi(k) = -\, V_{ij}^\pi(k)/(\mu^2 + k^2)^{1/2} \qquad (12)$$

whose expectation value can be taken with normal nuclear wave functions $|A\rangle$. Since $|A\rangle$ includes the effect of V_{ij}^π to all orders, this is not simply a perturbation estimate. Because the isobar plays an important role in πN interaction, we use the Argonne v28 model for our H, and the variationally calculated $|A\rangle$ for ^2H and nuclear matter to evaluate $\langle\delta n^\pi(k)\rangle$. Thus the important physical effects of the attractive $\pi N\Delta$ coupling and of short-range NN repulsion are brought together in a consistent manner. The local density approximation can then be used to estimate $\langle\delta n^\pi(k)\rangle$ in finite nuclei from the nuclear matter results. The results for $\langle\delta n^\pi(k)\rangle$ are shown in Fig.4 for a variety of nuclei; the total number of extra pions/nucleon is shown in parentheses. These results can then be used as input to the study of deep-inelastic lepton scattering[20] and other problems.

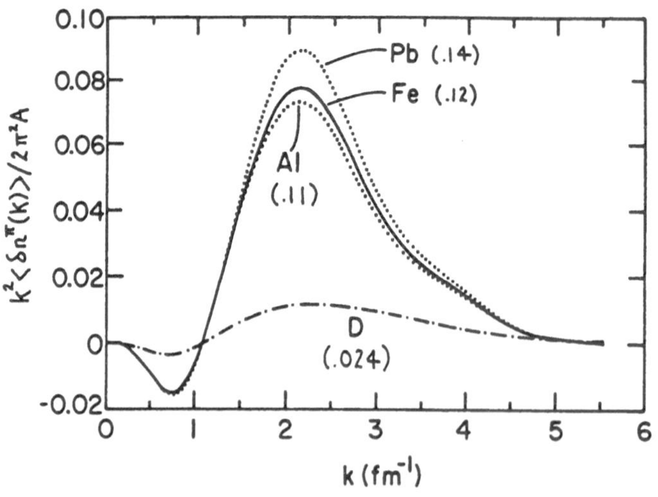

Fig.4

The work reviewed here was done in collaboration with T. L. Ainsworth, E. Berger, J. Carlson, F. Coester, J. L. Friar, B. L. Friman, V. R. Pandharipande and R. A. Smith, and was supported in part by the U. S. Dept. of Energy under contract W-31-109-ENG-38.

References

1. H. Kümmel, K. H. Lührmann and J. G. Zabolitzky, Phys. Rep. 36C, 1 (1978).
2. B. D. Day, Phys. Rev. Lett. 47, 226 (1981).
3. V. R. Pandharipande and R. B. Wiringa, Rev. Mod. Phys. 51, 821 (1979).
4. J. Lomnitz-Adler, V. R. Pandharipande and R. A. Smith, Nucl. Phys. A361, 399 (1981).
5. J. Carlson, V. R. Pandharipande and R. B. Wiringa, Nucl. Phys. A401, 59 (1983).
6. R. B. Wiringa, Nucl. Phys. A401, 86 (1983).
7. S. A. Coon, M. D. Scadron, P. C. McNamee, B. R. Barrett, D. W. E. Blatt and B. H. J. McKellar, Nucl. Phys. A317, 242 (1979).
8. R. B. Wiringa, R. A. Smith and T. L. Ainsworth, submitted to Phys. Rev. C.
9. I. E. Lagaris and V. R. Pandharipande, Nucl. Phys. A359, 331 (1981).
10. I. E. Lagaris and V. R. Pandharipande, Nucl. Phys. A369, 47 (1981).
11. Muslim, Y. E. Kim, and T. Ueda, Phys. Lett. 115B, 273 (1982).
12. A. Bomelburg, Phys. Rev. C28, 403 (1983); A. Bomelburg and W. Glockle, preprint (1983).
13. J. L. Friar, E. L. Tomusiak, B. F. Gibson and G. L. Payne, Phys. Rev. C24, 677 (1980).
14. F. Coester and W. N. Polyzou, Phys. Rev. D26, 1348 (1982).
15. F. Coester and R. B. Wiringa, to be published.
16. M. Lacombe, B. Loiseau, J. M. Richard, R. Vinh-Mau, J. Côté, P. Pirés and R. deTourreil, Phys. Rev. C21, 861 (1980).
17. R. A. Arndt and L. D. Roper, SAID program of the Center for Analysis of Particle Scattering, Dept. of Physics, V.P.I.&S.U.
18. J. J. Aubert et al., Phys. Lett. 123B, 275 (1983); A. Bodek et al., Phys. Rev. Lett. 50, 1431 (1983).
19. B. L. Friman, V. R. Pandharipande and R. B. Wiringa, Phys. Rev. Lett. 51, 763 (1983).
20. E. L. Berger, F. Coester and R. B. Wiringa, submitted to Phys. Rev. D.; F. Coester, proceedings of this conference.

Properties of Matter in Stellar Collapse

C. J. Pethick
Nordita,
Copenhagen, Denmark

and

Department of Physics
University of Illinois at Urbana-Champaign
Urbana, Illinois 61801, U.S.A

A brief review is given of the properties of matter in collapsing stars, with particular emphasis on the properties of nuclei.

In this talk I shall be interested in the properties of matter at densities between 10^7g cm^{-3}, where electrons first become relativistic in cold matter, and nuclear densities, about 3×10^{14}g cm^{-3}. The temperatures of interest range up to 10 ($\sim10^{11}$K) or more. This is the domain of ordinary low energy nuclear physics, and therefore we do not need to consider some of the more exotic states, such as quark-gluon plasmas discussed in other contributions at this meeting.

The stimulus to the studies I shall describe comes primarily from attempts to understand how collapse of a star can lead to a supernova, and, in some cases at least, neutron star formation, but the calculations are also relevant for the outer parts of neutron stars, and have implications for heavy ion reactions in the lab. (For more comprehensive reviews of work on stellar collapse and supernovae see Lattimer (1981), and Trimble (1982).) At the ends of their lives, stars with masses between about 10 and 25 solar masses develop cores of iron-peak elements, having densities of 10^9g cm^{-3} or more and masses of about one solar mass. In the surrounding envelope, which is much less dense, nuclear burning processes have not proceeded so far as in the core, and matter is made up of lighter elements. At low temperatures, which is the case for the infall stages of stellar collapse, most of the nucleons are locked up in nuclei and therefore contribute little to the pressure, which is provided mainly by the relativistic electrons, which are quite degenerate. The dominant contribution to the pressure is therefore $P \propto \rho^{4/3}$, since the electrons may be treated as free. A stellar core made up of such material is in a state of neutral equilibrium, since when the volume, V, of the core is altered, the internal energy, E, which is 3PV for a perfect relativistic gas, varies as R^{-1}, where R is the core radius. The gravitational energy also scales as 1/R, in Newtonian gravity. Consequently, if a core is in equilibrium it has zero total energy, and is in neutral equilibrium since the internal and gravitational energies scale is the same way with core radius. (See e.g. Landau & Lifshitz, 1980, Ch. XI). If P varies as ρ^γ, cores are stable if $\gamma > 4/3$ and unstable if $\gamma < 4/3$.

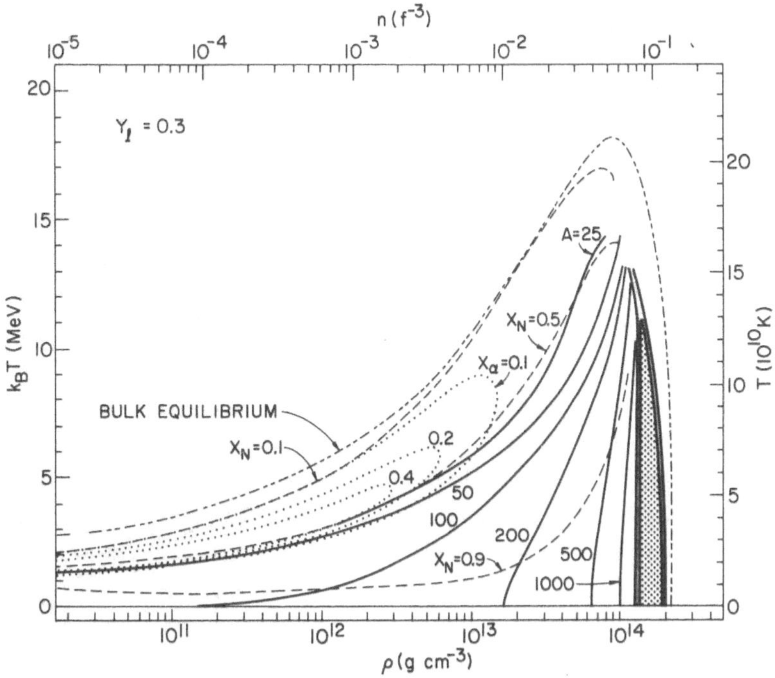

Fig. 1. Properties of hot dense matter at subnuclear densities for a lepton fraction Y_ℓ=0.3. The line labelled "bulk equilibrium" denotes the boundary of the two-phase region when surface and Coulomb effects are ignored. Finite nucleus effects were calculated using the methods of Lamb, Lattimer, Pethick, and Ravenhall (1978, 1984). Contours of constant nuclear mass number A are shown, as well as the mass fractions of nucleons in nuclei. X_α is the fraction of baryons in alpha particles, and X_N the fraction in heavy nuclei (including alpha particles). Bubbles are present in the dotted region, and mixtures (nuclei-bubbles, bubbles-uniform matter, or nuclei-uniform matter) in the surrounding grey region.

As one would expect, deviations of γ from 4/3 also play a crucial role in determining not only the stability of stellar cores, but also the outcome of stellar collapse. Small contributions to the pressure from sources other than electrons can therefore play an important role because they break the degeneracy of the problem.

In brief, the collapse of a core proceeds as follows. First, collapse is initiated by a slight pressure reduction due to the breakup of heavy nuclei. This reduces the energy available for thermal motion, and hence the thermal contribution to the pressure. As the collapse gets under way, the electron energy increases and some electrons are captured on protons, thereby producing neutrons and neutrinos. At densities of order 10^{12}g cm^{-3}, neutrinos become trapped in the star because the time for a neutrino to diffuse out of the stellar core exceeds the collapse time ($\lesssim$ 1s). Neutrinos become degenerate, and halt further beta decay by blocking final states. Thereafter the lepton fraction, Y_ℓ, in the core is fixed for the duration of the infall, and typically is ~0.35, most of which is electrons, compared with $Y_\ell = 0.46$ for ^{56}Fe matter with no neutrinos.

The first question we wish to address is when are nuclei present in hot dense matter. We note first of all, that both the neutrinos and the electrons may be treated as homogeneous ideal gases, the former because of the small interaction with other constituents, and the latter because the screening length is large compared with other microscopic lengths. We therefore need to consider in detail only the nucleons. If we neglect all finite nucleus effects, the possibility of having nuclei is equivalent to having equilibrium between a liquid and a gas, each of them consisting of a mixture of neutrons and protons. The condition for phase equilibrium is that the neutron chemical potentials in the two phases must be equal, as must the proton chemical potentials and the pressures. Results of such a calculation for the Skyrme I' nucleon-nucleon interaction are shown in Fig. 1. (Lattimer and Ravenhall, 1978), where the line marked "bulk equilibrium" is the boundary of the two phase region. At high temperatures the nucleons are in a single phase, while inside the boundary, liquid and vapor coexist. Note that at densities approaching nuclear density, one would expect nuclei to persist to temperatures of order 20 MeV.

We now turn to finite nucleus effects. These determine the nuclear size, but have little influence on the conditions under which one expects nuclei to be present. At low temperatures the most important finite size effects are the Coulomb and surface energies. The Coulomb energy of an isolated nucleus, if taken to be a uniformly charged sphere of radius r_N, is $(3/5)Z^2e^2/r_N$. For densities so high that the separation between nuclei is comparable with r_N the Coulomb interaction between nuclei must also be taken into account. In the Wigner-Seitz approximation, in which one calculates the energy of a spherical neutral sphere with the nucleus at its center, one finds the total energy to be $E_{Coul} = (3/5\, Z^2e^2/r_N)\left(1 - \frac{3}{2}\frac{r_N}{r_c} + \frac{1}{2}\left(\frac{r_N}{r_c}\right)^3\right)$, when $r_c = \left(\frac{3}{4\pi}\frac{Z}{n_e}\right)^{1/3}$ is the radius of the Wigner-Seitz cell, and n_e is the electron density. The second term in parentheses is the lattice energy for point nuclei, and the third term is a correction due to the finite nuclear size. Note that when nuclei fill all of space, the total Coulomb energy vanishes, since then the net charge is everywhere zero. The surface energy (or more correctly, the surface thermodynamic potential) is given by $E_{surf} = 4\pi r_N^2\sigma$, where σ is the surface tension. To determine the optimal nuclear size one minimizes the energy with respect to the nuclear radius, keeping the particle densities fixed, with the result $E_{surf} = 2E_{Coul}$. Putting in values for the coefficients one finds

$$A = \frac{12.5}{x^2}\,\frac{\sigma(x,T)}{\sigma(x=0.5)}\,\frac{n_i(x=0.5)}{n_i(x)}\,\frac{1}{1-\frac{3}{2}u^{1/3}+\frac{u}{2}},$$

where $u = (r_N/r_c)^3$ is the fraction of space filled by nuclei. If there is little matter outside nuclei, the total nucleon density is given by $n = un_i(x)$. Here x is the proton fraction of the matter in nuclei, and $n_i(x)$ is the nuclear saturation density. The result also applies at finite, but low, temperature if one allows for the temperature dependence of the surface tension. As the temperature increases the surface tension falls, and eventually it will vanish at the critical temperature, where there is no distinction between the matter inside nuclei and that outside. The expression above shows that with increasing density, A decreases, due to the negative Coulomb lattice energy reducing the Coulomb energy coefficient. With increasing temperature A decreases, due to the reduction of the surface tension. These trends are illustrated by the Z and A contours in Fig. 1 which were calculated by Lamb, Lattimer, Ravenhall and Pethick (1978, 1984) with finite nuclear size effects included. The figure also shows contours for X_H, the total fraction of nucleons in nuclei (including alpha particles), and X_α, the fraction of nucleons in alpha particles. Note that the region where a significant fraction of nucleons is in nuclei is within the two phase region given by the bulk equilibrium calculations.

So far we have assumed matter to consist of spherical nuclei. However it is easy to see that if nuclei fill half of space, ($u=1/2$), the energy is unchanged if nuclei are turned inside out, yielding a lattice of spherical bubbles like a sort of ordered Swiss cheese. At higher filling factors the bubbles state has less energy than the state with nuclei. Bubbles exist only if the nuclear matter is under tension, since both the Coulomb energy, which is minimized by making the charge density as uniform as possible, and the surface energy, which is minimized by reducing the total surface area, favor shrinking the bubble size, and give rise to a negative pressure on the nuclear matter. This stretching accounts for bubbles disappearing at densities somewhat below the nuclear saturation density. In Fig. 1 the region where bubbles exist is denoted by the dotted region.

Even the spherical bubble state is an oversimplification, since matter has a lattice structure, and consequently the Coulomb potential in the vicinity of a lattice site is not spherically symmetric. It has recently been shown that states with cylindrical and planar nuclei have lower energies than the state with spherical nuclei--we shall refer to these new states as spaghetti and lasagna, respectively. Ravenhall, Pethick and Wilson (1983) found that with increasing density matter went through the following sequence of states: spherical nuclei-spaghetti-lasagna-bubble spaghetti-spherical bubbles-uniform nuclear matter, with the filling factors at the transition being 0.22, 0.34, .61, 0.73, and 0.86 respectively for a proton fraction $Y_e = 0.3$. The physical reason for non-spherical nuclei being energetically favorable is that, with increasing density, the nuclear charge increases, due to the reduction of the Coulomb energy coefficient. The larger charge eventually results in nuclei being unstable to deformation, in essentially the same way as in the classic fission calculation of Bohr & Wheeler (1939). (See Pethick and Ravenhall, 1983.)

To illustrate effects on the pressure we show in Figs. 2 and 3 the pressure, P, and the adiabatic index $\Gamma = (\partial \ell n\, P/\partial \ell n\, n)_s$, where n is the baryon density. In the calculation neutrinos were neglected and the only states taken into account were spherical nuclei, spherical bubbles and uniform matter. (Pethick, Ravenhall, and

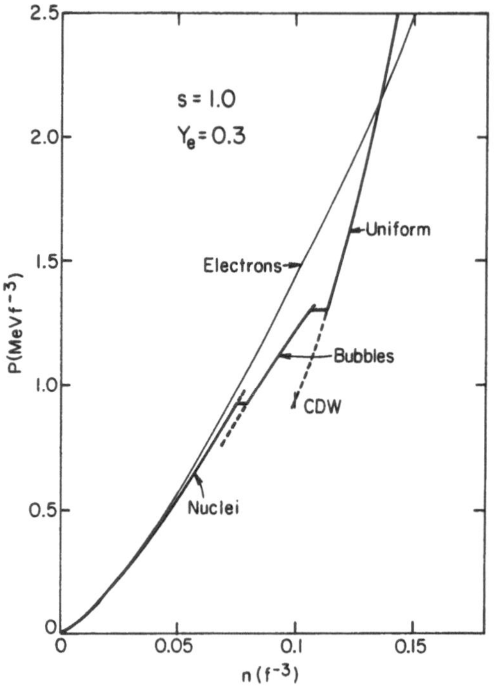

Fig. 2 Pressure v. density for matter with a proton fraction $Y_p = 0.3$ and entropy 1.0 k_B per nucleon. At the point denoted by CDW the uniform phase becomes unstable to formation of a charge density wave. Bubbles cannot exist in pressure equilibrium at densities greater than the uppermost density on the bubble-phase curve.

Lattimer, 1983.) The entropy is 1.0 k_B per nucleon, and the electron fraction is $Y_e = 0.3$, conditions which are typical of what is expected in stellar collapse. At low densities Γ is close to 4/3, but as the density increases, the lattice energy gives rise to an increasingly negative contribution to the pressure and to Γ. In the two-phase regions at the first-order phase transitions between the various states, Γ drops to low values. When the more exotic nuclear shapes are taken into account, the nuclei-bubbles transition will be replaced by a number of phase transitions, and Γ will be a smoother function of n. In the uniform phase Γ is much larger than 4/3, due to the large bulk modulus of nuclear matter. Stellar collapse will be halted, and the infall reversed, only when the core reaches densities above that of nuclear matter. (Bethe _et al._ 1979.)

Space does not allow me to go into details of the calculations, so I shall restrict myself to one aspect that illustrates the fruitful interplay between astrophysics and physics. Physical input is needed for astrophysical calculations, and astrophysics stimulates physics by expanding the range of conditions over which properties must be understood. The particular problem I wish to consider is that of calculating properties of nuclear surfaces at finite temperatures and for neutron

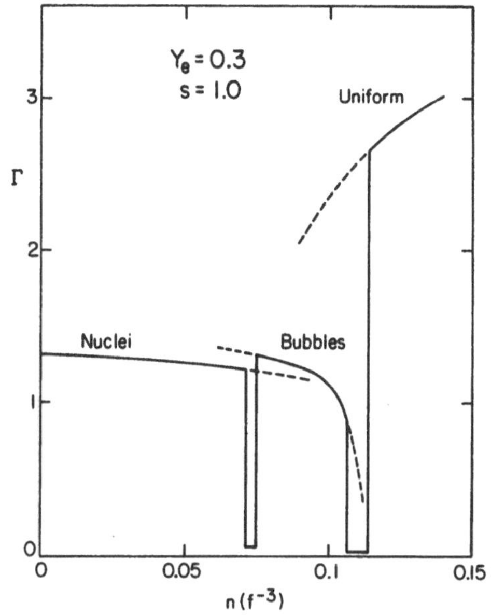

Fig. 3 The adiabatic index $\Gamma = (\partial \ln P/\partial \ln n)_s$ as a function of density for matter with a proton fraction $Y_p = 0.3$ and an entropy per nucleon $s = 1.0\ k_B$. Note that Γ is small but non-zero in the two phase regions. At low densities Γ is close to the value 4/3 of an ideal relativistic electron gas.

rich matter. As we have seen, nuclear surface properties are important for determining nuclear sizes in stellar collapse, and very neutron rich surfaces are of interest in neutron stars. In addition, they are also of interest for studies of heavy ion collisions in the laboratory. Calculations of the nuclear surface tension are to be found in Ravenhall, Bennett and Pethick (1972) and Ravenhall, Pethick and Lattimer (1983). A very important point is to start with the correct definition of the surface tension. Usually, at zero temperature, one defines a surface energy of a plane interface as the difference between the total energy and a bulk contribution. When nuclei are neutron rich, the neutron distribution extends further out than the proton one, and therefore there is ambiguity as to how one defines the bulk contribution to be subtracted. The most convenient quantity to define is the usual surface tension, which is, for many purposes, best thought of as the surface thermodynamic potential per unit area. (Landau and Lifshitz, 1980, Ch. XV.) The thermodynamic potential, Ω, is related to F, the Helmholtz free energy, by $\Omega = F - \sum_i \mu_i N_i$, where μ_i and N_i are the chemical potential and number of particles of species i. The advantage of working with the thermodynamic potential is that it does not depend on how one formally defines a location for the nuclear surface. In a two-component system, the surface tension is a function of the temperature and one other variable, which can be taken to be, among others, the proton concentration of the denser phase, the neutron chemical potential or the proton chemical potential. A natural choice, in view of the fact that the surface tension is a thermodynamic potential, is the neutron chemical potential, μ_n. The dependence of σ on μ_n gives information about the number, ν_{ns}, of excess neutrons per unit area of the surface:

$$\nu_{ns} = -\frac{\partial\sigma}{\partial\mu_n}(\mu_n,T).$$

In addition one can determine s_s, the surface contribution to the entropy by

$$s_s = -\frac{\partial\sigma}{\partial T}(\mu_n,T).$$

From this one can derive surface contributions to the specific heat and the density of states. We therefore see that the dependence of the surface tension on μ_n (or alternatively on proton concentration) and on temperature gives information valuable for nuclear physics applications - the size of the neutron skin, and the density of states.

In summary, studies of hot dense matter make use of a lot of input from nuclear physics, and since the nuclear properties required are not directly determined empirically it is frequently necessary to make use of theoretical calculations. Among the quantities of interest are the properties of bulk neutron rich nuclear matter at finite temperatures, and the properties of nuclear surfaces. Better microscopic calculations of these quantities would be extremely valuable. It is also unimportant to have microscopic calculations of complete nuclei, such as the finite temperature Hartree-Fock ones carried out by Bonche and Vautherin (1981,1982) and Wolff (1983). These incorporate a number of features, such as shell effects, not included in the simpler models.

I am grateful to my collaborators Don Lamb, Jim Lattimer and Geoff Ravenhall for their contributions to the calculations described above, and for many stimulating discussions. The work was supported in part by NSF grant NSF-PHY80-25605.

References

Bethe, H. A., Brown, G. E., Applegate, J., and Lattimer, J. M. 1979, Nucl. Phys. A **324**, 487.

Bohr, N., and Wheeler, J. A. 1939, Phys. Rev. **56**, 426.

Bonche, P., and Vautherin, D. 1981, Nucl. Phys. A **372**, 496.

Bonche, P., and Vautherin, D. 1982, Astron. and Ap. **112**, 168.

Lamb, D. Q., Lattimer, J. M., Pethick, C. J., and Ravenhall, D. G. 1978, Phys. Rev. Lett. **41**, 1623.

Lamb, D. Q., Lattimer, J. M., Pethick, C. J., and Ravenhall, D. G. 1983, Nucl. Phys. A (in press).

Lamb, D. Q., Lattimer, J. M., Pethick, C. J., and Ravenhall, D. G. 1984 (in preparation).

Landau, L. D., and Lifshitz, E. M. 1980, Statistical Physics, Part I (Pergamon, Oxford).

Lattimer, J. M. 1981, Ann. Rev. Nucl. Part. Sci. **31**, 337.

Lattimer, J. M., and Ravenhall, D. G. 1978, Ap. J. **223**, 314.

Pethick, C. J., and Ravenhall, D. G. 1983, in Numerical Astrophysics, ed J. Centrella, J. LeBlanc, M. LeBlanc, and R. L. Bowers, (in press).

Pethick, C. J., Ravenhall, D. G., and Lattimer, J. M. 1983, Properties of Warm Dense Matter at Low Entropies, submitted to Nucl. Phys. A.

Ravenhall, D. G., Bennett, C. D., and Pethick, C. J. 1972, Phys. Rev. Lett. **28**, 978.

Ravenhall, D. G., Pethick, C. J., and Lattimer, J. M. 1983, Nucl. Phys. A **407**, 571.

Ravenhall, D. G., Pethick, C. J., and Wilson, J. R. 1983, Phys. Rev. Lett. **50**, 2066.

Trimble, V. 1982, Rev. Mod. Phys. **54**, 1183.

Wolff, R. 1983, Thesis, T. U. München.

HYDRODYNAMICS OF ULTRA-RELATIVISTIC HEAVY ION COLLISIONS*

Bengt L.Friman**

Department of Physics
Åbo Akademi
SF-20500 Åbo 50
Finland

and

Gordon Baym

Department of Physics
University of Illinois
Urbana, Illinois 61801
U.S.A.

and

J.-P. Blaizot and M. Soyeur

Centre d'Etudes Nucléaires de Saclay
Service de Physique Théorique
F-91191 Gif-sur-Yvette Cedex
France

and

W. Czyz

Institute of Nuclear Physics, 31-342 Krakow
Poland

Abstract:

In central heavy-ion collisions at ultra relativistic energies, the central rapidity regime is expected to expand hydrodynamically. The hydrodynamic equations governing this behaviour are reviewed and numerical solutions are given for the ideal case of constant sound velocity. The cylindrical geometry, appropriate for central collisions, leads to a very rapid cooling of the matter, implying that a quark-gluon plasma formed in the collision will rapidly hadronize. Entropy production in the phase transition from a quark-gluon plasma to a hadron gas is also discussed.

*Work supported in part by U.S. N.S.F. Grants DMR-81-17182 and PHY81-21399

**Supported in part by the Academy of Finland.

Recent studies[1-5] suggest that in central heavy-ion collisions at center-of mass energies $\sqrt{s} > 25$ GeV per nucleon, large energy densities may be obtained, possibly leading to the formation of an extended quark-gluon plasma. At these ultra-relativistic energies nuclei are sufficiently transparent to nucleons that the two nuclei pass through each other, producing two highly excited nuclear fragmentation regions containing the net baryon number of the system. One expects these to be joined together by a central rapidity region, with negligible net baryon number but a substantial energy density[2] as in nucleon-nucleon collisions.[6] The geometry in the center-of mass frame after the collision is illustrated in fig. 1.

A model for the evolution of the central rapidity regime has recently been described by Bjorken.[2] At very short (proper) times following the collisions the degrees of freedom excited are, because of asymptotic freedom, weakly interacting. Only by a later proper time $\tau_0 \sim 1$ fm the interactions become sufficiently strong to establish thermodynamic equilibrium. Once thermodynamic equilibrium sets in, the evolution is described by Landau hydrodynamic model,[7] only with Lorentz-invariant boundary conditions inferred from the structure of the underlying nucleon-nucleon collisions together with the assumption of a central rapidity plateau.

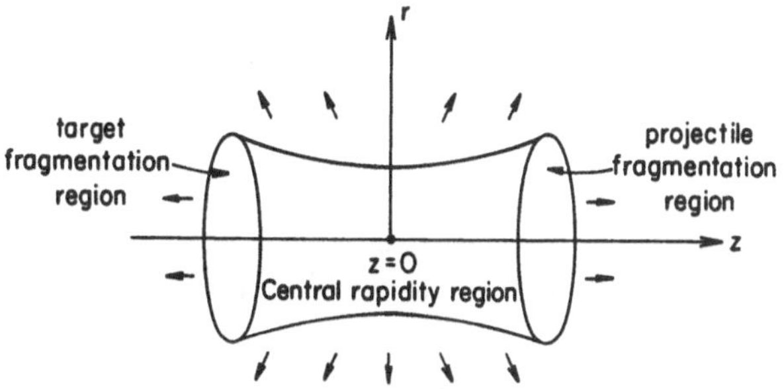

Fig. 1

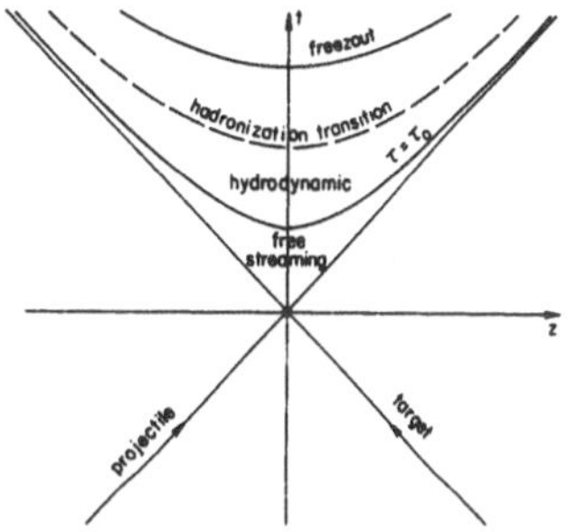

Fig. 2

The basic space-time picture is illustrated in the t,z-diagram in fig. 2. Here z is the coordinate along the collision axis, with z = t = 0 the central point of the collision. At very short proper times the weakly interacting degrees of freedom are "free streaming"; those reaching a point z,t from the collision region have velocity z/t and a local proper time (c = 1)

$$\tau = (t^2 - z^2)^{1/2}. \tag{1}$$

Assuming thermalization at $\tau = \tau_0$, the excitations are hydrodynamic thereafter. Thus the subsequent hydrodynamic behaviour can be specified by an initial boundary condition on the surface of constant $\tau = \tau_0$ (hyperbolas in fig. 2). With the assumption of a plateau in the final multiplicity versus rapidity the energy density ε is a constant ε_0 on this surface.

The initial energy density ε_0 can be estimated by extrapolating observed charged pion multiplicities in the central rapidity regime in pp and $\bar{p}$p collisions.[2] In the energy range 30-270 GeV per nucleon, the charged pion density per unit of rapidity is [8] $\sim$2-3. One expects, taking the neutral pions into account, a total pion multiplicity density a factor 3/2 larger. The energy per unit of rapidity, assuming a final pion energy $\sim$0.4 GeV, is thus $dE/dy \sim 1.2\text{-}1.8$ GeV. In a central A-A collision, the energy density is increased by a factor[9,10] (1-2)A, and thus the energy per unit volume is initially

$$\varepsilon \sim \frac{A}{\pi R_A^2} \left(\frac{dE}{dy}\right)_{pp} \frac{dy}{dz} , \tag{2}$$

where R_A is the nuclear radius 1.2 $A^{1/3}$ fm. Since the longitudinal velocity v_z is z/t, near the central slice (z = 0) dy/dz = 1/t. Thus at z = 0 we estimate that the energy density in the rest frame is

$$\varepsilon \sim (0.3 - 0.4) \frac{A^{1/3}}{t} \frac{\text{GeV}}{\text{fm}^3} , \tag{3}$$

with t in fm. At the time that thermal equilibrium sets in, t is simply τ_0. For $\tau_0 \sim 1$ fm, and $A \sim 238$, the initial energy density is $\sim 1.6\text{-}2.5$ GeV/fm^3. This is close to the energy density which, according to recent Monte Carlo simulations, is needed to produce deconfinement in a SU(3) lattice gauge theory with 2 flavours of massless quarks.[11] Thus, given the uncertainties in the estimates above, it is possible that the initial energy density is insufficient to carry the matter all the way through the deconfinement transition. However since, the Monte Carlo simulations, predict a first order transition, with a large latent heat, (1.5 GeV/fm^3) the system is in this case likely to develop a two-phase region, where the quark-gluon plasma coexists with a hadron gas.

One expects similar excitation energies to be achieved in the fragmentation region in a central collision.[1,12] However, because fragments continue to be produced in this regime, and the net baryon number is non-zero, the hydrodynamic description is somewhat more complex than in the central regime.[13-15]

If we assume that the matter at τ_0 consists of massless thermalized quarks with two flavours and neglect interaction effects, then an energy density 2 GeV/fm^3 corresponds to a temperature[16]

$$T \approx 160\, \varepsilon_0^{1/4} (\text{MeV}) \sim 200 \text{ MeV}, \tag{4}$$

where ε_0 is in GeV/fm^3. The total number of excited quanta is

$$n_0 = n_q + n_{\bar{q}} + n_g = 2.1\ \varepsilon_0^{3/4} (\text{fm}^{-3}) \sim 3.5\ \text{fm}^{-3}. \tag{5}$$

This large density of quanta implies that they have relatively short mean free paths

$$\lambda = \frac{1}{n_0 \sigma} \sim \frac{0.47}{\sigma_{fm} \varepsilon_0^{3/4}} (\text{fm}) , \tag{6}$$

where $\sigma_{fm}(\sim 1)$ is a mean scattering cross section in fm^2. Thus for $\varepsilon_0 \sim 2$ GeV/fm^3, λ is $\sim (1/4\sigma_{fm})$, and small compared with the transverse dimension $\sim A^{1/3}$ fm of the interaction volume. Therefore we expect a hydrodynamic description to be valid in this phase of the expansion, for sufficiently large nuclei.[5] The corresponding initial entropy density is

$$s_0 = 8.4\ \varepsilon_0^{3/4}\ \text{fm}^{-3} \sim 14\ \text{fm}^{-3} . \tag{7}$$

After these introductory remarks let us now discuss the hydrodynamic equations. Once local thermodynamic equilibrium is established in the central collision volume, the evolution is governed by the conservation laws for energy and momentum

$$\partial_\mu T^{\mu\nu}(x) = 0, \tag{8}$$

where in the absence of dissipation

$$T^{\mu\nu} = (\varepsilon + P) u^\mu u^\nu + P g^{\mu\nu} . \tag{9}$$

Here ε is the energy density, P the pressure, $g^{\mu\nu}$ the metric tensor,

$$u^\mu = \gamma(1, \underset{\sim}{v})$$

is the four-velocity, where v(x) is the local flow velocity and $\gamma = (1 - v^2)^{-1/2}$. In the central rapidity region, the local baryon density vanishes, so we need not include the baryon current conservation as a hydrodynamic equation. The equations of motion can, after some algebra, be written as[5] the entropy conservation law

$$\frac{\partial}{\partial t} (s\gamma) + \underset{\sim}{\nabla} \cdot (s\gamma \underset{\sim}{v}) = 0 \tag{10}$$

and the "acceleration" equation

$$\frac{\partial}{\partial t} (T\gamma \underset{\sim}{v}) + \underset{\sim}{\nabla}(T\gamma) = \underset{\sim}{v} \times \left(\underset{\sim}{\nabla} \times (T\gamma \underset{\sim}{v})\right). \tag{11}$$

The term on the right in (11) does not enter in the symmetric motions we consider here.

The dissipative terms, which are neglected in this description become significant when the temperature (and other thermodynamic variables) vary significantly within a mean free path. It is important to understand the entropy generating mechanisms in the expansion, since the total entropy of the matter is a diagnostic of the state produced in the initial collision volume;[17,18] to the extent that the entropy is conserved in the subsequent expansion, the final distribution of detected particles in phase space is related directly to the number of degrees of freedom excited in the initial collision.

In simple one-dimensional motion, including in particular the initial expansion along the collision axis, eqs. (10) and (11) reduce to

$$\frac{\partial}{\partial t}(s \cosh y) + \frac{\partial}{\partial z}(s \sinh y) = 0$$

$$\frac{\partial}{\partial t}(T \sinh y) + \frac{\partial}{\partial z}(T \cosh y) = 0, \qquad (12)$$

where

$$y = \tanh^{-1} v_z \qquad (13)$$

is the hydrodynamical rapidity variable.

The solution to eqs. (12) which satisfies the Lorentz invariant boundary conditions $\varepsilon = \varepsilon(\tau_0)$, or equivalently $s = s(\tau_0)$, $T = T(\tau_0)$ and $v_z = z/t$, is of the scaling form[2]

$$s(\tau) = s_0 \tau_0/\tau \ , \qquad v_z = z/t \ . \qquad (14)$$

This scaling of s results simply from the fact that the volume over which a given entropy is spread grows as τ. The scaling solution is stable with respect to small perturbations and, in particular, unlikely to develop a shock-discontinuity at the hadronization transition.[5]

The one-dimensional expansion of a semi-infinite slab of matter initially at temperature T_0 between $x = -\infty$ and $x = R$ (> 0) is described by the relativistic Riemann solution[5] to (12) (c_s = constant)

$$T = T_0\left(\frac{t-x+R}{t+x-R}\ \frac{1-c_s}{1+c_s}\right)^{c_s/2} ; \qquad v(x,t) = \frac{x-R+c_s t}{t+c_s(x-R)} \ . \qquad (15)$$

In the expansion of a finite slab and also in three dimensional expansions the leading edge of the matter is Riemann like.

One significant feature of the Riemann solution as that an entropy generating discontinuity or shock can develop in the expansion, when the condition $d(sc_s/T)/dT > 0$ is violated by the equation of state.[5] This happens in the transition from deconfined quark-gluon plasma to confined hadronic matter if this transition is first order.[5,11] However, the entropy produced in such a rarefaction shock is very small; on the order of a few percent of the initial entropy only.[19] Even though the shock is almost adiabatic , the phase transition might still have some effect on the final distribution of particles versus transverse rapidity.[18] This question requires further study.

Another source of entropy is the "freeze out" of the matter from collision dominated local equilibrium to free streaming particles. Work on this problem is in progress.[20]

Let us now turn to the question of the transverse expansion accompanying the longitudinal motion described by (14). Making use of the Lorentz invariance in the longitudinal direction and the cylindrical symmetry, we can reduce eqs. (10) and (11) to

$$\begin{gathered}\frac{\partial}{\partial t}(s\gamma) + \frac{\partial}{\partial r}(s\gamma v_r) + s\gamma\left(\frac{v_r}{r} + \frac{1}{t}\right) = 0, \\ \frac{\partial}{\partial t}(T\gamma v_r) + \frac{\partial}{\partial r}(T\gamma) = 0.\end{gathered} \tag{16}$$

which describe the transverse expansion of the central slice ($z = 0$). The hydrodynamic motion, in any other slice is obtained, by a Lorentz boost, from the solution in the central slice. In terms of the transverse rapidity variable $\alpha = \tanh^{-1} v_r$ these two equations can be written in a simple form

$$\begin{gathered}\frac{\partial}{\partial t}(rts \cosh\alpha) + \frac{\partial}{\partial r}(rts \sinh\alpha) = 0 \\ \frac{\partial}{\partial t}(T \sinh\alpha) + \frac{\partial}{\partial r}(T \cosh\alpha) = 0.\end{gathered} \tag{17}$$

We now turn to numerical results, obtained for the ideal equation of state ($c_s = 1/\sqrt{3}$). In figs. 3 and 4 we show the temperature and velocity distributions for a one dimensional expansion of a finite slab extending from $x = 0$ to $x = R$ and subject to the boundary condition $v(x = 0) = 0$. Each curve is marked by the corresponding value of t/R.

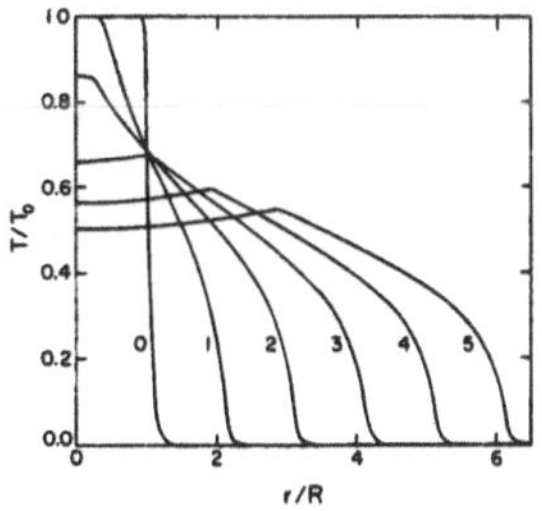

Fig. 3

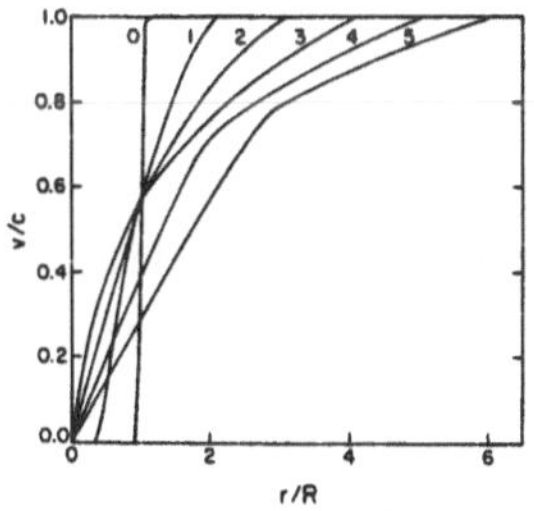

Fig. 4

The temperature and velocity distributions for the transverse expansion of cylindrically symmetric hot matter described by (17) are shown in figs. 5 and 6. Here we choose the time at which the initial conditions are specified $\tau_0 = 1$ fm. The coupling to the longitudical expansion qualitatively changes the hydrodynamic behaviour of the transverse motion compared to the one dimensional expansion (see e.g. figs. 3 and 5). The longitudinal expansion causes a cooling of the fluid, since it spreads the entropy over a constantly increasing longitudinal interval. In particular, the fluid cools uniformly at small r interior to the rarefaction front. This is seen in the temperature distribution, fig. 5.

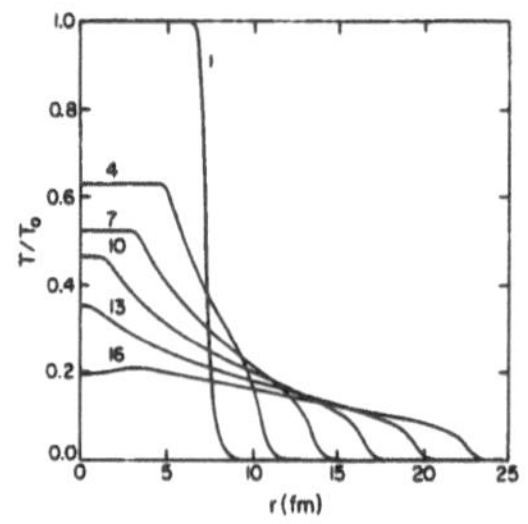

Fig. 5

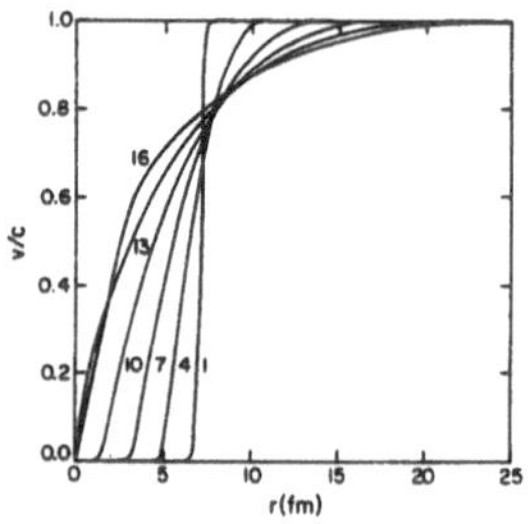

Fig. 6

The essential quantity one would like to determine from the hydrodynamic expansion is the final distribution of particle multiplicities and momenta. Here we give the results of a qualitative calculation of the expected multiplicities described im detail in ref. 5 (see fig. 7). The main assumption is that the freeze out occurs at a given temperature T_{fo}. A certain amount of the matter remains at rest and is frozen out before the rarefaction wave reaches it.[5] For R = 7 fm, $T_{fo} \sim 0.7$ and t_0 between 1 and 2 fm some 60 to 30 % of the initial matter remains unaffected by the transverse expansion prior to freeze out. Thus, due to the longitudinal expansion one expects for reasonable values of T_{fo}/T_0 only a small contribution to the transverse momentum from the hydrodynamic motion.[5] For comparison we show, in fig.8 the rapidity distribution obtained with a spherically symmetric,[21, 5] rather than cylindrical geometry. Since in this case the matter can cool only after the rarefaction front has passed one finds a larger contribution from the hydrodynamic motion.

Considerable work remains before quantitative predictions can be made with this model. One outstanding problem is a better treatment of the hadronization transition. In the freeze-out transition again it is necessary to include a good description of the hadronic mean free paths rather than to assume a discrete transition at a given freeze-out temperature, and to understand the entropy generation here.

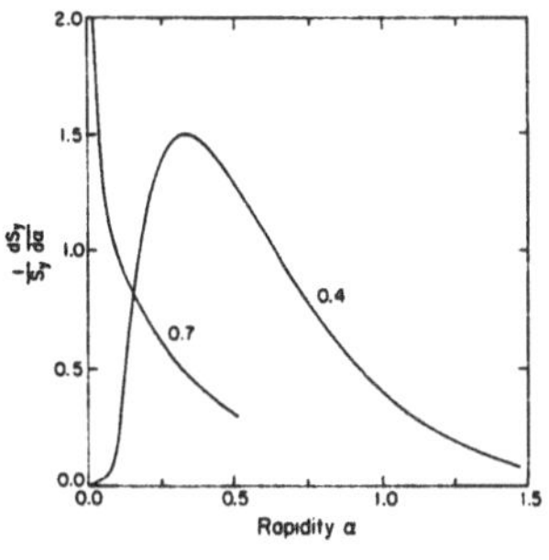

Fig. 7

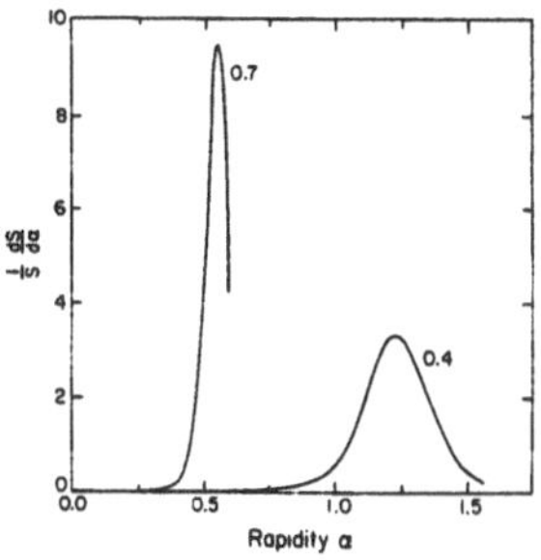

Fig. 8

References

1. R. Anishetty, P. Koehler and L. McLerran, Phys. Rev. D22 (1980) 2793.
2. J. D. Bjorken, Phys. Rev. D27 (1983) 140.
3. L.McLerran, in Quark Matter Formation and Heavy Ion Collisions (Proc. Bielefeld Workshop; ed. M. Jacob and H. Satz, World Scientific Publishing Co, Singapore, 1982), 63.
4. K. Kajantie, in Quark Matter Formation in Heavy Ion Collisions (Proc. Bielefeld Workshop; ed. M. Jacob and H. Satz, World Scientific Publishing Co, Singapore, 1982), 39.
5. G. Baym, B. L. Friman, J.-P. Blaizot, M. Soyeur and W. Czyz, Nucl. Phys. A (1983) in press.
6. J. D. Bjorken, Lecture Notes in Physics, 56 (Springer-Verlag, 1976) 93.
7. L. D. Landau, Izv. Akad. Nauk SSSR 17 (1953) 51; S.Z. Belen'kii and L. D. Landau, Usp. Fiz. Nauk 56 (1955) 309.
8. K. Alpgård et al., Phys. Lett. 107B (1981) 310; 112B (1982) 183.
9. A. Bialas, W. Czyz and L. Lesniak, Phys. Rev. D25 (1982) 2328.
10. J. E. Elias et al., Phys. Rev. D22 (1980) 13.
11. J. Kogut et al., Univ. of Illinois preprint ILL(TH)83-9; Phys Rev Lett. 50 (1983) 393.
12. J. Cleymans, M. Dechantsreiter and F. Halzen, Univ. of Wisconsin preprint 1982, MAD/TH/50.
13. K. Kajantie and L. McLerran, Phys Lett. 119B (1982) 203.
14. K. Kajantie and L. McLerran, Univ. of Helsinki preprint HU-TFT-82-30.
15. K. Kajantie and R. Raitio, Univ. of Helsinki preprint HU-TFT-82-52; K. Kajantie, R. Raitio and P. V. Ruuskanen, Nucl. Phys. B222 (1983) 152.
16. G. Baym, (Erice Lectures, April 1981) in Prog. in Part. and Nucl. Phys. 8 (1982) 73.
17. P. Siemens and J. Kapusta, Phys. Rev. Lett 43 (1979) 1486.
18. L. van Hove, Phys. Lett. 118B (1982) 138.
19. B. L. Friman, G. Baym and J.-P. Blaizot, Phys. Lett. (to be published).
20. G. Baym, B. L. Friman and S. Gavin, to be published.
21. F. Cooper, G. Frye and E. Schonberg, Phys. Rev. D11 (1975) 192.

VARIATIONAL TREATMENT OF π^0-CONDENSED NEUTRON MATTER IN A REALISTIC POTENTIAL MODEL

Omar Benhar
Istituto Nazionale di Fisica Nucleare, Sezione Sanità
Physics Laboratory, Istituto Superiore di Sanità
Viale Regina Elena 299, I-00161, Rome, Italy

Abstract

The results of a numerical study on the stability of neutron matter against neutral pion condensation are reported. Within the framework of a Jastrow-like variational approach, in which the Reid soft core potential has been modified to simulate Δ-resonance effects, the condensate phase turns out to be energetically favoured at densities larger than normal nuclear density by a factor 3÷4.

1. Introduction

There has been recently a growing effort aimed at firmly establishing the occurrence of pion condensates (for a review see Ref. 1) in nucleon matter within the framework of the existing microscopic many-body theories (2-8), namely the G-matrix perturbation theory (9) and the variational approach based on Jastrow-like correlated wave-functions (10). In fact, since the possibility of pion condensation was proposed, it has been realized that, to obtain an accurate estimate of the critical density ρ_c, at which the transition to the condensed phase takes place, the effect of strong short-range correlations between nucleons has to be properly taken into account.

It is well known that, owing to the P-wave pion-nucleon coupling, basically proportional to $(\underset{\sim}{\sigma}\cdot\underset{\sim}{k})$, a π^0 standing wave of wave vector $\underset{\sim}{k}_c$ gives rise to a spatially nonuniform spin-isospin ordered vacuum state of nuclear matter. In fact, the condensate field produces an attractive spin-isospin dependent periodic potential of wavelength π/k_c, felt by the nucleons, so that the total energy of the π^0-condensed system can be lowered by arranging the nucleons in a one-dimensional lattice-like structure, having lattice parameter π/k_c and a given spin-isopsin order.

In analogy with the case of a spatially nonuniform Coulomb system exhibiting a "longitudinal photon condensate", it is possible to describe the π^0-condensed ground state without explicitly including the pionic degrees of freedom, provided account is taken of the full nucleon-nucleon (NN) interaction. This is the basic assumption of the so-called potential model in which, the nonrelativistic nuclear Hamiltonian H being given, a comparison is made between the ground state energies of nuclear matter in the standard state and in the π^0-condensed state. It should be noticed that the potential approach seems to be quite well suited to study the onset of pion condensation for the following reasons:

i) it treats both phases of matter on the same footing;

ii) unlike the method based on the analysis of the pionic Green function in nuclear medium, it allows one to detect even a first order phase transition (for a discussion on the nature of the transition to the pion-condensed phase, see Ref. 11).

In this paper we report the results of an investigation on the stability of pure neutron matter (Z=0) against neutral pion condensation, in which the Jastrow-like variational approach has been employed. As for the NN force, an effective interaction based on the Reid soft core (RSC) potential (12) in the V6 form (13) has been constructed following the prescription of Ref. 4. The effect of the Δ-isobar mixing into the neutron states, which has been proved to be essential in producing a lowering of the critical density (1), has been simulated through a "renormalization" of the πN coupling constant in the one-pion-exchange (OPE) tail of the RSC interaction.

The description of the π^0-condensed ground state in terms of a Jastrow -like correlated wave-function and the method used in evaluating the expectation values of the nuclear Hamiltonian are discussed in Section 2, whereas in Section 3, the numerical results are analysed. Finally, in Section 4, the validity of the approximations employed and the possible improvements of the model are outlined.

2. Jastrow-like variational treatment of the π^0-condensed phase of neutron matter

Among the various configurations of neutron matter giving a nonvanishing expectation value of the π^0 field, we have selected the state proposed by Calogero and co-workers in Ref. 14-16, consisting of a determinant of Bloch-type single particle (sp) wave-functions characterized by a periodic spin-dependent localization along the direction of the z axis, while in the xy-plane the standard Fermi gas configuration survives. This state has been adopted as the model state Φ in constructing the correlated many-body wave-function $\Psi = F\Phi$. The sp states $\varphi_{\underset{\sim}{k},\lambda}$ (the wave-vectors $\underset{\sim}{k}$ belong to the Fermi sea $\{F\}$ and the index $\lambda = 1(2)$ denotes spin-up (down) neutrons) are defined as ($x \equiv \{\underset{\sim}{r}, \underset{\sim}{\sigma}\}$, Ω is the normalization volume):

$$\varphi_{\underset{\sim}{k},\lambda}(x) = \Omega^{-\frac{1}{2}}\ \chi_\lambda(z)\,\eta_\lambda(\sigma_z)\ \exp(i\ \underset{\sim}{k}\cdot\underset{\sim}{r}). \tag{1}$$

In eq. (1) η_λ is the spin state, whereas χ_λ produces a density modulation in the $\hat{z}$ direction with period ℓ and a spin dependent phase defined by the relation $\chi_\lambda(z) = \chi(z+\ell\delta_\lambda)$, where ($N(\alpha)$ is a normalization factor):

$$\chi(z) = N(\alpha) \sum_{n=-\infty}^{+\infty} \exp\left[-\frac{1}{2}\alpha^2(z-n\ell)^2\right] \tag{2}$$

From eqs. (1) and (2) it clearly follows that α can be regarded as an order parameter for the transition to the π^0-condensed phase, in the sense that as $\alpha \to 0$, $\chi \to 1$ and the standard uniform configuration of neutron matter is recovered.

The choice of a Fermi surface having a nonsymmetric z-dependence has been shown to be essential in producing π^0-condensation (5,7). In the present calculation a cylindrical shape with radius $k_{F\perp}$ and height $2k_{F}$ has been employed in the condensed phase, the values of $k_{F\perp}$ and k_{F_z} being related to the number density of the neutrons $\rho = N/\Omega$ by the expression $k_{F\perp}^2\ k_{F_z} = 2\pi^2\rho$.

As for the correlation factor, the simple form ($r_{ij} = |\underset{\sim}{r}_i - \underset{\sim}{r}_j|$)

$$F(1,\ldots,N) = \prod_{i<j} f(r_{ij}) \tag{3}$$

has been selected. The correlation function f is defined by the expression

$$f(r) = 1 - \exp\{-[\beta_\perp(x^2+y^2) + \beta_z\ z^2]\} \tag{4}$$

in the condensed phase, while in the standard phase a spherically symmetric f has been used taking $\beta_\perp = \beta_z = \beta$. It should be noticed that the simple correlation function employed, having no tensor component, seems to be suitable to describe pure neutron matter, since the potential energy of the standard phase is almost entirely given by the central singlet-even interaction, whereas in the condensed phase the effect of the tensor component in the NN interaction is taken into account by the model function Φ.

In the nuclear Hamiltonian

$$H = \sum_i \frac{p_i^2}{2m} + \sum_{i<j} v(ij) \tag{5}$$

an effective two-body interaction

$$v(12) = (\tilde{f}^2 - 1)\ \tilde{v}_{OPE}(12) + v_R(12) \tag{6}$$

has been used. In eq. (6), v_R is the V6 form (13) of the RSC potential (12). Following the approach of Ref. 4, the effect of Δ-resonance mixing into the neutron states has been included through the effective πN coupling constant $\tilde{f}$, whereas the modified OPE potential $\tilde{v}_{OPE}$ contains a medium range damping term which simulates the contribution of ρ-meson -exchange processes. It should be noticed that, as $\tilde{f}^2 \to 1$, corresponding to zero Δ-resonance percentage, $v \to v_R$.

The expectation value $\langle H \rangle = \langle \Psi | H | \Psi \rangle / \langle \Psi | \Psi \rangle$ has been evaluated by using the procedure described in Ref. 6. The one-body and two-body density matrices have been approximated in terms of the corresponding quantities associated with the correlation factor F and the model wavefunction Φ($N^{(1)}$ and $N^{(2)}$ are normalization factors)

$$p_{\Psi}^{(1)}(x,x') = N^{(1)} p_F^{(1)}(x,x') p_{\Phi}^{(1)}(x,x')$$
$$p_{\Psi}^{(2)}(x_1,x_2) = N^{(2)} p_F^{(2)}(x_1,x_2)\, p_{\Phi}^{(2)}(x_1,x_2) \qquad (7)$$

The calculation of $p_{\Phi}^{(1)}$ and $p_{\Phi}^{(2)}$ is straightforward, whereas $p_F^{(1)}$ and $p_F^{(2)}$ have been obtained by solving the hypernetted-chain integral equations for Bose liquids (17,18).

3. Numerical results

For any given value of the density ρ, the comparison of the binding energy per neutron in the two phases requires the following steps: i) the value of $\tilde{<}H>/N$ is minimized in the standard phase (i.e. with $\alpha = 0$, $\tilde{f}^2 = 1$ and spherical Fermi surface) with respect to the correlation parameter β; ii) the value of β giving the minimum energy in the standard phase being taken as $\beta_\perp$, the energy of the condensed phase is minimized with respect to α, β_z and k_{F_z}. As for the phase parameters, the set $\{\delta_1, \delta_2\} = \{0,1/2\}$, which maximizes the tensor attraction, has been employed.

The results of the variational search of the energy upper bounds are summarized in Table 1.

Table 1. Energy differences between the π^0-condensed and the standard phase of pure neutron matter (in MeV) for different effective coupling constant and densities. The parameter values corresponding to the minimum energy of the condensed phase are also listed (α and k_{F_z} expressed in fm^{-1}).

$\tilde{f}^2$	ρ/ρ_0	α	k_{F_z}	$\beta_z/\beta_\perp$	$E_\pi - E_N$
1	1	0.	1.4	1.0	2.2
	2	0.	1.8	1.0	4.1
	3	0.	2.0	1.0	5.3
	4	0.	2.1	1.0	5.5
2	1	0.	1.4	1.0	8.1
	2	0.	1.8	1.0	9.3
	3	0.	2.0	1.0	10.2
	4	0.	2.1	1.0	10.5
3	1	0.	1.4	1.0	29.7
	2	1.7	1.0	1.4	20.7
	3	1.7	1.0	1.2	8.5
	4	2.5	1.2	1.4	-11.4

It clearly appears that, as far as the bare Reid V6 interaction is adopted ($\tilde{f}^2=1$), the transition to the condensed phase does not occur,

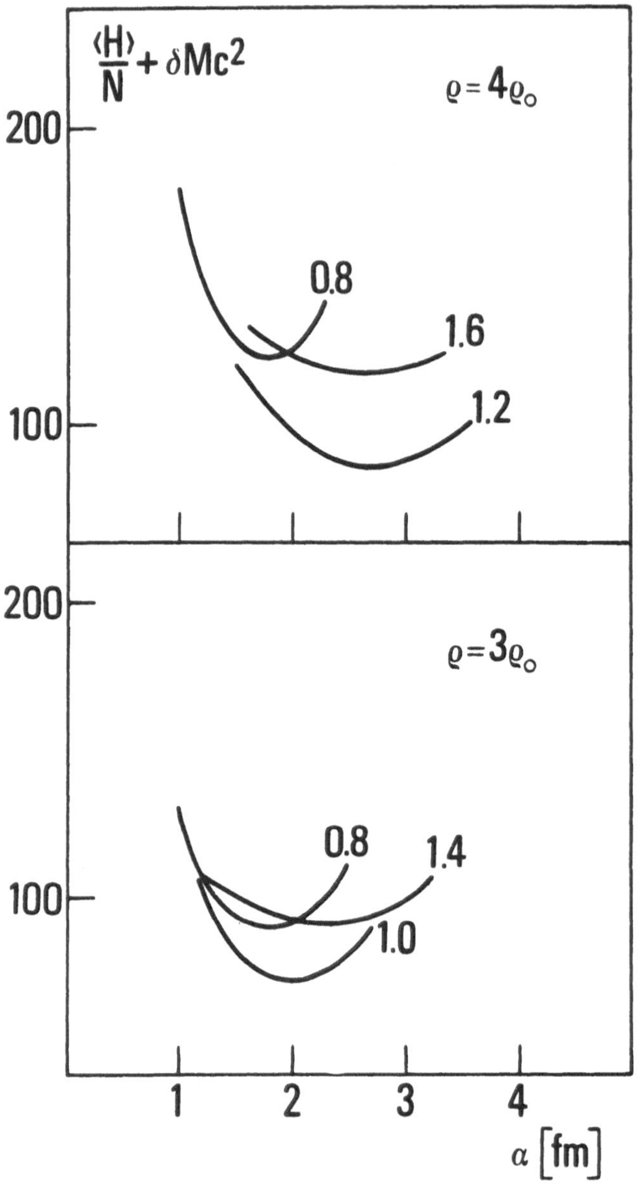

Fig. 1 Binding energies per particle (MeV) at $\tilde{f}^2 = 3$ versus the order parameter α. The curves are labelled according to the value of K_{Fz}, whereas the value of the ratio $\beta z/\beta_\perp$ is given in Table 1. The quantity ∂Mc^2 accounts for the mass difference between the neutron and the Δ-resonance.

up to densities $\rho \geq 4\rho_0$. For every value of k_{F_z} and β_z the binding energy turns out to be a monotonically increasing function of α, the gain in potential energy given by the tensor attraction being always overwhelmed by the kinetic energy needed to localize the particles along the z axis.

Taking $\tilde{f}^2=2$ the lowest energy configuration still corresponds to $\alpha=0$ but, for $\rho \geq 2\rho_0$ and $k_{F_z} \lesssim 1.4\ fm^{-1}$, the monotonical behaviour breaks down and secondary minima appear at $\alpha \neq 0$.

At $\tilde{f}^2=3$ (corresponding to a Δ percentage $v^2 \sim 0.106$ in the model of Ref. 4) and $\rho \geq 2\rho_0$ the lowest energy of the condensed phase is always achieved at $\alpha \neq 0$ and for $\rho = 4\rho_0$ the instability of standard neutron matter against π^0-condensation clearly appears.

In Fig. 1 the behaviour of the energy expectation value is shown, as a function in the order parameter α, for $\rho = 3\rho_0$, $\rho = 4\rho_0$ and different values of k_{F_z}.

As for the effect of the nonsymmetric z dependence of the correlation function of the condensed phase, it turns out that in the uniform configurations ($\alpha=0$) the ratio $\beta_z/\beta_\perp = 1$ is always favoured, while the onset of the one-dimensional localization produces a lowering of the correlation range in the $\hat{z}$ direction. However, it should be pointed out that the energy gain associated with the asymmetry of f is not very large (~ 1 MeV), at least for the values of α at which the energy minima occur.

4. Summary and conclusions

The results of the present calculations indicate that, at densities $\rho \sim 3\div4\rho_0$, the transition of pure neutron matter to the π^0-condensed phase takes place, provided the modification of the NN interaction associated with Δ-resonance mixing is taken into account.

These conclusions qualitatively agree with those of Tamiya and Tamagaki (4), who employed an effective interaction of the form (6), taking the full RSC potential as v_R, in a lowest-order (two hole-line) G-matrix calculation. In Ref. 4 the transition has been detected at $\rho_c \sim 2\div3\rho_0$ with an effective πN coupling constant $\tilde{f}^2=3$. The discrepancy between the G-matrix and variational estimates of the critical density can probably by ascribed to the following reasons: i) the V6 form of the RSC potential does not include the angular momentum dependence in the triplet-odd interaction, which has been proved to be very effective in producing the stabilization of the π^0-condensed phase (4); ii) in Ref. 4 a set of Bloch sp wave function constructed from a Wannier-type basis, suitable to describe well-defined localization, has been employed. It has been recently shown (19) that, within the Hartree-Fock approximation, this prescription for the sp states produces a significant lowering of the kinetic energy needed to localize the particles.

The approximated procedure adopted in the calculation of the expectation value <H> seems to be quite adequate as long as the short ranged nonovershooting correlation function defined in eq. (4) is employed. Nevertheless, a more accurate estimate of the critical density would probably require the use of the full hypernethed-chain procedure derived in Ref. 5, as well as some improvements of the variational wavefunction, both using a different sp basis in building up the model function Φ and allowing for a spin-dependence of the correlation functions.

Acknowledgments

The author is much endebted to Prof. C. Ciofi degli Atti for many clarifying discussions. Thanks are also due to Dr. G. Salmé for a critical reading of the manuscript.

References

1. A.B. Migdal, In "Mesons in Nuclei", Vol. 3, Eds., M. Rho and D.H. Wilkinson, (North-Holland, Amsterdam, 1979), p. 941.
   G.E. Brown and W. Weise, Phys. Rep. C27 (1976) 1.
2. D.J. Sandler and J.W. Clark, Phys. Lett. 100 B (1981) 213.
3. K. Tamiya and R. Tamagaki, Progr.Theor.Phys. 66 (1981) 948.
4. K. Tamiya and R. Tamagaki, Progr. Theor. Phys. 66 (1981) 1361.
5. O. Benhar, Lett. Nuovo Cimento 30 (1981) 517.
6. O. Benhar, Phys. Lett. 106 B (1981) 375.
7. O. Benhar, Phys. Lett. 124 B (1983) 305.
8. W.H. Dickhoff, A. Faessler, J. Meyer-ter-Vehn and H. Müther, Phys. Rev. C23 (1981) 1154.
9. B.D. Day, Rev. mod. Phys. 50 (1978) 495.
10. J.W. Clark, in "Progress in Particle and Nuclear Physics", Vol. 2, ed. D.H. Wilkinson (Pergamon, Oxford, 1979), p. 89.
11. A.M. Djugaev, JETP Lett. 22 (1975) 83.
    H. Kleinert, Lett. Nuovo Cimento 34 (1982) 133.
12. R.V. Reid, Ann. of Phys. 50 (1968) 411.
13. O. Benhar, C. Ciofi degli Atti, S. Fantoni and S. Rosati, Nucl. Phys. A328 (1979) 127.
14. F. Calogero, in "The Nuclear Many-Body Problem", Vol. 2, Eds. F. Calogero and C. Ciofi degli Atti, (Compositori, Bologna, 1973), p. 535.
15. F. Calogero and F. Palumbo, Lett. Nuovo Cimento 6 (1973) 663.
16. F. Calogero, F. Palumbo and O. Ragnisco, Nuovo Cimento 29A (1975)

509.

17. S. Fantoni and S. Rosati, Nuovo Cimento 25A (1975) 593.
18. S. Fantoni, Nuovo Cimento 44A (1978) 191.
19. E. Pace and F. Palumbo, Phys. Lett. 113 B (1982) 113.

Proton Mixing in Neutron Star Matter under $\pi°$ Condensation

T. Takatsuka

College of Humanities and Social Sciences,
Iwate University, Morioka 020

1. Introduction

It is known that protons become admixed by several % in the ordinary phase of neutron star matter where neutrons dominate in component.[1)] While pion condensation, an interesting new phase currently discussed by many autrhors, is now considered to occur at densities $\rho \gtrsim 2\rho_0$ (ρ_0 being the nuclear density).[2)-5)] This indicates that the pion-condensed phase comes into play, at least, in neutron star interiors, although its realization for finite nuclei is hard to expect. Then it becomes an interesting problem to ask how the admixture of protons is affected by the presence of pion condensation. The extent of proton contamination is important in connection with the superfluid properties,[6)7)] cooling processes [8)] and glitch models [9)] of neutron stars.

Among possible types of pion condensates, neutral pion ($\pi°$) condensation is of particular interest, since it causes a drastic structure-change of nucleon system characterized by a layer structure composed of the two-dimensional (2D) Fermi gas (FG) with specific spin-isospin ordering, in contrast with the 3D one in the ordinary phase.[10)]

In this talk, we discuss the proton mixing in neutron star matter under $\pi°$ condensation. We study the problem in the framework of the ALS model[10)] (Alternating Layer Spin) to describe suitably the system and with the use of the effective interaction approach, particularly intending to understand the underlying physics in a simple and transparent manner.

2. Outline of approach

ALS model

With assigning $\ell=0$ to a representative ($n\uparrow$, $p\downarrow$) layer, the configuration illustrated in Fig.1 is expressed as

$$\sigma_\ell \tau_\ell = -(-)^\ell, \quad \sigma_\ell(n) = (-)^\ell \quad \text{and} \quad \sigma_\ell(p) = -(-)^\ell, \tag{1}$$

where $\sigma_\ell(\tau_\ell)$ denotes the spin (isospin). Such aspect comes from that the condensed $\pi°$ field $\varphi_{\pi°}$ ($\propto \sin k_0 z$) generates a deep periodic potential $V_{\pi°}(\propto \tau_3 \sigma_z \nabla_z \varphi_{\pi°})$ with spin-isospin dependence, and hence nucleons arrange in the z-direction with the layer spacing $d=\pi/k_0$ so as to feel efficiently this potential. While the ALS configuration thus resulted provides the source function for π° field and maintain φ_{π° presumed. That is, the selfconsistent relation between the ALS structure and φ_{π° is fulfilled.

The single-particle basis are given by

$$\phi_\alpha(\vec{\xi})=\Omega_\perp^{-1/2}\exp(i\vec{q}_\perp\vec{r}_\perp)\phi_{w\ell}(z)\chi_{\sigma_\ell\tau_\ell}\text{ (spin, isospin)}, \qquad (2)$$

Fig.1.

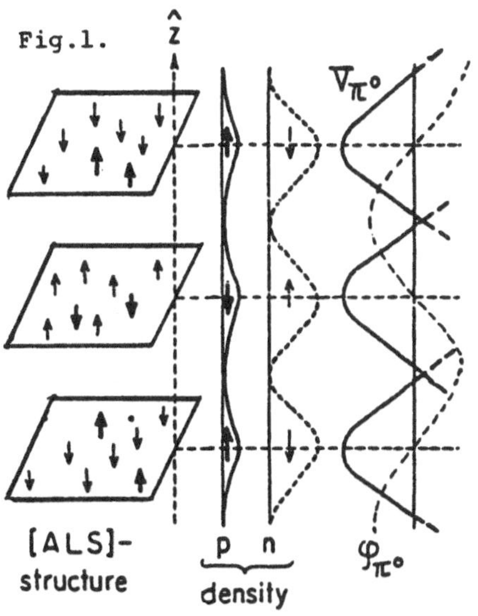

Where $\vec{\xi}=\{\vec{r}$, spin, isospin$\}$, $\vec{r}_\perp=\{x,y\}$, $\vec{q}_\perp=\{q_x, q_y\}$ and $\Omega_\perp$ is the 2D normalization volume. $\phi_{w\ell}(z)$ represent the Wannier type basis localized around the lattice cite ℓd and in most cases is well approximated by the gaussian form:

$$\phi_{w\ell}(z)\simeq\phi_\ell(z)$$

$$=(a/\pi)^{1/4}\exp[-a(z-d\ell)^2/2]. \qquad (3)$$

The nucleon ground state is given by the Slater determinant of $\{\phi_\alpha\}$, where the 2D FG state is occupied up to $|\vec{q}_\perp|\le q_{\perp F}^{(i)}$ with $q_{\perp F}^{(i)}$ being the 2D Fermi momentum; $q_{\perp F}^{(i)}=(4\pi\rho_i d)^{1/2}$. Here i=n(p) stands for neutron (proton).

Mechanism for proton mixing

In general, admixture of protons depends on the following factors. By the transition from neutron matter into the so called neutron star matter which consists of neutrons, protons and electrons to assure the charge neutrality (here we ignore the muon mixing for simplicity), there arises (i) decrease in the kinetic energy of nucleons, (ii) addition of the energy of relativistic electrons and (iii) energy gain from two-nucleon interaction since its effect is more attractive for n-p than n-n or p-p. The net effect (i)+(ii)+(iii) determines the extent of p-mixing.

Concerning these effects in the ALS(π°-condensed) phase, we remark the following points:

(a) The expressions of kinetic energy per nucleon for the ALS and the FG phases are given respectively as

$$\langle K.\ E.\rangle_{FG} = \frac{3}{5} E_F \{(1-y)^{5/3}+y^{5/3}\}, \tag{4}$$

$$\langle K.\ E.\rangle_{ALS} = \frac{1}{2} E_{\perp F} \{(1-y)^2+y^2\}+ \hbar a/4m_N \tag{5}$$

where $y=\rho_p/\rho$ denotes the p-mixing ratio ($\rho=\rho_n+\rho_p$), $E_F=\hbar^2(3\pi^2\rho)^{2/3}/2m_N$, $E_{\perp F}=2\hbar^2\pi\rho d/m_N$ and the second term in Eq. (5) represents the zero-point energy due to localization. From these, we can see that the effect (i) is enlarged in the ALS phase as compared with the FG one because the y-dependence is steeper in the former and also $\frac{1}{2}E_{\perp F} > \frac{3}{5}E_F$ results as a general tendency (for example[5]), $E_{\perp F}\simeq 190$MeV, $E_F\simeq 130$MeV at $\rho\simeq 3\rho_0$).

(b) Like in FG phase, we assume that in the ALS phase electrons are uniformly distributed as $\rho_e(r)=\rho_e=\rho_p$. Then the effect (ii) is the same in both phases. As for the coulomb energy, somewhat difference arises due to the density fluctuation of protons, $\rho_p(r)\simeq\rho_p\{1+2\exp(-\pi^2/ad^2)\times \cos(2\pi z/d)\}$ instead of $\rho_p(r)=\rho_p=\rho_e$ in the FG phase. But the contribution is found to be very small compared with the nucleonic energy quantities because of charge neutrality condition and therefore is neglected as in usual case.

(c) Oweing to the specific spin-isospin ordering in the ALS phase, the effect (iii) may acts in different manner from the usual case. To take this into account, it is indispensable to pay attentions to the state-dependence of two-nucleon interaction.

Based on (a)∿(c), we remark that protons are likely to mix in the ALS phase as far as the effect (iii) is not so different.

Effective Hamiltonian

we construct suitably an effective Hamiltonian $\tilde{H}$ to represent the the essentials of the system:

$$\tilde{H} = \Sigma_i T_N(i)+\Sigma_i T_e(i) + \frac{1}{2}\Sigma_{ij}\tilde{V}(ij), \tag{6}$$

$$\tilde{V}(ij) = \tilde{V}_T(ij) + \tilde{V}_C(ij), \tag{7}$$

where $T_N(T_e)$ denotes the kinetic (relativistic) energy part for nucleon (electron). The interaction potential $\tilde{V}$ is divided into two parts; the one is the effective tensor part $\tilde{V}_T$ which is just the driving force for the realization of the ALS phase and the other is the non-tensor part $\tilde{V}_C$ mainly composed of the central one which is the main ingredient for interaction energy in usual case.

For $\tilde{V}_T$, we use

$$\tilde{V}_T(1,2) = -\frac{1}{3}\eta^2 (f/m_\pi)^2 (\vec{\tau}_1\cdot\vec{\tau}_2) \int \frac{d\vec{k}}{(2\pi)^3} \frac{S_{12}(\vec{k})}{k^2+m_\pi^2} e^{i\vec{k}\vec{r}} \frac{m_\rho^2}{k^2+m_\rho^2} \tag{8}$$

with $S_{12}(\vec{k})=3(\vec{\sigma}_1\cdot\vec{k})(\vec{\sigma}_2\cdot\vec{k})-(\vec{\sigma}_1\cdot\vec{\sigma}_2)\vec{k}^2$, where the modification of the OPEP tensor force through its weakening due to ρ-meson effect (the appearence of $m_\rho^2/(k^2+m_\rho^2)$ with m_ρ the ρ-meson mass) and its enhancement due to the isobar Δ(1232) effect ($\eta^2>1$) are taken into account. As a typical example, we adopt $\eta^2=1.7$ [5] which well reproduces the results for neutron matter obtained by more fundamental approach.[3] It should be noted here that as far as the present problem are concerned with, the tensor force does not play a crucial role because its dominant direct part becomes independent of y and its exchange one is negligible.

As for $\tilde{V}_C$, we use the effective interaction obtained by solving the reaction matrix equation in the normal phase, assuming that the short-range correlation is not so different between two phases. For the calculations $\langle\frac{1}{2}\Sigma_{ij}\tilde{V}_C(ij)\rangle$ in both phases, we adopt the following approximations[1a)11)] as often used; the use of $\tilde{V}_C=\tilde{V}_C^{z=0}$ for n-n, the neglect of the contribution from p-p and the use of $\tilde{V}_C=\tilde{V}_C^{N=Z}$ for n-p, where $\tilde{V}_C^{z=0}$ ($\tilde{V}_C^{N=z}$) represents the effective interaction dependent on ρ for neutron (nuclear) matter.

As a suitable choice for $\tilde{V}_C^{N=z}$, we use the Go-force of Sprung and Banerjee given by [12]

$$\tilde{V}_C^{N=z}(r,\rho;\beta) = \Sigma_i \{a_i(\beta)+b_i(\beta)\sqrt{k_F}\} \exp[-\{r/\lambda_i(\beta)\}^2], \tag{9}$$

where $k_F=(3\pi^2\rho/2)^{1/3}$ and the parameters (a_i, b_i, λ_i) are referred to Ref. 12). β denotes the two nucleon states; $\beta={}^3O$, ${}^1E$, ${}^3E$, and ${}^1O$ according to (S, T)=(1, 1), (0, 1) (1, 0) and (0, 0) with S(T) denoting the total spin (isospin) of the pair. Also, For $\tilde{V}_C^{z=0}$, we use the modified version of Eq.(9) obtained by adjusting to the neutron matter results with the RSC potential:

$$\tilde{V}_C^{z=0}(r,\rho;\beta) = \Sigma_i \{a_i(\beta)+b_i(\beta)\sqrt{k_F}\} f(k_F) \exp[-\{r/\lambda_i'(\beta)\}^2],$$

$$\lambda_i' = \lambda_i/2^{1/3}, \tag{10}$$

$$f(k_F) = 4 \quad \text{for } \beta = {}^3O \quad \text{and} \quad (2.32-0.50k_F) \quad \text{for } \beta = {}^1E .$$

With the effective Hamiltonian thus represented, we are able to have the energy expressions where the roles of each components in $\tilde{H}$ become apparent. Parameters inherent there; y for FG phase and y, a and d for the ALS one, are determined by the energy minimization.

3. Numerical results and discussion

The total energies per nucleon E/A versus ρ of both phases are shown in Fig. 2 for neutron matter and neutron star matter. The results for neutron matter are similar to those from other approaches.[3)4)] The energy gain due to the p-mixing is known to be remarkable in the ALS phase.

In Fig. 3, y versus ρ for both phases are plotted. The results for the FG phase, $y \simeq (3-4)\%$ according to $\rho \simeq (1-4)\rho_0$, are considerably in agreement with those[1)] so far reported. We note here that the p-mixing becomes larger by a factor (1.5-2) according to $\rho \simeq (1.5-4)\rho_0$ in the π^0-condensed phase than in the ordinary one.

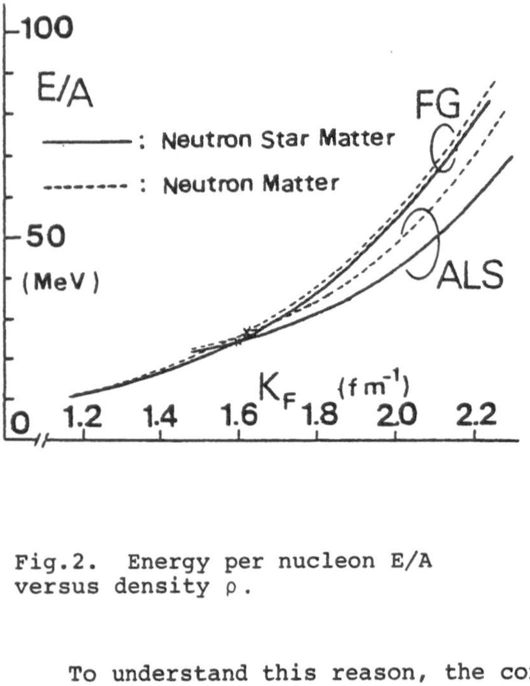

Fig.2. Energy per nucleon E/A versus density ρ.

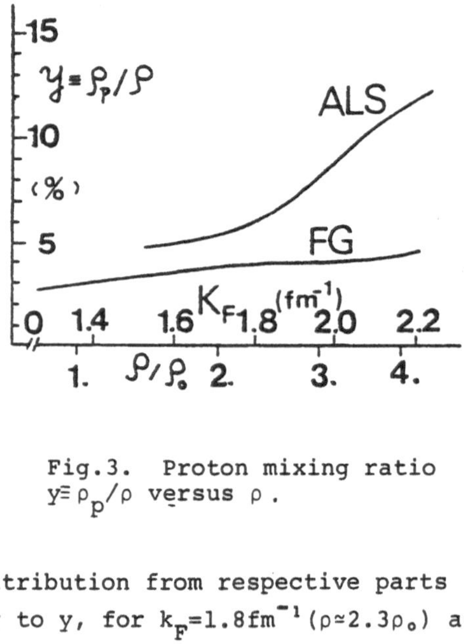

Fig.3. Proton mixing ratio $y \equiv \rho_p/\rho$ versus ρ.

To understand this reason, the contribution from respective parts in $\tilde{H}$ are illustrated in Fig.4 according to y, for k_F=1.8fm^{-1} ($\rho \simeq 2.3\rho_0$) as an example, where $\Delta\langle T_N\rangle$, $\Delta\langle T_e\rangle$, $\Delta\langle\tilde{V}_C\rangle$ and $\Delta\langle E\rangle$ denote the shifts in energies per nucleon from y=0 case; $\Delta\langle T_N\rangle = \langle T_N\rangle_y - \langle T_N\rangle_{y=0}$ and so on. Concerning the tensor part, $\Delta\langle\tilde{V}_T\rangle$ is very small in the ALS phase ($\Delta\langle\tilde{V}_T\rangle$ =0→0.4Mev for y=0→0.1) and vanishes exactly in the FG phase, as already mentioned in 2. It is clarified by the figure that the mechanism for the p-mixing lies in the energy gain from $\Delta\langle T_N\rangle + \Delta\langle\tilde{V}_C\rangle$ overwhelming the energy increase due to $\Delta\langle T_e\rangle$, as mentioned in the previous section.

Next, by comparing the results for two phases, it is pointed out that the large energy gain $\Delta\langle T_N\rangle$ is prominent in the ALS phase, while $\Delta\langle\tilde{V}_C\rangle$ is almost the same in both phases. Since $\Delta\langle T_e\rangle$ is the same, the resulting $\Delta\langle E\rangle(=\Delta\langle T_N\rangle+\Delta\langle T_e\rangle+\Delta\langle\tilde{V}_C\rangle+\Delta\langle\tilde{V}_T\rangle)$ has its minimum at larger y in the ALS phase, as denoted by the cross in the figure. This is the reason for larger p-mixing in the π°-condensed phase than in the ordinary one.

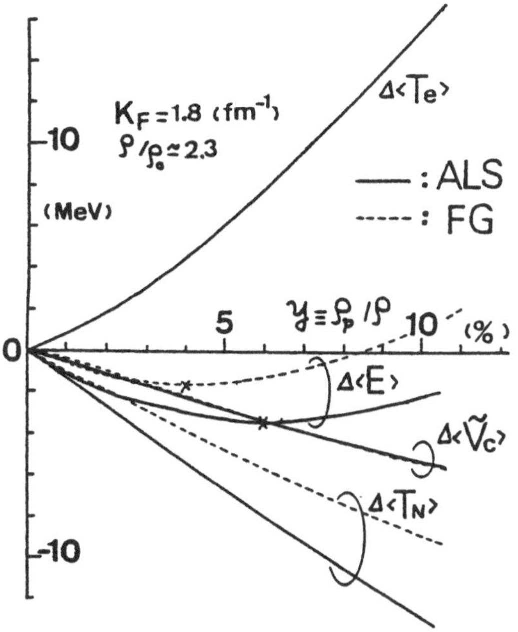

Fig.4. y-dependence of respective contributions in H. Notations are defined in the text.

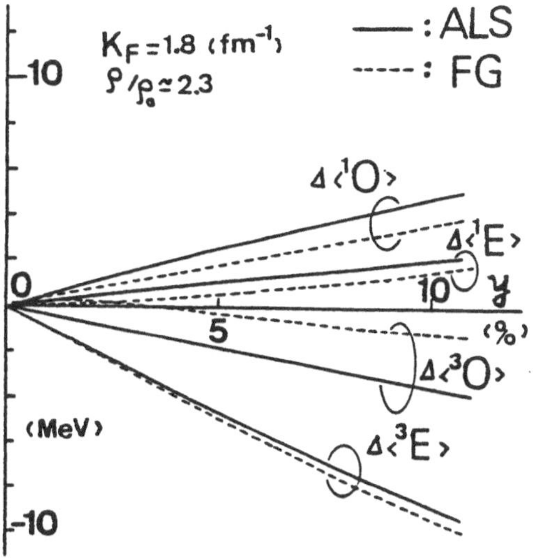

Fig.5. y-dependence of respective contributions in $\tilde{V}_C$. Notations are explained in the text.

In order to analize the result that the $\Delta\langle\tilde{V}_C\rangle$ is very close between two phases in spite of the distingnished difference of nucleon system, the contributions from various two-nucleon states are given in Fig. 5. It is seen that the contribution from ${}^1E+{}^1O+{}^3E$ states is less effective in the ALS phase than in the FG one. On the contrary, the energy gain from the ${}^3O$ state $\Delta\langle {}^3O\rangle$ is markedly larger in the former phase, which is because the ${}^3O$-repulsion becomes by far effective[5] for the ALS phase due to the spin alignment of the neutron pair in the same layer and this effect leads to be reduced by the proton contamination ($\Delta\langle {}^3O\rangle<0$ although $\langle {}^3O\rangle>0$). This ${}^3O$-effect almost compensates the disadvantage from the ${}^1E+{}^1O+{}^3E$ ones. As a result the net effect $\Delta\langle\tilde{V}_C\rangle(=\Delta\langle {}^1E\rangle+\Delta\langle {}^1O\rangle+\Delta\langle {}^3E\rangle+\Delta\langle {}^3O\rangle)$ becomes almost the same in two phases, which is understood as a consequence of the averaging over spin and isospin.

4. Concluding remarks

Admixture of protons is found to be also possible even when the developed π° condensation is at work in the high density region of neutron stars. The notable feature is in that the proton mixing ratio in the π°-condensed phase becomes remarkably larger as compared with the usual case without pion condensation. The reason can be attributed to the 2D nature of the FG state characteristic for nucleon system under π° condensate.

In the previous work[7], the proton $^1S_0$-dominant superfluidity under π° condensation was discussed by assuming the 5% mixing of protons at $\rho \simeq (1-3)\rho_o$. The present results support roughly the assumption and confirm the occurence of proton superfluid concluded there, although the investigation including ρ-dependence of y are needed for more quantitative results, especially at higher densities ($\rho \gtrsim 3\rho_o$). Further studies on the proton superfluid, together with its possible consequence on neutron star cooling, are our forthcoming subjects.

The author wish to thank Proffesor R. Tamagaki, Dr. T. Tatsumi and Dr. T. Kunihiro for their valuable discussions.

References

1) a) J. Nemeth and D.W.L. Sprung, Phys. Rev. 176 (1968) 1496.
b) S. Ikeuchi, S. Nagata, T. Mizutani and K. Nakazawa, Prog. Theor. Phys. 46 (1971) 95.
c) D. Ellis and D.W.L. Sprung, Can. J. Phys. 50 (1972) 2277.
2) As review articles,
A.B. Migdal, Rev. Mod. Phys. 50 (1978) 107;
G.E. Brown and W. Weise, Phys. Reports C27 (1976) 1;
G. Baym and D.K. Campbell, "Mesons in Nuclei" Vol.III (1979), chap.3.
3) T. Kunihiro and T. Tatsumi, Prog. Theor. Phys. 65 (1981) 613.
4) K. Tamiya and R. Tamagaki, Prog. Theor. Phys. 66 (1981) 948, 1361.
5) T. Takatsuka, Y. Saito and J. Hiura, Prog. Theor. Phys. 67 (1982) 254.
6) a) N.C. Chao, J.W. Clark and C.H. Yang, Nucl. Phys. A179 (1972) 320.
b) T. Takatsuka, Prog. Theor. Phys. 50 (1973) 1754 and 1755.
7) T. Takatsuka and R. Tamagaki, Prog. Theor. Phys. 65 (1981) 1333;
T. Takatsuka, "Lecture Notes in Physics" Vol. 142 (1981) 452.
8) S. Tsuruta, Phys. Reports 56 (1979) 237 and papers cited therein.
9) D. Pines, Proceedings of the 16 th Solvay Congress on Physics (1974 Editions de l' universite de Bruxells), P. 174.
10) T. Takatsuka, K. Tamiya, T. Tatsumi and R. Tamagaki, Prog. Theor. Phys. 59 (1979) 1933. AS a review article, R. Tamagaki, Nucl. Phys. A328 (1979) 352.
11) K.A. Brueckner and J. Dabrowski, Phys. Rev. 134 (1964) B722.
12) D.W.L. Sprung and P.K. Banerjee, Nucl. Phys. A168 (1971) 273.

TENSOR FORCES AND THE FERMI LIQUID PROPERTIES OF NUCLEAR MATTER

P. Haensel and A. J. Jerzak
Polish Academy of Sciences, N. Copernicus Astronomical Center,
Bartycka 18, PL-00-716 Warszawa, Poland

1. Introduction

The theory of normal Fermi liquids has been formulated by Landau in the fifties [1-3]. It has been originally devised for describing the properties of liquid ^{3}He near absolute zero (see Ref.[4] for a detailed review of present status of Fermi liquid theory of ^{3}He). In the sixties the basic idea of the Fermi liquid theory has been applied by Migdal and his collaborators to the description of atomic nuclei and nuclear matter [5].

The properties of a Fermi liquid are, in principle, determined by the elementary (bare) interaction between its particles. One of the main qualitative differences between the underlying two-body interaction in the normal ^{3}He and in nuclear matter is that in the former case tensor-term (resulting from the dipole-dipole interaction between nuclear spins) has negligible effects and can be safely neglected. On the contrary, the tensor and spin-orbit components of the nucleon-nucleon (N-N) interaction are known to be important. These terms are widely known to be to a large extent responsible for the fact that nuclear matter is a "nasty many-body system".

In the Fermi liquid theory the properties of the system are determined by the interaction between the quasiparticles (q.p.). In the case of the normal ^{3}He, the q.p. interaction is assumed to be central, in accordance with the properties of the bare two-body interaction. In the case of nuclear matter the tensor q.p. interaction, which couples q.p. momenta to their spins, results from the existence of strong tensor components in the bare two-body interaction [6-8]. In the standard Fermi liquid theory of nuclear matter, as initiated by Migdal, the tensor term in the q.p. interaction has been neglected. However, the study of Bäckman et al.[9] showed that the tensor forces could play an important role in the conditions for stability of the ground state of nuclear matter. Also, inclusion of tensor forces modifies the formulae of the Fermi liquid theory relative to the spin dependent excitations, as compared to the standard ones [9-11].

In this paper we briefly review our studies on the effect of tensor forces on the Fermi liquid properties of nuclear matter. Quantitative results are obtained using recent models of the q.p. interaction in nuclear matter, derived from realistic N-N potentials [9,12].

2. Quasiparticle interaction and stability of the ground state

The Fermi liquid theory establishes a one-to-one correspondence between low lying excitations in the system of quasiparticles and those in the real system. In what follows we consider for simplicity the case of static and space homogeneous excitations. The energy of the system is a complicated functional of the q.p. distribution matrix $n_{\vec{p}}$ (2×2 matrix in the spin and isospin space). However, for a small number of excited quasiparticles the excitation energy (per unit volume) implied by a small deviation $\delta n_{\vec{p}} = n_{\vec{p}} - n^{\circ}_{\vec{p}}$ from the ground state distribution matrix is given by a simple formula, including only terms quadratic in $\delta n_{\vec{p}}$,

$$\delta E = Tr_{\sigma} Tr_{\tau} \int \frac{d^3p}{(2\pi\hbar)^3} \left[e^{\circ}_{\vec{p}} + U_{\vec{p}}(\vec{\sigma},\vec{\tau}) - \mu \right] \delta n_{\vec{p}}(\vec{\sigma},\vec{\tau})$$

$$+ \frac{1}{2} Tr_{\sigma_1\sigma_2} Tr_{\tau_1\tau_2} \int \frac{d^3p_1}{(2\pi\hbar)^3} \int \frac{d^3p_2}{(2\pi\hbar)^3} f(\vec{p}_1,\vec{p}_2)\, \delta n_{\vec{p}_1}(\vec{\sigma}_1,\vec{\tau}_1)\, \delta n_{\vec{p}_2}(\vec{\sigma}_2,\vec{\tau}_2) \,, \tag{1}$$

where $e^{\circ}_{\vec{p}}$ is the q.p. energy in the absence of other elementary excitations, μ is chemical potential, and $U_{\vec{p}}$ describes the effect of possibly spin and/or isospin dependent external field (here time and space independent) applied to the system. The matrix structure of $\delta n_{\vec{p}}$ has been indicated using Pauli matrices in the spin and isospin spaces, $\vec{\sigma}_i$, $\vec{\tau}_i$. The quantity f is the q.p. interaction, which in symmetric nuclear matter is assumed to be of the form

$$N_0 f = F + F'\vec{\tau}_1\cdot\vec{\tau}_2 + G\vec{\sigma}_1\cdot\vec{\sigma}_2 + G'(\vec{\sigma}_1\cdot\vec{\sigma}_2)(\vec{\tau}_1\cdot\vec{\tau}_2)$$
$$+ q^2/p_F^2\,(H + H'\vec{\tau}_1\cdot\vec{\tau}_2)\, S_{12}(\hat{q}) \,, \tag{2}$$

where N_0 is the density of q.p. states at the Fermi surface, $N_0 = 2m^* p_F/\pi^2\hbar^3$ (m^* is the q.p. effective mass defined by $p_F/m^* = (\partial e^{\circ}_p/\partial p)_{p=p_F}$) The tensor operator is defined as

$$S_{12}(\hat{q}) = 3(\vec{\sigma}_1\cdot\hat{q})(\vec{\sigma}_2\cdot\hat{q}) - \vec{\sigma}_1\cdot\vec{\sigma}_2 \,, \tag{3}$$

where $\vec{q} = \vec{p}_1 - \vec{p}_2$. The tensor term, appearing in the second line of eq.(2), has been neglected in the original version of Migdal's theory. The q.p. momenta are restricted to the Fermi surface, $\vec{p}_i = p_F \vec{n}_i$. Notice, that the central part of f is scalar separately in momentum and spin spaces while tensor q.p. interaction is invariant only with respect to simultaneous rotation in both spaces (it is a contraction of the two rank two tensors).

The quantities F , F' , G , G' , H and H' depend only on the cosine of the angle between $\vec{p}_1$ and $\vec{p}_2$, $\cos\theta_{12} = \vec{n}_1\cdot\vec{n}_2$ This dependence is expanded in Legendre polynomials,

$$F(\vec{p}_1,\vec{p}_2) = \sum_l F_l\, P_l(\cos\theta_{12}) \,, \tag{4}$$

with similar expansions for the remaining functions.

So far no empirical information on the tensor q.p. interaction in nuclear matter is available. We have therefore used theoretical results for the spin dependent Fermi liquid parameters, calculated by Bäckman et al. [9] and Jackson et al. [12] starting from the realistic N-N potentials.

The Fermi liquid parameters of Ref. [9] have been calculated perturbationally starting from the Reid soft core (RSC) [13] N-N potential. The values of the spin dependent Fermi liquid parameters have been taken from table 2 of Ref. [9] (column "total"). In view of large uncertainties in perturbative results for m^* and the renormalization of the q.p. pole Z we made the simplest choice $m^* = m$ and $Z = 1$. We have put $G_l, G'_l = 0$ for $l > 4$ and $H_l, H'_l = 0$ for $l > 8$. This set of the Fermi liquid parameters will be referred to as the RSC(p) one.

In Ref. [12] the Fermi liquid parameters have been calculated using variational techniques from the RSC and Bethe Johnson (BJ) [14] N-N potentials. For technical reasons only lowest ($l < 3$ for the central part and $l < 2$ for the tensor part) parameters could be calculated with a sufficient precision. We have put $G_l, G'_l = 0$ for $l > 2$. However, putting $H_l, H'_l = 0$ for $l > 1$ would be unreasonable in view of the slow convergence of the Legendre expansion. Hence, the values of H_l, H'_l for $1 < l < 9$ have been calculated using the one-pion exchange (OPE) approximation (with a proper value of m^* in the density of states). As in the case of the RSC(p) model we put $H_l, H'_l = 0$ for $l > 8$. The corresponding sets of Fermi liquid parameters will be referred to as the RSC(v) and BJ ones, respectively.

In what follows we restrict ourselves to the case of excitations in the spin channel. The expansion of the $\delta n_{\vec{p}}$ matrix around the unperturbed Fermi surface reads then, including only first order term,

$$\delta n_{\vec{p}} = \frac{\partial n^0_{\vec{p}}}{\partial e_{\vec{p}}} w(\vec{n}) . \tag{5}$$

The hermitian traceless matrix w can be rewritten as

$$w(\vec{n}) = \vec{u}(\vec{n}) \cdot \vec{\sigma} = \sum_{\mu=-1}^{1} (-)^{\mu} u^{\mu}(\vec{n}) \sigma^{-\mu} . \tag{6}$$

Here, σ^{+1} and σ^{-1} are spin rising and spin lowering operators and $\sigma^0 = \sigma^z$. The part of the q.p. interaction relevant for the spin channel is

$$\Delta \mathcal{F} = G \vec{\sigma}_1 \cdot \vec{\sigma}_2 + q^2/p_F^2 \, H \, S_{12}(\hat{q}) . \tag{7}$$

We define a rank two spherical tensor

$$\begin{aligned} \Delta \mathcal{F}_{\mu\mu'} &= \tfrac{1}{4} \mathrm{Tr}_{\sigma_1 \sigma_2} \left[\Delta \mathcal{F} (-)^{\mu'} \sigma_2^{-\mu'} \sigma_1^{\mu} \right] = \\ &= \left(G - q^2/p_F^2 \, H \right) \delta_{\mu\mu'} + 3 H q^{\mu} q^{-\mu'} (-)^{\mu'} / p_F^2 . \end{aligned} \tag{8}$$

In view of the symmetry properties of the q.p. interaction and the presence of tensor terms it is particularly suitable to rewrite the equations of the Fermi liquid theory in the basis of the total angular momentum of the q.p.-q.hole pair, (this has been pointed out in Ref.[9]). The matrix elements of $\Delta\mathcal{F}$ read then

$$\Delta\mathcal{F}^{J}_{\ell\ell'} = \frac{1}{4\pi}\sum_{\substack{m\mu\\ m'\mu'}}(-)^{\mu+\mu'}(\ell m 1 -\mu|JM)(\ell' m' 1 -\mu'|JM)\cdot$$

$$\cdot\int d\vec{n}_1 \int d\vec{n}_2\, Y^{*}_{\ell m}(\vec{n}_1)\,\Delta\mathcal{F}_{\mu\mu'}\, Y_{\ell' m'}(\vec{n}_2). \tag{9}$$

The stability of the ground state at $T=0K$ in the spin channel implies that the energy has a <u>minimum</u> for $n_{\vec{p}} = n^{0}_{\vec{p}}$. We expand the angular dependence of u^{μ}, eq.(6), in spherical harmonics

$$u^{\mu}(\vec{n}) = \sum_{\ell m} u^{\mu}_{\ell m}\, Y_{\ell m}(\vec{n}) \tag{10}$$

and we pass to the J basis using the linear transformation

$$C^{M}_{\ell J} = \sum_{m\nu}(-)^{\nu}(\ell m 1 -\nu|JM)\,u^{\nu}_{\ell m} \tag{11}$$

This enables us to write down the following formula for δE

$$\delta E = \frac{3\rho}{8\pi m^{*}}\sum_{JM\ell\ell'}\left[\Delta\mathcal{F}^{J}_{\ell\ell'} + \delta_{\ell\ell'}\right] C^{M}_{\ell J}\, C^{M\,*}_{\ell' J} \tag{12}$$

where $\rho = 2k_F^3/3\pi^2$ and $k_F = p_F/\hbar$. Stability criteria are thus equivalent to the requirement that the stability matrix $\langle \ell J|A|\ell' J\rangle = \delta_{\ell\ell'} + \Delta\mathcal{F}^{J}_{\ell\ell'}$ be positive-definite for each value of J [9]. The condition for stability can be also stated as the requirement that the <u>lowest</u> eigenvalue of the $\langle \ell J|A|\ell' J\rangle$ matrix be positive.

The presence of tensor q.p. interaction may significantly lower the value of the $\langle 10|A|10\rangle$ element of the stability matrix; this is the lowest eigenvalue of A in the spin channel for the BJ and RSC q.p. interactions,

$$\langle 10|A|10\rangle = 1 - \frac{10}{3}H_0 + \frac{1}{3}G_1 + \frac{4}{3}H_1 - \frac{2}{15}H_2\,. \tag{13}$$

For both the RSC(v) and BJ parameters we find <u>negative</u> value of $\langle 10|A|10\rangle$ for $1.2\ \mathrm{fm}^{-1} < k_F < 2\ \mathrm{fm}^{-1}$. Let us mention that at the same time the system is stable with respect to the static, space homogeneous perturbations relevant for the definition of the static spin susceptibility and the spin symmetry energy of nuclear matter, discussed in the next section. The deformation of the Fermi surface is there restricted to the $J=1$, $\ell=0$, 2 channel.

The spin instability in the $J=0$, $\ell=1$ channel is driven by the tensor force. More precisely, it results from large, positive values of the parameter H_0. Quite large OPE values of H_ℓ for $\ell=1,2$ amplify further this effect. If one puts $H_\ell=0$ for $\ell>1$, the point of instability shifts towards higher value of k_F. Namely, nuclear matter becomes then unstable for $k_F>1.21\ \mathrm{fm}^{-1}$ in the case of the BJ model and for $k_F>1.255\ \mathrm{fm}^{-1}$ in the case of the RSC(v) model.

3. Spin susceptibility [10]

Consider static, space homogeneous perturbation of the form

$$H^{ext} = \vec{V}\,\vec{\sigma}(\vec{r}), \tag{14}$$

where $\vec{V}$ is a constant vector and $\vec{\sigma}(\vec{r})$ is the spin density operator. In what follows we assume that the $\vec{V}$ field is applied along the z-axis. To second order in V the formula for δE reads then

$$\delta E = \frac{3\rho}{8\pi m^*}\left(\sum_{JMll'} \langle lJ|A|l'J\rangle C^M_{lJ} C^{M*}_{l'J} + \frac{4\sqrt{\pi}}{v_F} C^0_{01} V\right), \tag{15}$$

where v_F is the q.p. velocity at the Fermi surface, $v_F = p_F/m^*$. The ground state of weakly spin polarized nuclear matter is determined from the condition

$$\delta E\,(C^M_{lJ}, C^{M'*}_{l'J'}) = \text{minimum}. \tag{16}$$

The solution to eq.(16) is surprisingly simple [10]. Only C^0_{01} and C^0_{21} do not vanish. Tensor force introduces thus a quadrupole deformation of the Fermi surface in polarized nuclear matter.

The spin susceptibility may be calculated as

$$\chi_\sigma = \lim_{V\to 0} \langle\sigma^0\rangle/V =$$

$$= N_0 \frac{\langle 21|A|21\rangle}{\langle 21|A|21\rangle\langle 01|A|01\rangle - \langle 01|A|21\rangle^2} \tag{17}$$

The condition $\chi_\sigma > 0$ is thus equivalent to the stability condition in the $J=1$, $S=1$, $T=0$, $l=0,2$ channel.

Using explicit formulae for the matrix elements $\langle l1|A|l'1\rangle$ we can rewrite the formula for χ_σ in a suitable form

$$1/\chi_\sigma = 1/\chi_{\sigma c}\,(1-\delta_{\sigma t}), \tag{18}$$

where $\chi_{\sigma c}$ corresponds to the standard case of purely central q.p. interaction,

$$\chi_{\sigma c} = N_0/(1+G_0), \tag{19}$$

and the effect of tensor q.p. interaction is given by

$$\delta_{\sigma t} = \frac{2(H_0 - \frac{2}{3}H_1 + \frac{1}{5}H_2)^2}{(1+G_0)(1+\frac{1}{5}G_2 - \frac{7}{15}H_1 + \frac{2}{5}H_2 - \frac{3}{35}H_3)} \tag{20}$$

The formula for the spin symmetry energy reads

$$\epsilon_\sigma = \epsilon_{\sigma c}\,(1-\delta_{\sigma t}). \tag{21}$$

The formulae for the spin-isospin channel (i.e. for $\chi_{\sigma\tau}$ and $\epsilon_{\sigma\tau}$) are obtained by a replacement of G_l, H_l by G_l', H_l'.

The calculations have been performed at the normal nuclear matter density corresponding to $k_F = 1.35\ \text{fm}^{-1}$. For the RSC(p) model of the q.p. interaction we get $\delta_{\sigma t} = 0.043$ and $\delta_{\sigma\tau t} = 0.005$. Hence, despite a strong tensor force in the spin channel (large H_l's) the value of $\delta_{\sigma t}$ is small.

This is due to exact cancellation of H_0 by $-\frac{2}{3}H_1$ in the numerator of expression (20). In the spin-isospin channel, where tensor forces are much weaker, the cancellation of H_0' by $-\frac{2}{3}H_1'$ is also very effective.

The results for $\delta_{\sigma t}$, $\delta_{\sigma\tau t}$ for the RSC(v) and BJ q.p. interaction are of only academic interest because of the instability in $J=0$, $\ell=1$, $S=1$, $T=0$ channel, but they may nevertheless be given for the sake of completeness. We get $\delta_{\sigma t}=0.27$ and 0.24 for the RSC(v) and BJ models, respectively. The effect in the spin-isospin channel is much weaker for both models.

4. Linear response and dynamic form factor [11]

We consider linear response of nuclear matter at $T=0K$ to a perturbation in the spin channel which is assumed to be of the following form in the momentum representation

$$H^{ext}(t) = \vec{V}\cdot\vec{\sigma}(-\vec{k})\, e^{-i\omega t} \quad + \text{hermitian conj.} \tag{22}$$

The spin-density-spin-density response tensor is then defined by

$$\sum_j \chi_{\ell j}(\vec{k}\omega)\, V^j = \langle\sigma^\ell\rangle_{\vec{k}\omega}\ , \tag{23}$$

where

$$e^{i(\vec{k}\cdot\vec{r}-\omega t)}\langle\sigma^\ell\rangle_{\vec{k}\omega} + \text{complex conj.} = \langle\Psi(t)|\sigma^\ell(\vec{r})|\Psi(t)\rangle \tag{24}$$

and the (perturbated) wave function $\Psi(t)$ is calculated using first order time dependent perturbation theory for $H = H^0 + H^{ext}(t)$. Here ℓ, j are cartesian indices. Using standard assumptions one can relate $\chi_{\ell j}(\vec{k}\omega)$ to the dynamic form factor for spin-density fluctuations, $S_{\ell j}(\vec{k}\omega)$, by the fluctuation-dissipation theorem,

$$\mathrm{Im}\,\chi_{\ell j}(\vec{k}\omega) = -\frac{\pi}{\hbar}\left[S_{\ell j}(\vec{k},\omega) - S_{\ell j}(\vec{k},-\omega)\right]. \tag{25}$$

In order to calculate $\langle\sigma^\ell\rangle_{\vec{k}\omega}$, equ.(23), we first calculate $\delta n_{\vec{p}}(\vec{r},t)$ implied by $H^{ext}(t)$, by solving the Landau kinetic equation in the collisionless regime ($T\cong 0K$). This can be done exactly in the long-wavelength limit ($k \ll k_F$).

Using the Ansatz

$$\delta n_{\vec{p}} = \frac{\partial n^0_{\vec{p}}}{\partial e_{\vec{p}}}\, w(\vec{n})\, e^{i(\vec{k}\cdot\vec{r}-\omega t)} \tag{26}$$

and assuming $\vec{k}$ to be our z-axis, we obtain an inhomogeneous integral equation for $u^\mu(\vec{n})$, eq.(6),

$$(s-z)\, u^\mu(\vec{n}_1) - z\sum_{\mu'}\int\frac{d\vec{n}_2}{4\pi}\,\Delta\mathcal{F}_{\mu\mu'}\, u^{\mu'}(\vec{n}_2) = -z V^\mu, \tag{27}$$

where $z = \vec{n}_1\cdot\hat{k}$ and $\Delta\mathcal{F}_{\mu\mu'}$ is given by eq.(8). To satisfy the proper initial condition at $t \to -\infty$ we have to put $s = \lambda + i\eta$, with real $\lambda = \omega/kv_F$ and $\eta \to +0$. We expand u^μ in spherical harmonics

$$u^{\mu}(\vec{n}) = \frac{z}{s-z} \sum_{lm} \sqrt{\frac{4\pi}{2l+1}}\, u^{\mu}_{lm} Y_{lm}(\vec{n}) \tag{28}$$

and pass to the C^{M}_{lJ} parameters using eq.(11). In the linear approximation the quantities $\bar{C}^{M}_{lJ} = C^{M}_{lJ} / V^{-M}$ do not depend on $\vec{V}$. In this way, we reduce eq.(27) to a system of linear equations for

$$\bar{C}^{M}_{lJ} - \sum_{J'l'l''} \Delta \mathcal{F}^{J}_{ll''} \alpha^{JJ'M}_{l''l'}(\lambda)\, \bar{C}^{M}_{l'J'} = -(-)^{M} \delta_{J1} \delta_{l0} \,. \tag{29}$$

Notice that matrices $\bar{C}^{M}_{lJ}$ corresponding to different M's are not coupled by the kinetic equation. The functions $\alpha^{JJ'M}_{ll'}(\lambda)$ are defined by eqs.(12) of Ref.[15]. The final formula for the diagonal elements of the χ tensor reads

$$\chi_{\mu\mu}(\lambda) = -(-)^{\mu} N_{0} \sum_{lJ} \alpha^{J1,-\mu}_{l0}(\lambda)\, \bar{C}^{-\mu}_{lJ} \,, \tag{30}$$

where $\bar{C}^{-\mu}_{lJ}$ are to be calculated from eq.(29). These solutions $\bar{C}^{-\mu}_{lJ}$ correspond to a fixed value of $M = -\mu = m - \nu$. With our assumptions the nondiagonal elements of χ vanish and $\chi_{+1,+1} = \chi_{-1,-1}$.

In what follows we consider the case $\lambda \geqslant 0$, relevant for nuclear matter at $T = 0$K. Then, for every $\lambda < 1$ there exists a nontrivial complex solution to eq.(29), yielding a nonvanishing imaginary part of $\chi_{\mu\mu}$. The corresponding contribution to $S_{\mu\mu}$, comming from the single pair excitations may be calculated using eq.(25).

For $\lambda > 1$ the only contribution to S can result from the existence of nontrivial solution to the homogeneous counterpart of eq.(29) at some $\lambda_0 > 1$. Such a nontrivial solution corresponds to an undamped zero-sound mode of collective excitation in the $M = -\mu$ channel, propagating at phase velocity $v_0 = v_F \lambda_0 > v_F$, its contribution to S being

$$S^{coll.}_{\mu\mu}(\lambda) = B_{\mu}\, \delta(\lambda - \lambda_0). \tag{31}$$

Let us remind once again that in the presence of tensor forces projections of orbital angular momentum, m, and spin, $-\nu$, on the z-axis are no longer good quantum numbers. The excitations are labelled by $M = m - \nu$. Solutions to eqs.(29) contributing to $S_{\mu\mu}$ are those from the $M = -\mu = 0, \pm 1$ channels, because only in these channels can the value $m = 0$ appear in expansion (28), giving rise to a nonvanishing expectation value $\langle \sigma^{\mu} \rangle$.

In view of the instability of the ground state of nuclear matter for the RSC(v) and BJ q.p. interactions, we restrict ourselves to the RSC(p) model. The dynamic form factors in the spin and spin-isospin channels, calculated at $k_F = 1.35$ fm^{-1}, are represented in Fig.1.

The main qualitative effect of tensor force is to remove degeneracy with respect to M. The only degeneracy left is that with respect to the sign of M. The effect of tensor force in the spin channel is stronger than that in the spin-isospin one. This is due to the fact that H_l's are usually 2÷3 times larger than H_l''s. The pole corresponding to an undamped collective excitation (spin or spin-isospin sound) splits into two poles: one corresponding to $M=0$ and the other corresponding to $M=\pm 1$.

Let us consider first the case of the spin channel. At a given value of k, tensor force shifts the energy of the $M=0$ mode down and that of the $M=\pm 1$ modes up with respect to the energy of spin sound mode in the absence of tensor forces. This is accompanied by a significant decrease of the relative strength of the $M=0$ mode.

In the spin-isospin channel the effect of tensor force is weaker and opposite to that seen in the spin channel. This is due to the fact that H_l' are negative while H_l were positive.

5. Conclusion

Inclusion of the tensor q.p. interaction modifies the formulae of the Fermi liquid theory of nuclear matter. The calculations performed using several sets of the spin dependent Fermi liquid parameters show that the effect of tensor forces may be important. In particular, for the q.p. interaction derived using variational methods from the RSC and BJ N-N potentials, tensor forces imply instability of the ground state of nuclear matter near and above normal nuclear matter density. The instability occuring in the $J=0$, $S=1$, $T=0$ channel may be (and most probably is) an artifact of the approximations made in the calculations. It seems to be reasonable to say that the appearence of this instability reflects large uncertainty in our theoretical knowledge of the q.p. interaction in nuclear matter.

References

[1] L.D.Landau, JETP (Sov. Phys.) 3, 920 (1956).
[2] L.D.Landau, JETP (Sov. Phys.) 5, 101 (1957).
[3] L.D.Landau, JETP (Sov. Phys.) 8, 70 (1959).
[4] G.Baym and C.J.Pethick, in Physics of liquid and solid helium, vol. II, p.115 (ed. by K.H.Bennemann and J.B.Ketterson) Wiley, N.York, 1978.
[5] A.B.Migdal, Theory of finite Fermi systems and applications to atomic nuclei, Interscience, London, 1967.
[6] P.Haensel and J.Dąbrowski, Nucl.Phys. A254, 211 (1975).
[7] J.Dąbrowski and P.Haensel, Ann.Phys. (N.Y.) 97, 452 (1976).
[8] G.E.Brown, S.-O.Bäckman, E.Oset and W.Weise, Nucl.Phys. A286,

191 (1977).

[9] S.-O.Bäckman, O.Sjöberg and A.D.Jackson, Nucl. Phys. A321, 10 (1979).

[10] P.Haensel and A.J.Jerzak, Phys. Lett. 112B, 285 (1982).

[11] P.Haensel and A.J.Jerzak, Acta Phys. Polonica (in press).

[12] A.D.Jackson, E.Krotscheck, D.E.Meltzner and R.A.Smith, Nucl. Phys. A386, 125 (1982).

[13] R.V.Reid, Ann. Phys. (N.Y.) 50, 411 (1968).

[14] H.A.Bethe and M.J.Johnson, Nucl. Phys. A230, 1 (1974).

[15] B.L.Friman and P.Haensel, Phys. Lett. 98B, 323 (1981).

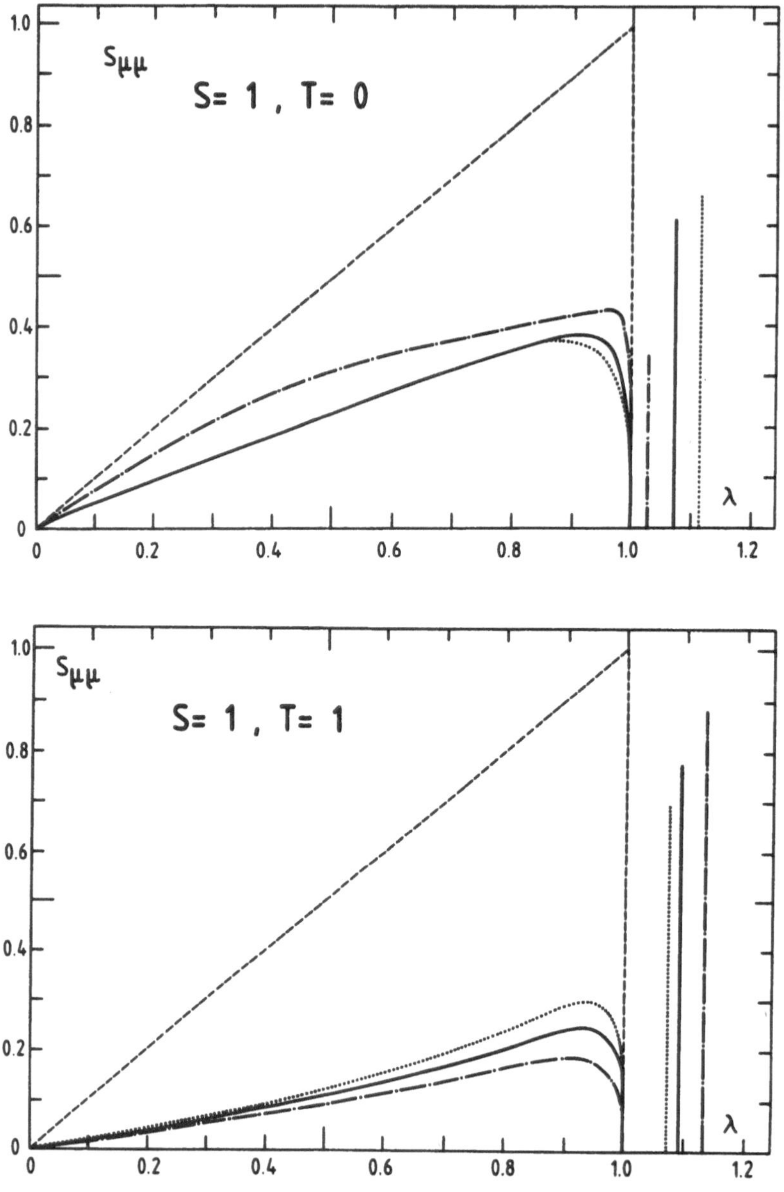

Fig.1. The plots of $S_{\mu\mu}(\lambda)$ (in the units of $m^* p_F/\pi^2\hbar^2$) calculated for the RSC(p) q.p. interaction at $k_F = 1.35$ fm^{-1}. Dotted line: $\mu = -M = \pm 1$, dashed line: $\mu = M = 0$. The contribution from collective modes is shown using vertical lines. The heights of vertical lines are proportional to the pole contribution to the integrated form factors. Results for the case of no tensor forces (solid line) and those for an ideal Fermi gas of particles of mass m^* (short dashes) are also shown.

THE EFFECTIVE MASS IN NUCLEAR MATTER AND IN NUCLEI

C. Mahaux,
Institut de Physique B5, Université de Liège, B-4000 Liège 1, Belgium

Abstract. Recent calculations of the effective mass are reviewed and their results are compared with experimental data.

1. INTRODUCTION

The effective mass plays an important role in nuclear physics : it is inversely proportional to the spacing between single-particle levels, is directly related to the parameters of the Landau interaction between quasiparticles, determines the excitation energy of several giant resonances, etc.

The effective mass is also of much interest for other many-body systems. In liquid ^{3}He for instance it is intimately related to the temperature dependence of the specific heat. Early[1]) as well as recent[2,3]) developments show a gratifying mutual fertilization between the theoretical understanding of the properties of the effective mass in liquid ^{3}He on the one hand and in nuclei on the other hand.

Here, we survey the progress which has been accomplished since a recent review[4]) on the effective mass in nuclear matter and in nuclei.

2. NUCLEAR MATTER

2.1. Definition

Let $V(e)$ denote the mean potential energy of a nucleon with energy e. The effective mass $m^*(e)$ is defined by

$$m^*(e)/m = 1 - dV(e)/de \quad . \tag{1}$$

When calculated in the Brueckner-Hartree-Fock approximation, $m^*(e)$ presents an enhancement peak whose maximum lies slightly above the Fermi energy and whose width decreases with decreasing density[5,6]). The Brueckner-Hartree-Fock approximation contains the Hartree-Fock field and "polarization" contributions like the one shown in the middle of Fig. 1 where a two particle-one hole intermediate state is excited. It does not contain the effect of "Pauli blocking", depicted by graph (c). It took several years before the latter contribution could be evaluated, first in the framework of the hard sphere Fermi gas model[7]) and of a schematic but more realistic model[8]).

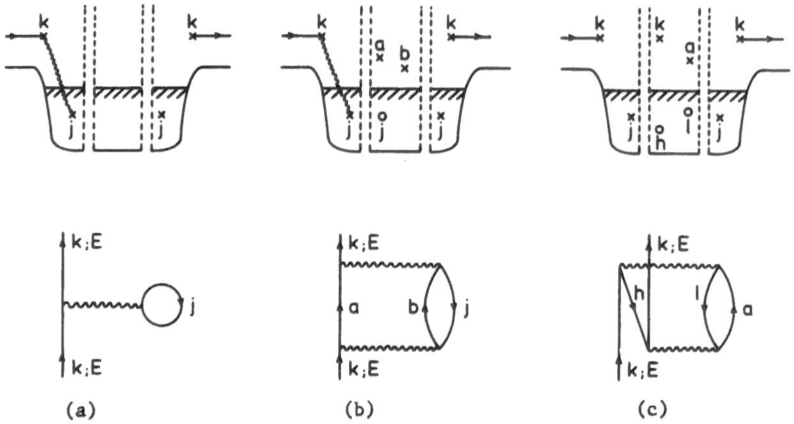

Fig. 1. Graph (a) represents the Hartree-Fock field and graph (b) the correlation contribution; both of these are included in the Brueckner-Hartree-Fock approximation. Graph (c) represents the blocking of those core correlations in which a nucleon is excited in the same state (k) as the external nucleon.

Still more realistic models were then studied in refs.[9-12], in which the sum of the three contributions shown in Fig. 1 have been calculated. These results are briefly described in sect. 2.4.

2.2. Mean potential energy[6]

Actually, a nucleon embedded in nuclear matter is labelled by two quantum numbers, namely its momentum k and its frequency $E/\hbar$. The corresponding single-particle field $\Sigma[k;E]$ is a function of both variables and is a complex quantity

$$\Sigma[k;E] = \mathcal{V}[k;E] + i\,\mathcal{W}[k;E] \quad . \tag{2}$$

In the quasi-particle (or optical model) approximation, it is assumed that a nucleon with given energy e is likely to have a momentum $k(e)$ given by the energy momentum relation

$$e = \hbar^2[k(e)]^2/2m + \mathcal{V}[k(e);e] \quad . \tag{3}$$

The quantity

$$V(e) =: \mathcal{V}[k(e);e] \tag{4}$$

is identified with the mean potential energy of a nucleon with momentum k .

2.3. k-mass and E-mass

If an approximation scheme gives $V[k;E]$, the corresponding effective mass can be calculated from eqs. (4) and (1). In the case of Fig. 1 for instance, one has

$$V[k;E] = V_{HF}(k) + V_{PO}[k;E] + V_{CO}[k;E] \quad , \tag{5}$$

where the three terms are the Hartree-Fock, polarization and Pauli blocking (or correlation) contributions, respectively. Note that the Hartree-Fock approximation only depends upon k while the other two terms depend upon both k and E . One can prove the identity

$$\frac{m^*(e)}{m} = \frac{\tilde{m}(e)}{m} \cdot \frac{\bar{m}(e)}{m} \quad , \tag{6}$$

where the k-mass $\tilde{m}$ and the E-mass $\bar{m}$ are defined by the relations

$$\tilde{m}(e)/m = \{1 + (m/\hbar^2 k)\ \partial V[k;E]/\partial k\}^{-1}_{k=k(e)} \quad , \tag{7}$$

$$\bar{m}(e)/m = 1 - \{\partial V[k;E]/\partial E\}_{k=k(e)} \tag{8}$$

2.4. Numerical results

The E-mass $\bar{m}(e)$ is found to have a maximum centered near the Fermi energy $\varepsilon_F = e(k_F)$, where k_F is the Fermi momentum. One example is shown in Fig. 2; the contribution of the polarization graph peaks slightly above ε_F and that of the correlation graph below ε_F ; their sum is practically centered on ε_F .

In the Hartree-Fock or in the Brueckner-Hartree-Fock approximation it is usually found that the k-mass $\tilde{m}(e)$ is a rather uninteresting smoothly increasing function of energy; an example is shown in Fig. 3. In the hard sphere Fermi gas model, however, it was found that $\tilde{m}(e)$ has a broad and shallow dip centered near the Fermi energy[7]). A similar finding appears in refs. [9,11]). The origin of this minimum is not clear. In ref.[9]) it already occurs in the Hartree-Fock approximation and is probably characteristic of the effective interaction

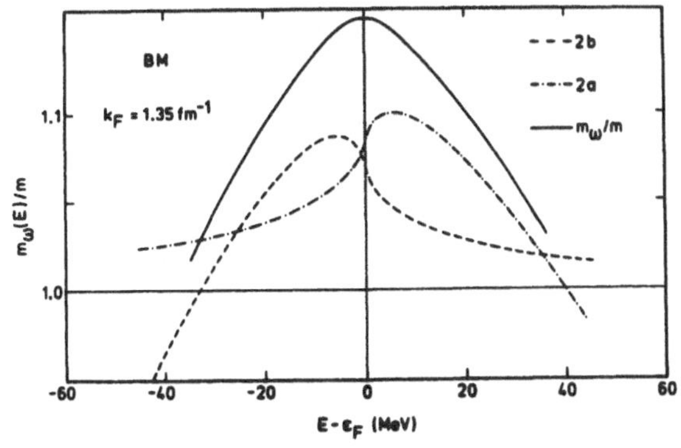

Fig. 2. Adapted from ref.[8]). The full curve shows the dependence of the E-mass $\bar{m}(E)$ (denoted $m_\omega(E)$ on the figure) upon the difference $E-\varepsilon_F$. The dash-and-dots correspond to the correlation graph and the dashes to the polarization graph of Fig. 1.

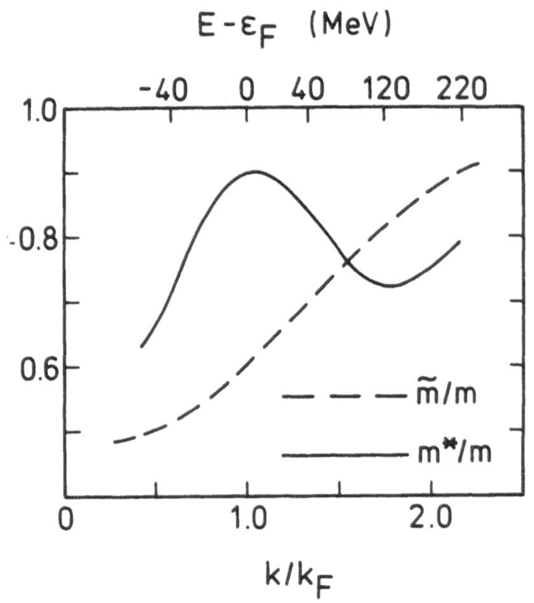

Fig. 3. Adapted from ref.[10]*). Dependence upon* $E-\varepsilon_F$ *of the quantities* $\tilde{m}(e)/m$ *(long dashes) and* $m^*(e)/m$ *(full curve) as evaluated in the correlated basis function approach. The k-mass* $\tilde{m}$ *is evaluated in the lowest order approximation, while the E-mass* $\bar{m}$ *is essentially derived from the two second-order diagrams of Fig. 1.*

used there. In ref.[11]) it seems to be due to the contribution to $\tilde{m}$ of the second-order diagrams shown in Fig. 1.

2.5. Discussion

The origin and the shape of the peak of $\bar{m}(e)$ near the Fermi energy can most easily be discussed in terms of the dispersion relation which relates the real part to the imaginary part of $\Sigma[k;E]$ and of the property that $\mathcal{W}[k;E]$ vanishes at $E=\varepsilon_F$ [13]). This dispersion relation can be used in conjunction with simple models for $\mathcal{W}[k;E]$. The resulting "schematic models"[14]) have proved useful in the case of liquid ^{3}He [2,3]) as well as of nuclear matter and nuclei[14-16]). These schematic models enable one to reproduce algebraically the result of microscopic calculations or of empirical fits, and to investigate the influence of various parameters or assumptions. It is unfortunate that the papers[9-12,17]) describing several recent microscopic calculations do not give separately the contributions to $\bar{m}$ of respectively the polarization and the correlation graphs of Fig. 1. Indeed each of these has a physical meaning in terms of the momentum distribution of the correlated ground state[6]). Moreover, a sum rule then enables one to check the accuracy of the numerical calculation.

There exists a detailed formal similarity between the expression of the polarization and correlation corrections derived from the perturbation (or from the hole line) expansion on the one hand, and from the correlated basis function approach on the other hand. However, the effective interaction which appears in the expression associated with the correlated basis function approach depends upon the frequency $E/\hbar$, see e.g. ref.[17]), while this is not the case in the hole-line (Brueckner) expansion. This difference should be better understood.

It would also be of interest to investigate the behaviour of the k-mass $\tilde{m}(e)$. In particular, one should study the origin of the broad and shallow minimum found in refs.[7,9,11]) in order to determine whether its existence is characteristic of specific interactions or approximation schemes, or else is a rather general property.

3. NUCLEI

In nuclear matter, the single-particle potential $\Sigma[k;E]$ is a function of only two coordinates; a Fourier transform yields the nonlocal potential $\Sigma[|\vec{r}-\vec{r}\,'|;E]$. Even if one excludes the spin and isospin degrees of freedom, the single-particle potential in a nucleus is a function $\Sigma[\vec{r},\vec{r}\,';E]$ of seven coordinates. Its properties are therefore difficult to study and to characterize, unless mother nature justifies the use of simple approximations.

In effect, most theoretical studies implicitly assume that the nonlocality of Σ is about the same in nuclei as in nuclear matter. More specifically, one usually assumes that one may write the real part (and similarly for the imaginary part) of Σ in the form

$$\mathcal{V}[\vec{r},\vec{r}\,';E] = (\beta\sqrt{\pi})^{-3}\, V(r;E)\, \exp(-\, s^2/\beta^2) \quad , \tag{9}$$

where $s = |\vec{r}-\vec{r}\,'|$. The assumption (9) is good in the Hartree-Fock approximation. Its validity when polarization and correlation corrections are included has only been studied at positive energy and appears to retain some validity although it becomes poorer[18-20]).

At negative energies one encounters the additional difficulty that $\mathcal{V}[\vec{r},\vec{r}\,';E]$ has a pole at each bound state energy and has no obvious meaning at other energies. This difficulty is circumvented by identifying the mean potential felt by a nucleon with energy E with the smoothly energy-dependent quantity $\Sigma[\vec{r},\vec{r}\,';E+iI]$, where I is an averaging interval larger than the separation between bound states or resonances[21]). This smooth quantity is the one which is discussed below, although we usually drop any explicit reference to I .

3.1. The k-mass

If assumption (9) is justified, the k-mass $\tilde{m}$ is determined by the nonlocality range β , and the E-mass $\bar{m}$ by the energy-dependence of $V(r;E)$:

$$\bar{m}(r;E)/m = 1 - dV(r;E)/dE \quad , \tag{10}$$

$$\tilde{m}(r;E)/m = \{1 - (2\hbar^2)^{-1}\, m\, \beta^2\, V(r;E)\}^{-1} \quad . \tag{11}$$

In practice, β is approximately independent of energy and $\tilde{m}(r;E)$ then has a Woods-Saxon like radial shape and is a smoothly increasing function of E . For Skyrme-type interactions $\tilde{m}(r;E)$ is independent of E in the Hartree-Fock approximation. Some information on the contribution to $\tilde{m}(r;E)$ of polarization and correlation effects at positive energy ($E \approx 10 - 50$ MeV) is implicitly contained in the results published in refs.[18-20] but has not been extracted yet. No information appears to be available on the value of $\tilde{m}$, or more generally on the nature of the nonlocality of $\Sigma[\vec{r},\vec{r}\,';E]$, at low and at negative energy when polarization and correlation corrections are included. Until now one has usually assumed that $\tilde{m}$ retains the form (11).

3.2. Energy-dependence of the E-mass

Most authors assume that Σ is diagonal in the $\{\varphi_{n\ell j}\}$ basis of the Hartree-Fock single-particle states. The problem then amounts to calculating $\Sigma_{n\ell j}(E)$, or equivalently the correction $\sigma_{n\ell j}(E)$ to the Hartree-Fock single-particle field. One usually performs an additional average and plots the quantity

$$\overline{V}(E) = \sum_{\ell j} (2j+1)\, \sigma_{n\ell j}(E) \quad . \tag{12}$$

It has recently been argued[22] that the average (12) is associated with the integral

$$\Delta(r;E) = \int \Sigma(\vec{r},\vec{r}\,';E)\, d^3s \quad . \tag{13}$$

In refs.[15,23,24], the quantities $\sigma_{n\ell j}(E)$ have been computed from particle-vibration coupling models and the E-mass has been defined by

$$\overline{m}(E)/m = 1 - d\overline{V}(E)/dE \quad . \tag{14}$$

Two examples are shown in Figs. 4 and 5. Two main differences exist between these two results. (i) In Fig. 5 $\overline{m}(E)/m$ shows a plateau for $|E| < 6$ MeV . The absence of this plateau in Fig. 4 is probably due to the fact that the authors did non include the particle-hole gap in their drawing. (ii) In Fig. 5 the quantity $\overline{m}(E)/m$ abruptly drops (to zero !) for $E \approx 8$ MeV . A similar feature was found in ref.[15]. Its absence in Fig. 4 may be due to the fact that the authors used a more realistic description of the core excitations[16].

An abrupt change of $\overline{m}(E)$ which would be limited to a narrow energy domain only reflects the existence of a kink in $\overline{V}(E)$, without implying a large change in the value of $\overline{V}(E)$. In the case of bound single-particle states, the quantity of interest appears to be the energy shift from the Hartree-Fock energy $\varepsilon_{n\ell j}$, namely

$$\delta\, \varepsilon_{n\ell j} \approx \sigma_{n\ell j}(\varepsilon_{n\ell j}) \quad . \tag{15}$$

It would thus be more meaningful to plot $\sigma_{n\ell j}(\varepsilon_{n\ell j})$ versus $\varepsilon_{n\ell j}$, to check whether

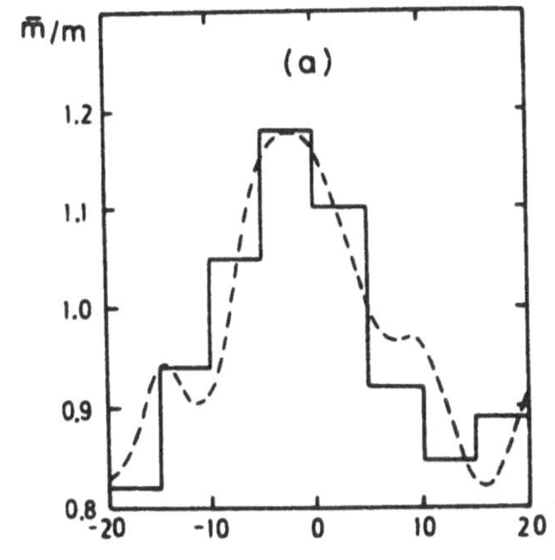

Fig. 4. Taken from ref.[23]*). Dependence upon* $E-\varepsilon_F$ *of expression (14) for neutrons in* ^{208}Pb .

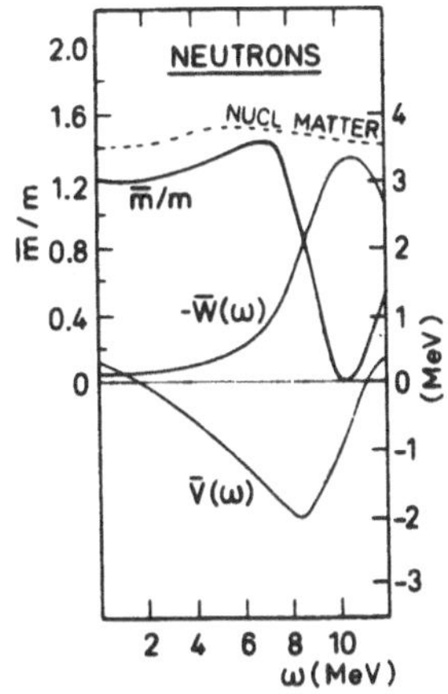

Fig. 5. Taken from ref.[24]*). Dependence upon* $\omega \equiv E-\varepsilon_F$ *of expressions (12) and (14) for neutrons in* ^{208}Pb *. The curve labelled* $-\bar{W}(\omega)$ *shows the imaginary part of the single-particle field and the dashed curve the E-dependence of* $\bar{m}(E)/m$ *in the case of nuclear matter.*

these values fall in the vicinity of a smooth curve $v(\varepsilon_{n\ell j})$ and to define the E-mass by eq. (14) with $\bar{V}(E)$ replaced by $v(E)$. This has recently been performed[25]) and yields results similar to those shown in Fig. 4, with due account for item (i) above.

In ref.[13]) it has been proposed to evaluate the E-mass from the dispersion relation which connects the real and the imaginary parts of $\Delta(r;E)$. The resulting average effective mass is shown in Fig. 6, where it is compared with the value deduced from the full curve drawn in Fig. 7 through the empirical values.

3.3. Radial dependence

The real part $\Delta V(r;E)$ and the imaginary part $\Delta W(r;E)$ of expression (13) are connected by a dispersion integral. In refs.[27-29]) $\Delta W(r;E)$ has been identified with the imaginary part $W(r;E)$ of the phenomenological optical-model potential. This enabled the calculation of the quantity

$$\bar{m}(r;E)/m = 1 - d[\Delta V(r;E)]/dE \quad . \qquad (16)$$

It was found that for r fixed the quantity $\bar{m}(r;E)$ has a maximum approximately centered on ε_F ; this peak is very broad in the nuclear interior and much narrower at the nuclear surface. For fixed E close to ε_F the E-mass $\bar{m}(r;E)$ has a peak for $r=R$, where R is slightly larger than the nuclear radius. Possible implications are described in refs.[30,31]). These results are confirmed by a recent microscopic calculation[22]), see Figs. 8 and 9.

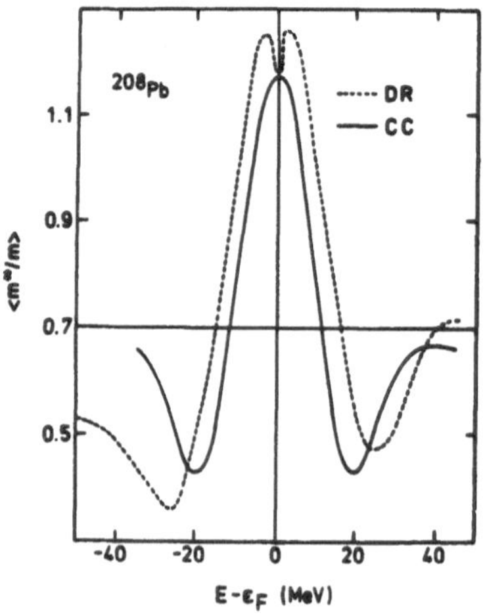

Fig. 6. Taken from ref.[26]). The dashed curve gives the energy dependence of the radial average of the effective mass in ^{208}Pb, as calculated from the dispersion relation[27]). The full curve is deduced from the empirical values shown in Fig. 7.

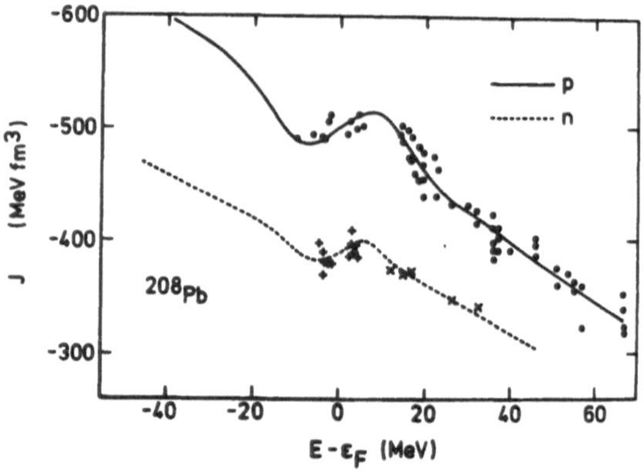

Fig. 7. Taken from ref.[26]). Dependence upon $E-\varepsilon_F$ of the empirical value of the volume integral per nucleon of the real part of the single-particle potential for protons (full dots) and neutrons (crosses) in ^{208}Pb.

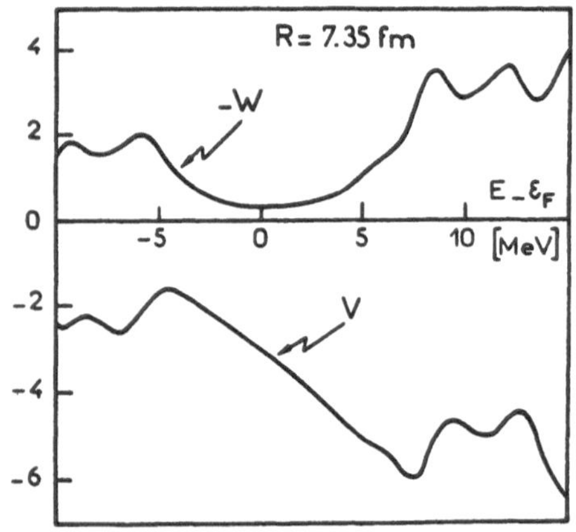

Fig. 8. Taken from ref.[22]). Dependence upon $E-\varepsilon_F$ of the real part $\Delta V(R;E)$ (labelled V on the drawing) and of the imaginary part $\Delta W(R;E)$ (labelled W here) of the quantity (13) for neutrons in ^{208}Pb and for $R = 7.35$ fm. The unit of the ordinate scale is MeV.

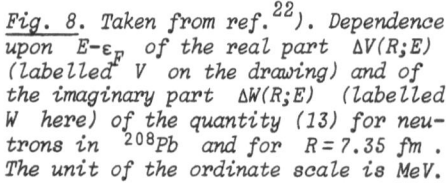

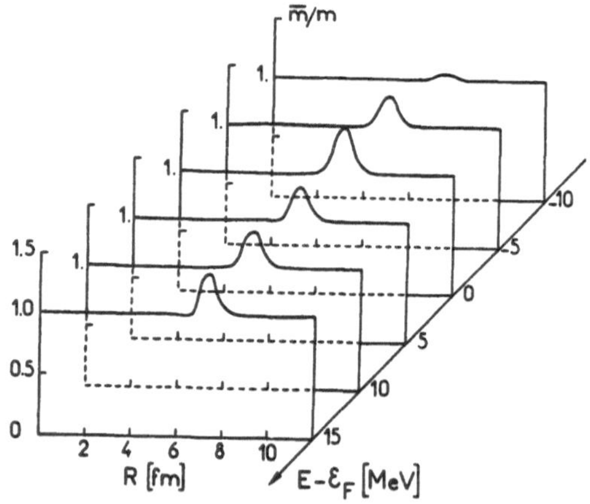

Fig. 9. Taken from ref.[22]). Dependence upon r and E of the quantity $\overline{m}(r;E)/m$ (see eq. (16)) for neutrons in ^{208}Pb.

4. DISCUSSION

Most of the existing calculations of the effective mass start from the assumption that it can be defined for negative as well as for positive energies. They do not contain internal checks whether this is possible indeed. In the case of ^{208}Pb, however, recent calculations[25]) and recent analyses of experimental data[28]) justify the approximate validity of introducing a state-independent effective mass or, equivalently, of including the polarization and correlation effects by adding a local energy-dependent correction to the Hartree-Fock field. One should study whether this also applies to lighter nuclei, e.g. ^{40}Ca.

REFERENCES

1. G.F. Bertsch and T.T.S. Kuo, Nucl.Phys. A112 (1968) 204
2. G.E. Brown, C.J. Pethick and A. Zaringhalam, J.Low Temp.Phys. 48 (1982) 349
3. S. Fantoni, V.R. Pandharipande and K.E. Schmidt, Phys.Rev.Lett. 48 (1982) 878
4. C. Mahaux, in "Nuclear Physics", edited by C.H. Dasso, R.A. Broglia and A. Winther (North-Holland Publ. Comp., Amsterdam, 1982) p. 319
5. J.P. Jeukenne, A. Lejeune and C. Mahaux, Phys.Lett. 59B (1975) 208
6. J.P. Jeukenne, A. Lejeune and C. Mahaux, Phys.Reports 25C (1976) 83
7. R. Sartor and C. Mahaux, Phys.Rev. 21C (1980) 1546
8. V. Bernard and C. Mahaux, Phys.Rev. C23 (1981) 888
9. J.P. Blaizot and B.L. Friman, Nucl.Phys. A372 (1981) 69
10. E. Krotscheck, R.A. Smith and A.D. Jackson, Phys.Lett. 104B (1981) 421
11. A.D. Jackson, E. Krotscheck, D.E. Meltzer and R.A. Smith, Nucl.Phys. A386 (1982)125
12. S. Fantoni, B.L. Friman and V.R. Pandharipande, Nucl.Phys. A399 (1983) 51
13. C. Mahaux and H. Ngô, Phys.Lett. 100B (1981) 285
14. G.E. Brown and M. Rho, Nucl.Phys. A372 (1981) 397
15. J. Wambach, V.K. Mishra and Li Chu-hsia, Nucl.Phys. A380 (1982) 285
16. J.P. Jeukenne and C. Mahaux, Nucl.Phys. A394 (1983) 445
17. E. Krotscheck and R.A. Smith, preprint (May 1982)
18. N. Vinh Mau and A. Bouyssy, Nucl.Phys. A257 (1976) 189
19. V. Bernard and Nguyen Van Giai, Nucl.Phys. A327 (1979) 397
20. A. Bouyssy, H. Ngô and N. Vinh Mau, Nucl.Phys. A371 (1981) 173
21. C.A. Engelbrecht and H.A. Weidenmüller, Nucl.Phys. A184 (1972) 385
22. Nguyen Van Giai and Pham Van Thieu, Phys.Lett. 126B (1983) 421
23. P.F. Bortignon, R.A. Broglia, C.H. Dasso and Fu De-ji, Phys.Lett. 108B (1982) 247
24. H.M. Sommermann, T.T.S. Kuo and K.F. Ratcliff, Phys.Lett. 112B (1982) 108
25. P.F. Bortignon, R.A. Broglia, C.H. Dasso and C. Mahaux, Communication to the 1983 Florence Conference and to be published
26. C. Mahaux and H. Ngô, Phys.Lett. 126B (1983) 1
27. C. Mahaux and H. Ngô, Nucl.Phys. A378 (1982) 205
28. C. Mahaux and H. Ngô, Nucl.Phys. (in press)
29. C. Mahaux and H. Ngô, Physica Scripta 27 (1983); and to be published
30. X. Campi and S. Stringari, Z.Phys. A309 (1983) 239
31. Z.Y. Ma and J. Wambach, Nucl.Phys. A402 (1983) 275.

DEFORMATIONS AND CORRELATIONS IN NUCLEI

E.Buendía and R.Guardiola
Departamento de Física Nuclear
and GIFT
Universidad de Granada, Granada (Spain)

MOTIVATION OF THIS WORK

In the 1981 edition of this conference, we presented an analysis of the ground states of nuclei with A=4n from ^{4}He to ^{40}Ca.[1] This study was based in the effective B1 potential of Brink and Boeker[2], and because of that the main purpose was to analyze de behaviour of the binding energy per nucleon in terms of the mass number, and not to get reliable values for that binding energy. Correlations were included via a Jastrow factor, and the ground states determined after a variation over the three parameters of the wave function, namely the harmonic oscillator parameter, and the range and depth of the correlation. It was observed that the spherical nuclei (He, O and Ca) had a binding energy per nucleon manifestly higher than its neighbours. In our opinion, this may be corrected by introducing a deformation in the shell model description with the appropriate angular momentum projection. This is the aim of this work.

We will describe below the wave functions used and the techniques for the angular momentum projection and the removal of the center of mass spurious motion as well as the practical formulae. These formulae will be used to determine the behaviour of the binding energy per particle with A, as well as the corresponging rotational bands.

WAVE FUNCTIONS

The nuclear states will be constructed by multiplying a Slater determinant times the Jastrow correlation factor corresponding to state independent correlations. The single particle states needed to construct the Slater determinant will be taken as eigenstates of a deformed harmonic oscillator with axial symmetry around the Z axis. The orbitals will be characterized by means of the three cartesian quantum numbers $(n_x\ n_y\ n_z)$, and each orbital will be occupied with the four spin and isospin orientations. Having in mind that we have to project on good angular momentum states, only those configurations with axial symmetry around the Z axis will be considered: this greatly simplifies the angular momentum projection. Accordingly with that restriction, the possible states are those

Nucleus	Configuration	Shape (d=1)
He-4	(000)	Spherical
Be-8	(000)(001)	Prolate
C-12	(000)(100)(010)	Oblate
O-16	(000)(100)(010)(001)	Spherical
Ne-20	(002)	Prolate
Mg-24	(101)(011)	Prolate
Si-28-o	(110)(200)(020)	Oblate
Si-28-p	(101)(011)(002)	Prolate
S-32	(110)(200)(020)(002)	Oblate
Ar-36	(110)(101)(011)(200)(020)	Oblate
Ca-40	(110)(101)(011)(200)(020)(002)	Spherical

Table 1.- The configurations with axial symmetry for the A=4n nuclei. Note that, with the exception of silicon, the configuration is unique. The two possible configurations of silicon are labelled with the letters o and p corresponding to the intrinsic shape, oblate and prolate. The intrinsic shape is the form corresponding to the configuration in the absence of deformation.

listed in Table 1. This table includes also the shape of the nucleus in the absence of deformations, and as we will see below, the deformation parameter will take values in agreement with the intrinsic shape.

There are some points of interest connected with Table 1. First of all, the condition of axial symmetry is very strong, and with the exception of silicon, it determines completely the configuration. In the s-d shell we may compare the adopted configurations of table 1 with the predictions of the SU(3) model: the configurations adopted for Ne, Ar and both of Si correspond to the lowest energy states of a quadrupole-quadrupole interaction. On the contrary, magnesium and sulfur prefer a wave function without axial symmetry.[3,4]

Once the configuration is chosen, the uncorrelated wave function is characterized by two parameters, R_s and R_z which are the harmonic oscillator lengths in the XY plane and in the Z axis, respectively. In the rest of the paper we will refer instead to the parameters $\alpha = 1/R_s^2$ and $d=R_s/R_z$. Values of d greater than 1 correspond to oblate nuclei, and values smaller than 1 to prolate nuclei. We will use also the parameter ρ given by

$$\rho = 1/R_z^2 - 1/R_s^2 .$$

ANGULAR MOMENTUM PROJECTION

The states we are interested in have a good value of J_z, equal to 0, but J is not a good quantum number. As it is well known, one may project states of good angular momentum by means of a operation consisting in rotating the intrinsic state (rotation characterized by the Euler angles Ω) and integrating over all angles after multiplying by the rotation matrix $D^J_{00}(\Omega)$. In this manner, for a given intrinsic state there results a rotational band just by varying J.

Because of the axial symmetry, it is only necessary to rotate over the axes Y and Z′, in that order. Such a rotation may be characterized by a unit vector $\vec{w}$ pointing along the symmetry axis of the nucleus. In the future we will use the following notation

$\vec{z}$ = A set of 3A-3 intrinsic (translationally invariant) coordinates

$\vec{R}$ = The center of mass coordinates

$F_J(\vec{z},\vec{R})$ = The angular momentum projected state

$F(\vec{z},\vec{R},\vec{w})$= The rotated Slater determinant including the Jastrow factor.

Then, the nuclear wave function is given by

$$F_J(\vec{z},\vec{R}) = \frac{1}{N_J} \int d\vec{w} \; D^{J*}(\vec{w}) \; F(\vec{z},\vec{R},\vec{w}) \tag{1}$$

where N_J is the normalization factor

$$N_J^2 = \int d\vec{w} \; d\vec{v} \; D^{J*}(\vec{w}) \; D^J(\vec{v}) \; O(\Omega) \tag{2}$$

and $O(\Omega)$ is the overlap integral

$$O(\Omega) = \langle F(\vec{z},\vec{R},\vec{w}) \mid F(\vec{z},\vec{R},\vec{v}) \rangle \tag{3}$$

which depends only on $\cos\Omega = \vec{v}.\vec{w}$. In the above formulae it is understood that the two lower quantum numbers of the rotation matrix D are both zero.

The calculation of matrix elements of operators is also simple. If we define

$$V(\vec{w},\vec{v}) = \langle F(\vec{z},\vec{R},\vec{w}) \mid V \mid F(\vec{z},\vec{R},\vec{v}) \rangle \tag{4}$$

then

$$\langle F_J | V | F_J \rangle = \frac{1}{N_J^2} \int d\vec{w} \; d\vec{v} \; D^{J*}(\vec{w}) \; D^J(\vec{v}) \; V(\vec{w},\vec{v}) \; . \tag{5}$$

For scalar operators $V(\vec{w},\vec{v})$ is again only a function of $\cos\Omega$. As a consequence, the four integrals which appear in (2) or (5) are trivially reduced to an integral over one variable, just $\cos\Omega$. There still remains the question of evaluating matrix elements of the kind (3) and (4), i.e. between rotated states. This corresponds to evaluate matrix elements between Slater determinants, but now the single particle states are no longer orthogonal due to the different rotations applied to the left and right wave functions. The reader may find more details on the evaluation of these matrix elements in Refs.[8,9].

THE REMOVAL OF THE SPURIOUS CENTER OF MASS MOTION

The last step in the construction of the nuclear wave function is the projection of the center of mass motion. In general we may define translationally invariant states by means of the operation

$$F_J(z) = \int dR\; G(R)\; F_J(z,R) \tag{6}$$

where G(R) is an arbitrary function of the center of mass coordinates. The only restriction is that G(R) must correspond to an L=0 state just to maintain the proper value for the nuclear angular momentum.

The freedom in choosing G(R) results in an ambiguous meaning for $F_J(z,R)$. There is a well known exception, namely when $F_J(z,R)$ is a Slater determinant built up with harmonic oscillator states (spherical). Then $F_J(z,R)$ factorizes into a center of mass state and an intrinsic, translationally invariant, wave function, so that any G(R) in (6) gives rise to the same intrinsic state.

The case of deformed harmonic oscillator states has some analogy with this case. Before the angular momentum projection the wave function is again factorable, but this property is lost by the angular momentum projection. Nevertheless, in a general expansion in center of mass states

$$F_J(z,R) = \sum_n C_n\, G_n(R)\; F_n(z) \tag{7}$$

we observe that in an appropriate basis there appears only one L=0 center of mass state, i.e. the spurious function contains only a translationally invariant component with the appropriate angular momentum quantum numbers.

The way to analyze the center of mass content of a shell model wave function has been described by Vincent [5] and applied to our problem in [6]. The center of mass states are determined by solving the integral equation

$$G(R) = g\int D(R,R')\; G(R')\; dR' \tag{8}$$

where

$$D(R,R') = \int dz \ \ F_J(z,R) \ \ F_J^*(z,R') \quad . \tag{9}$$

The solution of this integral equation in our case is given by [6]

$$G_L(R) = K_L^{-1/2} \exp(-\tfrac{1}{2}A\alpha R^2) \int_0^1 P_L(x) \exp(-\tfrac{1}{2}A\rho R^2x^2)\, dx \tag{10}$$

where K_L is the normalization constant, and P_L the Legendre polynomial. Equation (10) only includes the radial part of the center of mass state, and this radial part is independent of the angular momentum J at which the nucleus was projected.

Once the center of mass G_0 state has been determined, the matrix elements of physically interesting operators must be computed by means of the formula

$$\langle F_J(z)|V|F_J(z)\rangle = \int dR\, dR'\ G_0(R')\ F_J^*(z,R')\ V(z)\ F_J(z,R)\ G_0^*(R)\ dz \tag{11}$$

appart from a normalization constant related to the eigenvalue g of the integral equation (8). The r.h.s. of eq. (11) is rather complex to evaluate as it stands, but it may be managed so as to arrive to the simple form

$$\langle F_J(z)|V|F_J(z)\rangle = \frac{1}{M_J^2} \int dw\ dv\ D^{J*}(v)\ D^J(w)\sqrt{1+\lambda\sin^2\Omega}\ \ V(v,w) \tag{12}$$

where $\lambda = \frac{1}{4}\left(d - \frac{1}{d}\right)^2$.

Note the similarity between eq. (12) and eq. (5): the removal of the center of mass motion is just carried out by introducing the factor $\sqrt{1+\lambda\sin^2\Omega}$ in the angular integration. It is also interesting to observe that to compute $V(v,w)$ it is not necessary to work with the intrinsic coordinates. The normalization factor which appears in eq. (12) is computed like in eq. (2) but including the modifying factor $\sqrt{1+\lambda\sin^2\Omega}$.

ENERGY CALCULATIONS

All previous results are straightfordwardly applied to the case of correlated nuclei, the correlation factor being both rotationally and translationally invariant. In the calculations we have used the FAHT cluster expansion (see Ref. [7] and also [1] for further details). The wave functions contain four parameters: the harmonic oscillator parameter α , the deformation <u>d</u> and the range <u>b</u> and depth <u>a</u> characterizing the correlation factor $f(r) = 1 - a\exp(-br^2)$.

Nucleus	α	d	a	b	E(G.S.)	ΔE(Ang.Mom)	ΔE(3rd)
He-4	0.660	1.000	0.48	1.66	-36.55	---	3.49
Be-8	0.650	0.615	0.49	1.64	-69.15	-11.63	-4.37
C-12	0.330	1.440	0.49	1.61	-106.14	-10.63	-7.81
O-16	0.410	1.000	0.49	1.43	-156.88	---	-10.00
Ne-20	0.450	0.785	0.49	1.41	-191.72	-5.68	-10.92
Mg-24	0.375	0.920	0.49	1.62	-211.96	-0.77	-15.39
Si-28-o	0.310	1.285	0.50	1.37	-289.50	-6.62	-16.70
Si-28-p	0.415	0.805	0.49	1.49	-292.64	-5.26	-22.17
S-32	0.355	1.065	0.50	1.19	-292.09	-0.40	10.99
Ar-36	0.335	1.135	0.50	1.33	-412.73	-3.95	-21.43
Ca-40	0.360	1.000	0.51	1.60	-477.89	---	-13.37

Table 2.- Energy minima for the B1 Brink-Boeker force. The second column gives the value of the harmonic oscillator parameter (in fm^{-2}), the third is the deformation, the fourth is the depth of the correlation and the fifth the range (in fm^{-2}). The column labelled E(G.S.) is the value of the energy determined in second order with full angular momentum projection corrected for the third order of the cluster expansion. The importance of the angular momentum projection and the third order correction are shown in the two last columns.

The results of our calculations are shown in Table 2. The minima have been determined by computing the energy in the second order of the cluster expansion with projection on angular momentum and correction of the center of mass motion. The column labelled ΔE(Ang.Mom.) gives the value of the correction due to the angular momentum projection, and has been determined by computing the energy at second order of the cluster expansion in the minimum without angular momentum projection. The value of the correction at third order has been obtained by substracting the calculations at second and third order without angular momentum projection. The values given for the energy include these corrections.

To end up we will briefly comment the results. We may stress the following points:

i. There results a smooth variation in the parameters related to the correlation factor. Actually, the depth <u>a</u> of the correlation is practically constant. The variation of <u>b</u>, the range parameter, is probably related to the medium and long range effects induced by the correlation.

ii. The aparently lack of regularity in the harmonic oscillator parameter is due

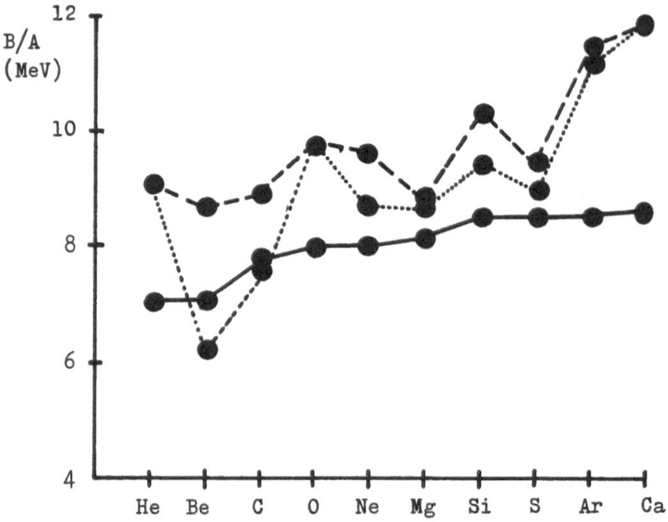

Figure 1.- Binding energy per nucleon versus the mass number. The continuous line refers to the experimental values. The dotted line joints the theoretical values determined in the absence of deformation, and the dashed line corresponds to the case where a deformation is allowed. Note that the dashed and dotted lines coincide at the spherical nuclei.

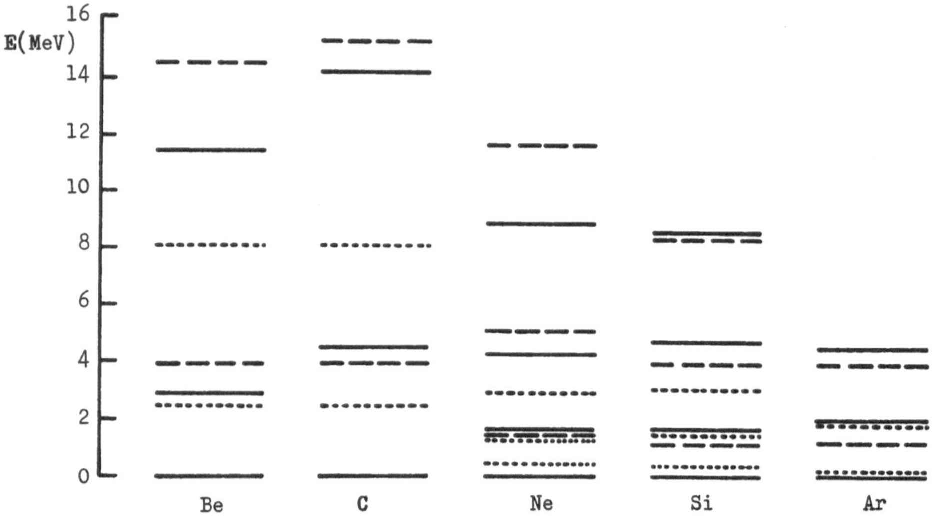

Figure 2.- The rotational spectrum. The continuous lines represent the experimental levels. The long dashed lines correspond to the levels determined in the presence of deformations. The short dashed lines are the results of the calculations with the constraint d=1. The quantum numbers of the levels have not been included to avoid confusing the figure. The levels are ordered as in the typical rotational bands, i.e. J=0, 2, 4 ...

to the presence of the deformation. This may be corrected with the introduction of an averaged harmonic oscillator length defined by the equation $R_o^3=R_s^2 R_z$.

iii. The convergence of the cluster expansion is quite good. Actually, the third order is rather small.

iv. There is an anomalous behaviour of the magnesium and sulfur nuclei. This is certainly due to the assumption of axial symmetry for that nuclei.

v. With regard to silicon, there is not too much difference between the two configurations, oblate and prolate. Actually, that two wave functions have, after the angular momentum projection, a large overlap.

vi. It is appealing the relation between the intrinsic shape and the values of the deformation, pointing both always in the same direction.

vii. There is an important effect of the deformation on the binding energy. This may be observed in the figure 1. If one forgets the anomalous nuclei Mg and S, there results an appealing pallelism between the theoretical results and the experimental values.

viii. Finally, the effect of the deformation on the rotational spectrum is very impressive, as shown in figure 2.

After these comments we conclude the important rôle of the deformation in these nuclei and the possibility of incorporating that effect in calculations including correlations.

REFERENCES

(1) R.Guardiola, in Recent Progress in Many Body Theories, pag. 398.
J.G.Zabolitzky, M.de Llano, M. Fortes and J.W.Clark editors. Springer (Heidelberg) (1981)
(2) D.M.Brink and E.Boeker Nucl.Phys. A91 (1967) 1
(3) G.Ripka, "The Hartree-Fock theory of deformed nuclei". Adv.Nucl.Phys. Vol. 1 (1968) 183
(4) J.P.Elliot in Selected Topics in Nuclear Theory, IAEA (Vienna) 1963
(5) C.M. Vincent Phys.Rev. C8 (1973) 929
(6) R.Guardiola Nucl.Phys. A253 (1975) 289
(7) J.W.Clark and P.Westhaus Jour.Math.Phys. 9 (1968) 131, 149
(8) J.Garcia-Roger y R.Guardiola Nucl.Phys. A267 (1976) 137
(9) D.M.Brink, The alpha particle model of light nuclei, in the Proc. Int. School Enrico Fermi, Course 36 (Academic Press, New York, 1966)

ACKNOWLEDGEMENTS

This work has been supported by the Comision Asesora de Investigación Científica y Técnica, Spain.

CORRELATED PAIRS NEAR THE FERMI SURFACE

W. Piechocki
Nuclear Theory Department
Institute of Nuclear Research
Hoza 69, PL-00-681 Warsaw, Poland

and

R.F. Bishop and G.A. Stevens
Department of Mathematics
UMIST, P.O.Box 88
Manchester M60 IQD, U.K.

One of the problems of the fifties, which occupied a lot of many body physicists dealing with strongly interacting Fermi systems of infinite extent like nucleon matter or liquid ^{3}He, was the occurence of a singularity at the Fermi surface in the Bethe-Goldstone (BG) equation. It was generally believed, at that time, that this singularity was associated with the existence of a superfluid state[1].

In the next two decades we can observe a few developments[2]. The approach I would like to discuss is the generalization of the BG equation. It includes, apart from particle-particle (pp) ladders, hole-hole (hh) ladders and all possible mixed pp and hh ladders. The generalized ladder equation of Mehta[3] and the Galitskii-Feynman (GF) equation investigated by the Manchester group[4] comprise the same type of ladder diagrams with the property that particles and holes are treated completely symmetrically. Metha's equation does not lose the BG equation singularity. The Manchester group, on the other hand, has not noticed any problem near the Fermi surface but has discovered that the GF T-matrix posses-ses a pole in the bound-state region which corresponds to the formation of bound-state pairs in the medium.

Results obtained by using symmetrical ladder equation are interesting because we know that bound composite clusters of fermions comprised of an even number of fermions (particles) and/or an even number of their superfluid-like phase[5]. As an example we have the Cooper pairs which are responsible for superconductivity in weakly interacting Fermi systems.

We have decided to examine a generalized ladder equation again, but in the context of the coupled-cluster (CC) method; the CC method is flexible and general enough to enable investigation of the pairing from different points of view and with better and better approximation. We have solved[6] the complete ladder (CLAD) equation of the CC method, corresponding to generalized ladder equations of the perturbation theory, for the separable two-body interaction. We have restricted ourselves to one-term S-wave interaction but the extension to both multi-term separable interactions and to arbitrary higher partial waves would be straightforward. The solution for the states close to the Fermi surface contains both the bound-pairs and the singularity. To understand that result better we have examined excited states[7]. Using the Emrich's ansatz for the excited state, we have solved the ladder type equation corresponding to the CLAD equation of the ground state. The solution is characterized entirely by the quantities which describe the pairing of the ground state. For the attractive interaction, the excited state has lower energy than the ground state which means that our reference state choosen to be a filled Fermi-sphere plane-wave determinant is not the best possible point of departure. For the repulsive interaction, on the other hand, we observe a real excitation for densities above a certain critical density and no excitation below that density; in the latter range of density we also do not observe any pairing in the ground state.

Both results, for the ground state and for the excited state, suggest the existence of the pairing in a many-fermion strongly interacting system. The question arises whether the observed pairing exists in the exact solution of the Schrödinger equation. To examine that, we should find a suitable truncation scheme of the CC method which would give a possibility to see the pairing in its successive levels[8]. We cannot apply the Bochum truncation scheme[9], because it was invented to describe the normal ground state of strongly interacting systems; by introducing a gap at k_F in the single-particle energy spectrum, this method relegates the pairing correlations near the Fermi surface to the terms of a very high order.

Presently we know how to treat the short-range repulsion of a two-body interaction, but we do not know how to describe properly the Fermi surface.

This research was supported in part by the Polish - U.S. Maria Sklodowska-Curie Fund and by the SERC of Great Britain.

References

[1] J.S. Bell and E.J. Squires, Adv.Phys.10 (1961) 211 and references therein

[2] B.H. Brandow, Phys.Rev. 152 (1966) 863; G.A. Baker Jr. and J.L.Gammel, Phys.Rev. C6 (1972) 403; J.P. Jeukenne, A. Lejeune and C. Mahaux, Nucl.Phys. A 245 (1975) 411.

[3] M.L. Mehta, Nucl.Phys. 20 (1960) 533.

[4]R.F. Bishop and M.R. Strayer, J.Phys. G, L13 (1977) and references therin.

[5]W. Kohn and D. Sherrington, Rev.Mod.Phys. 42 (1970) 1.

[6]R.F. Bishop, W. Piechocki and G.A. Stevens, to be published.

[7]R.F. Bishop, W. Piechocki and G.A. Stevens, to be published.

[8]W. Piechocki, Invited talk presented at 'The 7th (Pan-American) Workshop on Condensed Matter Theories', Altenberg, West-Germany, August 22 -27, 1983.

[9]B.D. Day and J.G. Zabolitzky, Nucl.Phys. A 366 (1981) 221.

EFFECTIVE INTERACTIONS AND ELEMENTARY EXCITATIONS IN ELECTRON AND HELIUM LIQUIDS

David Pines
Physics Department
University of Illinois at Urbana-Champaign
Urbana, IL 61801/USA

Introduction

In the first part of this talk I should like to describe some recent work which has had as one of its principal objectives the incorporation of very accurate calculations of the ground state energy and pair correlation functions into calculations of effective interactions and elementary excitation spectra. I shall describe work on electron liquids carried out in collaboration with Naoki Iwamoto and Eckhard Krotscheck,[1] and on liquid ^4He with Stratios Manousakis and Quatar Usmani.[2] In the second part of my talk, I should like to bring you up-to-date on some applications of polarization potential theory[3] to the helium liquids. This part may be regarded as a follow-up to my talk at Oaxtepec. I summarize briefly recent work on roton-roton interactions in superfluid ^4He,[4] effective interactions and transport properties of ^3He-^4He mixtures,[5] the mode-mode coupling between single particle (pair) and multiparticle (pair) excitations in quantum liquids,[6] and excitations and transport in liquid ^3He.

Electron-Hole Pseudopotentials

Perhaps the principal physical effect which is missing in the RPA is an account of the way in which charge-and-spin induced correlations between electrons act to modify, at short distances, the Coulomb interaction between electrons of parallel and antiparallel spin. Iwamoto and I (IP) have taken the point of view that just as was the case with the polarization potential theory of ^3He and ^4He,[3] it is both useful and instructive to describe the resulting effective interactions by means of configuration space and momentum space pseudopotentials.[1] Where the electron effective mass, $m^* \approx m$ (i.e., no backflow), which is the case for metallic electron densities, the spatial integral of these pseudopotentials is determined by the compressibility and spin susceptibility, which Vosko, Wilk, and Nusair[7] have determined from the Monte Carlo calculations of Ceperly and Alder.[8] Hence for a given choice of shape (in either configuration space or momentum space) the pseudopotential is determined uniquely. In this way IP arrive at a simple physical picture for the static local field correction, G(q), to the dielectric function, one which facilitates comparison with previous calculations.

IP write the coherent (density-density) and incoherent (spin-spin) response functions in the form,

$$\chi^{c}(q\omega) = \frac{\chi_{o}(q\omega)}{1 - \frac{4\pi e^{2}}{q^{2}}[1 - G^{s}(q)]\chi_{o}(q\omega)} \tag{1}$$

$$\chi^{I}(q\omega) = \frac{\chi_{o}(q\omega)}{1 + \frac{4\pi e^{2}}{q^{2}} G^{a}(q)\chi_{o}(q\omega)} \tag{2}$$

where $\chi_{o}(q\omega)$ is the free-electron (Lindhard) response function, and the spin-symmetric (anti-symmetric) local field corrections are given by

$$G^{s,a}(q) = - \frac{q^{2}}{4\pi e^{2}} f_{q}^{s,a} = - \frac{q^{2}}{4\pi e^{2}} \int dr \, [\frac{f^{\uparrow\uparrow}(r) \pm f^{\uparrow\downarrow}(r)}{2}] \, , \tag{3}$$

where $f^{\uparrow\uparrow}(r)$ and $f^{\uparrow\downarrow}(r)$ are the electron-hole pseudopotentials which determine the effective interactions between electrons,

$$v^{\uparrow\uparrow}(r) = \frac{e^{2}}{r} + f^{\uparrow\uparrow}(r) \tag{4a}$$

$$v^{\uparrow\downarrow}(r) = \frac{e^{2}}{r} + f^{\uparrow\downarrow}(r) \, . \tag{4b}$$

IP consider both Yukawa and a dielectrically screened pseudopotentials; with the former, one has

$$f^{\uparrow\uparrow}(r) = - \frac{e^{2}}{r} \exp - q_{\uparrow\uparrow}r; \qquad f_{q}^{\uparrow\uparrow} = - \frac{4\pi e^{2}}{q^{2}+q_{\uparrow\uparrow}^{2}} \, ; \tag{5a}$$

$$f^{\uparrow\downarrow}(r) = - \frac{e^{2}}{r} \exp - q_{\uparrow\downarrow}(r); \qquad f_{q}^{\uparrow\downarrow} = - \frac{4\pi e^{2}}{q^{2}+q_{\uparrow\downarrow}^{2}} \, . \tag{5b}$$

The screening wave vectors are determined uniquely by the compressibility, κ, and spin-susceptibility, χ_{p}, since

$$\frac{\kappa_{o}}{\kappa} - 1 = N^{o}(0) \, [\frac{f_{o}^{\uparrow\uparrow} + f_{o}^{\uparrow\downarrow}}{2}] \tag{6a}$$

$$\frac{\chi_{p}^{o}}{\chi_{p}} - 1 = N^{o}(0) \, [\frac{f_{o}^{\uparrow\uparrow} - f_{o}^{\uparrow\downarrow}}{2}] \, , \tag{6b}$$

where κ^{o} and χ_{p}^{o} are the free-electron susceptibility values and $N^{o}(0)$ the free electron density of states. In the high-density limit, IP find $q_{\uparrow\uparrow} = \sqrt{2}\, q_{F}$ and $q_{\uparrow\downarrow} = 0$; this is the Hubbard local field correction as modified by Geldart and

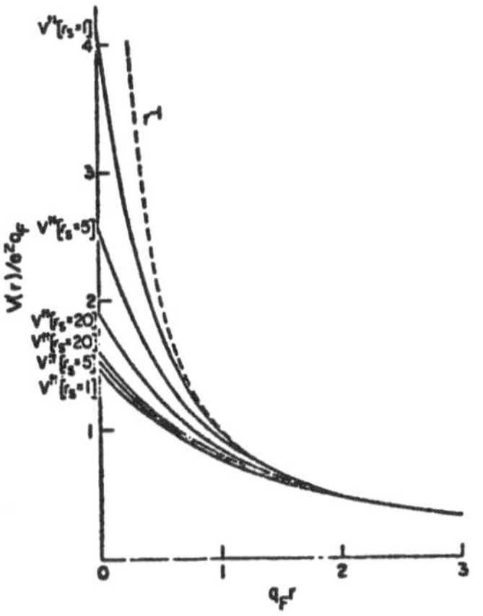

Fig. 1a

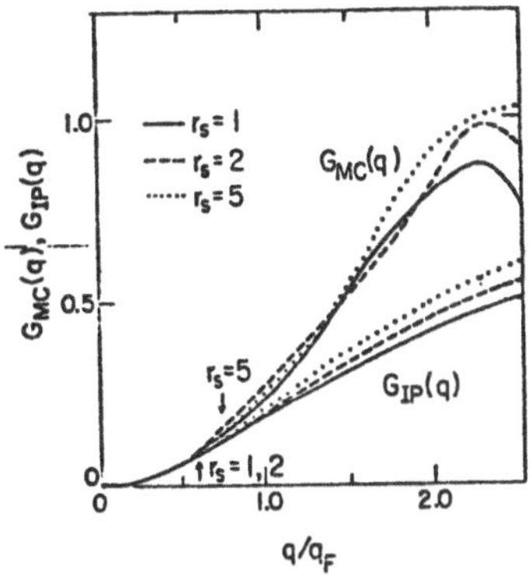

Fig. 1b

Fig. 1 (a) The IP[1] effective interactions between electrons of parallel and anti-parallel spin, $V^{\uparrow\uparrow}(r)$ and $V^{\uparrow\downarrow}(r)$, in units of $e^2 q_F$ for r_s=1, 5, and 20. The dotted line shows the bare Coulomb potential (b) Comparison of local field corrections.

Vosko, which is thus seen to be exact in this limit. At metallic (and lower) electron densities IP use the VWN values of χ_p and κ; their results for $V^{\uparrow\uparrow(\uparrow\downarrow)}(r)$ shown in Fig. 1a; those with the dielectrically screened pseudopotential are qualitatively similar.

Local field Corrections

The corresponding local field corrections are

$$G^{s,a}_{IP}(q) = \frac{1}{2}\left[\frac{q^2}{q^2 + q^2_{\uparrow\uparrow}} \pm \frac{q^2}{q^2 + q^2_{\uparrow\downarrow}}\right] ; \qquad (8)$$

and are exact in the long wavelength limit. This is but one of many possible choices for $G^{s,a}(q)$; a second choice, which also builds upon the Monte Carlo calculations of Ceperley and Alder,[8] is to take $\chi^{c,I}(q\omega)$ to have the form (1) and (2), but choose the corresponding local field corrections in such a way that when the appropriate frequency integral is carried out the resulting values of the spin symmetric and antisymmetric static form factors, $S^{s,a}(q)$, agree with the values calculated by Ceperley and Alder for $q \gtrsim \frac{q_F}{2}$. For smaller values of q, where there is some uncertainty in the Monte Carlo calculations of $S^c(q)$, one may use the exact long wavelength results obtained from the IP expression, Eq. (8), and interpolate smoothly to the larger q values. IKP[1] carry out this program with the aid of the mean spherical approximation.[9] Their result for $G^s(q)$, which I denote by $G_{MC}(q)$, is compared with other calculations of $G^s(q)$ in Fig. 1b.

IKP use the expression, Eq. (1), with $G^s(q) = G_{MC}(q)$, to calculate the contribution to the correlation energy from different momentum transfers, with results which agree to within 1% with the Monte Carlo values; they calculate as well the plasmon dispersion relation, with results which agree well with experiment for $\kappa(r_s \approx 5)$. However, IKP find that the corresponding calculations of the dynamic form factor at wavevectors $\gtrsim q_F$ (i.e., outside the plasmon regime) do not yield the two peak structure seen by Platzman and Eilenberger in Be, graphite, and Al.[10] Rather, one finds a single peak which is similar in shape to the RPA, but shifted to lower values of the energy. The physical reason for this disparity between theory and experiment may be traced to the role played by multipair excitations; the two peak structure reflects the presence of both single pair excitations and multipair excitations at a somewhat higher energy, in much the same way as these different excitations are responsible for the multipeak structure seen for corresponding wavevector values in neutron scattering experiments on ^3He.[3] IKP thus conclude that although the response function, Eq. (1) has the very attractive feature of yielding agreement with experiment for S_q; $E_{corr}(q)$, and plasmon dispersion, it is not the correct response function. A corollary of this conclusion is that the $G_{MC}(q)$ for $q \gtrsim q_F$, is not the correct value of the local field correction, because were it correct, there would be no "room" for multipair excitations to contribute to the static form factor and the f-sum rule. It is thus necessary to modify Eq. (1) in order to take into account multipair excitations, as well as the coupling between single pair and multipair excitations, a question to which I return later.

The ω^3 Sum Rule

A quantitative measure of the importance of multipair excitations comes from examination of the ω^3 sum rule. At high frequencies the exact density-density response function may be expanded in inverse square powers of ω, according to

$$\lim_{\omega \to \infty} \chi(q\omega) = \frac{Nq^2}{m\omega^2} + \frac{\langle\omega^3\rangle 2}{\omega^4} + \ldots \tag{9}$$

where

$$\langle\omega^3\rangle = \int_0^\infty d\omega\, \omega^3 S(q\omega) . \tag{10}$$

As first shown by Puff and Mihara,[11] $\langle\omega^3\rangle$ may be evaluated from a knowledge of the pair correlation function and the expectation value in the ground state of the kinetic energy. For the electron liquid their result may be written as

$$\langle\omega^3\rangle = \frac{Nq^2}{2m} \left\{\frac{q^4}{4m^2} + 2\langle KE\rangle \frac{q^2}{m} + \omega_p^2 (1-I_q)\right\} \tag{11}$$

where I_q can be evaluated from the static form factor and $\langle KE\rangle$ is the exact expectation value of the kinetic energy in the ground state.

If now one takes the density-density response function to be given by Eq. (1), one finds on carrying out the high frequency expansion,

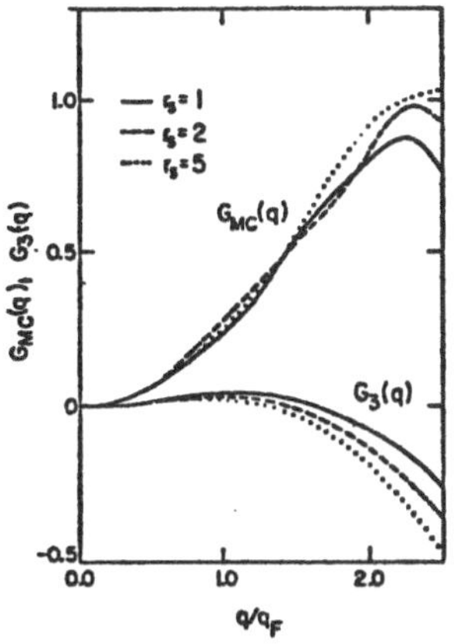

Fig. 2a

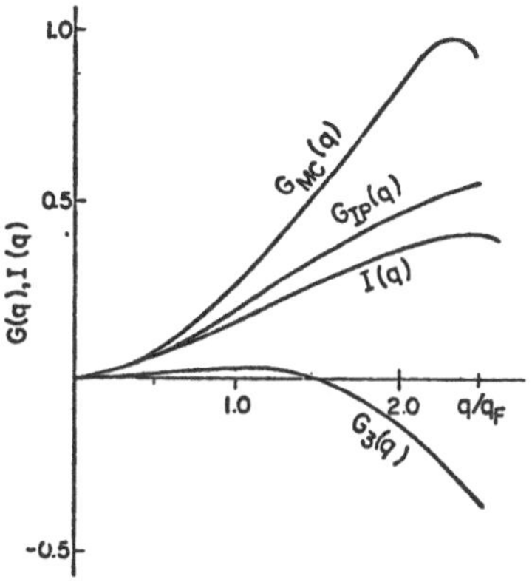

Fig. 2b

Fig. 2 (a)Comparison of local field corrections, $G_{MC}(q)$ and $G_3(q)$ for r_s = 1, 2, and 5; (b)Comparison of local field corrections in different theories for r_s = 2.

$$\langle\omega^3\rangle_{LF} = \frac{Nq^2}{2m}\left\{\frac{q^4}{4m^2} + 2\langle K.E.\rangle_o \frac{q^2}{m} + \omega_p^2\,[1-G^s(q)]\right\} \tag{12}$$

where $\langle KE\rangle_o$ is the average kinetic energy for the free electron gas. On comparing Eqs. (11) and (12), we see that one may choose a local field correction $G_3(q)$, which, when used in Eq. (12), will satisfy the ω^3 sum rule; it is

$$G_3(q) = I_q - \left(2q^2/m\omega_p^2\right)\left[\langle KE\rangle - \langle KE\rangle_o\right] . \tag{13}$$

IKP evaluate $G_3(q)$ from the Monte Carlo data; their results are shown in Figs. 2a,b, from which one may draw the following conclusions:

(1) There is a large difference between $G_{MC}(q)$ and $G_3(q)$ for all values of r_s between 1 and 5. The value of I(q) is about half of that of $G_{MC}(q)$ for the same r_s. I(q) and $G_{MC}(q)$ become close only for very small values of q ($\ll q_F$).

(2) I(q) is a monotonically increasing function of q; there is no peak structure near $q = 2q_F$ as found in $G_{MC}(q)$ for r_s = 1 and 2.

(3) For large values of q, where the second term on the right-hand side of Eq. (6.68) becomes appreciable, $G_3(q)$ takes on negative values.

The difference between $G_{MC}(q)$, which was chosen to yield agreement with S_q, and $G_3(q)$, which was chosen to yield agreement with the ω^3 sum rule, furnishes a measure of the importance of multipair excitations. Because these lie, on the average, at greater energies than the single pair plus plasmon energies, they make a much larger relative contribution to the ω^3 sum rule than to S_q. In the small q limit, their contribution to both quantities is of order q^4 (cf. Ref. 12): hence the agreement between all the various model calculations of G(q) in the long wavelength limit simply reflects the fact that, to lowest order in q, plasmons exhaust all the sum rules for $S(q,\omega)$.[12]

In both of the above model calculations of $G(q)$ multipair excitations are not taken explicitly into account; with $G_{MC}(q)$ one distorts the local field correction, and hence the pair + plasmon spectrum, in such a way as to yield the correct S_q, while with $G_3(q)$, one is distorting the pair + plasmon spectrum in such a way as to obtain the correct value of $\langle\omega^3\rangle$. To the extent that multipair excitations make a negligible contribution to both S_q and $\langle\omega^3\rangle$, $G_{MC}(q)$ and $G_3(q)$ should agree; as noted above, this is the case in the long wavelength limit. On the other hand, as one goes to larger values of q, neither local field correction is correct; the "true" local field correction must be such as to allow room for multipair excitations [through an appropriate modification of $\chi(q,\omega)$] to contribute to both S_q and $\langle\omega^3\rangle$; it may be expected to lie between $G_{MC}(q)$ and $G_3(q)$, as is the case for $G_{IP}(q)$.

Sum Rules and Multiparticle Excitation Spectra in Liquid ^4He

As first emphasized by Miller, Nozières and Pines,[13] the dynamic form factor, $S(q,\omega)$ of ^4He at very low temperatures may be written as the sum of two components:

$$S(q\omega) = Z_q\delta(\omega-\omega_q) + \zeta_q f(q\omega) \tag{14}$$

where the first term on the r.h.s. is the well known phonon-maxon-roton spectrum and corresponds to exciting a single quasiparticle out of the condensate, while the second, which forms a continuum, corresponds to exciting two or more quasiparticles from the condensate. The dynamic form factor satisfies the sum rules,[12]

$$\int_0^\infty d\omega\, \frac{S(q\omega)}{\omega} = -\frac{\chi(q,0)}{2} = \frac{Z_q}{\omega_q} + \zeta_q\langle\frac{1}{\omega_m}\rangle \tag{15a}$$

$$\int_0^\infty d\omega\, S(q\omega) = S(q) = Z_q + \zeta_q \tag{15b}$$

$$\int_0^\infty d\omega\, S(q\omega)\omega = \frac{Nq^2}{2m} = Z_q\omega_q + \zeta_q\langle\omega_m\rangle \;. \tag{15c}$$

If one ignores the contribution made by multiparticle excitations to $S(q,\omega)$, application of (15b) and (15c) yields at once the Feynman result for the single phonon spectrum, $\omega_q = q^2/2mS(q)$. Only when one takes the multiparticle excitations into account explicitly, and treats the mode-mode coupling between these and the single quasiparticle excitations, does one recover the experimental results for ω_q.

It is, however, quite difficult to calculate the multiparticle branch from first principles, while model calculations of the interplay between the single particle and multiparticle branch are limited by the fact that the moment conditions, Eqs. (15) are not sufficient to fix the multiparticle spectral density, $f(q\omega)$. To the extent however, that it is possible to calculate higher moments of $S(q\omega)$ this situation improves considerably. The next two moments are:[14]

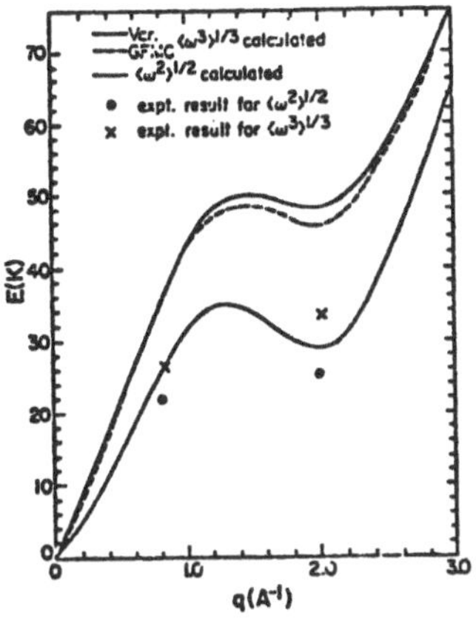

Fig. 3a

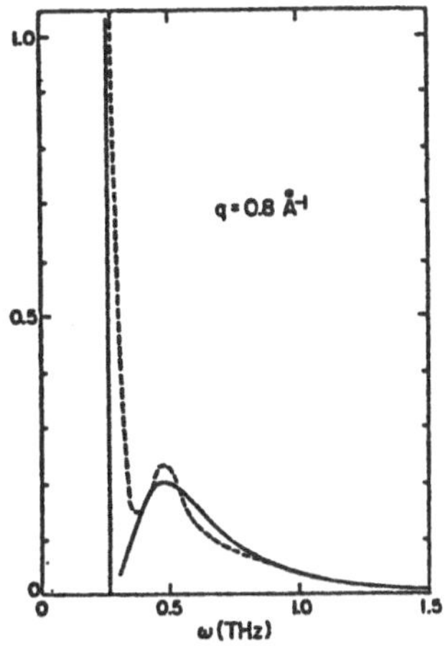

Fig. 3b

Fig. 3 (a) Theoretical calculations[2] of $\langle\omega^2\rangle^{1/2}$ and $\langle\omega^3\rangle^{1/3}$ compared with experimental results of Svennsen et al.[15]; (b) comparison of MPU two parameter fit to multiparticle spectral density with experiment (---).[15]

$$\int d\omega\, \omega^2 S(q\omega) = \frac{q^4}{4m^2}\left[2-S(q)\right] + \frac{q^2}{m^2} D(q) = Z_q\omega_q^2 + \zeta_q\langle\omega_m^2\rangle \qquad (16a)$$

$$\int d\omega\, \omega^3 S(q\omega) = \frac{q^2}{2m}\left[\frac{q^4}{4m^2} + \frac{2q^2}{m}\langle KE\rangle + \frac{N}{m}\int dr\, g(r)(1-\cos q\cdot r)\,\frac{(q\cdot\nabla)^2}{q^2}V(r)\right.$$

$$= Z_q\omega_q^3 + \zeta_q\langle\omega_m^3\rangle \qquad (16b)$$

where D(q) can be calculated from a knowledge of the ground state wave function, while g(r) is the pair distribution function. Manousakis, Usmani, and I (MPU)[2] have used the wave functions and pair distribution functions obtained by variational and Green's function Monte Carlo calculations to calculate the sum rules, (16a) and (16b); we then examine the constraints all five sum rules place on the multiparticle excitation spectrum.

The main conclusions reached by MPU are the following:

(1) There is comparatively little difference between using variational results (Jastrow or Jastrow-triplet) and the Green's function Monte Carlo results for g(r) in evaluating the ω^3 sum rule (Fig. 3a).

(2) The calculated results for $\langle\omega^2\rangle^{1/2}$ are in rather good agreement with the experimental results of Svennson et al.[15] at $q = 0.8\ A^{-1}$ and $q = 2.0\ A^{-1}$, while for $\langle\omega^3\rangle^{1/3}$ there is a substantial disagreement between theory and experiment. (Fig. 3a). If, however, one chooses a model spectral density which agrees with experiment up to the cut-off introduced in the analysis of the experimental data, but which contains as well a high energy tail, [(MPU choose $f(q\omega) \sim \omega^{-5}\exp\{\alpha_q[\omega-\omega_m(q)]^2/\omega^2\}$] this discrepancy is removed (Fig. 3b). MPU find that with this simple two-parameter spectral density one can fit all five sum rules. This consti-

tutes strong indirect evidence for the existence of a high energy tail (at $\omega > 60$ K for $q = 0.8\ \text{A}^{-1}$; $\omega > 90$ K for $q = 2.0\ \text{A}^{-1}$) in the multiparticle spectrum.

(3) The two new sum rules also help one disentangle the contribution to the multiparticle spectrum made by two-roton bound states or resonances from the remainder of the spectrum. As a first step in this direction, MPU split off a hypothesized two-roton bound state from the rest of the multiparticle spectral density; the sum rules, Eq. (16a) and (16b) then enable them to obtain the relative strengths of the two-roton and higher excitation modes. Since, however, the peaks observed by Svennson et al. fall at energies distinctly higher than that of a two-roton bound state, it is clear that a more sophisticated model for the two-roton states (and their coupling to the single phonon states) is required. I discuss below one possible approach to this problem.

The Roton Liquid and a Roton-Roton Pseudopotential

Above $T \sim 1$ K, rotons are the dominant elementary excitation in liquid He II. Just as ^{3}He atoms form a dilute normal Fermi liquid in ^{3}He-^{4}He mixtures, so may the roton excitations be regarded as forming a dilute Bose liquid in the background of superfluid ^{4}He. The interaction between rotons in this "roton liquid" determines the following properties of He II in this temperature regime:

- The temperature dependence of the roton energy, $\Delta(T)$.
- The temperature dependence of the normal fluid density, ρ_n, at $T \gtrsim 1.4$ K.
- The temperature dependence of the roton lifetime.
- Transport coefficients (viscosity, thermal conductivity...) at $T \gtrsim 1.4$ K.
- Neutron scattering experiments at momenta $\gtrsim 2.5\ \text{A}^{-1}$.

To describe the first two consequences of this interaction, Bedell, Fomin, and I developed a "roton liquid theory" which is analogous to Leggett's theory of the superfluid Fermi liquid.[16] We were able to derive stability conditions for the parameters which describe the effective roton-roton interaction, and to use one of these to place a bound on the variation with temperature of the roton energy. We obtained simple expressions for both the specific heat and normal fluid density as a function of temperature, and by fitting to experiment, were able to obtain the first two moments of the effective roton interaction.

To calculate these parameters from microscopic theory, and to obtain the remaining consequences of roton-roton interaction, it is necessary to calculate the roton-roton scattering amplitude as a function of the roton pair momentum and energy. Bedell, Zawadowski, and I (BPZ)[4] have carried out such a calculation by introducing a configuration space roton-roton pseudopotential, which is closely related to that used by Aldrich and Pines to describe the effective interaction between background particles in ^{3}He and ^{4}He.[3] According to Aldrich and Pines,[17] a roton is an excited quasiparticle of momentum $\sim p_0$, effective mass ~ 2.1 M (M is the mass of a bare ^{4}He

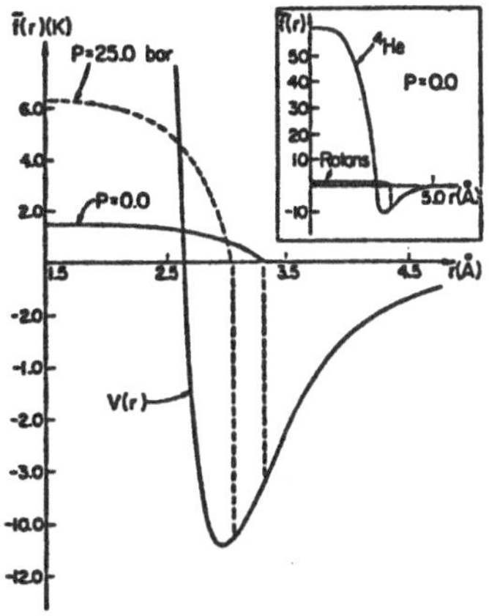

Fig. 4a

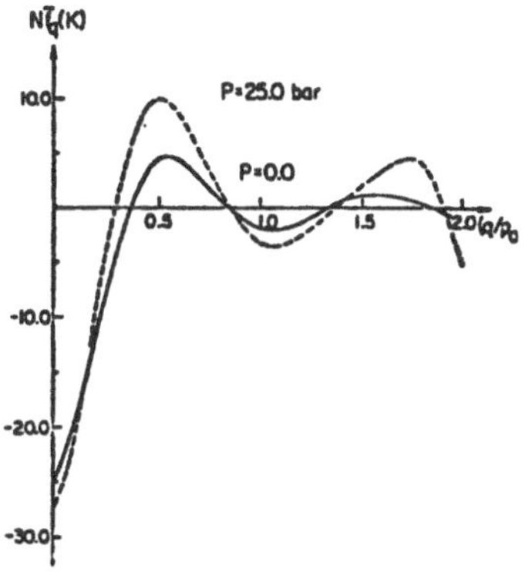

Fig. 4b

Fig. 4 BPZ configuration space (a) and momentum space (b) roton-roton pseudopotentials.

atom) moving in an attractive self consistent field (≃ -2 K) produced by the other quasiparticles in the liquid. BPZ therefore attempt a configuration-space description of roton interaction, $\tilde{f}(r)$, which is similar to the configuration-space interaction devised by Aldrich and Pines for the effective interaction between an excited ^{4}He atom and one in the ground state, viz: a long-range interaction which is identical to that between bare ^{4}He atoms, which becomes repulsive at some radius r_c; as with the ground state particles they argue that the short-range correlations brought about by the strong short-range repulsive part of the bare interaction will change the latter from its almost hard-core behavior to a soft-core repulsion, such that the over-all potential possesses a well-defined Fourier transform. To the extent that the transition from attraction to repulsion occurs over a distance small compared to r_c, and the exact form of the repulsion is of little importance, this potential, $\tilde{f}(r)$, may be characterized by two parameters; r_c, and a, the strength of the repulsive interaction at the origin. The BPZ model pseudopotentials in configuration space and momentum space are shown in Figs. 4a and 4b.

BPZ construct a scattering amplitude for these pseudopotentials and use that scattering amplitude to obtain a theory which is completely congruent with roton liquid theory. They calculate two-roton bound states, roton liquid parameters, roton life-times, and the hybridization of the two-roton bound state with excitations of higher and lower energy, and find excellent agreement between theory and experiment for the $\ell=2$ bound two-roton state at zero pair momentum, the roton life-time, the roton contribution to the normal fluid viscosity and normal fluid density, and the temperature variation of the roton energy.

Effective Interactions and Elementary Excitations in ^3He-^4He Mixtures

Dilute mixtures of ^3He in liquid ^4He represent an obvious target for application of polarization potential theory. Wei-chan Hsu, Kevin Bedell and I have been studying the properties of these dilute Fermi liquids,[5] and I wish to present a brief summary of our principal results:

- We have combined scaling arguments on the range of the repulsive part of the interaction with the results of Bardeen, Baym and Pines (BBP)[18] for the long wavelength limit of the direct ^3He-^3He interaction and the ^3He-^4He interaction, to obtain pseudopotentials to describe the above interactions. These are similar, but not identical to those of Aldrich and Pines for the quasiparticle interactions in pure ^4He and pure ^3He; the resulting non-local effective interaction between ^3He quasiparticles in the dilute mixtures is close to that of BBP, and provides physical insight into the origin and success of the BBP pseudopotential.
- Our results for the transport properties as a function of concentration are in better agreement with experiment than any previous calculations.
- By using a treatment of mode-mode coupling in which we incorporate changes in the ^3He quasiparticle-quasihole spectrum which result from ^3He quasiparticle-roton coupling, we are able to explain both the temperature and concentration dependence of the neutron scattering experiments on dilute ^3He-^4He mixtures.

Mode-Mode Coupling (Hybridization)

The coupling between multiparticle excitations, which form, in general, a continuum which may (or may not) peak at a comparatively large energy, and well-defined collective modes at lower energies plays an important role in the following systems:

- Electron liquids, where multipair excitations lower the plasmon energy
- ^4He, where hybridization, which results from the dynamic coupling between roton pair states (bound or resonant) and the phonon-maxon-roton spectrum, not only lowers the energies of the latter but may give rise to significant interference effects which are observable in, for example, the neutron scattering experiments of Svennson et al.[15]
- ^3He, where the interplay between this dynamic coupling and that of the single pairs to the collective modes is responsible for the maxon portion of the latter spectrum from $q \sim 1\ A^{-1}$ to $1.4\ A^{-1}$.

It seems appropriate to discuss hybridization at this meeting because while its physical consequences are often quite important, it has not been easy to develop a systematic approach to mode-mode coupling, so that its experimental consequences have not yet been fully explored for many situations of interest.

The approach which I shall describe has been developed in collaboration with Fred Zawadowski;[6] with Chuck Aldrich we have begun a preliminary examination of its consequences for ^{4}He. The basic idea is to separate the density (or spin density) fluctuation excitations into a single particle (pair) part, $\rho^{(1)}(q\omega)$ and a multiparticle (pair) part, $\rho^{(2)}(q\omega)$. In the presence of an external field, each responds to the sum of that field and induced polarization fields. Thus we write

$$\langle\rho^{(1)}(q\omega)\rangle = \chi_{sc}^{(1)}(q\omega)\{\phi_{ext} + V^{(1)}(q\omega)\langle\rho^{(1)}(q\omega)\rangle + V^{(1,2)}(q\omega)\langle\rho^{(2)}(q\omega)\rangle\} \tag{17}$$

$$\langle\rho^{2}(q\omega)\rangle = \chi_{sc}^{(2)}(q\omega)\{\phi_{ext} + V^{(2)}(q\omega)\langle\rho^{(2)}(q\omega)\rangle + V^{(1,2)}(q\omega)\langle\rho^{(1)}(q\omega)\rangle\} \tag{18}$$

from which one finds

$$\chi(q\omega) = \frac{\chi_o^{(1)}(q\omega) + \chi_o^{(2)}(q\omega) + 2V^{(1,2)}\chi_o^{(1)}(q\omega)\chi_o^{(2)}(q\omega)}{1 - [V^{(1,2)}(q\omega)]^2\chi_o^{(1)}(q\omega)\chi_o^{(2)}(q\omega)} \tag{19}$$

where

$$\chi_o^{(1)}(q\omega) = \frac{\chi_{sc}^{(1)}(q\omega)}{1 - V^{(1)}(q\omega)\chi_{sc}^{(1)}(q\omega)} \tag{20}$$

with a similar expression for $\chi_o^{(2)}(q\omega)$. Thus $\chi_o^{(1)}(q\omega)$ and $\chi_o^{(2)}(q\omega)$ are the response functions calculated in the absence of the mode-mode coupling described by $V^{(1,2)}(q\omega)$.

Consider for example, ^{4}He: $\chi_o^{(2)}(q\omega)$, in first approximation, is the roton-pair response function; the presence or absence of roton pair bound states depends on the roton-roton interaction, $V^{(2)}$, which can in turn be calculated from the roton-roton pseudopotential I discussed earlier. $\chi_o^{(1)}$ describes the phonon-maxon-roton spectrum, $\omega_o(q)$, calculated in the absence of the multiparticle excitations; thus $V^{(1)}(q\omega)$ is given by $f_q^s + (\omega^2/q^2)f_q^v$, where f_q^s and f_q^v are the strengths of the scalar and vector polarization fields previously determined by Aldrich and Pines.[17]
The dynamic coupling between $\chi_o^{(1)}$ and $\chi_o^{(2)}$ can then be described explicitly; where $\omega_o(q)$ becomes comparable to 2Δ, i.e., in the maxon region, a substantial reduction in the resulting maxon energy as a result of mode-mode repulsion is to be expected. On the other hand, the influence of the remaining part of the multiparticle response function, $\chi_o^{(2)}$, (corresponding to two maxon, four roton, two-maxon + two roton excitations, etc.) on $\omega_o(q)$ can be treated by taking the static limit of this contribution, since ω_o is small compared to the frequencies which characterize this part of $\chi_o^{(2)}$. The relative strengths of $\chi_o^{(1)}$ and $\chi_o^{(2)}$ are determined by requiring that in the absence of mode-mode coupling, $\chi(q\omega)$ satisfy the f-sum rule, while the term,

$2V^{(1,2)}\ \chi_o^{(1)}\chi_o^{(2)}$, describes the consequences of the interference between the single phonon and multiphonon modes, which is present whenever the two spectral densities overlap.

The above description of mode-mode coupling may seem at first sight overly complicated and overly laden with phenomenological parameters; however, it represents a minimum program for working out the physical consequences of this coupling in consistent fashion, and hence explaining, for example, the experimental results of Svennson et al.[15] For fermion systems, all physical effects of interest can be included in the screened response functions and the couplings $V^{(1)}$, $V^{(2)}$, and $V^{(1,2)}$; for ^{4}He there are a further set of mode-mode coupling terms which Zawadowski and I have identified as arising from the "anomalous" diagrams of the Bogoluibov-Beliaev-Hugenholtz-Pines theory, in which one excitation is directly coupled to two. We have been able to establish the connection between the above formalism and that developed by Ruvalds and Zawadowski for the consequences of hybridization in liquid ^{4}He;[19] our results reduce to those of Aldrich and Pines[17] for mode-mode coupling if we take $V^{(1,2)} = V^{(1)}$, an approximation which we do not expect to be generally valid.

To summarize, the expression, Eq. (19), should provide the basis for a quantitative theory of the consequences of mode-mode coupling in ^{4}He; because the couplings, $V^{(1)}(q\omega)$ and $V^{(2)}(q\omega)$ are known from previous work, and because the dynamic form factor must satisfy the five sum rules, Eqs. (15) and (16), the resulting calculation should require the introduction of few, if any, free parameters beyond those quantities which have been previously determined.

Excitations and Transport in Liquid ^{3}He

I should like to call your attention to three comparatively recent developments of interest.

- Bedell[20] has used the scattering amplitudes calculated by Bedell and me[21] from the AP polarization potentials to calculate the specific heat jumps in the superfluid phase transitions and the strong coupling corrections to the free energy. With no free parameters, he obtains excellent agreement with experiment. In similar vein, Hsu, Bedell, and Ono have calculated the transport properties of ^{3}He B, again obtaining good agreement with experiment. Both sets of calculations require that one take various angular averages of the BP scattering amplitude; the averages in question are, however, different for each of the quantities calculated. The agreement between theory and experiment for some nine distinct angular averages of the BP scattering amplitude would suggest that it must be very nearly the correct one.
- Pfitzner and Wölfle[22] have developed a more consistent way of calculating scattering amplitudes from the AP polarization potentials than the comparatively simple (and not self-consistent) approach made by Bedell and me.[21]

The resulting scattering amplitude differs somewhat from the BP result; it leads to quite reasonable agreement between theory and experiment for the transport properties, superfluid transition, etc. The agreement is, however, by no means as close as that found by BP who used a far less sophisticated (and more open-to-question) algorithm to calculate scattering amplitudes.

- Brown, Pethick, and Zaringalam[23] have pointed out that the experimental results on the specific heat of ^{3}He between, say, 40 mK and 1200 mK can be interpreted as requiring that the effective mass of a quasiparticle at or near the Fermi surface fall off rapidly with increasing temperature, becoming comparable to the free particle mass at ~ 1 K; they have suggested that this may be a consequence of decoupling of quasiparticles from the spin density fluctuation excitations as one goes away from the Fermi surface. Calculations by Fantoni, Pandharipande and Schmidt[24] tend to support this interpretation. On the other hand, apart from some broadening of the contribution made by the spin fluctuation excitations, there is essentially no change in the results of the neutron scattering experiments over this temperature range; hence the quasiparticle, quasihole pair effective mass, m_q^*, is essentially temperature independent. (It is, moreover, almost momentum independent up to $q \sim 2p_F$). How then can the effective mass of a quasiparticle (or a quasihole) vary dramatically with temperature, while that which characterizes quasiparticle-quasihole pairs remains nearly constant? One possible resolution is that the link between the strength of the induced current fluctuation responsible for the pair effective mass and the effective mass of a single quasiparticle might become broken as temperature increases. A second might be that while the paramagnon contribution to the effective mass of a single quasiparticle falls off rapidly with increasing temperature, as one reaches the comparatively high frequencies of interest in a neutron scattering experiment the zero sound contribution to the effective mass of a single quasiparticle (which might not be substantial at very low temperatures) enters in such a way as to compensate for the loss of the "paramagnon mass." Some recent calculations by Friman and Krotscheck[25] tend to support this point of view. Obviously more work needs to be done on this quite perplexing problem.

Acknowledgment

I should like to take this opportunity to thank my collaborators in this research, C. H. Aldrich, K. Bedell, W. C. Hsu, N. Iwamoto, E. Krotscheck, E. Manousakis, Q. Usmani, and A. Zawadowski, for helpful and stimulating discussions, to thank the Aspen Center for Physics for its hospitality during the preparation of this manuscript, and to acknowledge the support of the National Science Foundation (NSF Grant DMR82-15128).

References

1 N. Iwamoto and D. Pines, submitted to Phys. Rev. B. (hereafter referred to as IP); N. Iwamoto, E. Krotscheck and D. Pines, submitted to Phys. Rev. B. (hereafter referred to as IKP).

2. E. Manousakis, D. Pines and Q. Usmani, in preparation (hereafter referred to as MPU).

3. For a recent review, see D. Pines, Lecture at "Highlights of Condensed Matter Theory," the 89rd Varenna International School of Physics (to be published).

4. K. Bedell, D. Pines and A. Zawadowski, submitted to Phys. Rev. B. (hereafter referred to as BPZ).

5. K. Bedell, W.-C. Hsu and D. Pines, in preparation; W.-C. Hsu and D. Pines, in preparation.

6. D. Pines and A. Zawadowski, in preparation.

7. S. Vosko, L. Wilk and M. Nusair, Can. J. Phys. 58, 1200 (1980).

8. D. M. Ceperley and B. J. Alder, Phys. Rev. Lett. 45, 566 (1980).

9. E. Krotscheck, Phys. Rev. A26, 3536 (1982); R. F. Bishop and K. L. Lührmann, Phys. Rev. B26, 5523 (1982).

10. P. M. Platzman and P. Eisenberger, Phys. Rev. Lett. 33, 152 (1974).

11. R. D. Puff, Phys. Rev. 137A, 46 (1965); N. Mihara and R. D. Puff, Phys. Rev. 174, 221 (1968).

12. D. Pines and P. Nozières, Theory of Quantum Liquids, W. A. Benjamin, New York (1966).

13. A. Miller, D. Pines and P. Nozières, Phys. Rev. 127, 1452 (1962).

14. E. Feenberg, Theory of Quantum Fluids, Academic Press, New York (1969).

15. E. C. Svennson, P. Martel, V. F. Sears and A.D.B. Woods, Can. J. Phys. 54, 2178 (1976).

16. K. Bedell, D. Pines and I. Fomin, J. Low Temp. Phys. 48, 417 (1982).

17. C. H. Aldrich III and D. Pines, J. Low Temp. Phys. 25, 691 (1976).

18. J. Bardeen, G. Baym and D. Pines, Phys. Rev. 156, 207 (1967).

19. J. Ruvalds and A. Zawadowski, Phys. Rev. Lett. 25, 333 (1970).

20. K. Bedell, Phys. Rev. B26, 3747 (1982).

21. K. Bedell and D. Pines, Phys. Rev. Lett. 45, 39 (1980), hereafter referred to as BP.

22. M. Pfitzner and P. Wölfle, J. Low Temp. Phys. 51, 535 (1983).

23. G. E. Brown, C. J. Pethick and A. Zaringhalam, J. Low Temp. Phys. 48, 349 (1982).

24. S. Fantoni, V. R. Pandharipande and K. E. Schmidt, Phys. Rev. Lett. 48, 878 (1982).

25. B. L. Friman and E. Krotscheck, Phys. Rev. Lett. 49, 1705 (1982).

OLD DOGS AND NEW TRICKS: BEYOND THE GROUND STATE WITH CBF THEORY

J. W. Clark
McDonnell Center for the Space Sciences
and Department of Physics, Washington University
St. Louis, Missouri 63130, U.S.A.

E. Krotscheck
Institute for Theoretical Physics, University of California
Santa Barbara, California 93106, U.S.A.

I. INTRODUCTION

In this report we shall collect certain thoughts and outline key developments bearing on the next major thrust in the microscopic theory of strongly-interacting Fermi systems. The focus of activity is shifting from ground-state properties--where acceptable quantitative accuracy is either at hand or realizable in the near future--to elementary excitations and dynamical properties more broadly. En route one seeks to accomplish the microscopic construction of self-energies and quasi-particle interactions. As has been the case for static, ground-state properties, the method of correlated basis functions[1-5] (CBF) will offer a natural and convenient vehicle--and in many instances it should be the method of choice.

II. REVIEW: CBF SINGLE-PARTICLE ENERGIES AND EFFECTIVE INTERACTIONS

A number of recent publications (see, e.g., Refs. 4,6-18) have demonstrated how the most prominent methods of conventional many-body theory[19,20] may be recast, for application to strongly-coupled Fermi systems such as liquid ^{3}He, nuclear matter, and the electron gas, in terms of the matrix elements H_{mn}, N_{mn} of the Hamiltonian and unit operators in a correlated basis $\{|\psi_m\rangle = F|\Phi_m\rangle I_{mm}^{-\frac{1}{2}},\ I_{mm} \equiv \langle\Phi_m|F^\dagger F|\Phi_m\rangle\}$. (For the sake of economy we shall assume that the reader is familiar with the standard motivation, assumptions, and notations of CBF theory, which are explained in the cited articles, especially Refs. 5,7,9,14,15.)

It is convenient, formally as well as intuitively, to combine the given CBF matrix elements into off-diagonal quantities H'_{mn} and W_{mn} in accordance with

$$\begin{aligned} H'_{mn} &= H_{mn} - H_{oo}N_{mn} \\ &= W_{mn} + \frac{1}{2}(H_{mm} + H_{nn} - 2H_{oo})N_{mn} \qquad , \quad m \neq n \quad , \end{aligned} \tag{2.1}$$

and then to express W_{mn} and N_{mn} as

$$\begin{aligned} W_{mn} &\equiv \langle m_1 \ldots m_d|W_d(1\ldots d)|n_1 \ldots n_d\rangle_a \qquad , \\ N_{mn} &\equiv \langle m_1 \ldots m_d|N_d(1\ldots d)|n_1 \ldots n_d\rangle_a \qquad , \end{aligned} \tag{2.2}$$

where $W_d(1\ldots d)$ and $N_d(1\ldots d)$, $d = 1,\ldots A$, are suitable d-body "interaction" and "normalization" operators. (In (2.2), $\{m_1\ldots m_d\}$ denotes the set of orbitals present in the Slater determinant Φ_m but not in Φ_n, and conversely for $\{n_1\ldots n_d\}$.)

Now, the differences of diagonal matrix elements of H serve to define CBF single-particle energies e(k), e.g., in the case that Φ_m is chosen as a one-particle-one-hole (ph) state, we have

$$H_{mm} - H_{oo} = e(p) - e(h) \qquad , \tag{2.3}$$

where corrections down by O(1/A) are suppressed. From the operators W_d and N_d and the CBF single-particle energies e(k) one may form the CBF effective perturbation $H'_{mn} \equiv \langle m_1 \ldots m_d | V_d(1 \ldots d) | n_1 \ldots n_d \rangle_a$ and the CBF effective interaction $V_d(1 \ldots d)$. For example, if Φ_m and Φ_n are ph and p'h' states respectively, then

$$H'_{mn} = \langle ph' | V(12) | hp' \rangle_a = \langle ph' | W(12) | hp' \rangle_a + \tfrac{1}{2}[e(p)+e(p')-e(h)-e(h')]\langle ph' | N(12) | hp' \rangle_a \ , \tag{2.4}$$

the d = 2 subscript on the operators being implicit.

If one adopts a Jastrow correlation factor $F = \Pi_{i<j} f(r_{ij})$ (or more generally a Feenberg factor), the fundamental structural ingredients of e(k), W_d, and N_d, consisting of certain sums of cluster or FHNC diagrams which we denote collectively by X (see Refs. 5,13 for details), are in fact generated automatically upon solving the FHNC and FHNC' integral equations which arise in the optimal variational theory pursued in Ref. 21. While there are some uncertainties in this approach to the evaluation of X, associated with the treatment of "elementary" diagrams, it is certainly superior to evaluation by brute-force cluster expansion. On the other hand, the latter has greater flexibility, being readily generalized to the wider context of state-dependent correlations. Be that as it may, both pedestrian cluster expansion and the "horse-and-buggy" phase of stepwise inclusion or fudging of elementary-diagram effects within FHNC, will ultimately be supplanted by Monte Carlo technology.

In the culminating step of a CBF calculation, the matrix elements H_{mn}, N_{mn} (or rather, e(k), W(12), and N(12)) are fed into some appropriately generalized many-body machinery to grind out the desired static or dynamic properties of the strongly-coupled system. The "bottom line" is that the CBF formalism serves to transform the problem of a system of bare particles with strong two-body interactions into a problem involving dressed particles with weak two-body, three-body,... effective interactions.[2,10] With due attention to the nonorthogonality of the basis, standard textbook recipes for weakly interacting systems may be adapted to treat the ground and excited states of the transformed problem. The next two sections provide important examples of explicit realization of this strategy.

III. CORRELATED-BASIS PERTURBATION THEORY

Various means are available[4,8,22] for developing a perturbation expansion of the (exact) ground-state energy in the nonorthogonal correlated basis $\{|\psi_m\rangle\}$. The first few terms of the raw expansion in the effective perturbation H'_{mn} read:

$$E = H_{oo} - \Sigma_m H'_{om} H'_{mo}/(H_{mm}-H_{oo}) + \Sigma_{m,n} H'_{om} H'_{mn} H'_{no}/(H_{mm}-H_{oo})(H_{nn}-H_{oo}) + \ldots \quad . \tag{3.1}$$

The correlated-coupled-cluster (CCC) procedure,[8] wherein the usual exp S method is modified to deal with the composite ansatz $|\psi\rangle = Fe^S|\Phi_0\rangle$ for the exact ground state, offers an economical route to (3.1). This procedure also has the special advantages that (a) it leads in a natural way to a linked expansion for E in terms of generalized Goldstone diagrams built from vertex elements representing the N_d and W_d (or N_d and V_d) operators, with propagators and insertions involving the e(k), and (b) it suggests natural partial resummations of big classes of such diagrams (e.g., those with ring or ladder topology).

As a simple but valuable illustration, the leading term of (3.1) in the effective two-body interaction $V_2(12) = V(12)$ takes the familiar form

$$E^{(2)}_{2p2h} = -\frac{1}{4}\sum_{pp'hh'} \frac{|\langle pp'|V(12)|hh'\rangle_a|^2}{e(p)+e(p')-e(h)-e(h')} = \tag{3.2}$$

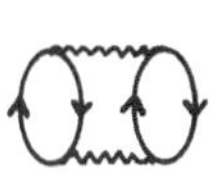

with

$$\langle pp'|V(12)|hh'\rangle_a = \langle pp'|W(12)|hh'\rangle_a + \tfrac{1}{2}[e(p)+e(p')-e(h)-e(h')]\langle pp'|N(12)|hh'\rangle_a . \tag{3.3}$$

As is true in ordinary, weak-interaction many-body theory, it is possible to get much more than just the ground-state energy out of an expansion such as (3.1) or its linked-diagrammatic counterpart. Upon introduction of Fermi-sea occupation numbers n(k) to explicate the various sums over particle and hole states, reinterpretation of these as quasiparticle occupation numbers, and functional variation of the expansion once or twice with respect to the n(k), one may obtain corresponding diagrammatic expansions of the (on-shell) self-energy or of the quasiparticle interaction. With care, an off-shell extension of the former construction may be made, yielding an expansion for the wave-vector and frequency dependent self-energy $\Sigma(k,\omega)$. In particular, the CBF versions of the familiar "correlation" and "polarization" contributions[23] to Σ may be extracted in this way from the second-order term (3.2).

It is useful at this point to adopt an historical perspective and identify three generations[17] of CBF calculations, according to whether the CBF inputs e, W_d, N_d are evaluated to low finite order or (partially) to infinite order in their diagrammatic cluster expansions and to whether perturbation theory in the correlated basis is carried to low finite or (partially) to infinite order in the CBF effective perturbation. First-generation CBF calculations employ low-cluster-order approximations for the off-diagonal elements N_{mn}, W_{mn} (and usually also for the expectation values H_{mm}), and are confined to low perturbative order. Such a treatment may be adequate for some properties of nuclear systems. The distinguishing mark of second-generation CBF calculations, currently in full swing, is the inclusion of cluster contributions to infinite order (in particular, by FHNC resummation), while proceeding again only to low finite order in the CBF perturbation expansion. This level of treatment has been applied primarily to liquid ^{3}He, where cluster expansions converge poorly. Finally, the <u>third generation</u> of CBF calculations builds on the successes of the second by reaching to all orders in the CBF effective perturbation. We are now entering a period in which third-generation calculations are both (a) technically feasible and (b) demanded by the nature of the quantities being studied. Quantities which require such treatment include the self-energy in polarized ^{3}He (Ref. 17) and the quasiparticle interaction in liquid ^{3}He (Ref. 16).

IV. CORRELATED RANDOM-PHASE APPROXIMATION

A generalization of the random-phase approximation has recently been formulated[7,12] which promises a semi-quantitative microscopic description of elementary excitations in nuclei, nuclear matter, liquid ^{3}He, the electron gas, and other strongly-interacting Fermi systems.

The derivation of Ref. 12 is patterned after the conventional one based on time-dependent Hartree-Fock theory,[24] but with the essential strong short-range correlations built in from the outset: The least action principle

$$\delta\int_{t_1}^{t_2} \langle\Psi(t)|H - i\hbar\, \partial/\partial t|\Psi(t)\rangle dt = 0 \tag{4.1}$$

is applied to a class of time-dependent correlated states

$$|\Psi(t)\rangle = F|\Phi(t)\rangle/\langle\Phi(t)|F^\dagger F|\Phi(t)\rangle^{\frac{1}{2}} \quad , \tag{4.2}$$

where F is a predetermined time-independent correlation operator and

$$|\Phi(t)\rangle = \exp(-iH_{oo}t/\hbar)\exp\left(\sum_{ph} c_{ph}(t)a_p^\dagger a_h\right)|\Phi_o\rangle \ . \tag{4.3}$$

The recipe (4.1)-(4.3) is pursued in the context of (a) the correlated Brillouin condition, meaning $H'_{mo} = 0\ \forall\ m = ph$, and (b) the small-amplitude (small $|c_{ph}|$) limit. On making a harmonic decomposition, there results a set of equations, called the correlated RPA (CRPA) equations, which have the same structure as the standard RPA equations,[20,25] but with a suitable redefinition of the matrices $A = (A_{ph,p'h'})$ and $B = (B_{ph,p'h'})$ in terms of the CBF effective interaction V(12), and with a nontrivial metric matrix $M = (M_{ph,p'h'})$. To be explicit, the CRPA equations take the form

$$\begin{bmatrix} A & B \\ B^* & A^* \end{bmatrix}\begin{bmatrix} x \\ y \end{bmatrix} = \hbar\omega\begin{bmatrix} M & 0 \\ 0 & -M^* \end{bmatrix}\begin{bmatrix} x \\ y \end{bmatrix} \ , \tag{4.4}$$

where, for the uniform medium,

$$\begin{aligned} A_{ph,p'h'} &= [e(p)-e(h)]\delta_{pp'}\delta_{hh'} + \langle ph'|V(12)|hp'\rangle_a \quad , \\ B_{ph,p'h'} &= \langle pp'|V(12)|hh'\rangle_a \quad , \\ M_{ph,p'h'} &= \delta_{pp'}\delta_{hh'} + \langle ph'|N(12)|hp'\rangle_a \quad . \end{aligned} \tag{4.5}$$

The equations of the theory thus involve only CBF inputs already introduced, namely e(k), W(12), and N(12).

The formulation has been extended to treat linear response of the dynamically correlated system to a weak external perturbation,[12] and a corresponding Green's function has been constructed which may be used to describe the propagation of particle-hole pairs. This approach to the elementary-excitation problem is currently being implemented for various systems, particularly by Sandler and Kwong.[26] An alternative version[14] of CRPA theory, rooted in an analysis of ring diagrams in the perturbation expansion (3.1), will be sketched below.

V. RELATIONS BETWEEN FHNC-CLUSTER DIAGRAMS AND CBF-GOLDSTONE DIAGRAMS

The chief message of the preceding sections is that the most prominent methods of conventional many-body theory may be rewritten, without excessive complications, in terms of the interaction and normalization matrix elements W_{mn}, N_{mn} of CBF theory, along with the diagonal quantities H_{oo}, $H_{mm} - H_{oo}$. We turn now to the relations between the diagrams of the variational estimate of the desired physical quantity and the diagrams of CBF perturbation theory. An understanding of these relations has proven crucial in cases where it is necessary to go to infinite-order CBF theory to achieve a correct description. For simplicity, we shall operate within the context of a Jastrow correlation operator.

Attention is directed to the decomposition (2.1) of the effective perturbation H'_{mn} into the "interaction" matrix element W_{mn} and the "energy-numerator" term $\frac{1}{2}(H_{mm} - H_{oo} + H_{nn} - H_{oo})N_{mn}$ (see also (2.4) and (3.3)). We can use this decomposition in a systematic analysis of CBF perturbation series like (3.1); carrying out all possible cancellations of energy numerators against identical energy denominators, one is left with a class of contributions <u>without any energy denominators</u>. In these "propagator-free" contributions, all Fermi-sea summations may be performed explicitly, and for each such summation we will obtain the familiar exchange or Slater function ℓ as a factor. Consequently, the CBF perturbation-theoretic contributions without energy denominators may be represented in terms of cluster diagrams of FHNC theory.

We hasten to note that such an analysis often "splits small quantities into large pieces." For instance, in the calculation of energy corrections, the W and N terms of H'_{mn} cancel to a high degree (cf. Ref. 27). Great care is therefore needed to assure consistent approximations for the W and N parts of the effective interaction.

The association of FHNC diagrams with some contributions to the CBF perturbation series suggests that certain FHNC diagrams (or combinations of FHNC graphs) may actually be thought of as <u>approximations</u> to Goldstone diagrams of an ordinary perturbation series. This idea has been developed quite thoroughly for the HNC variational theory of Bose systems where it is found[28,29] that optimal Jastrow HNC (more precisely, HNC/0) in fact gives all ring and ladder diagrams exactly, while the iterations between the ring and the ladder graphs are treated approximately. Before proceeding with the application of variational-CBF theory to excited states, we need to illuminate further the relations between FHNC and Goldstone diagrams.

Operating in an heuristic mode, let us compare the phase-space restrictions imposed by the Pauli principle, in FHNC theory and in Goldstone-type perturbation theory. In preparation for the applications to come, we concentrate on RPA-like diagrams; the same sort of analysis can be carried through for ladder-like diagrams. For simplicity, we restrict explicit considerations to the normalization operator N(12). (The interaction W(12) is of identical topological structure, and can be derived from N(12) by the diagrammatic differentiation technique introduced in Ref. 5.) As a first example we study the <u>chain diagrams</u> of FHNC theory. Consider, in particular, the two leading chain diagrams, drawn in part (a) of Figure 1, where we adhere to the accepted configuration-space diagrammatic conventions[30,5] of FHNC technology. We may introduce an alternative, Goldstone-like diagrammatic representation of the contribution of these diagrams to the matrix elements of N(12), according to the following scheme:

(i) Horizontal dashed lines represent matrix elements of the correlation bond $h(r) \equiv f^2(r) - 1$.

(ii) Upward-going lines represent particles, and downward-going lines represent holes.

(iii) <u>No energy denominators appear</u>.

The Goldstone-like representation of the matrix elements of the (sum of) cluster diagrams appearing in part (a) of Fig. 1 is shown in part (b) of the figure. The family resemblance of the chain diagrams of FHNC theory to the ring diagrams of the RPA becomes quite apparent, though we still have to find out where all the energy denominators went.

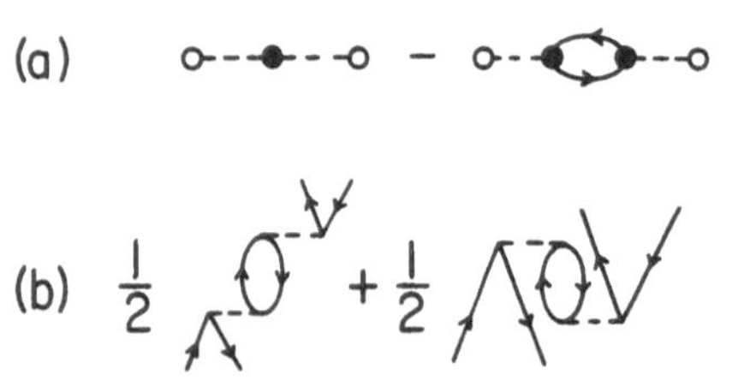

Figure 1
(a) FHNC chain diagrams contributing to N(12) and (b) Goldstone-like counterpart. Note that in (b) the two time orderings are displayed to achieve a symmetrical representation; they are analytically identical.

To expand on our investigation of diagrammatic analogies and correspondences, let us next look at the first-order corrections to the "particle-hole propagator." In Fig. 2 we depict the contributions from the two possible "time orderings," in both the FHNC diagrammatic language (to the left of the equal sign) and the Goldstone-like representation (to the right). For the first time ordering (shown in part (i)) the bare propagator is drawn; in the case of the second (shown in part (ii)), the propagator is attached to external dashed lines. More importantly, note that in the case of the second time ordering there are factorizable particle- and hole-line insertions present. On reverting to the FHNC graphology and invoking the convolution property[30] of the Slater function ℓ, it is seen that cancellations take place within (ii) such that the sum of these diagrams, with dangling dashed lines removed, coincides with the sum of diagrams (i). Such cancellations are no longer complete if the correlation bond h is spin- or isospin-dependent, this being the origin of the commutator diagrams appearing in state-dependent variational theories.[31,32]

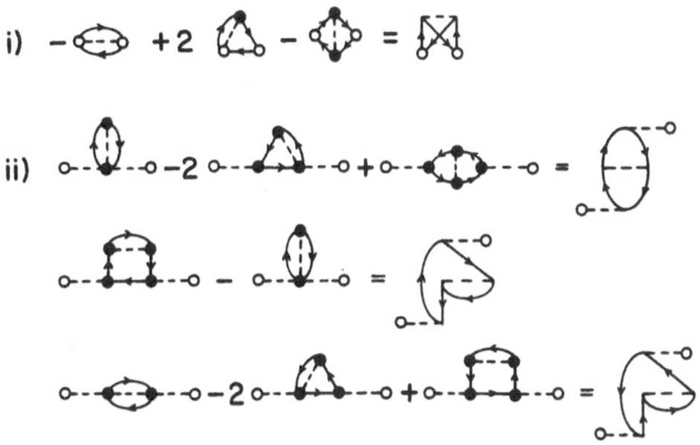

Figure 2
First-order corrections to the "particle-hole propagator," in FHNC and Goldstone-like versions.

VI. IRREDUCIBLE INTERACTIONS, CORRELATED RPA, AND CORRELATED RINGS

The connections between FHNC and Goldstone diagrams noted so far have been based on optical identification. Systematic extension of these considerations to larger classes of diagrams was carried forth in Ref. 14. To set up a formal, analytical definition of what one might call a "particle-hole-irreducible diagram" in FHNC theory, we follow the lead of that work and introduce a "correlation matrix"

$$C = \begin{bmatrix} (\langle ph'|N(12)|hp'\rangle_a) & (\langle pp'|N(12)|hh'\rangle_a) \\ (\langle hh'|N(12)|pp'\rangle_a) & (\langle hp'|N(12)|ph'\rangle_a) \end{bmatrix} = \begin{bmatrix} (C_{ph,p'h'}) & (C_{ph,h'p'}) \\ (C_{hp,p'h'}) & (C_{hp,h'p'}) \end{bmatrix} \tag{6.1}$$

and a matrix of "particle-hole-irreducible" diagrams

$$X = \begin{bmatrix} (X_{ph,p'h'}) & (X_{ph,h'p'}) \\ (X_{hp,p'h'}) & (X_{hp,h'p'}) \end{bmatrix} \tag{6.2}$$

which are related through

$$C = X + \frac{1}{2}CX \quad . \tag{6.3}$$

The matrices C and X are to be understood as supermatrices, the notation being rather schematic in that each is a 4×4 matrix in which the entries are matrices with rows labeled by ph, columns by p'h'. Thus the element in the (ph)th row and the (p'h')th column of the matrix forming the 22 element of C is $C_{hp,h'p'} = \langle hp'|N(12)|ph'\rangle_a$, etc.

It is easily verified that the matrix elements appearing in X do not contain any cluster diagrams which can be visually identified, according to the heuristic considerations of the preceding section, as "particle-hole reducible." In particular, the chain diagrams contributing to the operator N(12) do not occur in the elements of X. In situations where the notion of particle-hole reducibility is not so obvious, the relation (6.3) is taken as definition; this relation is to be used, in a diagrammatic expansion, to locate and eliminate diagrams contributing to C, but not to X. Though the matrix X should not be confused with the X-sets of non-nodal diagrams,[30] the similarity of notation is intentional--indeed the leading, direct part of the X matrix is determined by the matrix elements of $X_{dd}(r)$.

Our formal prescription for identifying "particle-hole reducible" diagrams is readily extended to the vertex function W(12). Writing the matrix elements of W(12) in both particle-hole channels in the same supermatrix form as (6.1), with $N \rightarrow W$, we define the "particle-hole-irreducible" piece X' through

$$W = \left(1 + \tfrac{1}{2}C\right) X' \left(1 + \tfrac{1}{2}C\right) \quad , \tag{6.4}$$

where X' is a supermatrix of the same form as X.

We are now ready to transform the CRPA equations (4.4) into new equations in terms of the "irreducible vertex functions" X, X'. To this end, we construct a diagonal 4 x 4 supermatrix Ω with nontrivial components $\Omega_{11} = ([e(p)-e(h)-\hbar\omega-i\eta]\delta_{pp'}\delta_{hh'})$ and $\Omega_{22} = ([e(p)-e(h)+\hbar\omega+i\eta]\delta_{pp'}\delta_{hh'})$, and observe that (4.4) can be written as $[\Omega + W + \frac{1}{2}\Omega C + \frac{1}{2}C\Omega]\begin{bmatrix} x \\ y \end{bmatrix} = 0$. Introducing $\begin{bmatrix} \hat{x} \\ \hat{y} \end{bmatrix} = \left[1 + \frac{1}{2}C\right]\begin{bmatrix} x \\ y \end{bmatrix}$ and the new supermatrix

$$U(\omega) = X' - \frac{1}{4} X\Omega X \tag{6.5}$$

representing the particle-hole interaction, we may now recast the CRPA equations in the form

$$\left[\Omega + U(\omega)\right]\begin{bmatrix} \hat{x} \\ \hat{y} \end{bmatrix} = 0 \quad . \tag{6.6}$$

Though exact, the manipulations of this section have been quite formal. It is therefore instructive to study their implications within the context of a simple example. For this purpose, we adopt the chain approximation to the particle-hole matrix elements and neglect the exchange terms. All manipulations can then be carried out analytically, yielding

$$U_{ph,p'h'} = A^{-1}\left[\tilde{X}'_{dd}(q) - \frac{\hbar^2 q^2}{4m}\tilde{X}_{dd}(q)\right]\delta(\vec{p} + \vec{p}\,' - \vec{h} - \vec{h}\,') \quad , \tag{6.7}$$

where $q = |\vec{p} - \vec{h}|$. In fact, in the more elaborate case that exchange diagrams are included, (6.7) is the local portion of the general particle-hole interaction (6.5), depending only on the momentum transfer q of the particle-hole pair.

This archetypal example shows how, by elimination of the chain diagrams from N(12) and W(12), we may arrive at a formulation of the RPA in the correlated basis which is structurally identical to the conventional RPA. In particular, the remaining energy dependence, i.e., that of the supermatrix $U(\omega)$, is totally **eradicated.** For the more general case of nonlocal correlations, the energy dependence of $U(\omega)$ has been discussed in some detail in Ref. 14. Suppressing or neglecting any residual energy dependence present in $U(\omega)$, we henceforth refer to this supermatrix simply as U.

We may now proceed with the formal developments and reformulate the CBF perturbation series for the energy, or the self-energy, or any other physical quantity of interest, in terms of the irreducible vertex U. However, it must be reiterated that serious pitfalls are in the offing unless one not only retains all diagrams of a chosen topological class, but also treats the "interaction" (W) and "normalization" (N) terms in such a manner as to assure certain crucial cancellations. An explicit rearrangement of the ground-state energy along these lines has been given in Ref. 14, the end result being the sum of all ring diagrams of CBF theory, expressed in the form

$$E^{CBF}_{ring} = (\Delta E)_0 + E_{RPA}[U] \quad . \tag{6.8}$$

Here, $E_{RPA}[U]$ is the sum of all RPA ring diagrams in terms of the irreducible vertex U, while $(\Delta E)_0$ collects those ring diagrams of the CBF perturbation series in which all energy denominators have been canceled against corresponding energy numerators. The latter quantity is given by

$$(\Delta E)_0 = -\frac{1}{4} \sum_{ph,p'h'} \Big\{ [e(p)-e(h)] X_{ph,h'p'} X_{hp,p'h'} + X'_{ph,h'p'} C_{hp,p'h'} + C_{ph,h'p'} X'_{hp,p'h'} \Big\} \quad . \tag{6.9}$$

Similarly, one may develop a (correlated) random-phase approximation for the self-energy in terms of the irreducible interaction U (see Refs. 14,17). Essentially, textbook equations may be carried over. The only additional wrinkle is that the Hartree-Fock-like single-particle energies e(k) must be modified to take account of the cancellations between the FHNC chain diagrams present in e(k) and similar diagrams generated by the variational derivative of $(\Delta E)_0$ with respect to the particle number.[17]

VII. LOCAL APPROXIMATIONS

At this point in the development it is merely a matter of concerted numerical effort to solve the RPA equations as they stand for some pristine CBF input and to calculate, for example, ground-state energy corrections or self-energies. For nuclear-matter problems one would thereby effectively be calculating all operator-chain diagrams and the most important commutator diagrams in an efficient and transparent way. We wish here to pursue a slightly different avenue which, at some sacrifice of realism, makes the content of CBF theory and the underlying approximations of the Jastrow form of the wave function (and more generally, the Feenberg form), somewhat clearer. For this purpose we search for the best local approximation of the operators N(12), W(12), and U(12), local in the sense that the matrix elements of these operators in the particle-hole channel are functions of momentum transfer $q=|\vec{p}-\vec{h}|$ alone. Guidance as to the best choice is provided by the fact that the hole-state averages[14,33] of these matrix elements can be related to the static structure factor S(k) and the variational derivative T(r) of the ground-state energy expectation value,

$$T(r) \equiv \delta H_{oo}/\delta \ln f^2(r) \quad . \tag{7.1}$$

Within such an approximation, the interaction matrix U is specified by

$$U_{ph,p'h'} = A^{-1}\,U(q)\,\delta(\vec{p}+\vec{p}\,'-\vec{h}-\vec{h}\,') \quad , \tag{7.2}$$

and we are led to choose

$$U(q) = \tilde{T}(q)S^{-2}(q) + (\hbar^2q^2/4m)[S^{-2}(q) - S_F^{-2}(q)] \quad , \tag{7.3}$$

where $S_F(q)$ is the static structure factor of the noninteracting Fermi gas.

Let us examine the implications of this representation of the particle-hole interaction, within some important settings. First, insert the local approximation (7.3) into the expression (6.8) for the sum of all CBF ring diagrams. For consistency a noninteracting single-particle spectrum must be used. The result may be written

$$E_{ring}^{CBF} = \Delta E_{opt} + E_{RPA} - E_{RPA}^{coll} \quad . \tag{7.4}$$

The first term,

$$\Delta E_{opt} = -\frac{1}{2}\sum_{\vec{q}} \frac{\hbar^2q^2}{4mS(q)}\left\{\sqrt{1+\frac{4m\tilde{T}(q)}{\hbar^2q^2}} - 1\right\} \quad , \tag{7.5}$$

reflects the fact that the CBF perturbation series also takes care of the optimization of the local two-body correlations. This term is zero if optimal two-body correlations are used to generate the CBF series. The second term is the familiar sum of the RPA ring diagrams,

$$E_{RPA} = -\frac{i\hbar}{2}\sum_{\vec{q}}\int \frac{d\omega}{2\pi}\left\{\log[1-U(q)\Pi_o(q,\omega)] - U(q)\Pi_o(q,\omega)\right\} \quad , \tag{7.6}$$

in which $\Pi_o(q,\omega)$ is the bare particle-hole propagator, i.e., the Lindhard function. In the third term E_{RPA}^{coll} is identical with the expression (7.6), except that the full

Lindhard function is replaced by its "collective" (or "mean-spherical") approximation

$$\Pi_0^{coll}(q,\omega) = \frac{\hbar^2 q^2/m}{\hbar^2\omega^2 - (\hbar^2 q^2/2mS_F(q))^2} \quad . \qquad (7.7)$$

Next, we calculate the static structure factor of the interacting system via the fluctuation-dissipation theorem from the density-density response function obtained with the particle-hole interaction U(q). Invoking in addition the collective approximation for the Lindhard function, one finds, starting from a variational structure factor $S_{var}(q)$, a corrected structure function of the form

$$S_{CBF}(q) = \frac{S_{var}(q)}{\sqrt{1 + 4m\tilde{T}(q)/\hbar^2 q^2}} \quad . \qquad (7.8)$$

In this relation we recover the first iteration in the optimization of the correlation factor f through the paired-phonon analysis.[34,3]

The summation of CBF ring diagrams is seen to have two effects: (i) It provides the first step in the optimization of the two-body correlation factor by the inclusion of virtual phonons,[34] and (ii) it replaces the "collective" particle-hole propagator by the correct Lindhard function. If the two-body correlations have been optimized already, only the propagator corrections survive, and the sum of all CBF ring diagrams vanishes when the particle-hole propagator is replaced by its collective approximation.

The above insights into the nature of correlations introduced by the variational wave functions and CBF perturbation furnish vital clues for a satisfactory microscopic treatment of the quasiparticle interaction. It is clear that simple variation of the energy expection value with respect to the quasiparticle occupation number must lead to wrong results, especially since it is not clear how to "undo" the frequency sums. While an energy-averaged particle-hole propagator can be acceptable for ground-state calculations, it is definitely incorrect if one looks at a specific $\omega \to 0$ limit. This is the basic reason why purely variational calculations of Fermi-liquid parameters yield generally poor results[35,9] and experience substantial corrections in CBF theory.[35,11] It also points to a cure. The inclusion, to infinite order, of CBF ring diagrams serves to replace the collective propagator by an inherently correct (or nearly correct) particle-hole propagator, and should therefore substantially improve the quality of predictions for the quasiparticle interaction.[16]

The above explications have also documented the consistency of our local effective interactions: for optimal correlations, the corrections to static properties of the ground state reduce to propagator corrections and are, therefore, as small as they can possibly be.

An essential aspect of the variational ansatz for the wave function that has come to light in this section is the simulation of the particle-hole propagator by one in which the single-particle spectrum is replaced by a collective mode of energy

$\hbar\omega_q = \hbar^2 q^2/2mS_F(q)$. One can expect that such an approximation is justified in cases where the elementary excitations of the system are in fact dominated by a collective mode. This is the situation in the density channel in liquid ^{3}He, and in the electron gas. That it is not so in the spin channel is the underlying reason why the collective correlations give unacceptable results for spin excitations in ^{3}He, and, in particular, the notorious spin instability.

VIII. APPLICATIONS TO LIQUID ^{3}He

Our numerical applications of the CBF-RPA theory to liquid ^{3}He are based on the optimized FHNC/C theory of Ref. 21, with the Aziz potential[36] HFDHE2 used as the bare two-body interaction. We shall not quote ground-state energies, since these can today be more accurately determined with Monte Carlo techniques. Rather, we consider the most important virtue of an integral-equation-based computational apparatus to be that it provides the raw material necessary for the calculation of the CBF effective interactions. We expect that in due course the FHNC evaluation of most of the ingredients of CBF theory will be superceded by more accurate Monte Carlo simulations.

The particle-hole interactions $U_\bullet(q)$ in normal, unpolarized liquid ^{3}He and in the fully-spin-polarized phase (^{3}He$\updownarrow$) are shown in Fig. 3. These interactions should be identified with the Aldrich-Pines pseudopotentials,[37] though we stress that they are based here on a purely microscopic theory. Note also that $\tilde{T}(q) \neq 0$ in the spin channel. Results of a calculation of energy corrections, separately for the density and the spin channels in the normal liquid, are presented in Table I. We see that propagator corrections contribute roughly 0.4 K of additional binding energy in this

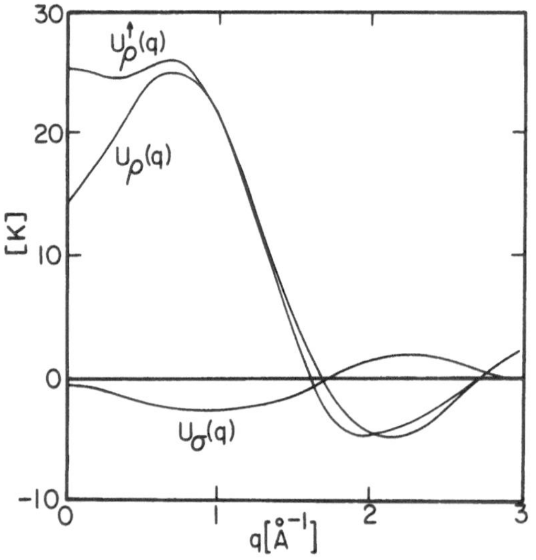

Figure 3

Local particle-hole interaction for unpolarized liquid ^{3}He, as appropriate to the density channel (U_ρ) and the spin channel (U_σ), at the equilibrium density $\rho = 0.0166\ \mathring{A}^{-3}$. Also shown is the density-channel potential ($U_\rho^\uparrow$) for the fully-spin-polarized liquid at the same density.

phase--to be appended to the variational estimate H_{oo}. Such corrections may be understood also as the effect of momentum-dependent correlations (occasionally confused with backflow). The propagator corrections for the fully-polarized liquid are only half their unpolarized-phase values, which indicates that $^3He\downarrow$ is more strongly dominated by a collective mode.

Table I. Propagator corrections in the density channel (E_{pp}) and spin-correlation corrections (E_{sp}) in the ground-state energy of normal and fully polarized liquid ^3He.

ρ $[A^{-3}]$	Normal E_{pp}	Normal E_{sp}	Polarized E_{pp}
0.0100	-0.15	-0.25	-0.06
0.0112	-0.19	-0.28	-0.07
0.0130	-0.25	-0.32	-0.11
0.0142	-0.30	-0.34	-0.14
0.0148	-0.32	-0.36	-0.16
0.0166	-0.41	-0.39	-0.21
0.0180	-0.48	-0.43	-0.26
0.0200	-0.59	-0.48	-0.34

Of rather more interest than the propagator corrections in the density channel are the spin correlations. There is no collective mode in the spin channel in (normal) ^3He; in fact, the system is known to be close to an instability. It should therefore come as no surprise that collective propagators do not work well in the spin channel: using the collective approximation (7.7) there, one underestimates the additional binding energy by typically 30%. The propagator corrections in the density and spin channels, together with the somewhat larger effect of static three-body ("triplet") correlations in normal ^3He, bring this phase safely into the regime of stability against spin fluctuations.

As a second application we calculate the self-energy of a ^3He atom in (normal) ^3He, a quantity currently under intensive theoretical[38,13,39] and experimental[40,41] investigation. Within the local model for the particle-hole interaction, the random-phase approximation for the self-energy can be written in the form[17,20] (ignoring $\hbar$'s)

$$\Sigma(k,E) = U_F(k) + i(2\pi)^{-4}\int d^3q d\omega\; G^0(\vec{k}-\vec{q},E-\omega)U^2(q)\Pi(q,\omega)_\rho^{-1}, \tag{8.1}$$

where $U_F(k)$ is the Fock term of the particle-hole interaction $U(q)$, G^0 is the non-interacting one-body Green's function, and $\Pi(q,\omega)$ is the density (respectively, the spin-density) response function, here taken as

$$\Pi(q,\omega) = \frac{\Pi_0(q,\omega)}{1 - U(q)\Pi_0(q,\omega)} \quad . \tag{8.2}$$

We regard spin summations, where appropriate, as implicit.

The effective mass of unpolarized ^3He as obtained from the on-shell derivative of the self-energy[13,17] is shown, for three different densities, in Fig. 4. We find the expected enhancement peak due to the attractive spin-channel interaction.

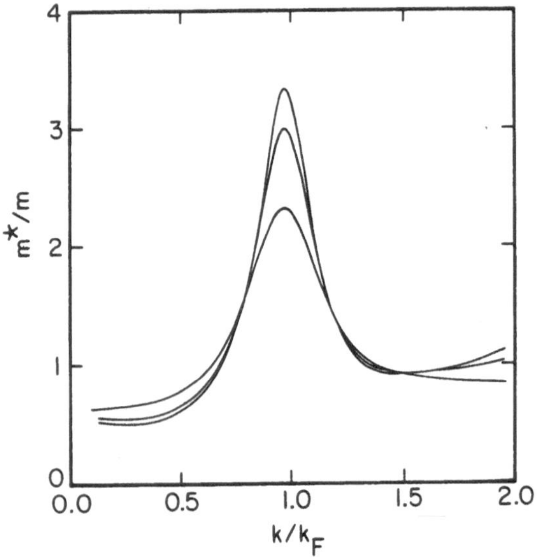

Figure 4

The "on-shell" effective mass in normal liquid ^3He as a function of wave number for three densities, ρ = 0.0148 $Å^{-3}$, 0.0166 $Å^{-3}$ and 0.0180 $Å^{-3}$. The peak of m* increases with density.

Some caution must, however, be applied in assessing the accuracy of these results: the size of the enhancement peak depends very sensitively on the strength of the spin-channel interaction, and a change of a few percent can have quite dramatic effects. It is presently not clear whether such an accuracy can be attributed to the FHNC evaluation of the spin-channel interaction. Another point of concern is the local approximation to the intrinsically nonlocal spin-spin interaction.

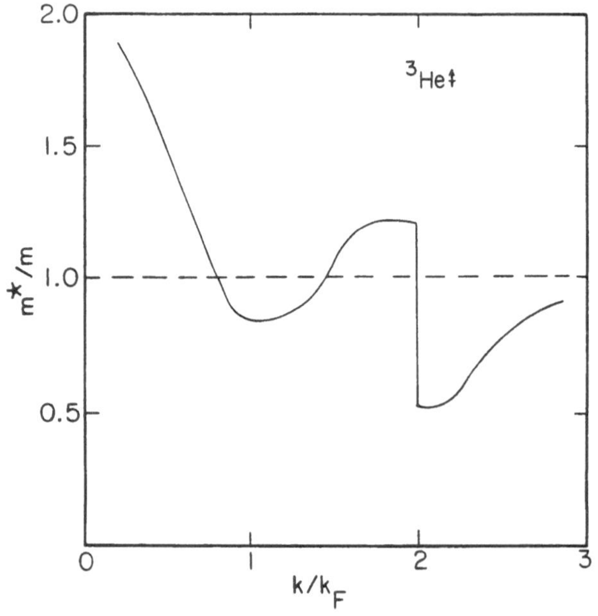

Figure 5

The "on-shell" effective mass in polarized ^3He over a broad range of wave number, at ρ = 0.0166 $Å^{-3}$. The discontinuity is due to the coupling of the single-particle excitation to the zero-sound mode.

For more complete picture, we include, in Fig. 5, results for the "on-shell" effective mass of $^3He\ddagger$.

IX. APPLICATIONS TO THE ELECTRON GAS

A thorough account of progress on the electron gas within the CBF theory has been given recently.[18] We confine ourselves to a brief outline of the findings of that work. In particular we refrain from presenting any data on the ground-state energy. The results for H_{oo} obtained in Ref. 18 are similar to those of earlier variational studies[42,43] and furnish, if compared with results of variational Monte Carlo evaluation,[44] an estimate of the typical accuracy one should expect from an integral-equation treatment. The juxtaposition of ground-state energies as calculated in second-order CBF theory with Green's function Monte Carlo results[45] is also illuminating.[18]

The formalities of applying CBF theory to dynamical properties of the electron-gas closely parallel those sketched for the Fermi helium liquids. As a matter of presentation, local models U(q) of the particle-hole interaction are commonly discussed in terms of a <u>local screening function</u>[46,47] G(q) related to U(q) by

$$U(q) \equiv \tilde{v}(q)[1 - G(q)] \quad , \tag{9.1}$$

where $\tilde{v}(q)=4\pi\rho e^2/q^2$ represents the bare Coulomb potential. Adopting the approximation (7.3) with optimal two-body correlations (hence $T(r) \equiv 0$), we have simply

$$G(q) = 1 - \frac{\hbar^2 q^2}{4m\tilde{v}(q)}\,[S^{-2}(q) - S_F^{-2}(q)] \quad . \tag{9.2}$$

This quantity gives a sensitive measure of errors in the static structure function S(q). A comparison of the screening function (9.2) based on an optimized FHNC/C evaluation[18] of S(q) with (9.2) based on Green's function Monte Carlo (GFMC) data[48] for S(q) is shown in Fig. 6. The overall agreement is satisfactory, in view of the fact that the correlations involved in the FHNC/C calculation and the GFMC calculation are not quite the same.

Within the model (9.2) for the local screening function, the dynamic structure factor $S(q,\omega) = -\pi^{-1}\,\mathrm{Im}\,\Pi(q,\omega)$ and the self-energy $\Sigma(k,E)$ were evaluated in correlated RPA, i.e., using (8.1)-(8.2). The plasmon dispersion relation determined by $S(q,\omega)$ is displayed in Fig. 7, at three different densities. There is little point in displaying the single-pair part of the dynamic structure factor, which is of the typical RPA shape, but does not show the double-peak structure discussed by Green *et al.*[49] and Awa *et al.*[50] It remains to be seen whether the double-peak structure is due to dynamic screening effects or due to the structure of the imaginary part of the self-energy.

Finally, let us focus on aspects of the self-energy. A comparison of the various effective masses (k-mass, E-mass, and "on-shell" mass) derived from our results for $\Sigma(k,E)$ is presented in Table II. Another piece of information contained in the self-energy is the Z-factor

$$Z_F = [1 - \partial\Sigma(k_F,E)/\partial E]^{-1}_{E=\hbar^2 k_F^2/2m} \tag{9.3}$$

measuring the quasiparticle pole strength. We include in Table II a comparison of the results of several independent estimates of this factor.

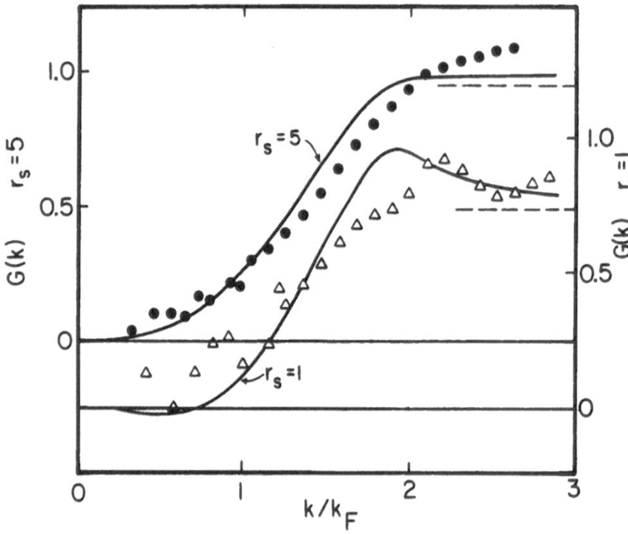

Figure 6

The electron-gas local screening function G(k) of (9.2) as derived from optimized FHNC/C calculations (solid curves) and from Green's function Monte Carlo results (dots and triangles). The horizontal dashed lines indicate the corresponding values of g(0+).

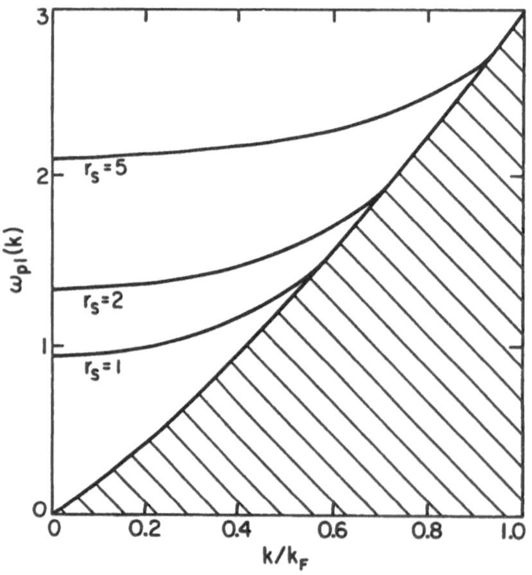

Figure 7

Plasmon dispersion relation for the electron gas, based on optimized FHNC/C calculations. The plasmon frequency is given in units of $\hbar^2 k_F^2/2m$. The shaded region is the particle-hole continuum.

Table II
Comparison of several versions of the effective mass and of the quasiparticle pole strength Z_F, in the electron gas. Here m_k^*, m_E^*, and m_{os}^* are the k-mass, E-mass, and "on-shell" effective mass, respectively. Columns 6-9 give Z_F values from (a) the present treatment, (b) Lantto,[42] (c) the Hubbard model,[51] and (d) RPA.[51]

r_s	m_k^*	m_E^*	$m_k^* m_E^*$	m_{os}^*	$Z_F^{(a)}$	$Z_F^{(b)}$	$Z_F^{(c)}$	$Z_F^{(d)}$
1	0.85	1.12	0.95	0.95	0.89	0.89	0.86	0.87
2	0.78	1.22	0.95	0.95	0.82	0.81	0.77	0.77
3	0.73	1.31	0.96	0.95	0.76	0.76	0.70	0.70
4	0.70	1.39	0.97	0.95	0.72	0.71	0.65	0.63
5	0.67	1.45	0.97	0.96	0.69	0.66	0.60	–

X. SUMMARY

Since the mid-sixties we have witnessed the development of CBF theory from the simplest, variational calculations of ground-state energies, through pedestrian, order-by-order nonorthogonal perturbation theory, up to a point where infinite-order CBF theory is being brought to bear for the correct description of salient physical effects. Depending on the questions asked, one may stop at any level: Ground-state-energy calculations can be performed with some confidence within a one-shot variational treatment. Where this is not sufficient, a low-order perturbation treatment will usually do the job. In contrast, Fermi-liquid parameters cannot be reliably obtained within a purely variational theory. As long as the F_ℓ are much smaller than $(2\ell+1)$, finite-order CBF theory is again sufficient,[11] but improvements along the lines of the Babu-Brown theory[52] must be implemented for strongly-coupled systems like ^{3}He. Infinite-order CBF theory is also generally necessary for the description of the dynamic response of a system to external perturbations[12] and for mode-mode coupling effects.

The CBF scheme puts at our disposal a comprehensive theory of <u>static</u> effective interactions. It is up to the user which interaction is considered static: In finite-order CBF theory, the effective interaction V(12) (cf. (2.4), (3.3)) is regarded as static. At a second stage of development elaborated herein, we went a step further and derived a static microscopic model for pseudopotentials and local screening functions. In future work, one may choose to go still further[16] and develop a static model for the direct interaction of the Babu-Brown theory.[52]

The first author acknowledges research support from the National Science Foundation under Grant No. DMR-8304213. The second author acknowledges support from the Deutsche Forschungsgemeinschaft through a Heisenberg Fellowship and from the National Science Foundation under Grant Nos. DMR-8114556 and PHY-7727084, supplemented by funds from the National Aeronautics and Space Administration and Office of Naval Research Contract No. N00014-82-K-0626. Informative discussions with David Pines are appreciated.

References

1. J. W. Clark and E. Feenberg, Phys. Rev. 113, 388 (1959).
2. J. W. Clark and P. Westhaus, Phys. Rev. 141, 833 (1966).
3. E. Feenberg, Theory of Quantum Fluids (Academic, New York, 1969).
4. J. W. Clark, L. R. Mead, E. Krotscheck, K. E. Kürten, and M. L. Ristig, Nucl. Phys. A328, 45 (1979).
5. E. Krotscheck and J. W. Clark, Nucl. Phys. A328, 73 (1979).
6. E. Krotscheck and J. W. Clark, Nucl. Phys. A333, 77 (1980).
7. J. W. Clark, Lecture Notes in Physics 138, 184 (1981).
8. E. Krotscheck, H. Kümmel, and J. G. Zabolitzky, Phys. Rev. A 22, 1243 (1980).
9. E. Krotscheck, R. A. Smith, and J. W. Clark, Lectures Notes in Physics 142, 270 (1981).
10. E. Krotscheck, R. A. Smith, and A. D. Jackson, Phys. Rev. B 24, 6404 (1981).
11. A. D. Jackson, E. Krotscheck, D. Meltzer, and R. A. Smith, Nucl. Phys. A386, 125 (1982).
12. J. M. C. Chen, J. W. Clark, and D. G. Sandler, Z. Physik A 305, 223 (1982).
13. E. Krotscheck and R. A. Smith, Phys. Rev. B 27, 4222 (1983).
14. E. Krotscheck, Phys. Rev. A 26, 3536 (1982).
15. J. W. Clark, in Proceedings of the 3rd International Conference on Nuclear Reaction Mechanisms, Varenna, June 14-19, 1982, ed. E. Gadioli (University of Milano, 1982), Suppl. no. 28, p. 464.
16. E. Krotscheck, invited talk at the Symposium on Quantum Fluids and Solids, Sanibel, Florida, to appear in AIP Conference Proceedings (1983).
17. E. Krotscheck, J. W. Clark, and A. D. Jackson, submitted to Phys. Rev. B.
18. E. Krotscheck, submitted to Ann. of Phys.
19. A. L. Fetter and J. D. Walecka, Quantum Theory of Many-Particle Systems (McGraw-Hill, New York, 1971).
20. G. E. Brown, Many Body Problems (North Holland, Amsterdam, 1972).
21. E. Krotscheck, R. A. Smith, J. W. Clark, and R. M. Panoff, Phys. Rev. B 24, 6383 (1981).
22. S. Fantoni, Phys. Rev. B, to be published; and these Proceedings.
23. J. P. Jeukenne, A. Lejeune, and C. Mahaux, Phys. Reports 25, 83 (1976).
24. A. K. Kerman and S. E. Koonin, Ann. of Phys. 100, 332 (1974).
25. D. J. Thouless, The Quantum Mechanics of Many-Body Systems, 2nd ed. (Academic Press, New York, 1972).
26. D. G. Sandler and N.-H. Kwong, these Proceedings; and to be published.
27. J. W. Clark, P. M. Lam, and W. J. Ter Louw, Nucl. Phys. A255, 1 (1975).
28. H. K. Sim, C.-W. Woo, and J. R. Buchler, Phys. Rev. A 2, 2024 (1970).
29. A. D. Jackson, A. Lande, and R. A. Smith, Phys. Reports 86, 55 (1982).
30. J. W. Clark, in Progress in Nuclear and Particle Physics, ed. D. H. Wilkinson (Pergamon, Oxford, 1979), Vol. 2, p. 89.
31. M. L. Ristig, W. J. Ter Louw, and J. W. Clark, Phys. Rev. C 3, 1504 (1971); C 5, 695 (1972).
32. V. R. Pandharipande and R. B. Wiringa, Rev. Mod. Phys. 51, 821 (1979).

33. E. Krotscheck, in Proceedings of the VIth Pan-American Workshop on Condensed Matter Theories, eds. J.M.C. Chen, J. W. Clark, and P. Suntharothok-Priesmeyer (Washington University, St. Louis, 1983).

34. C. E. Campbell and E. Feenberg, Phys. Rev. 188, 396 (1969).

35. H.-T. Tan and E. Feenberg, Phys. Rev. 176, 360 (1968).

36. R. A. Aziz, V. P. S. Nain, J. C. Carly, W. J. Taylor, and G. T. McConville, J. Chem. Phys. 70, 4330 (1979).

37. C. H. Aldrich, III and D. Pines, J. Low Temp. Phys. 32, 689 (1978).

38. G. E. Brown, C. J. Pethick, and A. Zaringhalam, J. Low Temp. Phys. 48, 349 (1982).

39. B. L. Friman and E. Krotscheck, Phys. Rev. Lett. 49, 3536 (1982).

40. D. S. Greywall and P. A. Busch, Phys. Rev. Lett. 49, 146 (1982).

41. D. S. Greywall, Phys. Rev. B 26, 2747 (1980).

42. L. J. Lantto, Phys. Rev. B 22, 1380 (1980).

43. J. G. Zabolitzky, Phys. Rev. B 22, 2353 (1980).

44. D. Ceperley, Phys. Rev. B 18, 3126 (1978).

45. D. M. Ceperley and B. J. Alder, Phys. Rev. Lett. 45, 566 (1980).

46. J. Hubbard, Proc. Roy. Soc. (London) A 240, 539 (1957).

47. N. Iwamoto and D. Pines, Phys. Rev. B, in press.

48. D. Ceperley, private communication.

49. F. Green, D. N. Lowy, and J. Szymanski, Phys. Rev. Lett. 48, 638 (1982).

50. K. Awa, H. Yasuhara, and T. Ashai, Phys. Rev. B 25, 3687 (1982).

51. G. D. Rice, Ann. of Phys. 31, 100 (1965).

52. S. Babu and G. E. Brown, Ann. of Phys. 78, 1 (1973).

SOLUTION OF THE ORNSTEIN-ZERNIKE EQUATION FOR NON-UNIFORM SYSTEMS

M. D. Miller*, M. L. Ristig, and N. Schulz
Institute for Theoretical Physics
University of Cologne
5000 Cologne 41
WEST GERMANY

The solution to the Ornstein-Zernike Equation for a liquid with a free surface is formulated in terms of an integral eigenvalue problem. We discuss the properties of the integral kernel and its associated eigenfunctions and eigenvalues. Our method is then applied to the zero-temperature free surface of liquid ^{4}He.

I. Introduction

In recent years, the development of powerful numerical and analytical tools has yielded great advances in our understanding of the bulk properties of uniform states of matter. The equations of state and distribution functions for simple classical liquids [1] or quantum liquids [2] (at zero temperature) can be accurately calculated by stochastic methods or by solving approximate integral equations. The state of the art in the case of non-uniform systems is not nearly as advanced. The complications introduced by generalizing the successful uniform system approaches to non-uniform systems often render the techniques intractable. For example, in a system with a planar surface, the surface two-particle distribution function is a function of three variables. A straightforward iterative solution of a non-linear integral equation for a function of three variables is generally intractable. Thus, a popular approach to this problem consists of simply replacing the surface distribution function by either the bulk radial distribution function or by a set of radial distribution functions calculated over a distribution of densities (the local density approximation). There is no microscopic basis for the local density approximation (to be referred to henceforth as the LDA). It is usually justified post hoc as being a not unreasonable way to turn an intractable problem into a tractable one. The LDA

---

*Present address: Department of Physics, Washington State University, Pullman, WA 99164, USA

surface distribution function is constructed to have the correct bulk limit, unfortunately, its accuracy in the surface region is unknown (although one expects that for a system with a diffuse, slowly varying surface profile, the LDA should be adequate).

The use of an uncontrolled approximation within the integral equation approach to non-uniform fluids [3] is unfortunate and, as we hope to point out in the body of this paper, unnecessary. In this context, we shall study the zero temperature surface distribution function and surface tension for the quantum liquid ^{4}He. In Sec. II, we write down the surface tension in terms of the one and two-particle surface distribution functions using a variational ansatz. We then discuss the solution of the Ornstein-Zernike (OZ) equation for the two-particle function, $g(\rho_{12}, z_1, z_2)$, by relating the OZ equation to an equivalent integral eigenvalue problem. In Sec. III, we present our results for $g(\rho_{12}, z_1, z_2)$ using a surface profile determined by Edwards [4]. Section IV is the Conclusion.

II. Surface Distribution Function.

We consider a system of N ^{4}He atoms in a box of volume $2L^3$ such that the ^{4}He atoms fill the lower (negative-z) half space with a Gibbs surface located at $z = 0$ (i.e. the bulk number density $\bar{\rho} = N/L^3$). The Hamiltonian for the system can be written

$$H = \sum_{i=1}^{N} -\frac{\hbar^2}{2m}\nabla_i^2 + \sum_{i<j} v(r_{ij}) \quad , \tag{1}$$

where the specific form used for the He-He interaction, v(r), will be noted below. Following Shih and Woo [5] and Chang and Cohen [6], we introduce a trial wave function in the form

$$\psi(1,2\ldots N) = \prod_i e^{\frac{1}{2}t(z_i)} \prod_{i<j} e^{\frac{1}{2}u_{ij}} \quad , \tag{2}$$

and we note that the two-particle (Jastrow) part includes the possibility of coordinate dependencies more general than simply scalar relative distance, r_{ij}. (This possibility had previously been discussed by Saarela, Pietilainen and Kallio [7] in their *hypernetted chain (HNC)* discussion of the helium surface.) The energy expectation can now immediately be written

$$E = -\frac{\hbar^2}{8m} A \int dz_1 \, \rho_1 \frac{d^2}{dz_1^2}(\ln \rho_1) + \frac{1}{2} \int d1\, d2 \; \rho_{12}^{(2)} v_{12}^* , \qquad (3)$$

where $A = L^2$ is the area of the system, $\rho_1 \equiv \rho(z_1)$ is the single particle distribution function (i.e., the density profile)

$$\rho_1 = \frac{1}{N} \frac{\int d2 \cdots dN \; |\Psi|^2}{\int d1 \cdots dN \; |\Psi|^2} , \qquad (4)$$

$d2 \equiv dr_2$, etc. and $\rho_{12}^{(2)} = \rho_1 \rho_2 g_{12}$ is the two particle distribution function and is defined analogously to Eq. (4). Thus $g_{12} \equiv g(\rho_{12}, z_1, z_2)$ is the surface equivalent of the bulk radial distribution function. The quantity v_{12}^* in Eq. (3) is an effective potential function obtained by using the Jackson-Feenberg form of the correlational kinetic energy,

$$v_{12}^* = v(r_{12}) - \frac{\hbar^2}{4m} \left[\nabla_1^2 u_{12} + \vec{\nabla}_1 (\ln \rho_1) \cdot \vec{\nabla}_1 u_{12} \right] . \qquad (5)$$

The surface tension, α, is obtained from Eq. (3) by use of the Gibbs prescription:

$$\alpha = \lim_{A \to \infty} \left(\frac{E - E_0}{A} \right) , \qquad (6)$$

where E_0 is the energy of a fictitious bulk system which at constant density, $\bar{\rho}$, fills the lower half space up to the Gibbs surface.

In principle, we now consider ρ_1 and u_{12} to be the independent functions which are varied to minimize . The problem then is to generate g_{12} for a given ρ_1 and u_{12}. The OZ equation relates g_{12} to a new function c_{12}, the direct correlation function:

$$h_{12} = c_{12} + \int d3 \; \rho_3 \, h_{13} \, c_{32} , \qquad (7)$$

where $h_{12} = g_{12} - 1$. The HNC equation then closes the loop by relating c_{12} back to g_{12}:

$$g_{12} = \exp\left(u_{12} + N_{12} + E_{12}\right), \tag{8}$$

$$\text{where} \quad N_{12} = h_{12} - c_{12} , \tag{9}$$

and, in the HNC/0 approximation, the elementary diagrams $E_{12} = 0$. We can take advantage of translational invariance in the xy plane and rewrite Eq. (7) as

$$h_\kappa(z_1,z_2) = c_\kappa(z_1,z_2) + \int dz_3\, h_\kappa(z_1,z_3)\, c_\kappa(z_3,z_2) , \tag{10}$$

where $h_\kappa(z_1,z_2)$ is given by the Hankel transform

$$h_\kappa(z_1,z_2) = \rho_1^{1/2}\, \rho_2^{1/2}\, 2\pi \int d\rho_{12}\, \rho_{12}\, J_0(\kappa\rho_{12})\, h_{12} , \tag{11}$$

and $c_\kappa(z_1,z_2)$ is defined analogously. Equation (10) can immediately be solved by treating $h_\kappa(z_1, z_2)$, say, as the kernel of an integral eigenvalue problem, namely

$$\int dz_2\, h_\kappa(z_1,z_2)\, \phi_\mu(z_2) = \lambda_\mu\, \phi_\mu(z_1) . \tag{12}$$

Then we immediately find

$$c_\kappa(z_1,z_2) = \sum_\mu \left(\frac{\lambda_\mu}{1+\lambda_\mu}\right) \phi_\mu^*(z_1)\, \phi_\mu(z_2) , \tag{13}$$

or from Eq. (9)

$$N_\kappa(z_1,z_2) = \sum_\mu \left(\frac{\lambda_\mu^2}{1+\lambda_\mu}\right) \phi_\mu^*(z_1)\, \phi_\mu(z_2) . \tag{14}$$

The summations in Eqs. (13) are generalized sums since the eigenvalue spectrum can be both discrete and continuous. Thus for a given choice of ρ_1 and u_{12} the solution of the HNC/0 equation for a non-uniform system is equivalent to being able to solve the integral eigenvalue Eq. (12).

In the next section we discuss the solution of the HNC/0 equation for the zero temperature surface of liquid ^{4}He. In the remainder of this section, we shall discuss some properties of Eq. (12).

1. The kernel $h_\kappa(z_1,z_2)$ is real and symmetric. Thus, the eigenvalues are

real, the eigenfunctions can be chosen to be real and eigenfunctions belonging to different eigenvalues are orthogonal.

2. From translational invariance in the limit $z \rightarrow -L$ ($|L| \rightarrow \infty$), the continuous spectrum eigenfunctions must behave asymptotically like

$$\phi_\mu \sim e^{i k_\mu z} \qquad (15)$$

Their eigenvalues can be immediately obtained:

$$\lambda_\mu = S(k) - 1 + O\left(\frac{1}{L}\right), \qquad (16)$$

where $k^2 = \kappa^2 + k_\mu^2$ and $S(k)$ is the bulk liquid structure factor.

3. The kernel is square integrable. Proof: The $\mathcal{L}^2$ norm of a kernel is given by

$$\|h_\kappa\| = \left[\int dz_1 \int dz_2 \; |h_\kappa(z_1,z_2)|^2\right]^{\frac{1}{2}} \qquad (17)$$

After inserting the definition of $h(z_1,z_2)$ and using the inequality $\int |h_\kappa| dz \geq |\int h_\kappa dz|$ in (17), we immediately find

$$\|h_\kappa\| \leq \left|\int d1 \, \rho_1 h_{12}\right|, \qquad (18)$$

but using a sequential relation [8],

$$\int d1 \, \rho_1 h_{12} = -1, \qquad (19)$$

and therefore

$$\|h_\kappa\| \leq 1, \qquad (20)$$

which completes the proof. Indeed, in the sense of Smithies [9] $h_\kappa(z_1,z_2)$ is an $\mathcal{L}^2$ kernel. (In addition, a <u>lower</u> bound on $\|h_\kappa\|$ can be established from Bessel's inequality $\|h_\kappa\| \geq \sum_\mu \lambda_\mu^2$.)

4. Since $h_\kappa(z_1,z_2)$ is a real symmetric $\mathcal{L}^2$ kernel, bilinear series such as Eqs. (13) and (14) are at least mean-convergent [10]. For the more physically

interesting situations (i.e. non-pathological u_{12}'s) one also expects $h_{\kappa}(z_1,z_2)$ to be continuous. This raises the possibility of applying Mercer's Theorem [10] which would guarantee uniform and absolute convergence of the bilinear series. If, for example, the series converges uniformly it is straightforward to show that the sequential relation, Eq. (19), can be rewritten as a constraint on the $\kappa = 0$ eigenvalues. In particular, at $\kappa = 0$ $\lambda_\mu = -1$ $(\forall \mu)$. (This result is the surface equivalent of the bulk statement that S(k) = 0 at k = 0.)

5. The considerations in the above sub-section do not require that the set $\{\phi_\mu\}$ be complete in $\mathcal{L}^2$.

6. An upper bound for the energy of the ripplon spectrum can be obtained from a Feynman type approach and these energies depend directly on the eigenvalues $\{\lambda_\mu\}$. Thus, using a trial function in the form [11]

$$\psi_{\vec{k}}(1,2\cdots N) = \sum_i e^{\ell z_i} e^{i\vec{k}\cdot\vec{\rho}_i} \psi_0 \quad , \tag{21}$$

where ψ_0 is the ground state eigenfunction of ^{4}He with a free surface and $\ell = \ell(\kappa)$ is a variational parameter, the excitation energies can immediately be written

$$\epsilon_\kappa = \frac{\hbar^2(\kappa^2+\ell^2)}{2m\,\Sigma_\ell(\kappa)} \quad . \tag{22}$$

In Eq. (21) we have defined

$$\Sigma_\ell(\kappa) = 1 + a_\ell \sum_\nu \lambda_\nu \rho_\nu^2 \quad , \tag{23}$$

where $$a_\ell^{-1} = \int dz\, e^{2\ell z} \rho(z) \tag{24}$$

$$\rho_\nu \equiv \rho_\nu(\kappa,\ell) = \int dz\, e^{\ell z} \rho(z)^{\frac{1}{2}} \phi_\nu(z) \quad . \tag{25}$$

It is simple to show that one solution of Eq. (22) is at $\ell = 0$ with $\Sigma_0(\kappa) = S(\kappa) + O(\frac{1}{L})$ (Feynman phonon). Presumably there is a second solution with $\ell \neq 0$ (the ripplon) which lies below the phonon dispersion curve. We note that Eq. (23) is only valid to the extent that we can integrate the sum over eigenvalues term by term (i.e.

uniform convergence).

III. Results for ^{4}He

In the following, we wish to set the stage for the examination of the validity of the LDA as applied to the problem of the ^{4}He free surface. Thus, we shall not try to vary the shape of the surface profile in order to minimize α , but instead, we shall fix $\rho(z)$ in the functional form determined by Edwards and Fatouros /4/ to be consistent with their atomic scattering data. Thus,

$$\rho(z) = \frac{\bar{\rho}}{\left[1 + \varepsilon \exp\left(\beta z + \gamma + \frac{\lambda}{4\beta}\left(\frac{1}{z^2 + \delta^2}\right)\right]^2}$$

where $\bar{\rho} = 0.0218$ Å^{-3}, $\gamma = -2.5$, $\delta^2 = 8.5$ Å^2 , $\beta = \sqrt{2m E_0}/\hbar = 1.1$ Å^{-1} and $\lambda = 2mC_6/\hbar^2 = 20$ Å. The potential function was chosen to be the HFD HE2 form /12/. The pair function u_{12} was chosen in a parameterized form to depend on both the relative scalar seperation, r_{12} , and the center of mass z-coordinate $Z_{12} = \frac{1}{2}(z_1 + z_2)$, viz.,

$$u_{12} = -\left(\frac{b_{12}}{r_{12}}\right)^5 , \qquad (27)$$

where $b_{12} = b[\rho(Z_{12})]$ is the value of the variational parameter in a bulk system of density $\rho(Z_{12})$.

In Figs. 1 and 2 we show the preliminary results obtained from our method. Figure 1 shows g_{12} in layers parallel to the surface at z_1 , z_2 = -2,-2; 0,0; and +2,+2 Å. In Figure 2, we show g_{12} for the atoms being in different layers at $z_1 = -2.5$ Å and z_2 = -5; -4.5 and -2.5 Å . It is clear that we have generated within a tractable procedure the full range of behavior to be expected from surface distribution functions.

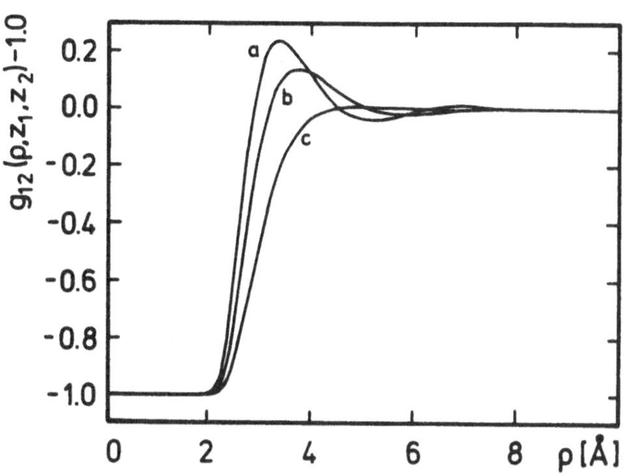

Fig. 1:

The distribution function g_{12} in layers parallel to the surface.
Curve a: $z_1=z_2=-2Å$;
curve b: $z_1=z_2=0Å$;
curve c: $z_1=z_2=+2Å$.

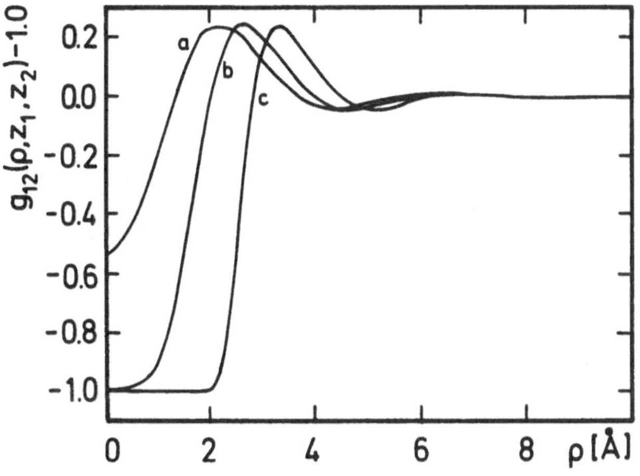

Fig. 2:

The distribution function g_{12} in different layers. z_1 is fixed at -2.5Å.
Curve a: z_2= -5Å;
curve b: z_2= -4.5Å;
curve c: z_2= -2.5Å.

IV. Conclusion

In the above we have set forth a straightforward approach for solving the Ornstein-Zernike equation for non-uniform systems. In Figs. 1 and 2 we presented some preliminary surface distribution functions which were obtained without the uncontrolled LDA or any other approximation schemes.

There are a number of directions which we shall take in further work. First, we shall make quantitative comparisons between our exact surface distribution functions and those obtained from the LDA (the complication being that the g(r)'s used for the LDA cannot be an arbitrary set of solutions to the HNC equation but they must be obtained using the same numerical algorithms as the surface functions). Second, we shall examine the ripplon spectrum in the hope that our results will show some roton-like bending which would be in better agreement with a phenomenological form obtained by fitting the measured temperature dependence of the surface tension. Next, we shall examine some of the remaining mathematical uncertainties with regards the integral eigenvalue problem. Does the bilinear form converge uniformly? Are the $\{\phi_\mu\}$ complete in $\mathcal{L}^2$? In the infinite volume limit, is there a point spectrum as well as a continuous one? [For the numerical work the system is placed on a grid in a finite sized box, i.e. a finite dimensional space. Thus, these questions are not practical problems. The basic numerical problem is to understand in what sense our sequence of degenerate kernels approximates the continuum system kernel in the infinite box limit.] Finally, we would like to apply our procedure to the classical liquid free surface where perhaps in combination with a density functional approach [13] for the surface profile a tractable yet quantitative theory of the temperature dependent distribution functions may result.

Acknowledgment: We would like to acknowledge the support of the Deutsche Forschungsgemeinschaft.

References

1. See, for example, J. P. Hansen and I. R. McDonald, Theory of Simple Liquids (Academic, New York, 1976) and references therein.
2. See, for example, M. H. Kalos, M. A. Lee, P. A. Whitlock, and G. V. Chester, Phys. Rev. 24 (1981) 115.
3. For a discussion of the theory of non-uniform fluids see J. K. Percus in: Studies in Statistical Mechanics, Vol. VIII, Eds. E. W. Montroll and J. L. Lebowitz (North Holland, Amsterdam, 1982).
4. D. O. Edwards and P. P. Fatouros, Phys. Rev. B17 (1978) 2147. V. U. Nayak, D. O. Edwards, and N. Masuhara, Phys. Rev. Lett. 50 (1983) 990.
5. Y. M. Shih and C.-W. Woo, Phys. Rev. Lett. 30 (1973) 478.
6. C. C. Chang and M. H. Cohen, Phys. Rev. A8 (1973) 3131.
7. M. Saarela, P. Pietilainen, and A. Kallio, Phys. Rev. B27 (1983) 22.
8. E. Feenberg, Theory of Quantum Fluids (Academic, New york, 1969), p. 3.
9. F. Smithies, Integral Equations (Cambridge London, 1958), ch. 1.
10. J. A. Cochran, The Analysis of Linear Integral Equations (McGraw-Hill, New York, 972), ch. 13.
11. C. C. Chang and M. H. Cohen, Phys. Rev. B11 (1975) 1059.
12. R. A. Aziz, V.P.S. Nain, J. S. Carley, W. L. Taylor, and G. T. McConville, J. chem. Phys. 77 (1979)4330.
13. See, for example, A. Evans, Adv. Phys. 28 (1979) 147.

JASTROW-SLATER TRIAL ENERGY FOR THE LOW DENSITY HARD SPHERE FERMI GAS

L.W. Bruch

Department of Physics
University of Wisconsin-Madison
Madison, WI 53706/USA

Abstract. The ability of a variational calculation with a Jastrow-Slater trial function to yield the leading terms in the ground state energy of a low density hard sphere fermi gas is discussed. A cluster expansion of the variational trial energy is developed which fully includes exchange processes between particles in the cluster and the other particles of the medium. The result, however, is that the spin-degeneracy of the Lee-Yang term is not recovered; this result was previously obtained with analyses in which the series of exchange terms was truncated.

The numerical evaluation of variational trial energies continues to be an important method of deriving quantitative information on the quantum mechanical many body problem. One test of the adequacy of the functional form assumed for the trial wave function has been[1-5] to analyze the resulting trial energy in limiting cases where analytical results are available. For the ground state energy of the low density hard sphere fermi and bose gases, the leading interaction (Lenz) term can be recovered and the functional form of the next term obtained. For the hard sphere bose gas a Jastrow trial function generates this next term (proportional to the 3/2-power of the density) completely[3,4]. However, for the second interaction term in the ground state energy of the hard sphere fermi gas, first obtained by Lee and Yang[6], there is a long standing question[1,2,5] whether the Jastrow-Slater trial function is sufficiently general to yield the correct dependence on exchange processes. Previous treatments of this question have been in the context of theories which truncate the series of exchange terms. In view of the importance of the statistical correlations in the low density fermi gas, emphasized by Brueckner[7], it seems appropriate to reconsider this question with a cluster expansion in the direct particle interactions which maintains the full set of exchange processes with particles outside the clusters. This is done in the present note; the result, however, is that the discrepancy in the spin-degeneracy dependence of the Lee-Yang term remains even in this analysis.

The N fermions in three-dimensional volume Ω with number density $\rho = N/\Omega$ and spin degeneracy S have a fermi wave number k_F

$$k_F = (6\pi^2 \rho/S)^{1/3} . \tag{1}$$

When they interact by a hard-sphere potential

$$\phi(r) = \begin{cases} \infty & r < a \\ 0 & r > a \end{cases} , \tag{2}$$

the ground state energy of the low density gas is[6]

$$E_0/N = (\hbar^2 k_F^2/2m)\ \{(3/5) + (2/3\pi)(S-1)(k_F a) + \gamma(k_F a)^2 + \ldots\}, \qquad (3)$$

where $\hbar$ is the reduced Planck constant and m is the particle mass. The coefficient γ is given by Lee and Yang[6] and by Abrikosov and Khalatnikov[8] as

$$\gamma = (S-1)(4/35\pi^2)(11-2\ln 2) \simeq 0.1113(S-1)\ . \qquad (4)$$

In the perturbation theory calculations the fact that γ is proportional to S-1 follows immediately from the structure of the exchange terms. In the variational calculations it is much less apparent that the variational estimate γ_t must be proportional to (S-1) and in fact this feature is not recovered from the calculations[1,2,5].

The Jastrow-Slater trial function for the many-fermion system has the form

$$\Psi_t = (\prod_{i<j} f_{ij}) S_N\ , \qquad (5)$$

where S_N denotes the Slater determinant ground state (fermi sphere) of a noninteracting system of density ρ. Such a trial function has explicit pair correlations and no explicit three-particle correlations, as in the many-boson calculations[3,4]. In the present work, two forms for the pair factor f are used

$$f_1(r) = \begin{cases} 0 & r < a \\ [1-(a/r)]/[1-(a/R)] & a < r < R \\ 1 & R < r \end{cases} \qquad (6a)$$

where the variational parameter R denotes the separation at which f is scaled to 1[5] and

$$f_2(r) = \begin{cases} 0 & r < a \\ 1-(a/r)\exp[-\lambda(r-a)] & a < r \end{cases} \qquad (6b)$$

where the damping factor λ is the variational parameter[2]. Both these forms permit application of the cluster expansion formalism; the volume integral of f-1 is finite.

For the hard sphere system, the trial energy reduces to the expectation value of the kinetic energy operator

$$K = \sum_{i=1}^{N} p_i^2/2m \qquad (7)$$

in the state Ψ_t. The cluster expansion of the trial energy begins with the formal expansion of the pair product

$$\prod_{i<j} f_{ij} = 1 + F\ , \qquad (8)$$

with

$$F = h_I + h_{II} + \cdots \tag{9}$$

$$h_I = \sum_{i<j} (f_{ij}-1)$$

$$h_{II} = 1/2 \sum_{i<j} \sum_{k<\ell} (f_{ij}-1)(f_{k\ell}-1)\Big|_{(i,j)\neq(k,\ell)}$$

If the Slater determinant S_N is denoted by a ket $|0\rangle$

$$K|0\rangle = \varepsilon_o |0\rangle \ , \qquad \varepsilon_o = N(3\hbar^2 k_F^2/10\ m) \ , \tag{10}$$

the trial energy E_t is written in terms of a double commutator:

$$E_t - \varepsilon_o = (\tfrac{1}{2})\langle 0|[F,[K,F]]|0\rangle / \langle 0|(1+F)^2|0\rangle \ . \tag{11}$$

Expansion in clusters gives for the leading terms

$$E_t - \varepsilon_o = (g_1/2) + \{g_2 - g_3 g_1\} + \cdots \tag{12}$$

with

$$g_1 = \langle 0|[h_I,[K,h_I]]|0\rangle$$

$$g_2 = \langle 0|[h_I,[K,h_{II}]]|0\rangle \tag{13}$$

$$g_3 = \langle 0|h_I|0\rangle \ .$$

Eq. (12) provides a different starting point for the analysis of the low density fermi gas than that used in previous work: exchange terms of the many-fermion system have not been truncated. Expectation values with respect to the Slater determinant are readily evaluated by the use of second quantized operators in a plane wave basis.

The explicit forms for the terms g_1, g_2, and g_3 in first quantized representation are

$$g_1 = (\hbar^2/m)\langle 0| \sum_{ijk} (\nabla_i f_{ij})\cdot(\nabla_i f_{ik})|0\rangle_{i\neq j, i\neq k}$$

$$g_2 = (\hbar^2/2m)\langle 0| \sum_{ijmk\ell} (f_{k\ell}-1)(\nabla_i f_{ij})\cdot(\nabla_i f_{im})|0\rangle$$

$$[i\neq j,\ i\neq m,\ k\neq \ell,\ (i,j)\neq(k,\ell)]$$

$$g_3 = 1/2 \langle 0|\sum_{ij} (f_{ij}-1)|0\rangle_{i\neq j} \ . \tag{14}$$

The term g_1 has contributions involving two and three distinct particle labels while the leading piece of g_2, which is evaluated here, has m=j and involves the expectation values of three and four body operators.

Some fourier transforms which arise in the second quantization of the operators are

$$V(\underset{\sim}{k}) = \int d\underset{\sim}{r}\ (\nabla f(r))^2\ e^{i\underset{\sim}{k}\cdot\underset{\sim}{r}}$$

$$H(\underset{\sim}{k}) = \int d\underset{\sim}{r}\ [f(r)-1]\ e^{i\underset{\sim}{k}\cdot\underset{\sim}{r}} \tag{15}$$

$$\underset{\sim}{G}(\underset{\sim}{k}) = \int d\underset{\sim}{r}\ \nabla f(r)\ e^{i\underset{\sim}{k}\cdot\underset{\sim}{r}}\ .$$

A sum over the fermi sphere of plane waves is defined

$$\ell(k_F r) = \sum_k n_k\ e^{i\underset{\sim}{k}\cdot\underset{\sim}{r}}/\sum_k n_k = (3/z^3)\{\sin z - z\cos z\}\Big|_{z=k_F r}\ , \tag{16}$$

where n_k is 1 for $k < k_F$ and 0 for $k > k_F$.

The term g_1 is

$$g_1/(\hbar^2/m) = N(k_F^3/6\pi^2)V(0)(S-1) + N(k_F^3/6\pi^2)\int d\underset{\sim}{r}(\nabla f)^2[1-\ell^2(k_F r)]$$
$$+ 2S\ \Sigma_\ell\ n_\ell\ |\Sigma_{\underset{\sim}{k}}\underset{\sim}{G}(\underset{\sim}{k}-\underset{\sim}{\ell})n_k/\Omega|^2 - (NS/\Omega)\ \Sigma_k\ |\underset{\sim}{G}(\underset{\sim}{k})|^2(\Sigma_\ell\ n_{\underset{\sim}{\ell}}n_{\underset{\sim}{\ell}+\underset{\sim}{k}}/\Omega) \tag{17}$$

The part of the second term which contributes to the coefficient γ is

$$(g_2 - g_3 g_1)/(\hbar^2/m) \simeq (2/\Omega^2)\Sigma_{pqt}n_p n_q n_t\{SV(\underset{\sim}{t}-\underset{\sim}{q})H(\underset{\sim}{p}-\underset{\sim}{q}) - S^2V(\underset{\sim}{q}-\underset{\sim}{t})H(\underset{\sim}{t}-\underset{\sim}{q})\}$$
$$+ (1/\Omega^2)\Sigma_{pqr}n_p n_q n_{p+r}n_{q-r}V(r)\{S^2H(\underset{\sim}{r})-SH(\underset{\sim}{p}-\underset{\sim}{q}+\underset{\sim}{r})\}$$
$$\simeq S(S-1)V(0)\ \Sigma_k H(k)(\Sigma_\ell\ n_\ell n_{\ell+k}/\Omega)^2$$
$$- 2N(S-1)V(0)(1/\Omega)\Sigma_k H(k)(\Sigma_\ell n_\ell n_{\ell+k}/\Omega)\ . \tag{18}$$

In obtaining the last form the estimate

$$V(k) \simeq V(0) + O(k_F a^2)$$
$$\simeq 4\pi a + O(k_F a^2) \tag{19}$$

has been used to isolate the terms contributing to γ_t. This is satisfied for the functions in Eq. (6); similar estimates have been used to identity the clusters contributing to γ_t. The first term on the right-hand side of Eq. (17) gives the Lenz term, linear in the sphere diameter a. The remaining terms of Eqs. (17) and (18), with 1/R and γ proportional to k_F, contribute to the coefficient γ_t.

In bringing the variational estimate for γ_t to its final form, the following summation is performed:

$$\Sigma_\ell\ n_\ell n_{\ell+k}/\Omega = (k_F^3/6\pi^2)\ F(k/k_F)\ ,$$

$$F(x) = \begin{cases} 1 - (3x/4) + (x^3/16) & x < 2 \\ 0 & x > 2 \end{cases}\ .$$

Then, if the variational parameters R and λ are assumed to be independent of S, the variational estimate for γ_t is

$$\gamma_t = (S-1)\,\sigma_1 + \sigma_2 \tag{20}$$

with

$$\sigma_1 = (2/3\pi k_F a)[(V(0)/4\pi a) - 1] + (4/3\pi^2) \int_0^2 dx\, h(x)F(x)[4 - 2F(x) - h(x)] \tag{21}$$

and

$$\sigma_2 = [2/3\pi(k_F a)^2] \int_0^\infty x^2\, dx\, (df/dx)^2[1 - \ell^2(x)] - (4/3\pi^2) \int_0^2 dx\, h^2(x)F(x) + (8/3\pi^2) \int_0^1 x^2 dx\, [\int_0^\infty dy\, g(y)\, j_1(xy)]^2 . \tag{22}$$

The functions h(x)and g(y) are

$$h(x) = -(k_F x)^2 H(k_F x)/4\pi a \tag{23}$$

and

$$g(y) = y^2(df/dy)\ell(y)/(k_F a)$$

and $j_1(z)$ is the spherical bessel function of order one.

For $S = 2$, the minimum value of γ_t with the function f_1 is 0.244 at $R = 1.65/k_F$ with $\sigma_2/\sigma_1 = 0.362$. For the function f_2, the $S = 2$ minimum of γ_t is 0.218 at $\gamma = 0.95\ k_F$ with $\sigma_2/\sigma_1 = 0.306$. Both variational estimates are considerably greater than the analytical value of 0.111 ...; the exponential damping used to impose the cluster condition for f_2 gives a somewhat lower energy than the sharp cutoff used for f_1. For neither trial function does the σ_2 term, which violates the spin degeneracy form of the Lee-Yang result, vanish. In fact, if the pair factor f is independent of S, a proof which was used previously for truncated exchange expansions[5] also applies here and shows that σ_2 is positive. The present work therefore shows that the failure to obtain the Lee-Yang degeneracy factor is not an artifact of truncated exchange expansions.

Acknowledgement

This work was supported in part by the National Science Foundation through grant DMR-8214518.

References

1. J.B. Aviles, Ann. Phys. (NY) 5 (1958) 251.
2. A. Temkin, Ann. Phys. (NY) 9 (1960) 93.
3. H.A. Gersch and V.H. Smith, Phys. Rev. 119 (1960) 886.
4. E.H. Lieb, Phys. Rev. 130 (1963) 2518.
5. L.W. Bruch, Physica 94A (1978) 586.
6. T.D. Lee and C.N. Yang, Phys. Rev. 105 (1957) 1119.
7. K.A. Brueckner, Phys. Rev. C14 (1976) 1196.
8. A.A. Abrikosov and I.M. Khalatnikov, Sov. Phys. JETP 6 (1958) 888.

VARIATIONAL MONTE CARLO APPROACH ON ATOMIC IMPURITIES IN ^4He

K.E. Kürten

Courant Institute of Mathematical Sciences
New York University
New York, New York 10012

A great deal of theoretical and experimental effort has gone into trying to understand the behavior of ions and other impurities in bulk liquid helium.[1,2] Studies of two component systems, particularly in the limit of small concentrations, form the basis for understanding the physics of rather complicated systems and offer an intermediate step in developing an adequate microscopic theory for surface and interface problems. Also, theoretical studies of the impurtiy problem as a zero-concentration mixture anticipate the description of quantum fluid mixtures such as ^3He-^4He or Hydrogen-Helium at low temperatures.[3-5] Moreover, a study of the internal structure of a ^3He impurity or other impurities in ^4He is of interest in its own right. Such descriptions here reached a very sophisticated level, since recent GFMC results for the pure ^4He system are almost indistinguishable from the experimental data,[6] whereas at the present stage the ^3He system has not been treated on the same level of accuracy.[7]

We consider a homogeneous system of (N-1) identical background particles of mass m_1 and one foreign particle of mass m uniformly distributed in a cubic box of volume Ω and density N/Ω with periodic boundary conditions. Starting from the Hamiltonian

$$(1) \quad H = -\frac{\hbar^2}{2m_1}\sum_{i=1}^{N-1}\Delta_i - \frac{\hbar^2}{2m}\Delta_N + \sum_{i<j}^{N-1} v_{11}(r_{ij}) + \sum_{k=1}^{N-1} v_{12}(r_{kN})$$

with realistic interaction potentials $v_{\alpha\beta}(r)$, $\alpha\beta = 11,12$.
As a trial function we choose the symmetric form

$$(2) \quad \psi = \exp\left(-\frac{1}{2}\sum_{i<j}^{N-1} u_{11}(r_{ij}) - \frac{1}{2}\sum_{k=1}^{N-1} u_{12}(r_{kN})\right)$$

For a ^3He-impurity we may ignore the fermion statistics, since a single fermion does not have any exchange correlation with the background particles. In eq. (2) $u_{11}(r)$ and $u_{12}(r)$ represent the standard Jastrow pseudopotentials, which may differ from one another in order to reflect possible different correlations.

The Monte Carlo algorithm

It is well known that the variational energy

(3) $$E_o \leq E_T = \frac{\langle \Psi_T | H | \Psi_T \rangle}{\langle \Psi_T | \Psi_T \rangle}$$

is an upper bound to the ground state energy E_o and attains that value when $\Psi_T = \Psi_O$, the fundamental eigenfunction. The best estimate of the energies is that obtained by minimizing E with respect to numbers of different classes of trial functions.

The task now is to evaluate the multidimensional integral in order to determine the variational energy expectation value.

Let $\underline{R}_i$ be a set of points drawn from the probability distribution

(4) $$p(\underline{R}) = \frac{|\Psi_T(\underline{R})|^2}{\int |\Psi_T(\underline{R})|^2 d\underline{R}},$$

where the integral in the denominator serves to normalize $p(\underline{R})$. Then the central limit theorem of probability gives that

(5) $$F_M = \frac{1}{M} \sum_{i=1}^{M} f(\underline{R}_i)$$

$$\text{with} \qquad f(\underline{R}) = \frac{H \, \Psi_T(\underline{R})}{\Psi_T(\underline{R})}$$

is an approximation to E_T associated with a statistical error proportional to $\frac{1}{\sqrt{M}}$ for large M.

The Monte Carlo algorithm $M(RT)^2$,[8] which has been invented to calculate properties of classical statistical systems, is an extremely powerful way to compute multidimensional integrals. This $M(RT)^2$ algorithm is a biased random walk in configuration space. As usually carried out, each particle is moved one after another to a new position uniformly distributed inside a cube of side s. That move is either accepted or rejected depending on the magnitude of the trial function at the new position compared with the old position.

A possible move from the old position $\underline{R}'$ to the new position $\underline{R}$ is accepted automatically if $|\Psi_T(\underline{R})|^2 > |\Psi_T(\underline{R}')|^2$ and if not, it is accepted with probability

(6) $$q = |\Psi_T(\underline{R})|^2 / |\Psi_T(\underline{R}')|^2.$$

If the move is not accepted, the configuration is returned to $\underline{R}'$.

It has been shown that under certain very general conditions, the points of the random walk have $|\Psi_T(\underline{R})|^2$ as their density asymptotically as the number of steps increases.

In general the algorithm is very simple to program and test, and follows very closely a Monte Carlo simulation of a classical system. Brute force application of the algorithm would lead to large statistical errors for quantities related to the impurity. If each particle of the system is moved, one after another, one pass gives (N-1)(N-2) pieces of informations about the ^{4}He particles compared to 2(N-1) ones about the impurity atom. This fact leads to rather poor statistics for quantities related to the impurity atom.
Thus we modify the algorithm such that each move of a background particle is followed by an attempt to move the impurity so that half the attempts are moves of the impurity atom.

Another very efficient way to improve the statistics for quantities related to the impurity atom is by moving the particles in the vicinity of the impurity atom more often.

The chemical potential difference

For a finite system of N particles, containing in addition one impurity, the difference of the chemical potentials of the differing constituents is

$$(7) \quad \mu_F = \frac{E_i - E_p}{1/N}$$

where E_p represents the energy expectation value per particle of N ^{4}He atoms and E_i represents the energy expectation value per particle. of (N-1) ^{4}He atoms and one impurity atom.

Since the energy difference $E_i - E_p$ in (7) is proportional to 1/N for large N, the evaluation of (7) is somewhat delicate for isotopic systems. A straightforward analysis of the statistical error $\delta\mu_F$ of the quantity μ_F evaluated in a Monte Carlo approach gives

$$(8) \quad \delta\mu_F = N \sqrt{(\delta E_i)^2 + (\delta E_p)^2}$$

with δE_p and δE_i being the absolute statistical errors associated with the quantities E_p and E_i, respectively. Thus the relative statistical error is

$$(9)\quad \left|\frac{\delta\mu_F}{\mu_F}\right| = \frac{\sqrt{(\delta E_i)^2 + (\delta E_p)^2}}{|E_i - E_p|}$$

If the constituents and their mutual interactions are quite different from one another, as is the case for Xe or Cs dissolved in ^{4}He,[3] the denominator $(E_i - E_p)$ in (9) might be large enough in order to guarantee a sufficiently small relative statistical error for the difference of the chemical potentials.

However, for isotopic components the energy difference is necessarily very small and consequently the relative statistical error might be even larger than the difference of the chemical potentials. There are a number of techniques in order to circumvent these difficulties such as reweighting methods, parametrization of the Hamiltonian and Baym's approximation.

Reweighting method

If one has available configurations generated from a trial function Ψ and a Hamiltonian H, it is possible to calculate properties, such as the energy,of a slightly different trial function $\tilde{\Psi}$ and a Hamiltonian $\tilde{H}$.
The energy expectation value E of the Hamiltonian H with respect to the trial function Ψ can be written as

$$(10)\quad \int \frac{\tilde{\Psi}\,\tilde{H}\,\tilde{\Psi}}{\int|\tilde{\Psi}|^2 d\underline{R}}\, d\underline{R} = \int \frac{\Psi^2(\underline{R})}{\int|\Psi|^2 d\underline{R}}\, \frac{\tilde{H}\,\tilde{\Psi}(\underline{R})}{\tilde{\Psi}(\underline{R})}\, W(\underline{R})\, d\underline{R}$$

with
$$W(\underline{R}) = \frac{\int|\Psi|^2 d\underline{R}\,\tilde{\Psi}^2(\underline{R})}{\int|\tilde{\Psi}|^2 d\underline{R}\,\Psi^2(\underline{R})}$$

One samples from the probability contribution $\dfrac{\Psi^2(\underline{R})}{\int|\Psi|^2 d\underline{R}}$ and calculates averages from the function

$$(11)\quad f(\underline{R}) = \frac{\tilde{H}\,\tilde{\Psi}(\underline{R})}{\tilde{\Psi}(\underline{R})}\, W(\underline{R})$$

Parametrization of the Hamiltonian

Another possibility still to be explored is to parametrize the Hamiltonian

$$(12)\quad H(\lambda) = \frac{\hbar^2}{2m_4}\sum_{i=1}^{N}\Delta_i + \sum_{i<j}^{N} v(r_{ij}) - \lambda\,\frac{\hbar^2}{2m_4}\left(\frac{m_4}{m} - 1\right)\Delta_N .$$

The choice $\lambda = 0$ corresponds to the Hamiltonian of the background system consisting of N ^{4}He atoms, the choice $\lambda = 1$ corresponds to the Hamiltonian which exactly treats the ^{4}He-impurity problem. Using the Hellman-Feynman theorem the energy difference (7) can then be written as

$$(13)\quad \frac{E_i - E_p}{1/N} = \int_0^1 \frac{\partial E(\lambda')}{\partial \lambda} d\lambda' = \left(\frac{m_4}{m} - 1\right) \int_0^1 T(\lambda')\, d\lambda'$$

where $E(\lambda)$ represents the energy expectation value of the Hamiltonian (12) and $T(\lambda)$ the kinetic energy, respectively.

This method has the advantage that we only have to deal with kinetic energies and no delicate energy differences have to be calculated.

Baym's approximation

The choice $\lambda = 1$ in eq. (12) gives the Hamiltonian

$$(14)\quad H = \frac{\hbar^2}{2m_4} \sum_{i=1}^{N} \Delta_i + \sum_{i<j}^{N} v(r_{ij}) - \frac{\hbar^2}{2m_4} \left(\frac{m_4}{m} - 1\right) \Delta_N$$

If the mass ratio $\frac{m_4}{m}$ is close to unity we expect that the approximation $u_{11}(r) = u_{12}(r)$ in evaluating the energy expectation value is very good. This assumption leads to Baym's formula[9] for the difference of the chemical potentials

$$(15)\quad \mu_F = T\left(\frac{m_4}{m} - 1\right), \text{ where}$$

T is the kinetic energy p.p. of the pure ^{4}He system.

^{3}He - ^{4}He impurity problem

Table 1:

Approximation	Baym	state dependent calculation	interaction
PPA/HNC/O	5.04 (0.65)	4.92 (0.53)	LJ
GFMC	4.54 (0.15)	n.a.	LJ
PPA/HNC/O	5.38 (0.99)	5.20 (0.81)	HFDHE 2
GFMC	4.67 (0.28)	n.a.	HFDHE 2

In Table 1 we have listed in various approximations the difference of the chemical potentials and the discrepancy from experiment associated with the corresponding approximation for the LJ and the HFDHE 2 He-He-interactions.[10] (All energies are in K)

Column 1 gives the difference of the chemical potentials adopting Baym's formula and kinetic energies derived from various theoretical calculations. Column 2 shows the result of a state dependent PPA/HNC/O calculation, which assumes different correlations between the ^{4}He and the ^{3}He particles. State dependent correlations in the wave function (2) are responsible for lowering the difference of the chemical potentials by 0.12 K for the LJ-interaction and by 0.18 K for the HFDHF 2-interaction. It is certainly plausible to expect an energy shift of the same order upon performing a GFMC calculation with a Hamiltonian (1), which exactly treats the ^{3}He-^{4}He impurity problem. If this were true one would obtain a chemical potential difference within about 1% of the experimental value.

Xenon-^{4}He impurity problem

Although Baym's formula (15) might be generalized for nonisotopic systems by adding the term

$$(16)\quad \langle v_{12}\rangle - \langle v_{11}\rangle = \frac{\rho}{2}\int (v_{12}(r) - v_{11}(r))\, g_{11}(r)\, d\underline{r},$$

resulting from the fact that the interaction potentials are different, it will be a rather poor approximation if there is a large difference in the core sizes and perhaps the well depths, too. Consequently the pseudopotentials $u_{11}(r)$ and $u_{12}(r)$ have to be chosen differently. Moreover, since the mass ratio of a ^{4}He and Xe atom is 4:132 the Xe atom will almost behave like a classical particle. Thus this system will show effects quite different from the ^{3}He-^{4}He impurity problem. As a first step, we adopt the standard McMillan choice of the pseudopotentials

$$(17)\qquad u_{ij}(r) = \left(\frac{b_{ij}}{r}\right)^5, \quad ij = 11,12$$

for the wave function with b_{ij} optimized within the HNC/O approximation.

Alternatively we employ more refined pseudopotentials, taken from functional optimization of the Jastrow wave function (2) by the two component Paired Phonon-Analysis (PPA) method,[11] where one solves the extremum condition

$$(18)\qquad 0 = \frac{\delta}{\delta u_{ij}} \frac{\langle\Psi|H|\Psi\rangle}{\langle\Psi|\Psi\rangle}$$

in the HNC/O approximation in order to find the best variational $u_{ij}(r)$'s. As interaction potentials we adopt the widely used

HFDHE2 interatomic potential of Aziz et.al. for the ^4He-^4He interaction, whereas for the Xe-^4He interaction, which is not as well known, we follow a suggestion by Buck[12], which is of LJ 6-12 form with the parameters:

$$\varepsilon_{12} = 25.18\text{K} \text{ and } \sigma_{12} = 3.697\text{Å}$$

In table 2 we have listed the numerical results on the difference of the chemical potentials of the Xe-impurity in the background for two choices of pseudopotentials. For comparison $\left.\frac{\partial E}{\partial x}\right|_{x=0}$ is given in the PPA/HNC/O approximation for an infinite system and the corresponding Monte Carlo quantity $\left.\frac{\Delta E}{\Delta x}\right|_{x=\frac{1}{N}}$ has been calculated for 63 ^4He and one Xe atom for two choices of pseudopotentials. As found in the Monte Carlo treatment of the pure ^4He system there are only very small size effects[13]. We could find no essential differences between a 32 and a 108 particle system.

Table 2 Chemical potential difference for a Xe-impurity in ^4He in various approximations (in Å)

$\mu_{ij}(r)$ $ij=11,12$	$\left.\frac{\partial \langle H\rangle}{\delta x}\right\|_{x=0}$	$\frac{E_i-E_p}{1/N}$
	HNC/O	Variational MC
McMillan	-212.5	-230.0 ± 5.5
PPA	-287.0	-290.0 ± 7.0

Figs. 1 and 2 display the radial distribution functions $g_{12}(r)$ and the associated structure functions $S_{12}(k)$ for the McMillan choice and the PPA choice respectively. The ^4He-Xe impurity problem is only poorly described by the McMillan choice, Eq. (13), which is not flexible enough to take care of the core and the considerably large well depth of the Xe-^4He interaction. The $g_{12}(r)$ for the McMillan form shows no more structure than the well known $g_{11}(r)$ of the ^4He background, corresponding to the fact that this functional form is only able to take care of the core. This is also reflected in the energies of the optimal PPA/HNC/O choice of the pseudopotentials lowering the total energy considerably about 25 % (Table 2).

The strong oscillatory behavior of the optimal radial distribution function resulting from the immense attraction felt by the ^4He-atoms causes large cancellations within and among the elementary diagrams neglected in the HNC/O results. This indicates that the accuracy of the HNC/O approximation for the optimal Jastrow results is excellent.

The results on the mixed structure function also agree quite well with the Monte Carlo results except at smaller wave numbers due to the finite box size of 15A for 64 particles adopted in the Monte Carlo treatment.

The pronounced peak in the optimal pseudopotential $u_{12}(r)$ [13] (figure 3) and the pronounced shell structure 3 around the impurity might be an indication that the simple Jastrow result does not represent the correct ground state wave function for the Xe-^{4}He impurity system. We find about 14-15 Helium atoms in the first shell around the Xe-atom suggesting that there are strong angular correlations among the ^{4}He-atoms in the vicinity of the Xe impurity.

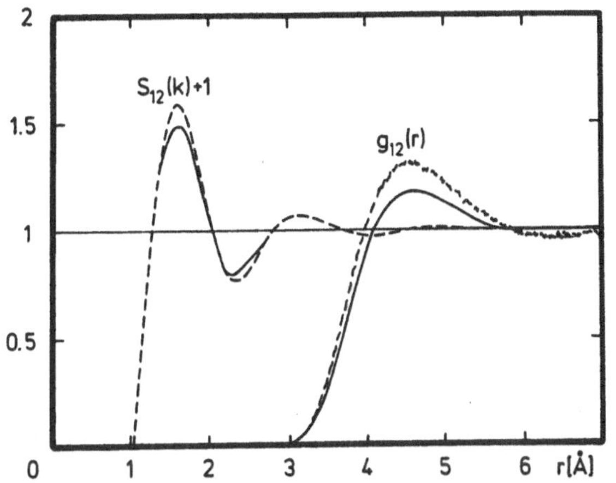

Fig.1: Radial distribution function $g_{12}(r)$ and the associated structure function for the McMillan choice at density ρ=0.02185 $Å^{-3}$ for a Xe-impurity in ^{4}He

Fig.2 Radial distribution function $g_{12}(r)$ and the associated structure function for the PPA choice at density ρ =0.02185 $Å^{-3}$ for a Xe-impurity in ^{4}He

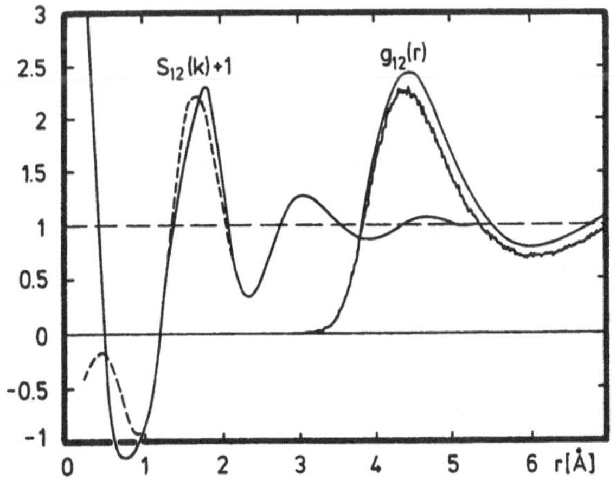

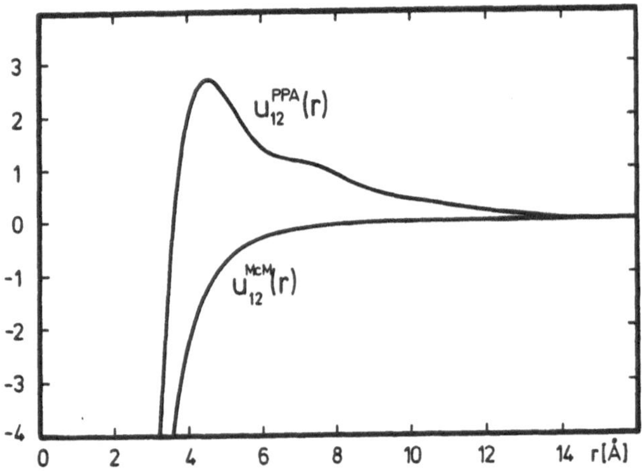

Fig.3:
Jastrow pseudopotential $U_{12}(r)$ for the PPA- and McMillan choice at density $\rho=0.02185\ A^{-3}$ for a Xe-impurity in ^4He

Only a Green'sfunction Monte Carlo analysis will be able to determine the importance and influence of such angular dependences in the exact ground state.

I wish to thank G.V.Chester,J.Gspann,M.H.Kalos,M.D.Miller, M.L.Ristig and P.A.Whitlock for discussions and valuable information. This work has been supported by the US Dept. of Energy under contract DE-AC 0276 ERO-3077 and by the NSF under grant DMR-77-18329.

1. A.L. Fetter, The Physics of Liquid and Solid Helium, ed.by K.H. Bennemann and J.B. Ketterson (Wiley, New York, 1976).
2. G. Baym and C. Pethick, The Physics of Liquid and Solid Helium, ed. by K.H. Bennemann and J.B. Ketterson (Wiley, New York, 1976).
3. K.E. Kürten and M.L. Ristig, Phys.Rev. B 27, 9 (1983).
4. M.D. Miller,Phys.Rev. B 18, 4730 (1978).
5. M.D. Miller, Ann.Phys. (N.Y.) 127, 367 (1980).
6. M.H. Kalos, M.A. Lee, P.A. Whitlock and G.V. Chester, Phys.Rev. B 24, 115 (1981).
7. M.A. Lee, K.E. Schmidt, M.H. Kalos and G.V. Chester, Phys.Letters 46, 11, 728 (1981).
8. M. Metropolis, A.W. Rosenbluth, M.N. Rosenbluth, G.H. Teller, and E. Teller, J. Chem. Phys. 21, 1087 (1953).
9. G. Baym, Phys.Rev. Letters 17, 952 (1966).
10. R.A. Aziz, V.P.S. Nain, J.S. Carley, W.L. Taylor and G.T. McConville, J.Chem.Phys. 70, 4330 (1979).
11. K.E. Kürten and C.E. Campbell, Phys.Rev. B 26, 124 (1982).
12. W. Buck, Adv.Chem.Phys. 30, 368 (1975).
13. E. Feenberg, Low Temp.Phys. 16, 112, 125 (1974).

DENSITY-FLUCTUATION SPECTRA OF ^{3}He-HeII MIXTURES AT T=0 K

A. Szprynger

Institute of Low Temperature and Structure Research, Polish Academy of Sciences, 50-950 Wrocław, P.O. Box 937, Poland

I. Introduction

The density excitation spectrum of a dilute mixture of ^{3}He in HeII is an interesting experimental problem and a challenge for many-body theories of strongly interacting fermions and bosons. Since Landau and Pomeranchuk /1/ proposed for a quasiparticle moving in HeII a parabolic dispersion with an enlarged effective mass m^* reflecting the HeII backflow much effort has been invested to calculate m^* /2-4/ and to determine the dispersion outside the small-momentum region /4-10/. Using the method developed by Götze and Lücke /11/ for pure ^{4}He we demonstrate how to solve these problems within a mode-coupling theory.

The presence of macroscopically many ^{3}He atoms changes the ^{4}He density-fluctuation spectrum. The few theoretical attempts /10,12/ were focused on evaluating roton energy shifts produced by hybridization with ^{3}He quasiparticle-quasihole excitations. Common to them is that the shift of the ^{4}He single-mode excitation energy is due to the repulsion of ^{4}He and ^{3}He levels; i.e. it is upwards to the left of the crossing where $\varepsilon_4 > \varepsilon_3$ and downwards to the right where $\varepsilon_4 < \varepsilon_3$. This disagrees, at least partly, with neutron scattering data /13,14/ since the more important effect of the changed ^{4}He structure in the mixture on the restoring force against a ^{4}He density fluctuation has been neglected. The structurally induced frequency shifts turn out to have the opposite sign of the level repulsion and are mostly larger than the level repulsion /15/. We evaluate these and other features of the density-fluctuation spectra $S_{ij}(k,\omega)$ $(i,j=3,4)$ for a mixture of concentration $x=N_3/(N_3+N_4)$.

II. Density response of ^{3}He-^{4}He mixtures

In order to describe dynamics of the particle densities $\varrho_i(k,t)$ we use the representation /16/ expressing the response-function matrix

$$\chi(k,z) = -\left[z^2-\Omega^2(k)+z\Sigma(k,z)\right]^{-1}k^2m^{-1} \tag{1}$$

in terms of self-energies /$(A|B)=\chi_{AB}(k)$ is a scalar product of A,B/:

$$\Sigma_{ij}(k,z) = \left(\rho_i(k)\mathcal{L}^2\, Q|(Q\mathcal{L}Q-z)^{-1}|Q\mathcal{L}^2\rho_j(k)\right)m_j/k^2 \tag{2}$$

and characteristic frequencies of static restoring forces:

$$\Omega^2(k) = k^2m^{-1}\chi^{-1}(k) = \frac{1}{1-W^2(k)}\begin{pmatrix}\omega_3^2(k); & -\sqrt{\frac{4}{3}}\,W(k)\omega_3(k)\omega_4(k)\\ -\sqrt{\frac{3}{4}}\,W(k)\omega_3(k)\omega_4(k); & \omega_4^2(k)\end{pmatrix} . \tag{3}$$

The diagonal matrix of masses m_3, m_4 is denoted by m. The projector Q is orthogonal to densities $\rho_i(k)$ and longitudinal currents $\mathcal{L}\rho_i(k)$. The restoring forces of ^{3}He and ^{4}He density fluctuations $\omega_i^2(k) = k^2/m_i\chi_{ii}(k)$ are coupled via the off-diagonal elements of $\Omega^2(k)$ with the strength

$$W(k) = \chi_{34}(k)/\sqrt{\chi_{33}(k)\chi_{44}(k)} \;, \tag{4}$$

proportional to the square-root of the mean densities $\sqrt{n_3n_4}$.
The spectral functions $\chi''_{ij}(k,\omega)$ related via the fluctuation-dissipation theorem to the dynamical structure factors $S_{ij}(k,\omega)$ fulfil the sum rules

$$\int d\omega\; \chi''_{ij}(k,\omega)\omega/\pi = \delta_{ij}k^2/m_i \;, \qquad \int d\omega\, \chi''_{ij}(k,\omega)/\pi\omega = \chi_{ij}(k) \;, \qquad \frac{1}{\pi}\int_0^\infty d\omega\; \chi''_{ij}(k,\omega) = s_{ij}(k) \;. \tag{5}$$

The total neutron scattering intensity neglecting small incoherent contribution is proportional to

$$S_{tot}(k,\omega) = (1-x)\sigma_4\; S_{44}(k,\omega)+2\sqrt{x(1-x)\sigma_3\sigma_4}\; S_{34}(k,\omega)+x\sigma_3\; S_{33}(k,\omega), \tag{6}$$

where σ_i are coherent cross sections.

III. Single ^{3}He atom in HeII bath

In the limit of zero ^{3}He concentration the spectral function $\chi^{o\,''}_{33}(k,\omega)$ is given by /4/

$$\chi^{o\,''}_{33}(k,\omega) = \frac{\omega\Sigma^{o\,''}_{33}(k,\omega)\; k^2/m_3}{\left[\omega^2-\Omega^{o2}_{33}(k)+\omega\Sigma^{o\,'}_{33}(k,\omega)\right]^2+\left[\omega\Sigma^{o\,''}_{33}(k,\omega)\right]^2} . \tag{7}$$

All quantities referring to this limit are marked by a superscript "o". The undamped quasiparticle dispersion is determined by the equations

$$\varepsilon_3^{o2}(k)-\Omega^{o2}_{33}(k)+\varepsilon_3^o(k)\Sigma^{o\,'}_{33}(k,\varepsilon_3^o(k)) = 0, \qquad \Sigma^{o\,''}_{33}(k,\varepsilon_3^o(k)) = 0 \;. \tag{8}$$

Thus the spectral function (7) consists of the δ-function at $\varepsilon_3^o(k)$

$$\chi_{33}^{o''}(k,\omega) = \pi f_3^o(k)\ \delta(\omega-\varepsilon_3^o(k))+\chi_{33}^{o''}(k,\omega)|_{con} \tag{9}$$

with the spectral weight

$$f_3^o(k) = \left[1+\Omega_{33}^{o2}(k)/\omega^2+\partial\Sigma_{33}^{o'}(k,\omega)/\partial\omega\right]^{-1}_{\omega=\varepsilon_3^o(k)}\ k^2/m_3\varepsilon_3^o(k)\ , \tag{10}$$

and the broad continuum part $\chi_{33}^{o''}(k,\omega)|_{con}$.

The imaginary part of the self-energy describes processes whereby a ^{3}He atom with energy ω and wave vector k transfers energy and momentum to the HeII bath. We take into account only dominant processes in which ^{3}He mode decays into uncorrelated HeII and ^{3}He density fluctuations. Then Eq. (2) leads to

$$\omega\Sigma_{33}^{o''}(k,\omega) = (m_3/2(2\pi)^4k^2n_4)\int d\underline{p}\int d\varepsilon\ V(\underline{k},\underline{p})^2\left[\chi_{44}^{o''}(p,\varepsilon)\chi_{33}^{o''}(|\underline{k}-\underline{p}|,\omega-\varepsilon)+\right.$$
$$\left.+(33\leftrightarrow 44)\right]\left[\mathrm{sgn}(\varepsilon)+\mathrm{sgn}(\omega-\varepsilon)\right] \tag{11}$$

Owing to a lowest-order correlation approximation used to calculate the vertex $V(\underline{k},\underline{p})$ it can be expressed in terms of the static structure factors $s_{44}^o(k)$, $s_{34}^o(k)$ and the bare masses m_i /4/. The real part $\Sigma_{33}^{o'}(k,\omega)$ renormalizing via Eq. (8) the free dispersion $\varepsilon_3^{oF}(k)=\Omega_{33}^{oF}(k)$ can be found from the Kramers-Kronig relation

$$\Sigma'(k,\omega) = (2\omega/\pi)\int_0^\infty d\varepsilon\ \Sigma''(k,\varepsilon)/(\varepsilon^2-\omega^2)\ . \tag{12}$$

To solve numerically the closed set of equations (5,7,11,12) one needs: 1. The ^{4}He density spectrum $\chi_{44}^{o''}(k,\omega)$ taken from /11/; 2. The structure factors $s_{44}^o(k)$ /17/ and $s_{34}^o(k)$ /18/; 3. The bare masses m_i and particle density n_4^o.

Below a critical wave number k_c which is close to the roton momentum k_o the spectrum $\chi_{33}^{o''}(k,\omega)$ is of the form (9). In the continuum part two peaks are evident. For small k the quasiparticle dispersion $k^2/2m^*$ ($m^*=2.35\ m_3$) is parabolic in agreement with the Landau-Pomeranchuk prediction /1/. With increasing momentum $\varepsilon_3^o(k)$ is subjected to the level repulsion from the pair continuum and lowers below the parabola. Near the roton minimum the dispersion slightly bends over (without forming a minimum), enhancing the density of quasiparticle states which agrees with the earlier experimental results /8,19/. A comparison of our $\varepsilon_3^o(k)$ with the neutron scattering data /14/ and high-precission specific heat measurements /9/ allows to conclude that the self-energy Eq. (11) expresses rather accurately polarization of the HeII bath caused by the ^{3}He atom which surrounds itself by a "cloud" of density excitations.

The undamped ^{3}He dispersion terminates at k_c where $\varepsilon_3^o(k)$ approaches the treshold curve u(k) for roton emission. Above k_c $\Sigma_{33}^{o''}(k,\varepsilon_3^o(k)) \neq 0$ and the dispersion continues as the lowest resonance in $\chi_{33}^{o''}(k,\omega)|_{con}$. The contribution to the self-energy due to this quasiparticle decay has a square-root singularity /20/ $\omega\Sigma_{33}^{o''}(k,\omega)|_{th} \sim \sqrt{u(k)-\omega}\,\Theta(u(k)-\omega)$. Eq. (8) leads then to conclussions that $\varepsilon_3^o(k)$ merges tangentially with the threshold $u(k)$ and the excitation strength $f_3^o(k)$ drops abruptly to zero when k approaches k_c.

IV. ^{4}He single-mode excitations and quasiparticle-quasihole continuum

In a mixture /our numerical calculations are performed for a 6% mixture/ the quasiparticles modify the phonon-roton spectrum e.g. generating an energy shift $\delta\varepsilon_4(k)$. To describe this shift properly it is necessary to include structural changes of ^{4}He due to the presence of ^{3}He component. We propose other approach /15/ than the self-consistent treatment in Sec.III. On account of the dilution the ^{3}He subsystem can be described as an ideal Fermi gas (FG) of quasiparticles with a mean effective mass $\bar{m}^* = 2.65\ m_3$. So we write

$$\omega_3^2(k) \simeq \omega_{FG}^2(k) = k^2/\bar{m}^*\chi_{FG}(k)\ , \qquad s_{33}(k) \simeq s_{FG}(k)\ ,$$

$$z^2+z\Sigma_{33}(k,z) \simeq \omega_{FG}^2(k)\left[1+\pi_{FG}^{-1}(k,z)\right] \quad . \tag{13}$$

The reduced polarization operator $\pi_{FG} = -\chi_{FG}(k,z)/\chi_{FG}(k)$ describes quasiparticle-quasihole excitations.

Since the roton decay rate measured /21/ for a 6% mixture at T = 0.25 K is only about 0.06 K we neglect $\Sigma_{44}''(k,\omega)$ as well as the off-diagonal terms of the self-energy matrix $\Sigma_{34}(k,z)$. Thus the only damping mechanism we admit here is decay of density fluctuations into quasiparticle-quasihole continuum determined by $\pi_{FG}''(k,\omega)$ vanishing outside the band

$$k^2/2\bar{m}^*-kv_F \leq \omega \leq k^2/2\bar{m}^*+kv_F \quad . \tag{14}$$

The real part $\Sigma_{44}'(k,\omega)$ is accounted for by replacing the bare frequency $\omega_4(k)$ by a renormalized energy $\varepsilon_4(k)$. In pure ^{4}He $\omega_4^o(k)$ and $\varepsilon_4^o(k)$ are related by Eq.(8) with index "4" substituted for "3" /$\Omega_{44}^o(k) = \omega_4^o(k)$/. In the mixture the bare restoring force $\omega_4^o(k)$ is shifted to $\omega_4 = \omega_4^o + d\omega_4$. If we ignore a change in $\Sigma_{44}'(k,\omega)$ the corresponding shift in energy is of the form

$$d\varepsilon_4(k) = (\omega_4^o(k)/\omega_F^o(k))f_4^o(k)\ d\omega_4(k) \quad , \tag{15}$$

where $\omega_F^o(k)=k^2/2m_4 s_{44}^o(k)$ is the Feynman energy. For $\varepsilon_4^o(k)$, $\omega_4^o(k)$, and $f_4^o(k)$ we use the results obtained by Götze and Lücke /11/. Thus the restoring force is replaced by the renormalized energy

$$\omega_4(k) \longrightarrow \varepsilon_4(k) = \varepsilon_4^o(k)+\left[\omega_4(k)/\omega_4^o(k)-1\right] f_4^o(k)\omega_4^{o2}(k)/\omega_F^o(k) \tag{16}$$

with the quotient $\omega_4(k)/\omega_4^0(k) \simeq s_{44}^0/s_{44}$ evaluated within a Feynman approximation. This faithfully reflects the structural effects that are mainly responsible for the change of the restoring forces. The generalized Feynman model /15,22/ assuming $\Sigma(k,z) = 0$ yields a simple relation $\chi(k) = 4s(k)ms(k)/k^2$ between static-susceptibility and structure-factor matrices which determines the coupling function $W(k)$ Eq. (4). The quotient $s_{44}(k)/s_{44}^0(k)$ was extracted within a lowest order concentration approximation from the total x-ray scattering intensity $s_{tot}(k)$ /23/ and $s_{34}(k)$ was taken from /18/ with some modifications in the region of smaller k.

Inserting the above $\Omega_{ij}(k)$ and $\Sigma_{ij}(k,z)$ into the density response matrix Eq.(1) one gets a formula of a RPA type for $\chi_{33}(k,z)$

$$\chi_{33}(k,z) = (\bar{m}^*/m_3)\chi_{FG}(k,z)\left[1-V_{eff}(k,z)\Pi_{FG}(k,z)\right]^{-1} \tag{17}$$

with a k,z dependent effective potential caused by the coupling $W(k)$

$$V_{eff}(k,z) = \gamma^2(k)z^2\left[z^2-\tilde{\varepsilon}_4^2(k)\right]^{-1} ,$$

$$\gamma^2(k) = W^2(k)/(1-W^2(k)) \sim n_3 n_4 . \tag{18}$$

The potential is induced and mediated by the exchange of undamped ^{4}He excitations with energy $\tilde{\varepsilon}_4(k) = \varepsilon_4(k)/\sqrt{1-W^2(k)}$ and is an analogous to the Fröhlich phonon-exchange potential between electrons /24/.
The ^{4}He density response takes the form

$$\chi_{44}(k,z) = -(k^2/m_4)\left[z^2-\tilde{\varepsilon}_4^2(k)\left[1+\gamma^2(k)\Pi(k,z)\right]\right]^{-1} . \tag{19}$$

The polarization potential

$$\Pi(k,z) = \Pi_{FG}(k,z)\left[1-\gamma^2(k)\Pi_{FG}(k,z)\right]^{-1} , \tag{20}$$

appears in (19) due to the coupling $W(k)$ and reflects a possibility of ^{4}He density fluctuation decay into quasiparticle-quasihole continuum.

$$\chi_{34}(k,z) = -\sqrt{3/4}\,\gamma(k)\chi_{33}(k,z)\omega_{FG}(k)\tilde{\varepsilon}_4(k)\left[z^2-\tilde{\varepsilon}_4^2(k)\right]^{-1} \tag{21}$$

is "almost" a product of $\chi_{33}(k,z)$ and $\chi_{44}(k,z)$.

In our model, all density fluctuations are damped only within the band (14) /thin lines in Fig.1/. Outside the band all response spectra $\chi''_{ij}(k,\omega)$ display a delta function spike at the frequency where $\mathrm{Det}\left[\chi(k,z)\right]^{-1} = 0$ /dots outside the band/. While the δ function carries practically the total spectral weight of $\chi''_{44}(k,\omega)$, the weight of $\chi''_{33}(k,\omega)$ is negligible outside the band, so we drew open triangles denoting the upper peak positions of $\chi''_{33}(k,\omega)$ only within the band. The lower peak /closed triangles / of $\chi''_{33}(k,\omega)$ carries for small k almost the total spectral weight of ^{3}He density fluctuations. Approaching the roton momentum its dispersion bends over, and more and more weight is

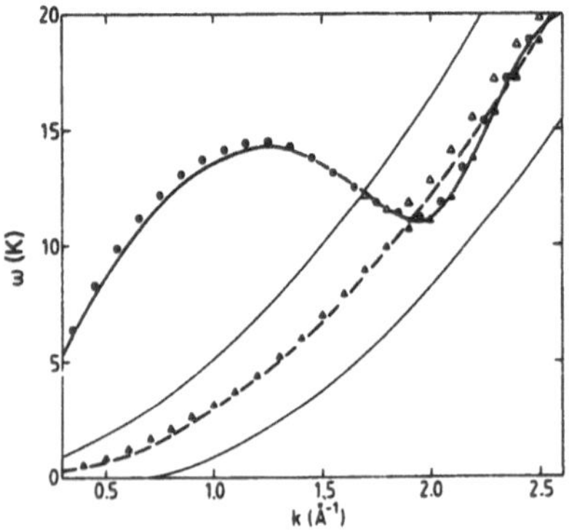

Fig.1 Peak position (closed circles) of the total neutron scattering intensity /Eq.(6)/ compared with $\varepsilon_4^o(k)$ of Ref. /11/ (full thick curve).

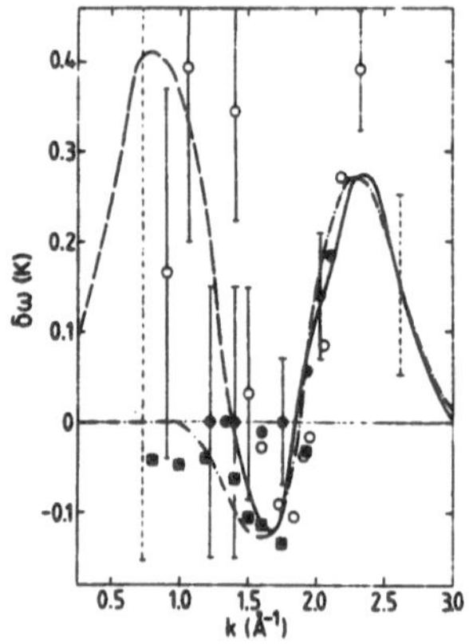

Fig.2 Shift of the peak position of $S_{tot}(k,\omega)$ with respect to $\varepsilon_4^o(k)$ (full curve and dashed curve) compared with neutron scattering results. The dash-dot curve denotes $d\varepsilon_4(k)$ /Eq.(15)/.

transferred to the upper peak. The upper (lower) peak remains always above (below) the energy $\tilde{\varepsilon}_4(k)$ where $\chi''_{33}(k,\omega)$ has a zero because of the diverging phonon-roton exchange potential (18). This peak splitting /25/ which is about 1.2 K around k_o, is due to the hybridization of a ^{3}He quasiparticle-quasihole excitation with a ^{4}He excitation.

The spectrum $\chi''_{44}(k,\omega)$ has practically only one peak, indicated (approximately) by solid circles in Fig.1. For all k shown there, the vertex $\gamma(k)$ for decay of ^{4}He density fluctuations into ^{3}He excitations is too small to induce an additional zero of the denominator in (19) at real frequency other than that at $\omega \simeq \tilde{\varepsilon}_4(k)$. Only for smaller k, where that zero is outside the band, does $\chi''_{44}(k,\omega)$ have a small side maximum in the band somewhat above its center. However, as soon as the former enters the band around k=1.7 Å^{-1} the side maximum disappears. The polarization potential $\gamma^2(k)\pi'(k,\omega)$ is too small to cause a detectable hybridization of a roton with the quasiparticle excitations /25/.

In Fig.2 the shift of the peak position of $S_{tot}(k,\omega)$ with respect to the single-mode excitation energy of pure ^{4}He is compared with neutron scattering data of Hilton et al. /14/ obtained for a 6% mixture at T=0.75 K (closed circles), T=0.6 K (squares), and of Rowe et al. /13/ (open circles) obtained for x=5% at T=1.6 K. Error bars of the experiments (solid vertical lines) and from uncertainties of input data entering theory (dashed vertical lines) are shown as well. The uncertainties of our input parameters - experimental error of $s_{44}(k)$ and an assumed uncertainty of $s_{34}(k)$ of about 50% - cause uncertainties of our theoretical results exceeding 100% for smaller wave numbers.

The energy shift is caused by various, partly competing effects (1)-(4) listed below in the order of their importance. The first two are of static and the remaining of dynamic origin.

(1) The first effect due to the different structure of ^4He in the mixture dominates the other three. In the range 1-2.2 $Å^{-1}$ the ratio $d\varepsilon_4(k)/d\omega_4(k)$ turns out to be almost k independent. Thus the wave number dependence of $d\varepsilon_4(k)$ (dash-dot line in Fig.2) is dictated by the relative difference $s^o_{44}(k)/s_{44}(k) - 1$ which changes sign around $k \simeq 1.9$ $Å^{-1}$. Above $k \simeq 1.5$ $Å^{-1}$ $d\varepsilon_4(k)$ is almost identical to the peak shift of $S_{tot}(k,\omega)$.

(2) The coupling $W(k)$ to the ^3He restoring force enhances $\varepsilon_4(k)$ towards $\tilde{\varepsilon}_4(k)$ thus inducing upwards energy shifts which are typically less than 0.05 K. However, below $k \simeq 1.25$ $Å^{-1}$, where $W^2(k)$ as well as $\varepsilon_4(k)$ is relatively large, this enhancement is large enough to dominate the other effects.

(3) The frequency $\tilde{\varepsilon}_4(k)$ is further renormalized by the polarization potential $\gamma^2(k)\Pi(k,z)$. Its real part causes a level repulsion between $\tilde{\varepsilon}_4(k)$ and the center frequency $k^2/2\bar{m}_3^*$ of the quasiparticle excitation band: For k smaller (larger) than about 1.9 $Å^{-1}$, where both levels cross, $\tilde{\varepsilon}_4(k)$ is larger (smaller) than $k^2/2\bar{m}_3^*$ and consequently the position of the main peak of $\chi''_{44}(k,\omega)$ is pushed towards a larger (smaller) frequency than $\tilde{\varepsilon}_4(k)$. Thus the energy shift induced by the repulsion between ^4He fluctuation energies, and ^3He quasiparticle excitation levels is opposite to the structurally induced shift $d\varepsilon_4(k)$. However, the repulsion effect is small - at most 0.06 K - compared to the effect (1).

(4) The last effect is due to the cross spectrum $\chi''_{34}(k,\omega)$ being positive below the frequency $\tilde{\varepsilon}_4(k)$ and negative above /25/. That causes $S_{tot}(k,\omega)$ to be slightly asymmetric and also to have a peak at a lower frequency than the peak position of $\chi''_{44}(k,\omega)$. The downward shift is quite small, mostly about 0.01 K with a maximum of 0.03 K at $k=2.2$ $Å^{-1}$.

Acknowledgment

This lecture is based upon work done together with M. Lücke and W. Götze.

References

/1/ L.D. Landau and I.Ya. Pomeranchuk, Dokl.Akad.Nauk USSR 59, 669 (1948); I.Ya. Pomeranchuk, Zh.Eksp.Teor.Fiz. 19, 42 (1949).
/2/ N.R. Brubaker, D.O. Edwards, R.E. Sarwinski, P. Seligmann, and R.A. Sherlock, Phys.Rev.Lett. 25, 715 (1970); J.Low Temp.Phys. 3, 619 (1970).
/3/ T.B. Davison and E. Feenberg, Phys.Rev. 178, 306 (1969); C.W. Woo, H. Tan, and W.E. Massey, Phys.Rev. 185, 287 (1969); V.R. Pandharipande and N. Itoh, Phys.Rev. A8, 2564 (1973).
/4/ W. Götze, M. Lücke, and A. Szprynger Phys.Rev. B19, 206 (1979).
/5/ R.N. Bhatt, Phys.Rev. B18, 2108 (1979).
/6/ M.J. Stephen and L. Mittag, Phys.Rev.Lett. 31, 923 (1973); C.M. Varma, Phys.Lett. 45A, 301 (1973).
/7/ B.N. Esel'son, V.A. Slyusarev, V.I. Sobolev, and M.A. Strzhemechnyi, JETP Lett. 21, 115 (1975).
/8/ R.B. Kummer, V. Narayanamurti, and R.C. Dynes, Phys.Rev. B16, 1046 (1977).
/9/ D.S. Greywall, Phys.Rev. B20, 2643 (1979).
/10/ J. Ruvalds, J. Slinkman, A.K. Rajagopal, and A. Bagchi, Phys.Rev. B16, 2047 (1977).
/11/ W. Götze and M. Lücke Phys.Rev. B13, 3825 (1976); M. Lücke, in Lecture Notes in Physics, edited by A. Pekalski and J. Przystawa (Springer, Berlin, 1980), Vol. 115.
/12/ D.L. Bartley, J.E. Robinson, and V.K. Wong, J.Low Temp.Phys. 12, 71 (1973); D.L. Bartley, V.K. Wong, and J.E. Robinson, ibid. 17, 551 (1974).
/13/ J.M. Rowe, D.L. Price, and G.E. Ostrowski, Phys.Rev. Lett. 31, 510 (1973).
/14/ P.A. Hilton, R. Scherm, and W.G. Stirling, J.Low Temp.Phys. 27, 851 (1977).
/15/ M. Lücke and A. Szprynger, Phys.Rev. B26, 1374 (1982).
/16/ H. Mori, Prog.Theor.Phys. 33, 423 (1965); 34, 399 (1965).
/17/ E.K. Achter and L. Meyer, Phys.Rev. 188, 291 (1969); R.B. Hallock, ibid. A5, 320 (1972).
/18/ W.E. Massey, C.W. Woo, and H.T. Tan, Phys.Rev. A1, 519 (1970).
/19/ B.N. Esel'son, Yu.Z. Kovdrya, and V.B. Shikin, Sov. Phys.JETP 32, 37 (1971); V.I. Sobolev and B.N. Esel'son, ibid. 33, 132 (1971).
/20/ L.P. Pitaevskij, Sov.Phys. JETP 9 830 (1959); 12, 155 (1961).
/21/ P.A. Hilton, R. Scherm, and G.W. Stirling (unpublished).
/22/ C.E. Campbell, J.Low Temp.Phys. 4, 433 (1971).
/23/ M. Suemitsu and Y. Sawada, Phys.Lett. 71A, 71 (1979).
/24/ N.W. Ashcroft and N.D. Mermin, Solid State Physics (Holt, Rinehart and Winston, New York, 1976).
/25/ M. Lücke and A. Szprynger, Physica 107B, 33 (1981).

QUANTUM-MECHANICAL CALCULATIONS OF THE PROPERTIES OF LIQUID He DROPLETS

Steven C. Pieper

Argonne National Laboratory, Argonne, Illinois 60439

We have made[1] calculations of the ground-state properties of droplets of liquid ^{4}He, and are presently involved in calculations for droplets of liquid ^{3}He. The already published[1] ^{4}He results will be summarized here; the ^{3}He results to be presented are preliminary.

The system we are considering consists of N atoms of He with a Hamiltonian given by

$$H = -\sum_{i=1}^{N} \frac{\hbar^2}{2m} \nabla_i^2 + \sum_{i<j}^{N} V(r_{ij}) + \frac{\hbar^2}{2Nm} \left(\sum_{i=1}^{N} \vec{\nabla}_i\right)^2 ,$$

where the last term removes the kinetic energy of the center of mass (if H is applied to a translationally invariant wave function, the last term is zero). The potential V is the HFDHE2 potential of Aziz, et al.[2] Two methods for finding the ground-state properties of this system are used in the present study: variational Monte Carlo (VMC) and Green's Function Monte Carlo (GFMC).

For the VMC calculations, the trial wave function contains one-, two- and three-atom correlation terms. The single atom correlation is the bound-state wave function of a Woods-Saxon potential; for the case of fermions, a Slater determinant of such wave functions is used. The two- and three-atom correlations have the same functional form as that used in the best variational calculations[3,4] of infinite liquid He, and do not depend on the positions of the atoms in the droplet but only on the distances between pairs of atoms. The parameters in the various correlations were separately varied for each droplet (value of N) considered. The Metropolis random walk is used to compute the expectation of H in the trial wave function.

Table I shows some of the computed ground-state energies for ^{4}He droplets. The statistical errors of the GFMC values are shown in parentheses; except for these errors, the values are exact solutions of the given Hamiltonian. As can be seen, the VMC values are some 3% to 4% higher than the GFMC values.

Table I. Energy per atom for droplets of ^{4}He

N	GFMC (K)	VMC (K)	Difference (K)	Difference %
8	-0.6165(6)	-0.597	.02	3.1
20	-1.627(3)	-1.570	.06	3.5
40	-2.487(3)	-2.396	.09	3.7
70	-3.12(4)	-3.02	.10	3.2
112	-3.60(1)	-3.52	.08	2.2
240	---	-4.19	---	---
728	---	-4.95	---	---
∞	-7.11(2)	-6.88	.23	3.2

Liquid-drop expansions of the droplet energies may be used to extrapolate to the infinite liquid. For example, a fit to the GFMC energies for $20 \leq N \leq 112$ gives (in K)

$$E/N = -7.02 + 18.8\, x - 11.2\, x^2 ,$$

where

$$x = N^{-1/3} .$$

The constant term is just the extrapolated value of the bulk energy per atom and should be compared with the experimental (and GFMC) value of -7.11K. The linear term is related to the surface tension of the liquid and gives

$$t = 0.30 \text{ K Å}^{-2}$$

compared to the experimental value of 0.27 K $Å^{-2}$. Finally the large quadratic term shows that the curvature energy cannot be neglected in liquid-drop expansions of these energies.

The density profiles of the ^4He droplets are shown in the figure. The GFMC and VMC profiles show no significant differences. We find that as the number of atoms in the droplet is increased, the central density of the droplet steadily increases until (for about 40 atoms) it reaches the density of the infinite liquid. Contrary to simple hydrodynamical predictions, the small droplets do not have central densities greater than that of the infinite liquid. Larger drops consist of a central region with the infinite liquid density and a surface whose thickness for drops of more than 100 atoms is independent of the number of atoms. From this asymptotic surface profile we predict that the surface thickness of the semi-infinite liquid is 7 Å. The surface structure is monotonic with no density overshoot.

It is interesting to note that although our VMC two- and three-atom correlations do not depend on the individual atomic positions, but only on the distances between pairs of atoms, the VMC wave functions do a good job of reproducing the exact GFMC energies and densities.

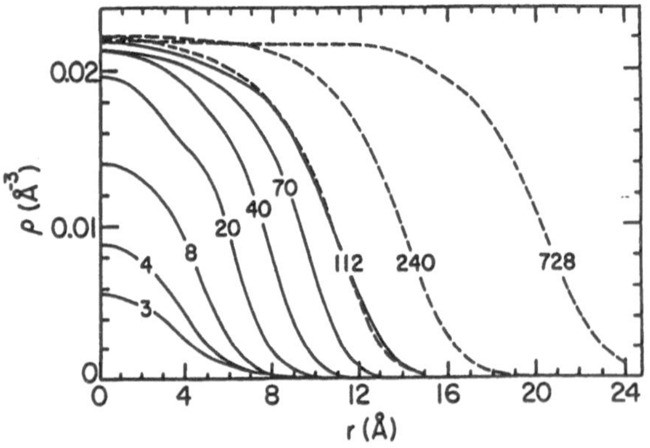

Density distributions of ^4He droplets computed by GFMC (solid lines) and VMC (dashed lines). The curves are labeled with the number of atoms.

The rms radii of the very small droplets decrease with increasing N; the minimum radius occurs for N = 6. For larger drops the radius increases approximately as $N^{1/3}$. Although the picture is somewhat confused by shell effects, nuclear rms radii show similar effects with a minimum occuring around N = 11. The radius parameter, defined as

$$r_0(N) = \sqrt{\tfrac{5}{3}}\ \langle r^2\rangle^{1/2}\ N^{1/3} \quad ,$$

may be extrapolated to $N = \infty$ to get a prediction of the saturation density of the infinite liquid. The result agrees to within 5% with the experimental value.

We turn now to preliminary results of calculations of ^3He droplets. So far we have used only VMC. The trial wave function has the same form as that used for ^4He except that the single-atom correlation factor is now a Slater determinant of bound-state solutions of the Schrödinger equation with a Woods-Saxon potential. Of course all of the variational parameters are readjusted to find the minimum ground-state energy.

Table II shows results for the cases of 18 and 40 atoms for which we have made calculations for the four possible permutations of mass (^4He, ^3He) and statistics (Bose, Fermi). We see that 1K to 1.5K of binding is lost due to the change from mass 4 to mass 3. This is of course due to the increased kinetic energy of the lighter atoms. The change in statistics adds about 1K to E/N for both droplets. These two effects add so that both droplets are unbound (or at least we have only a positive variational bound on the ground-state energy) for physical ^3He.

Table II - E/N for Systems of 18 and 40 Atoms (in K)

	18 Atoms		40 Atoms	
	Bosons	Fermions	Bosons	Fermions
^4He	-1.45	-0.50	-2.40	-1.30
^3He	-0.46	~ +0.5	-0.92	~ +0.1

Table III also shows results for droplets of 70 and 112 atoms; for these we find that the ^3He droplets are bound. The calculations for these droplets are, however, still unsatisfactory for the following reason. The density profiles show a large (6 Å radius for N = 70) central region of constant density but, as can be seen in Table III, the density is only 75% that of the infinite liquid. This does not seem reasonable and suggests that we must find better trial wave functions that will yield a higher central density.

Table III - VMC results for E/N and the central density of droplets of ^{4}He (Bose) and ^{3}He (Fermi).

N	E/N (K)			ρ(r=0) (Å^{-3})	
	^{4}He	^{3}He	Δ	^{4}He	^{3}He
18	-1.45	+0.5	2.0	0.018	---
40	-2.40	+0.1	2.5	0.022	---
70	-3.10	-0.10±.01	3.0	0.022	0.012
112	-3.52	-0.27±.01	3.25	0.023	0.013
∞	-6.9	-2.4	4.5	0.023	0.017

The Feynman-Cohen backflow has been found to be important in variational calculations[4] of infinite ^{3}He. In this case it lowers the ground-state energy by 0.5K per atom and slightly increases the density. We have made very preliminary calculations in which the increased kinetic energy due to the backflow correlation is ignored; thus we obtain a definite overestimate of the benefit of the backflow correlation. These calculations suggest that the upper bounds for 70 and 112 ^{3}He atoms given in Table III might be reduced by 50%. We are now working on a complete calculation including the backflow correlation.

Perhaps the most interesting experimental question raised by this work is that of the minimum number of ^{3}He atoms that will form a bound droplet (all ^{4}He clusters for $N>3$ are predicted to be bound). It appears that this number will be around 40.

The research reported here was done in collaboration with V. R. Pandharipande, J. G. Zabolitzky, R. B. Wiringa and U. Helmbrecht and was partially supported by the U. S. Department of Energy under Contracts No. W-31-109-ENG-38 and No. DE-AC02-76ER01198 and by the Deutsche Forschungsgemeinschaft.

References

1. V. R. Pandharipande, J. G. Zabolitzky, S. C. Pieper, R. B. Wiringa and U. Helmbrecht, Phys. Rev. Lett. 50, 1676 (1983).
2. R. A. Aziz, V. P. S. Nain, J. S. Carley, W. L. Taylor and G. T. McConville, J. Chem. Phys. 70, 4330 (1979).
3. Q. N. Usmani, S. Fantoni and V. R. Pandharipande, Phys. Rev. B26, 6123 (1982).
4. E. Manousakis, S. Fantoni, V. R. Pandharipande and Q. N. Usmani, "Microscopic Calculations of Normal and Polarized Liquid ^{3}He," submitted to Phys. Rev. B, (U. of Illinois preprint ILL-(TH)-83-11, 1983).

VARIATIONAL APPROACH TO
TWO-COMPONENT COULOMB LIQUIDS

P. Pietiläinen, L. Lantto and A. Kallio
University of Oulu
Department of Theoretical Physics
Linnanmaa, SF-90570 Oulu 57
Finland

I. Introduction

In this paper we intend to discuss the extension of the standard Jastrow approach to the case of multi-component fermion mixtures. In particular we shall present numerical results which characterize the ground state structure and energetics of a two-component plasma. The physical systems which we have in mind are, for example, the liquid phases of the highly compressed hydrogen and deuteron[1] at low temperature and electron-hole liquids[2] in semiconductors. The two-component plasma model to be considered here consists of equal number N of oppositely charged particles with masses m_1 and m_2 contained in a box of volume V with the mean density $\rho = N/V$. At T=0 this system is characterized by two parameters: the coupling strength $r_s = (3/4\pi\rho)^{1/3}/a_0$ and the mass-ratio m_2/m_1. Here, in order to allow direct comparisons with one-component plasma, we define the Bohr radius in terms of mass $m_1 : a_0 = \hbar^2/m_1 e^2$. Typical electron-hole liquids have $r_s \sim 1\text{-}4$ and mass-ratios rather close to unity, while for liquid metallic hydrogen $r_s \approx 1$ and $m_2/m_1 \sim 2 \times 10^3$.

Recently several authors have employed the Jastrow variational approach to study ground state properties of binary boson-boson and fermion-boson mixtures. Fabrocini and Polls[3] have generalized the Fantoni-Rosati[4] fermion hypernetted-chain equations (FHNC) to the case of fermion-boson mixture and they computed energetics of the ^{3}He-^{4}He mixture using simply parametrized Jastrow correlation factors. For boson mixtures, Chakraborty[5] has constructed Euler-Lagrange equations to optimize the trial wave function in conjunction with the HNC-approximation and Kürten and Campbell[6] have extended the paired-phonon treatment to multi-component boson mixtures. Fermion-fermion mixtures have been recently studied by Chakraborty et al.[7] who adopted the so-called Lado approximation to simplify the treatment of the antisymmetry. In the present work we continue along these lines to study the structure of the two-component plasma. Also, in order to obtain more reliable estimates of the ground state energy we shall generalize the fermion hypernetted-chain theory to the case of multi-component fermion mixtures and use this method to calculate the correlation energies of the two-component fermion plasma.

II. Ground state structure

The ground state wave function of the two-component fluid is approximated by the standard form

$$\psi = F_{11}\, F_{12}\, F_{22}\, \phi_1 \phi_2 \tag{1}$$

where each $F_{\alpha\beta}(\alpha,\beta = 1,2)$ is a product of pairwise correlation factors and ϕ_α is a plane wave Slater determinant. In the Lado approximation[8] one replaces the square of the Slater determinant by a Jastrow type product: $|\phi_\alpha|^2 \approx \exp\left[\sum_{i<j} u_{0\alpha}(r_{ij})\right]$ and chooses u_0 such that the boson HNC approximation produces the ideal fermion gas pair-distribution function. Then the exchange-correlation energy can be written as

$$\begin{aligned} E^L_{xc} &= \sum_{\alpha\beta} \frac{1}{2} \int \rho\, d\bar{r} \left\{ g_{\alpha\beta} v_{\alpha\beta} + \frac{\hbar^2}{8\mu_{\alpha\beta}} \nabla g_{\alpha\beta} \cdot \nabla \left[\ln g_{\alpha\beta} - N^B_{\alpha\beta} \right] \right\} \\ &+ \sum_{\alpha} \frac{1}{2} \int \rho\, d\bar{r}\; \frac{\hbar^2}{4m_\alpha}\, g_{\alpha\alpha} \left[2\nabla^2 u_{0\alpha} + (\nabla u_{0\alpha})^2 \right] \\ &+ \sum_{\alpha} \int \rho^2 d\bar{r}\, d\bar{r}'\; \frac{\hbar^2}{4m_\alpha}\, g^{(3)}_{\alpha\alpha\alpha}(\bar{r},\bar{r}',\bar{r}-\bar{r}') \left[\nabla u_{0\alpha}(\bar{r}) \cdot \nabla u_{0\alpha}(\bar{r}') \right]. \end{aligned} \tag{2}$$

$N^B_{\alpha\beta}$ in the above expression are conveniently given in momentum space as matrix elements of

$$N^B(k) = S^{-1} - I + S - I \tag{3}$$

where the partial structure functions (matrix elements of S) are defined by

$$(S - I)_{\alpha\beta} = \int \rho\, d\bar{r}\; e^{i\bar{k}\cdot\bar{r}} \left[g_{\alpha\beta}(r) - 1 \right] . \tag{4}$$

If we neglect, for the moment, the three-body term on the third line of eq. (2) and make the remaining energy stationary with respect to small variations of $g_{\alpha\beta}$, we end up with

$$\left[-\frac{\hbar^2}{2\mu_{\alpha\beta}} \nabla^2 + v_{\alpha\beta}(r) + w^B_{\alpha\beta}(r) + w^x_{\alpha\beta}(r) \right] g^{1/2}_{\alpha\beta}(r) = 0. \tag{5}$$

Here $w^x_{12} = 0$ and $w^x_{\alpha\alpha}(r) = \frac{\hbar^2}{4m_\alpha}\left[2\nabla^2 u_0 + (\nabla u_0)^2\right]$ for $\alpha = 1,2$. In the present work we have equal number of both particle species, hence u_0 is the same for both species. The induced potentials $w^B_{\alpha\beta}$ depend on $g_{\alpha\beta}$ and they can be expressed as a k-space matrix

$$w^B(k) = -\frac{\hbar^2 k^2}{4} \left[MS + SM - 3M + S^{-1} MS^{-1} \right] \tag{6}$$

where $M = \mathrm{diag}\,(1/m_\alpha)$.

Let's assume that physically reasonable solutions to eqs. (5)-(6) exist. Then the small-k behaviour of these solutions may be derived by rewriting the eqs. (5)-(6) in the form

$$S(k) M^{-1} S(k) = \left\{ M + \frac{4}{\hbar^2 k^2} \left[v + R \right] \right\}^{-1} \tag{7}$$

with

$$R_{\alpha\beta}(r) = w^x_{\alpha\beta} + \frac{\hbar^2}{2\mu_{\alpha\beta}} \left(\frac{1}{2} \nabla^2 g_{\alpha\beta} - \frac{\nabla^2 g^{1/2}_{\alpha\beta}}{g^{1/2}_{\alpha\beta}}\right). \tag{8}$$

By solving eq. (7) for partial structure functions $S_{\alpha\beta}$ one can see that they vanish linearly all with the same slope: $S_{\alpha\beta}(k) = a\,k$, as $k \to 0$. The slope a is given in terms of solutions $g_{\alpha\beta}$ as

$$a = \left\{ \frac{m_1 + m_2}{4\hbar^2} \int \rho \, d\bar{r} \left(\sum_{\alpha\beta} R_{\alpha\beta}\right)\right\}^{-1/2} . \tag{9}$$

Further, the perfect screening condition assume the form: $S_{11}(k) + S_{22}(k) - 2\,S_{12}(k) = \hbar^2 k^2/2\mu_{12}\omega_p$, as $k \to 0$, where ω_p is the plasma frequency.

The numerical results of the present work are consistent with the above observations. These results are obtained by two different methods. The first one is based on a straightforward generalization of the linearization scheme introduced in ref. (9). In the second method[10)] the linearization is avoided by adopting an iterative procedure in which each iteration starts by calculation of $R_{\alpha\beta}$ using the given estimates of $g_{\alpha\beta}$ in eq. (8). Then $S_{\alpha\beta}$ and $w^B_{\alpha\beta}$ are evaluated from eqs. (7) and (6), respectively, and finally eq. (5) is numerically integrated to obtain improved $g_{\alpha\beta}$ .The converged solutions $g_{\alpha\beta}$(or $S_{\alpha\beta}$) of both methods agree within numerical accuracy. However, we were able to find properly converged solutions only for a rather limited range of parameters r_s and m_2/m_1 . For example, at $r_s = 1$ the maximum mass-ratio for which the solutions were found was about 25, while for $m_2 = m_1$ the largest stable r_s value was only slightly more than 2. The range of stability depends, of course, on the choice of induced potential $w^x_{\alpha\beta}$ and the numbers quoted here are based on the simplest Lado approximation, in which the variation of the three-body term on the third line of eq. (2) was neglected in the calculation of $w^x_{\alpha\beta}$. The inclusion of this term with a convolution approximation for $g^{(3)}$ extends the stability range to considerably larger r_s values but is not adequate to make the equations stable for the mass-ratio suitable for the liquid metallic hydrogen. We note, however, that although the Euler-Lagrange equations cannot be accurately solved outside the stability range, approximate solutions can still be constructed and used in the calculation of the energy. Our studies of the nature of the instabilities seem to indicate that they are caused by the fact that the energy becomes insensitive (flat) with respect to small variations of $g_{\alpha\beta}$. Thus the structure functions cannot be very accurately determined while reasonable estimates for the energy may still be obtained.

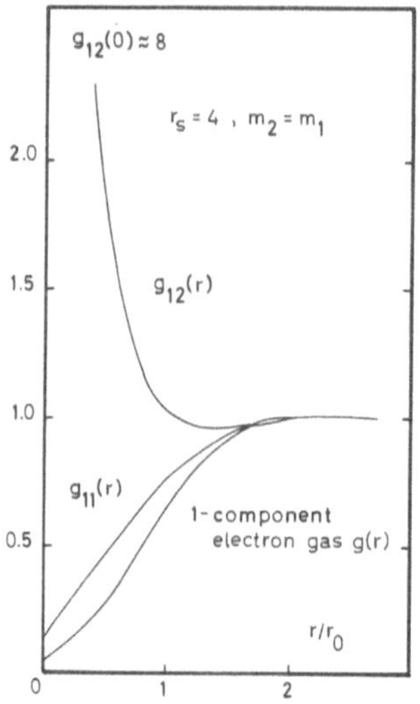

Fig. 1

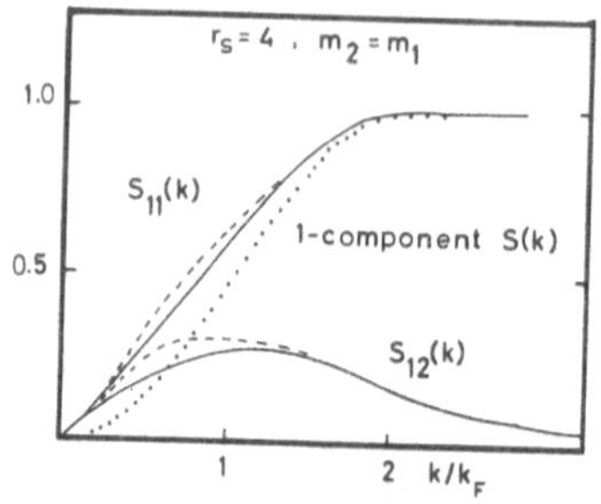

Fig. 2

Fig. 1 shows numerically calculated pair-correlation functions $g_{\alpha\beta}(r)$ for $r_s = 4$, $m_2 = m_1$. $g_{12}(r)$ has a strong peak at the origin, while $g_{11}(r)$ does not deviate very much from the corresponding one-component electron gas correlation function.

Fig. 2 shows partial structure functions at $r_s = 4$, $m_2 = m_1$, compared with the 1-component S(k). Solid lines represent solutions in the case when the above mentioned three-body term was included in $w^x_{\alpha\beta}$, while dashed lines correspond to "approximate solutions" constructed without the three-body term. The reason for showing these "approximate solutions" is that they give lower energy when it is computed in the FHNC-approximation to be discussed below.

In Fig. 3 $S_{12}(k)$ is shown as a function of the mass-ratio m_2/m_1. We observe that the maximum of $S_{12}(k)$ at $k \approx k_F$ first increases with increasing m_2/m_1, then at about $m_2/m_1 \approx 10$ it saturates and starts decreasing again for $m_2/m_1 \gtrsim 20$.

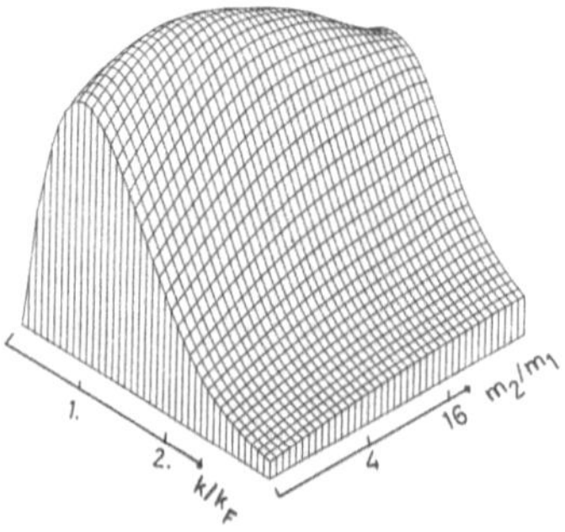

Fig. 3

III. Ground-state energy

In this section we intend to determine the ground-state energy using the pair-distribution functions $g_{\alpha\beta}$ of the previous section as an input to the FHNC calculation to be described below. The present FHNC scheme is actually a straightforward extension to multi-component systems of the method which earlier has been applied to various one-component systems.[11] In this scheme we treat $g_{\alpha\beta}$(or $S_{\alpha\beta}$) as input quantities and solve the FHNC equations "backwards" to obtain $f_{\alpha\beta}$ and other functions needed in the energy calculation. We use the standard notation[12]

$$g_{\alpha\beta}(r) = 1 + \Gamma^{dd}_{\alpha\beta} + \Gamma^{de}_{\alpha\beta} + \Gamma^{ed}_{\alpha\beta} + \Gamma^{ee}_{\alpha\beta} \tag{10}$$

where each $\Gamma^{(i)} = N^{(i)} + X^{(i)}$. Quantities N and X are related by (Örnstein-Zernike) convolution equations, i.e. each $N^{(i)}$ has the form

$$N^{(j)} = \sum_{\text{allowed } (i,i')} \Gamma^{(i)} \overset{*}{} X^{(i')} \quad , \quad (i,j = dd,de,ed,ee,cc) \tag{11}$$

where $*$ indicates an r-space convolution integration. Here "allowed" refers to the graphical rules of the FHNC expansion. In the multi-component case these equations may be conveniently expressed using matrix notation in k-space. Thus, for example, components $N^{de}_{\alpha\beta}$ of the matrix N^{de} obey equation

$$N^{de}(k) = \Gamma^{dd} X^{de} + \Gamma^{de} X^{de} + \Gamma^{dd} X^{ee} \tag{12}$$

and $N^{ed} = (N^{de})^T$. Similar relations hold for N^{dd}, N^{ee} and N^{cc}. In this work we solve these equations in the following way. Let's introduce matrices

$$\begin{cases} b(k) = (I + X^{ee})^{-1} \\ c(k) = (I - X^{de}) \end{cases} .$$

After some manipulation we may write:

$$\begin{cases} N^{de}(k) = Scb - Scbcb + b - I - X^{de} & (13a) \\ N^{ee}(k) = S - 2Scb + Scbcb - b + I - X^{ee} & (13b) \\ N^{cc}(k) = (X^{cc} X^{cc} - \ell/\nu\, I)(I - X^{cc})^{-1} & (13c) \end{cases}$$

where ℓ/ν is the step function $\ell/\nu = \theta(k-k_F)$. The direct correlation functions $X^{(i)}$, which appear on the r.h.s. of eqs. (13) are in turn calculated in r-space as

$$\begin{cases} X^{de}_{\alpha\beta}(r) = \Gamma^{dd}_{\alpha\beta} N^{de}_{\alpha\beta} & (14a) \\ X^{ee}_{\alpha\beta}(r) = \Gamma^{dd}_{\alpha\beta} N^{ee}_{\alpha\beta} + (1 + \Gamma^{dd}_{\alpha\beta})\left[N^{de}_{\alpha\beta} N^{ed}_{\alpha\beta} - \delta_{\alpha\beta}\nu(N^{cc}_{\alpha\beta} - \ell/\nu)^2\right] & (14b) \\ X^{cc}_{\alpha\beta}(r) = \delta_{\alpha\beta}\, \Gamma^{dd}_{\alpha\beta}\, (N^{cc}_{\alpha\beta} - \ell/\nu). & (14c) \end{cases}$$

Here $1 + \Gamma^{dd}_{\alpha\beta}(r) = g_{\alpha\beta}/\left[1 + N^{de}_{\alpha\beta} + N^{ed}_{\alpha\beta} + N^{ee}_{\alpha\beta} + N^{de}_{\alpha\beta} N^{ed}_{\alpha\beta} - \delta_{\alpha\beta}\nu(N^{cc}_{\alpha\beta} - \ell/\nu)^2\right]$. We solve the eqs. (13)-(14) by a straightforward iteration. All explicit elementary or bridge diagrams are neglected, but so-called "Fermi-cancellation" (FHNC/C) is imposed as proposed by Krotscheck.[13] Then we are ready to compute the energy using the following expression[14]

$$E_{xc} = \sum_{\alpha\beta} \frac{1}{2} \int \rho d\bar{r}\, g_{\alpha\beta}\, v_{\alpha\beta} + \frac{\hbar^2}{8\mu_{\alpha\beta}} g_{\alpha\beta} \cdot \nabla \left[\ell n\, (1 + \Gamma^{dd}_{\alpha\beta}) - N^{dd}_{\alpha\beta}\right]$$
$$- \sum_{\alpha} \frac{\hbar^2}{8m_\alpha} \int \rho d\bar{r}\, \Gamma^{dd} \left[\nabla^2 \ell^2/\nu - 2N^{cc}_{\alpha\alpha} \nabla^2 \ell\right]$$
$$- \sum_{\alpha} \frac{\hbar^2}{2m_\alpha} \frac{1}{\nu^2} \iint \rho^2 d\bar{r}\, d\bar{r}'\, \Gamma^{dd}_{\alpha\alpha}(\bar{r})\Gamma^{dd}_{\alpha\alpha}(\bar{r}')\Gamma^{cc}_{\alpha\alpha}(\bar{r}-\bar{r}')\nabla\ell(k_F\bar{r})\cdot\nabla\ell(k_F\bar{r}') \qquad (15)$$

The quantities $N^{dd}_{\alpha\beta}$ are computed using a k-space formula similar to eqs. (13a-c)

$$N^{dd}(k) = Scbcb + (Scb)^T - Scb - b - X^{dd} \qquad (16)$$

where $X^{dd}(k) = cbc^T - S^{-1}$. We note that in the case of Bose statistics both c and b become equal to unity and eq. (16) reduces to N^B of eq.(3). For consistency in the energy calculation we evaluate also Γ^{dd} in k-space using the fact that $\Gamma^{dd} = N^{dd} + X^{dd}$. The third line in eq. (15) represents a small three-body contribution.

In Table 1 we show the correlation energies of a model plasma ($m_2 = m_1$) computed with eq. (15). The input pair-correlation functions $g_{\alpha\beta}$ are (approximate) solutions to eqs. (5)-(6) with the simple Lado approximation for $w^x_{\alpha\beta}$. The choice of this form of $w^x_{\alpha\beta}$ is motivated by the experience that in the case of one-component electron gas, it very nearly optimizes the corresponding FHNC energy. The second solumn in Table 1 contains the correlation energies obtained via eq. (2) with the same input for $g_{\alpha\beta}$. For comparison we also show corresponding results of some earlier calculations.[15-16] Finally, the last column in Table 1 lists the calculated enhancement factors $g_{12}(0)$.

Table 1. Correlation energies for a plasma with $m_1 = m_2$. The energy unit is $1Ry = m_1e^4/2\hbar^2$. The columns labeled VBS, RPA and Hubbard are taken from ref. (16). The last column shows the enhancement factor $g_{12}(0)$.

r_s	Present FHNC	Present Lado	VBS	RPA	Hubbard	$g_{12}(r = 0)$
1	-.48	-.43	-.46	-.50	-.45	1.8
2	-.38	-.36	-.38	-.38	-.34	3.
3	-.33	-.32	-.33	-.32	-.29	5.
4	-.30	-.30	-.31	-.28	-.25	8.

In Fig. 4 we compare the correlation energies of the present work with several other calculations. For reference, the corresponding one-component results of the present FHNC method are compared with the exact Green function Monte Carlo results due to Ceperley and Alder[17] We note that our calculation overestimates the correlation energy at small r_s while in the most interesting region of $r_s \approx 3 - 4$ our results are slightly above those of GFMC. If the similar situation holds for the two-component system, as we may expect, then the close agreement between our results and those of ref. (16) (VBS) become even better. Thus a key result to emerge from the present work is that it seems to support the results at ref. (16) and hence it should remove the apparent controversy between the variational and other methods recently discussed in the literature[18] in the context of the ground-state energies of the electron-hole liquids.

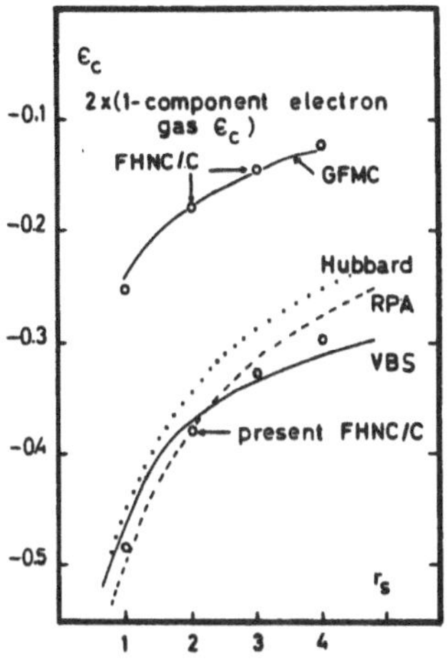

Fig. 4

References:

1) See e.g. N.W. Ashcroftt and J. Oliva, these proceedings.

2) For a review, see T.M. Rice, The Electron-Hole Liquid in Semiconductors: Theoretical Aspects, in Solid State Physics, vol. 32,p.1 (1977).

3) A. Fabrocini and A. Polls, Phys. Rev. B25, 4533 (1982).

4) S. Fantoni and S. Rosati, Nuovo Cimento A25, 593 (1975).

5) T. Chakraborty, Phys. Rev. B25, 3177 (1982).

6) K.E. Kürten and C.E. Campbell, Phys. Rev. B26, 124 (1982).

7) T. Chakraborty, A. Kallio, L. Lantto and P. Pietiläinen, Phys. Rev. B27, 3061 (1983); T. Chakraborty and P. Pietiläinen, Phys. Rev. Lett. 49, 1034 (1982).

8) J.G. Zabolitzky, Adv. Nucl. Phys. 12, 1 (1982); F. Lado, J. Chem. Phys. 47, 5369 (1967).

9) L.J. Lantto, A.D. Jackson and P.J. Siemens, Phys. Lett. 68B, 311 (1977).

10) P. Pietiläinen and A. Kallio, Phys. Rev. B27, 224 (1983).

11) L.J. Lantto and P.J. Siemens, Nucl. Phys. A317, 55 (1979); L.J. Lantto, Phys. Rev. B22, 1380 (1980).

12) J.W. Clark, in Progress in Particle and Nuclear Physics (Pergamon,

Oxford and New York 1979) vol. 2.
13) E. Krotscheck, Nucl. Phys. A317, 149 (1979).
14) S. Fantoni and S. Rosati, Phys. Lett. B84, 23 (1979).
15) W.F. Brinkman and T.M. Rice, Phys. Rev. B7, 1508 (1973).
16) P. Vashista, P. Bhattacharyya and K.S. Singwi, Phys. Rev. B10, 5108 (1974).
17) D.M. Ceperley and B.J. Alder, Phys. Rev. Lett. 45, 566 (1980).
18) P. Vashista, R.K. Kalia and K.S. Singwi, Phys. Rev. Lett. 50, 2036 (1983).

SPIN POLARIZED ^{3}He

C. Lhuillier
Laboratoire de Spectroscopie Hertzienne de l'E.N.S.
24, rue Lhomond
F - 75231 Paris Cedex 05

ABSTRACT

The effects of a strong nuclear polarization on the macroscopic properties of gaseous and liquid ^{3}He at low temperatures are discussed. They are all pure consequences of the atom indistinguishability.

In the gas phase, these effects can completely be calculated from first principles. We discuss the changes of the heat conduction and viscosity originating from the strong reduction of the effective interaction between polarized fermions and the existence of coupling terms between heat conduction and longitudinal spin diffusion. Another consequence of particle indistinguishability is the appearance of transverse spin waves in a dilute, highly polarized gas at low temperatures. In the degenerate liquid phase, the nuclear polarization has a double role : it modifies the effective interaction between atoms and also reduces the phase space available to the system. We give some illustrations of this last effect in a variational framework, and discuss the accuracy of a variational calculation of unpolarized and polarized ^{3}He. We finally give a brief description of the experiments actually in progress at the E.N.S.

1. INTRODUCTION

Spin polarized quantum fluids provide us with attractive physical systems that can exhibit many fascinating features in their equilibrium properties ([1]) (i.e. Bose-Einstein condensation in the gaseous phase of H↑, presumably new phase diagrams for the two Fermi systems D↑ and ^{3}He↑ and the various spin polarized mixtures of helium and hydrogen isotopes) as well as in their transport properties.

All these systems possess nuclear spin (I = 1/2 or 1) and the nuclear spin polarization -through its controlling of indistinguishability effects- can lead to a better understanding of the exact microscopic structure of these systems. In fact, the experimental situation is rather different with the hydrogen isotopes as compared to ^{3}He. Due to molecular recombination, the polarization of any sample of hydrogen or deuterium rapidly relaxes towards a very high value ([2]) and experiments at different nuclear polarizations may be rather difficult to perform ([3]). At variance, very long nuclear relaxation times can be achieved in the fluid phase of ^{3}He. Due to the weakness of the dipolar interaction, the relaxation time in the dilute phases (gaseous ^{3}He or low concentration of ^{3}He in liquid ^{4}He) is always controlled by the wall relaxation; the coating of these walls either with H_2 solid ([4])([5]) or with superfluid ^{4}He minimizes this extrinsic relaxation and can lead to relaxation times T_1 ranging from 10^3 s to 10^5 s ([6])([7]). In the dense phase, by some similar elimination of wall relaxation, long nuclear relaxation times (several minutes) have already been measured ([8]) and it seems well established that the intrinsic value of T_1 will remain very long even in highly polarized liquid ^{3}He ([9]), the only possible departure from the low magnetization behaviour being an eventual non exponential decay of the polarization. On a time scale much smaller than T_1 (but long enough to perform experiments !), the nuclear magnetization appears as a new thermodynamic variable that can be varied continuously from zero to one.

From a macroscopic point of view, one can be tempted to think of ^{3}He as a mixture of two chemical species (spin up and spin down) and use for this system the usual concepts of the thermodynamics of mixtures. It will be shown in the following that this point of view allows to describe a large body of interesting new phenomena but is too restrictive for a complete description of these new phases.

2. TRANSPORT IN DILUTE, NON DEGENERATE GASES

Several characteristic lengths govern the transport phenomena of quantum fluids : the range a of the interaction potential (typically a few angströms), the average distance between particles, their thermal wavelength λ_{DB}, the mean free path of the (quasi)-particles and finally the macroscopic experimental length scale. In this paper, we shall be concerned with temperatures low enough for the de Broglie wavelength λ_{DB} to be comparable with the range a of the potential. In this case, whatever is the density n of the fluid , the collisions must be treated quantum mechanically and interference effects will appear as a consequence of the indistinguishability of particles [10]. In this paragraph, we focus the analysis on the simple situation of gaseous, non degenerate systems that can be exactly studied from first principles, but it is quite clear that many of the effects encountered in this situation can equally occur in dilute mixtures of ^3He in liquid ^4He [11] as well as in the liquid ^3He phase.

2.a. A quantum mechanical Boltzmann equation

In the dilute non degenerate systems ($n\lambda^3_{DB} < 1$), the free propagation of atoms between two binary collisions can be described classically. In so far as the interest is focussed on phenomena occurring on a length scale much larger than λ_{DB} and a, the evolution of the system can be described by a Boltzmann-like equation with a classical drift term and a collision term involving all the quantum mechanical effects. However, due to the presence of the spin variables, the description of the state of the system cannot be reduced to the knowledge of the distribution function f(r,p,t) of the atoms, but must equally involve (in the spin 1/2 case) the distribution function of the magnetization $\vec{\mathcal{M}}(\vec{r},\vec{p},t)$. In the general case of a nuclear spin I, all these informations can be described by a (2I+1)×(2I+1) density matrix $\rho_I(\vec{r},\vec{p},t)$, which is the Wigner transform of the quantum Liouville-Von Neuman density operator σ. Waldmann [12] and Snider [13] were the first to write such a quantum mechanical Boltzmann equation for $\rho_I(\vec{r},\vec{p})$. Initially not taken into account, particle indistinguishability was subsequently introduced through the use of symmetrized cross-sections for bosons and antisymmetrized ones for fermions. If such an heuristic approach is perfectly justified for systems with no internal degree of freedom (i.e. spin 0 bosons, or totally polarized fermions), it is not obvious to generalize it to partially polarized systems and moreover this approach is of no help when dealing with situations where the polarization can vary in magnitude and direction. In fact, Boltzmann and Waldmann's original line of argument can be readily developped together with a complete quantum mechanical description of the collisions [14] : the only problem with respect to the indistinguishability of particles arise from the *formulation of the molecular chaos assumption*. The classical form of this assumption which links the reduced two body operators $\sigma^{(2)}$ to the one-body operators $\sigma^{(1)}$ is not consistent with the symmetrization principle and must be replaced by the correctly symmetrized form :

$$\sigma^{(2)}_{(1,2)} = \frac{1}{2}\,(1 + \varepsilon P_{12})\,\sigma^{(1)}_{(1)}\,\sigma^{(1)}_{(2)}\,(1 + \varepsilon P_{12}) \tag{1}$$

where the symmetrization operator

$$\frac{1}{\sqrt{2}}\,(1 + \varepsilon P_{12}) \qquad \text{with} \qquad \begin{array}{l} \varepsilon = +1 \text{ for bosons} \\ \varepsilon = -1 \text{ for fermions} \end{array}$$

acts on both internal (i.e. spin) variables and external variables. From that point the developpment of the argument is straightforward [14] and leads to an equation (equ.(32) of ref. 14) somewhat similar to the classical Boltzmann equation but with a more complex collision term, depending on four independant real 'cross sections" : $\sigma_k(\theta)$, the unsymmetrized differential cross section, and $\sigma_k^{ex}(\theta)$, $\tau_k^{ex}(\theta)$ and τ_{fwd}^{ex} three independant real functions of the scattering amplitude which arises through exchange effects. We don't want to discuss here the multiple characteristics of this quantum Boltzmann equation (see ref. 14) but it is possible to summarize the essentials of its physical content and consequences by the use of two ideas :

(i) in the simple case where the directions of the nuclear spins are parallel everywhere, the medium can be described as a chemical mixture of (2I+1) species, the cross section for collisions of atoms in different spin states being simply σ_k (standard "distinguishable cross section") and that for atoms in the same spin state being the properly symmetrized cross section $\sigma_k + \varepsilon\sigma_k^{ex}$.

(ii) At variance if the magnitude and direction of the magnetization is not constant in the medium, it appears a quite new phenomenon due to the strong correlations introduced by the symmetrization principle between spin and kinetic variables. Although the real interaction between particles is considered as purely electrostatic during collision, due to symmetry effects, the spins of the colliding particles precess around each other, the total spin of the pair being conserved : it is what we called "identical spin rotation effects"; they are at the origin of unusual oscillatory phenomena in dilute, non degenerate quantum gases. It must be noticed that a somewhat similar transport equation for quasi-particles in degenerate Fermi-liquids immersed in a magnetic field was derived by V.P. Silin ([17]) from the Landau theory. In a certain sense, the "identical spin rotation effect" can be thought of as the precursor of the molecular field of the Landau-Silin theory.

The theory so far developped is valid for bosonic as well as for fermionic systems; in the following, we will concentrate on the fermionic systems and more precisely on ^{3}He, the case of H↑ and D↑ being studied in ref. 16.

2.b. Transport properties

The quantum Boltzmann equation can be studied by standard techniques ([14])([15]), without any approximation on the polarization. Due to the exchange contributions to the collision terms, the various transport coefficients strongly depend on nuclear polarization in a non linear fashion. This is illustrated in fig. 1 for the heat conduction coefficient (which qualitatively behaves like the viscosity) and in fig. 3 for the transverse spin diffusion coefficient. The only exception is the longitudinal spin diffusion which in a first approximation is independant on the polarization because it only depends on collisions between atoms in different spin states (i.e. distinguishable) ([15])([18]). The exchange effects also introduce new non linear couplings between heat conduction and longitudinal spin diffusion that could lead to some new hydrodynamic behaviours ([15])([16]).

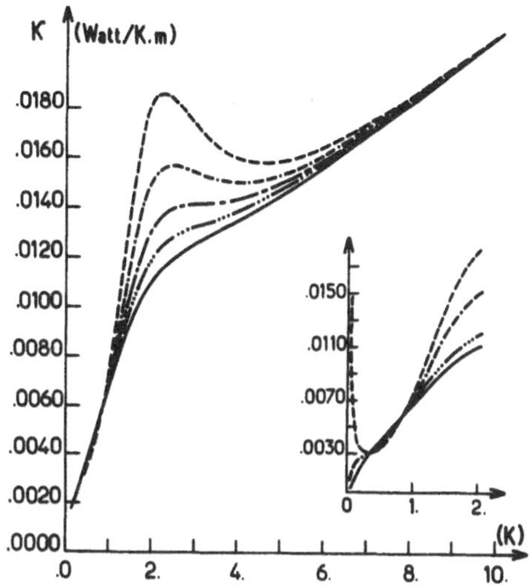

Fig. 1 : Heat conductivity in ^3He versus temperature and nuclear polarization M

—	M = 0;
—·· ·—	M = 0.4;
— · —	M = 0.6;
·—·——	M = 0.8;
---	M = 1.

The low temperature divergence of heat conductivity for fully polarized systems is a characteristic of fermionic systems. Due to the exclusion principle, low temperature effective interaction between atoms in the same spin state goes to zero, and the mean free path for heat (and impulsion) transfers diverges. The "bump" about 2K is a dynamical feature explained in ref. 16. (by courtesy of J. de Physique).

The "identical spin rotation effect" is at the origin of an unexpected oscillatory behaviour of the transverse magnetization current; this effect has two main consequences : first the possibility of observation of transverse spin waves in a dilute, non degenerate system (the ratio between the angular frequency of the oscillation and its damping constant can be calculated to be $\varepsilon\mu M$ where μ is a dimensionless parameter illustrated in fig. 2, M being the local polarization); secundly this effect is responsible for a strong decrease of the transverse spin diffusion as illustrated in fig. 3. (Similar effects were predicted in dense ^3He by Legett and Rice ([19]) and demonstrated by Corrucini, Osheroff, Lee and Richardson ([20]).)

These special features associated with the "identical spin rotation effects" cannot be understood within a model of spin up and spin down mixtures. They are likely to be most important in a great number of experimental situations where field gradients are unavoidable; in particular, they could deeply affect some relaxation processes as illustrated in ref. 21.

3. THE LIQUID PHASE

The evolution of the liquid phase with the nuclear polarization offers a still greater challenge to the theorists. It is widely admitted that the binding energy of the liquid will not vary drastically with the magnetic polarization. The binding energy of ^3He is known experimentally to be of the order of -2.48K. Extrapolation of its low magnetic properties would give an energy of the order of -2.2K for $^3He\uparrow$. Microscopic variational calculations led to upper bounds for the energy of the nor-

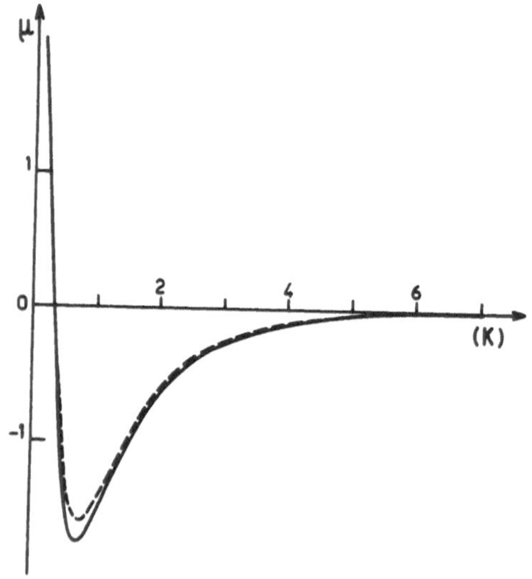

Fig. 2 : "Identical spin rotation" coefficient μ versus temperature. The full line has been calculated with the Lennard-Jones potential, the dashed one with the Aziz potential.

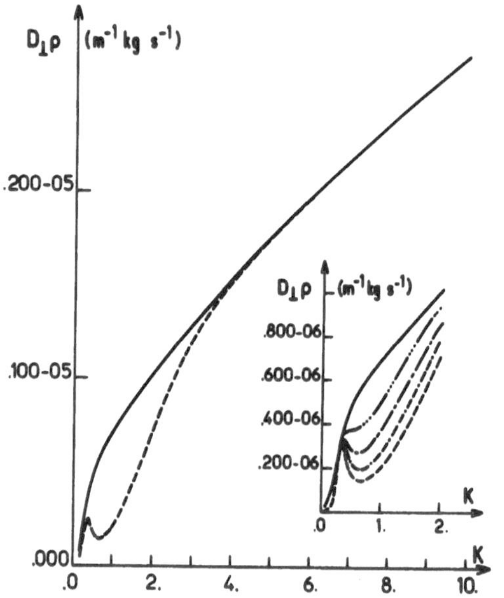

Fig. 3 : Transverse spin diffusion coefficient. Same legend as in fig. 1 for nuclear polarization. The full line curve is a good first approximation to the longitudinal spin diffusion coefficient for any value of the polarization.

mal liquid ranging from -1.9K (Monte Carlo calculations with wavefunctions including triplet and backflow correlations of Schmidt, Lee, Kalos and Chester (22)) to -2.2K (G.F.M.C. calculations of the same authors (23)) while a somewhat cruder Monte Carlo calculation of totally polarized ^3He yield an upper bound of -1.55K for the binding energy of $^3He\uparrow$ (24). In a very recent and accurate study of both normal and spin polarized liquid ^3He, with F.H.N.C. summation techniques, Manousakis, Fantoni, Pandharipande and Usmani (25) obtained -2.35K for the binding energy of normal ^3He and -2.1K for $^3He\uparrow$.

It would be unwarranted to deduce from these results that spin polarized helium three will be quite comparable to normal helium three; in fact, the very framework of the above variational calculations is rather in favour of the opposite conclusion. All the present variational approaches for the fundamental (and excited states) of both ^3He and ^4He were performed within the framework drawn years ago by Bijl (26), Dingle (27) and Jastrow (28), Feynman (29) and other contributors (30). In this approach, it is supposed that the short range correlations of the helium liquids (^4He, ^3He, $^3He\uparrow$) are roughly the same in their fundamental state as well as in their first excited states. This assumption leads to orbital eigenfunctions of the helium N particles hamiltonian in the factorized form

$$\psi(\vec{r}_1 \ldots \vec{r}_N) = \phi_S . \theta \, , \tag{2}$$

involving a symmetrical correlation factor ϕ_S (Jastrow factor with eventually three-body correlations ...) multiplied by a function θ intended to select the fundamental (or excited states) of appropriate permutation symmetry. All the calculations presently achieved seem to prove that a choice of θ as the solution of the non interacting particles problem is a reasonable first approximation [$\theta = 1$ for the fundamental state of liquid ^{4}He, θ is a phonon-like state for the first excitation of ^{4}He and is a Slater determinant (resp. the product of two) for the fundamental state of ^{3}He↑ (resp. ^{3}He)] ; the use of wavefunctions including backflow, or of more general correlated basis functions can then be thought of, as a perturbative correction to this first approximation.

Let us now consider the variational space $\mathcal{E}$ spanned by the (unphysical) functions

$$\Phi = \phi_S \prod_{j=1,N} e^{i\vec{k}(j).\vec{r}_j}$$

where the $\vec{k}(j)$ are determined by the periodic boundary conditions in a box ($k_\alpha = n_\alpha \frac{2\pi}{L}$; $n_\alpha = 1,2$) and are to be chosen amongst the N first k vectors. The space encompasses all the above-mentioned variational wavefunctions.

Let us now try to count the number of states of a given symmetry in this space. This problem can be made simpler if we select among the possible ϕ the functions ϕ_o for which the set { k(j) } is a permutation of the N first eigenvectors $\vec{k}_1 \ldots \vec{k}_N$. In this restricted space $\mathcal{E}_o$ (of dimension N !), there is only :

i) one completely symmetrical wavefunction that can be used to describe an assembly of N bosons

ii) one completely antisymmetrical wavefunction that can describe an assembly of N polarized fermions

iii) and $\alpha = N\,! \,/\, (\frac{N!}{2})^2$ a priori distinct eigensubspaces (with degeneracy α) that span the set of states physically accessible to an assembly of N fermions of spin 1/2 with a total spin I=0. (For N large enough, α is of the order of N).

If we now relax the constraint upon the set { $\vec{k}$ } to recover the whole $\mathcal{E}$ space, we find new states accessible to an assembly of bosons or unpolarized fermions but no more that can describe an assembly of ^{3}He↑.

This simple result may be the starting point for two kinds of deductions :

i) firstly, it explains why it seems to be much simpler to have a good microscopic description of ^{4}He or ^{3}He↑ in contrast with the case of normal ^{3}He (the improvement in the wavefunctions associated with triplet and backflow correlations are very important for the description of normal ^{3}He and of much less importance in the description of $^3He\uparrow$ [24][25]).

ii) secondly, if the variational space $\mathcal{E}$ happens to be the proper one for describing the low lying levels of the helium hamiltonian, then the density of first excitations must be very low in $^3He\uparrow$ as compared to normal ^3He. As a result, the low temperature properties of $^3He\uparrow$ should be very different from that of normal ^3He. The nuclear polarization would then monitor a decrease of the heat capacity, an increase of the sound velocity, large modification of transport properties and perhaps an increase of the range of superfluidity. But if the variational approach with space $\mathcal{E}$ is not adequate, then different properties and structures may be expected for $^3He\uparrow$.

Spin fluctuations and paramagnous theories [31][32][33] seem to support the idea of scarcity of the excited states of $^3He\uparrow$ while the Hubbard-Anderson-Brinkman model of ^3He emphasizes on the quasi solid structure of ^3He which should be reinforced by nuclear polarization [9][34]. Oviously, there is a great deal to be learnt from the experimental study of liquid ^3He at various polarizations.

4. THE EXPERIMENTS ON SPIN POLARIZED ^3He AT THE E.N.S.

Several methods have been proposed to produce high nuclear polarization in ^3He. One method suggested by Castaing and Nozières [9] is based on the very fast melting of highly polarized solid ^3He to get a metastable liquid with roughly the same polarization. With this technique, several groups [35][36] obtained highly polarized liquid, the polarization relaxing to its equilibrium value in a few minutes.

The method chosen at the E.N.S. is quite different in its principle : it involves the optical pumping of gaseous ^3He which transfers polarization from a suitable photon beam to the nuclear spin of the atoms. The optical pumping of ^3He was first realized by Colegrove, Schearer and Walters [37]. It was only with the recent development of coulour center lasers that it became possible to get large nuclear polarizations in gaseous ^3He. Nuclear polarizations of 70% were recently obtained at room temperature [38] (for comparison purposes, thermodynamic equilibrium polarization at 1K in 10 Tesla is only 1%) (see fig. 4). By spin diffusion to a low temperature cell, polarization rates as high as 50 percent have been observed in gaseous ^3He between 1.7K and 4.2K. The T_r relaxation time was about 6 hours at 4.2K, of the order of 30 minutes near 2K [39]. First evidences of the "identical spins rotation effect" have been observed using a technique of relaxation in weak magnetic field gradients [21]. By spin diffusion techniques, it should now be possible to polarize a drop of liquid as was first demonstrated by H.H. McAdams [40].

Acknowledgements. Part of this work was carried out in collaboration with Franck Laloë and I want to thank him and Dominique Levesque for numerous stimulating discussions.

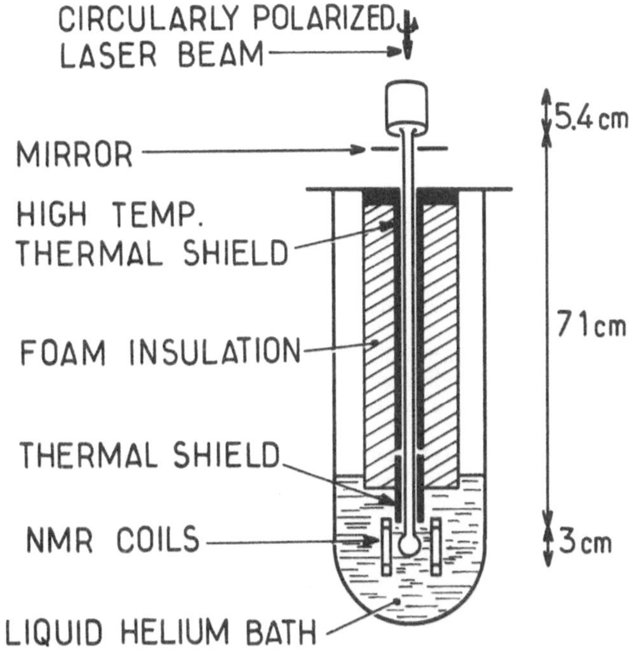

Fig. 4 : Sketch of the double cell experiment used for polarizing ^3He at low temperature. The upper cell at 300K is sbmitted to optical pumping. The nuclear polarization is transferred by spin diffusion to the low cell immersed in a liquid helium bath (M. Leduc, P.J. Nacher, S.B. Crampton, F. Laloë).

REFERENCES

(1) J. Phys. Coll., 41, C-7 (1980); "Spin polarized quantum systems", and references contained.
(2) S.B. Crampton, T.J. Greytak, D. Kleppner, W.D. Phillips, D.A. Smith, A. Weinrib, Phys. Rev. Lett. 42, 1039 (1979)
(3) J.T.M. Walraven, J.F. Silvera and A.P.M. Matthey, Phys. Rev. Lett. 45, 449 (1980)
M. Morrow, R. Jochemsen, A.J. Berlinsky and W.N. Hardy, Phys. Rev. Lett. 46, 165 (1981)
R.W. Cline, T.J. Greytak and D. Kleppner, Phys. Rev. Lett. 47, 1195 (1981)
R. Sprik, J.T.M. Walraven, G.H. Van Yperen and J.F. Silvera, Phys. Rev. Lett. 49, 153 (1982)
B. Yurke, J.S. Denker, B.R. Johnson, L.P. Lévy, D.M. Lee and J.H. Freed, preprint
(4) R. Barbé, F. Laloë, J. Brossel, Phys. Rev. Lett. 34, 1488 (1975)
(5) V. Lefèvre-Seguin, F. Laloë, to be published in J. Physique (1983)
(6) M.A. Taber, LT-15, J. Physique, C-6, 39, 192 (1978)
(7) M. Himbert, V. Lefèvre-Seguin, P.J. Nacher, J. Dupont-Roc, M. Leduc and F. Laloë, J. Physique Lett. 44, 523 (1983)
(8) H. Godfrin, G. Frossati, D. Thoulouze, M. Chapellier and W.G. Clark, J. Phys. Coll. 39, C6-287 (1978)
H. Godfrin, G. Frossati, B. Hebral, D. Thoulouze, J. Phys. Coll. 41, C7-275 (1980)
(9) B. Castaing, P. Nozières, J. Physique, 40, 257 (1979)
(10) This elementary quantum mechanical effect is illustrated for the present situation in ref. (1) page 51.
(11) E.P. Bashkin and A.E. Meyerovich, Advances in Physics, 30, 1-92 (1981)
Some of the effects predicted by these authors have already been observed :
D.S. Greywall and M.A. Paalanen, Phys. Rev. Lett. 46, 1292 (1981)
W. Gully, B.A.P.S. (April 1983).
(12) L. Waldmann, Z. Naturforsch. 13a, 609 (1958)
(13) R.F. Snider, J. Chem. Phys. 32, 1051 (1960)
(14) C. Lhuillier and F. Laloë, J. Phys. 43, 197 (1982)
(15) C. Lhuillier and F. Laloë, J. Phys. 43, 225 (1982)
(16) C. Lhuillier, J. Phys. 44, 1 (1983)
(17) V.P. Silin, J.E.T.P. 33, 1227 (1957); Sov. Phys. JETP 6, 945 (1958)
(18) V.J. Emery, Phys. Rev. 133A, 661 (1964)
(19) A.J. Legett, M.J. Rice, Phys. Rev. Lett. 20, 586 (1968)
A.J. Legett, J. Phys. C 12, 447 (1970)
(20) L.R. Corrucini, D.D. Osheroff, D.M. Lee and R.C. Richardson, Phys. Rev. Lett. 27, 650 (1971); J. Low Temp. Phys. 8, 229 (1972)
(21) V. Lefèvre-Seguin, P.J. Nacher, and F. Laloë, J. Physique, 43, 737 (1982)
(22) K.E. Schmidt, M.A. Lee, M.H. Kalos, G.V. Chester, Phys. Rev. Lett. 47, 807 (1981)
(23) M.A. Lee, K.E. Schmidt, M.H. Kalos, G.V. Chester, Phys. Rev. Lett. 46, 728 (1981)
(24) C. Lhuillier, D. Levesque, Phys. Rev. B 23, 2203 (1981)
(25) E. Manouzakis, S. Fantoni, V.R. Pandharipande, Q.N. Usmani, preprint submitted to Phys. Rev. B, April 1983
(26) A. Bijl, Physica, 7, 869 (1940)
(27) R.B. Dingle, Phil. Mag. 40, 573 (1949)
(28) R. Jastrow, Phys. Rev. 98, 1479 (1955)
(29) R.P. Feynman, Phys. Rev. 94, 262 (1954)
R.P. Feynman and M. Cohen, Phys. Rev. 102, 1189 (1956)
M. Cohen and R.P. Feynman, Phys. Rev. 107, 13 (1957)
(30) See for instance the review of :
C.W. Woo in "Microscopic Calculations for Condensed Phases of Helium" in "the Physics of Liquid and Solid Helium", K.H. Bennemann and J.B. Ketterson editors (J. Wiley & Sons, 1976).
(31) K.S. Bedell and K.F. Quader, Physics Lett. 96A, 91 (1983)
(32) T.L. Ainsworth, K.S. Bedell, G.E. Brown and K.F. Quader, J. Low Temp. Phys. 50, 319 (1983)
(33) B.L. Friman and E. Krotschek, Phys. Rev. Lett. 49, 1705 (1982)
(34) D. Vollhardt, preprint, Mai 1983)

(35) G. Schumacher, D. Thoulouze, B. Castaing, Y. Chabre, P. Segranson and J. Joffrin, J. Physique Lett. 40, L143 (1979)
(36) M. Chapellier, G. Frossati, F.B. Rassmussen, Phys. Rev. Lett. 42, 904 (1979)
(37) F.D. Colegrove, L.D. Schearer, and G.K. Walters, Phys. Rev. 132, 2561 (1963)
(38) P.J. Nacher, M. Leduc, G. Trénec and F. Laloë, J. Physique Lett. 43, L-525 (1982)
(39) M. Leduc, P.J. Nacher, S.B. Crampton and F. Laloë, in the Proceedings of the Conference on "Quantum Fluids and Solids", in Sanibel, Florida (1983)
(40) H.H. MacAdams, Phys. Rev. 170, 276 (1963)

THE PROPERTIES OF PAULI ENHANCED NORMAL FERMI LIQUIDS IN FINITE MAGNETIC FIELDS

Kevin S. Bedell
Physics Department
State University of New York
Stony Brook, NY 11794, USA

and

NORDITA
Blegdamsvej 17
DK-2100 Copenhagen Ø, Denmark

Abstract

The qualitative features of arbitrarily polarized Pauli enhanced Fermi liquids are discussed. These are presented within the context of the Landau theory of a polarized Fermi liquid and the induced interaction model. Some general properties of the polarized Fermi liquids regarding such things as the field dependence of the specific heat, effective masses, and scattering amplitudes will emerge.

1. Introduction

The study of highly polarized Pauli enhanced normal Fermi liquids can lead to a better understanding of the rôle of the Pauli principle and spin fluctuations in these liquids. Their contributions to the behaviour of the systems such as ^3He, Pd, $TiBe_2$, etc. are masked by the presence of a large singlet component in the quasiparticle interaction and scattering amplitude. By polarizing these systems, the strong singlet scattering and the spin fluctuations will be 'frozen' out, leaving a system driven by a weaker triplet interaction. This should lead to important modifications in those properties that depend most sensitively on statistics and the spin fluctuations, e.g. transport coefficients and the effective mass.

In this article, I would like to discuss the extent to which theory can give us some insight into the rôle of statistics and spin fluctuations in the Pauli enhanced systems. In particular, I will concentrate on the qualitative aspects of the results obtained from the Landau theory for a polarized Fermi liquid and the induced interaction model.[1,2,3] Some very general features of the polarized Fermi liquids regarding such things as the field dependence of C_V, the specific heat, the effective mass and the scattering amplitudes will emerge.

2. Landau Theory of a Polarized Fermi Liquid

Many of the properties of the elementary excitations and collective modes of a normal Fermi liquid can be expressed in terms of Landau theory[4]. The generalization to arbitrarily polarized systems can be developed along the same lines as used by Landau. Some of the results have appeared already in several places[3,5,6] and I will only discuss some of the key features of the theory. Here we will only consider excitations and interactions with spin projection zero in the particle-hole channel, longitudinal fluctuations. Thus, we will not consider such things as spin-flip interactions, transverse fluctuations, that carry spin one in the particle-hole channel. Since the spin projection in this channel is conserved, these excitations will decouple.

The energy density, taking into account longitudinal fluctuations, to second order is given by

$$\epsilon = E/V = \epsilon_0 + \sum_{\vec{p}\sigma} \varepsilon^0_{\vec{p}\sigma}\delta n_{\vec{p}\sigma} + \tfrac{1}{2}\sum_{\vec{p}\sigma,\vec{p}'\sigma'} f^{\sigma\sigma'}_{\vec{p}\vec{p}'}\delta n_{\vec{p}\sigma}\delta n_{\vec{p}'\sigma'} , \tag{1}$$

where the quasiparticle energy is defined by

$$\tilde{\varepsilon}_{\vec{p}\sigma} = \varepsilon^0_{\vec{p}\sigma} - \vec{\sigma}\cdot\vec{B} + \sum_{\vec{p}'\sigma'} f^{\sigma\sigma'}_{\vec{p}\vec{p}'}\delta n_{\vec{p}'\sigma'} , \tag{2}$$

with $\delta n_{\vec{p}\sigma} = n_{\vec{p}\sigma} - n^0_{\vec{p}\sigma}$. Here $n^0_{\vec{p}\sigma}$ is the equilibrium distribution function in the presence of the magnetic field $\vec{B}$. The quasiparticle interaction with spin projection zero $f^{\sigma\sigma'}_{\vec{p}\vec{p}'}$ has three components, $f^{\uparrow\uparrow}_{\vec{p}\vec{p}'}$, $f^{\uparrow\downarrow}_{\vec{p}\vec{p}'}$, and $f^{\downarrow\downarrow}_{\vec{p}\vec{p}'} \neq f^{\uparrow\uparrow}_{\vec{p}\vec{p}'}$, each depending on the polarization. From eqs.(1) and (2) we can obtain the various equilibrium properties of the system. To obtain the scattering amplitudes and the collective modes in the limit of long wavelengths and low frequencies, we use the linearized kinetic equation[3,8].

One of the most important quantities in Fermi liquid theory is the effective mass which enters into all measurable quantities. In a polarized Fermi liquid we have two effective masses m^*_σ for $\sigma = \uparrow$ or $\downarrow$. By invoking Galilean invariance we can relate m^*_σ to the $\ell = 1$ moments of $f^{\sigma\sigma'}_{\vec{p}\vec{p}'}$, which yields

$$\frac{m^*_\sigma}{m} = \frac{1}{1 - \frac{1}{3}N^0_\sigma(0)\left[f^{\sigma\sigma}_1 + \left(\frac{k_F^{-\sigma}}{k_F^\sigma}\right)^2 f^{\uparrow\downarrow}_1\right]} , \tag{3}$$

where $N_\sigma(0) = \frac{k_F^\sigma m_\sigma^*}{2\pi^2}$ and $N_\sigma^0(0) = \frac{k_F^\sigma m}{2\pi^2}$. The interaction has been expanded as follows:

$$f_{\vec{p}\vec{p}'}^{\sigma\sigma'} = \sum_\ell f_\ell^{\sigma\sigma'} P_\ell(\hat{p}\cdot\hat{p}') \quad . \tag{4}$$

In the unpolarized case $m_\uparrow^* = m_\downarrow^* = m^*$ and eq.(3) reduces to the standard results.[4,8] In an unpolarized Fermi liquid the effective mass can be obtained from specific heat measurements. For an arbitrarily polarized system, C_V measurements for $T \ll \min\left(k_F^{\uparrow 2}/2m_\uparrow^*\ ,\ k_F^{\downarrow 2}/2m_\downarrow^*\right)$ will yield the combination

$$C_V/T \sim m_\uparrow^* k_F^\uparrow + m_\downarrow^* k_F^\downarrow \quad . \tag{5}$$

We then see that C_V can only determine a combination of the two effective masses.

From eqs.(3) and (5) some general features of the field dependence of the effective mass and C_V emerge. The first thing we note is that in the expression for $m_\downarrow^*$ there is a term of the form $f_1^{\uparrow\downarrow}/k_F^\downarrow$. When $k_F^\downarrow \to 0$, this term might appear to diverge; however, from general arguments it remains finite. The momenta $\vec{p}$ and $\vec{p}'$ are the relative momenta of the incoming and outgoing particle-hole pair (analogous to the relative momenta in the particle-particle channel). From quantum mechanics we know that when the relative momentum of the incoming (or outgoing) pair of particles vanishes, only s-wave scattering is possible. This is also true in the particle-hole channel. Thus, if $\vec{p}$ and $\vec{p}'$ are on their respective Fermi surfaces, then $f_\ell^{\uparrow\downarrow} = 0$ and $f_\ell^{\downarrow\downarrow} = 0$ for $\ell \geq 1$ when $k_F^\downarrow = 0$. From this general observation we have that

$$f_1^{\uparrow\downarrow} = k_F^\downarrow \lambda\left(k_F^\uparrow, k_F^\downarrow\right) \xrightarrow[k_F^\downarrow \to 0]{} k_F^\downarrow \lambda\left(k_F^\uparrow, k_F^\downarrow = 0\right) \quad , \tag{6}$$

and thus $m_\downarrow^*$ will approach the value for a single down-spin impurity in a liquid of up-spin fermions when $k_F^\downarrow \to 0$. We are, of course, assuming that the normal liquid remains so after it is polarized.

Qualitatively, we expect the effective masses to have the behaviour shown in fig.1(a). From very general arguments we know that m_σ^* should have a term linear in the field. On the other hand, from thermodynamic arguments C_V can have no linear term. Thus, the coefficients of the linear terms in $m_\uparrow^*$ and $m_\downarrow^*$ must be the same but with opposite signs. In ^{3}He we would expect from the sign and size of f_1^s that $m_\uparrow^*$ would decrease and $m_\downarrow^*$ would initially increase with increasing polarization. In the fully polarized limit, it might be amusing to make a short speculative trip regarding $m^*_\uparrow$ and $m^*_\downarrow$ in ^{3}He.

Let us first consider the effective mass of a single down-spin particle in a liquid of fully polarized ^{3}He, denoted by ^{3}He$\uparrow$, at $T=0$. In this case the analogy with a single ^{3}He atom in superfluid ^{4}He is appropriate. If a classical backflow argument is used for a single ^{3}He atom in ^{4}He, we get a mass enhancement of 0.85.[9] At s.v.p. in ^{4}He (which has the same density as ^{3}He at $P \simeq 21$ bar) this number is off roughly by a factor of 1.5 from experiment. The corrections arise since the classical results ignore quantum and interaction effects. A single down-spin ^{3}He atom in polarized ^{3}He has a classical component since up and down spins, like a ^{3}He and ^{4}He atom, are distinguishable. From the classical arguments, the mass enhancement in this case would be $\frac{1}{2}$. If at the same density in ^{3}He as ^{4}He at s.v.p., i.e. $P \simeq 21$ bar, the corrections are comparable, this would yield a mass enhancement around 0.75 and $m^*_\downarrow < m^*$. Even though this argument is somewhat naive, it is very suggestive and probably correct.

For the case of the effective mass of an up-spin particle in ^{3}He$\uparrow$, the classical arguments can not be applied. In this case, the particles are indistinguishable and statistics will play an important rôle. In particular, the interactions that would give rise to the effective mass must be anti-symmetrized. This will eliminate the singlet component from the effective mass, leaving it to be determined by much weaker triplet interactions. Moreover, the spin-flip contribution to $m^*_\uparrow$, which is responsible for the mass enhancement at the Fermi surface in ^{3}He, is 'frozen' out. We would then expect that $m^*_\uparrow$ is considerably reduced from its unpolarized value. This is what was found in BQ and in some more recent microscopic calculations.[10,11]

The qualitative features of $m^*_\uparrow$ and $m^*_\downarrow$ described above and shown in fig.1(a) would lead to the behaviour for $\Delta C = (C_V^0 - C_V)/C_V^0$, where C_V^0 is the specific heat in zero field, shown in fig.1(b). The field denoted by B_c separates the regions where $m^*_\uparrow$ and $m^*_\downarrow$ work against each other and where $m^*_\downarrow$ begins to decrease and work with $m^*_\uparrow$. For larger values of B, the spin fluctuations are no longer very effective in changing the masses, leading then to the levelling off of C_V. The scale over which C_V will change depends on the particular details of the interaction.

If we are to observe some of the features of C_V in ^{3}He, it is important that B_c is not very large. Since B_c will depend on the quasiparticle interactions in a non-linear manner, it can not be readily estimated. We can, however, make some speculation based on the similarity of ^{3}He to some of the Pauli enhanced metals. Within the approach of

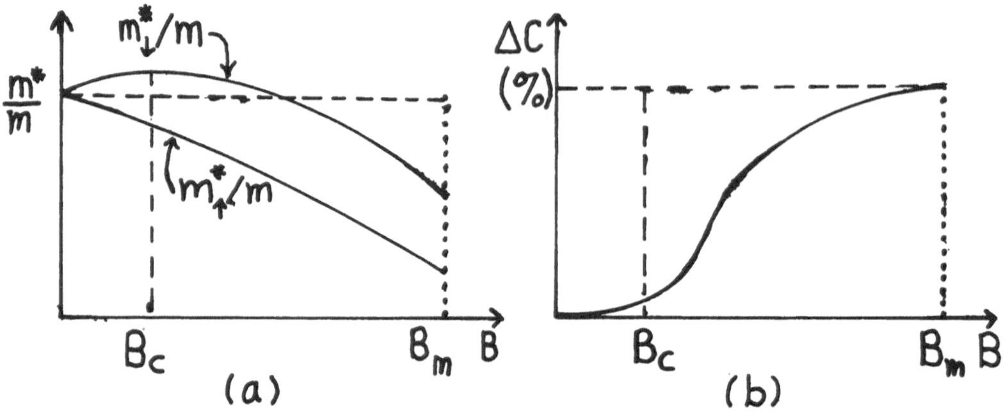

Fig.1. In (a) is shown the qualitative shape of m^*_σ based upon the speculations made in the text. The field B_m is the maximum field required to fully polarize the liquid and B_c is defined in the text. In (b) is the expected shape for $\Delta C = (C_V^0 - C_V)/C_V^0$ given fig.(a). These are not drawn to scale and serve only as a qualitative guide.

ABBQ, these systems are essentially the same. For example, in $TiBe_2$ the field B_c is around 4 or 5 T.[12] Generally speaking, changes in the Pauli enhanced systems occur on a scale given by the spin-fluctuation temperature T_{SF}. In $TiBe_2$ the field corresponding to this energy scale is $B_{SF} = \mu\, T_{SF} \simeq 50$ T;[12] thus $B_c \simeq .1\, B_{SF}$. If we simply assume that the relationship between B_c and B_{SF} is the same in ^{3}He, then we find $B_c \sim 30$ T in ^{3}He at melting pressure. The precise value of B_c should not be taken too seriously, however. That $B_c \ll B_{SF}$ is probably correct and may even be experimentally accessible.

Some additional insight into the properties of polarized Fermi liquids can be gained from a study of the scattering amplitudes. To obtain the scattering amplitudes with spin projection zero in the particle-hole channel, we use the kinetic equation defined in BQ. Here I wish to comment only on the sum rules, so the details of the solutions will not be given.

In a finite field B, the degeneracy of the triplet state is lifted. The argument from quantum mechanics that only s-wave scattering is possible when the relative momentum $\vec{k}$ (or $\vec{k}'$) of the incoming (or outgoing) particles vanishes still holds for each sub-state. In principle then there should be three sum rules for $m_s = 0$ and $m_s = \pm 1$ in the particle-particle channel. If we only consider particles on the Fermi surfaces, then the triplet scattering amplitude with $m_s = 0$ will not vanish since it involves states on different Fermi seas ($|\vec{k}| \neq 0$ and $|\vec{k}'| \neq 0$ for

$m_s = 0$). For the situations in which $m_s = \pm 1$ we get two distinct sum rules,

$$\sum_\ell a_\ell^{\sigma\sigma} = 0 \tag{7}$$

for $\sigma = \uparrow$ or $\downarrow$, and where[7]

$$a_\ell^{\uparrow\uparrow} = \frac{f_\ell^{\uparrow\uparrow}\left(1 + \frac{N_\downarrow(0) f_\ell^{\downarrow\downarrow}}{2\ell+1}\right) - N_\downarrow(0) f_\ell^{\uparrow\downarrow^2}/2\ell+1}{\left(1 + N_\uparrow(0)\frac{f_\ell^{\uparrow\uparrow}}{2\ell+1}\right)\left(1 + N_\downarrow(0)\frac{f_\ell^{\downarrow\downarrow}}{2\ell+1}\right) - N_\uparrow(0) N_\downarrow(0)\left(f_\ell^{\uparrow\downarrow}/2\ell+1\right)^2} \quad . \tag{7a}$$

To get $a_\ell^{\downarrow\downarrow}$ just interchange $\uparrow$ and $\downarrow$ in eq.(7a). In passing, we note that one of the stability conditions requires the denominator of eq.(7a) to be positive.

In the fully polarized limit, the sum rules lead to interesting consequences. If we combine the sum rule for $m_s = -1$ and the fact that $f_\ell^{\downarrow\downarrow} = 0$ and $f_\ell^{\uparrow\downarrow} = 0$ for $\ell \geq 1$ when $k_F^\downarrow = 0$, we have

$$f_0^{\downarrow\downarrow} - \frac{N_\uparrow(0) f_0^{\uparrow\downarrow^2}}{1 + N_\uparrow(0) f_0^{\uparrow\uparrow}} = 0 \quad ,$$

which yields

$$\frac{m_\uparrow^*}{m} = \frac{2^{2/3}\pi^2}{mk_F} \frac{f_0^{\downarrow\downarrow}}{f_0^{\uparrow\downarrow^2} - f_0^{\uparrow\uparrow} f_0^{\downarrow\downarrow}} = \frac{1}{1 - N_\uparrow^\bullet(0) f_1^{\uparrow\uparrow}/3} \quad . \tag{8}$$

This is a rather interesting result, since the effective mass is determined exactly by the $\ell = 0$ moments of the Landau interaction. In addition to being an interesting result it also acts as a check on calculations of the Landau parameters in the fully polarized system.

The sum rule for $m_s = +1$ when $k_F^\downarrow = 0$ yields

$$\sum_\ell \frac{f_\ell^{\uparrow\uparrow}}{1 + N_\uparrow(0) f_\ell^{\uparrow\uparrow}/(2\ell+1)} = 0 \quad . \tag{9}$$

This is more restrictive than the sum rule in the unpolarized phase which involves both the symmetric and anti-symmetric interactions. Here if we truncate eq.(9) at $\ell = 1$ we can conclude that either $f_0^{\uparrow\uparrow} < 0$ or $f_1^{\uparrow\uparrow} < 0$, but they can not have the same sign. In $^3He\uparrow$, if we try to fit the compressibility calculated by Luhillier and Levesque[13] and truncate eq.(9) at $\ell = 1$, it is found that $f_0^{\uparrow\uparrow} > 0$ and $f_1^{\uparrow\uparrow} < 0$, i.e. $m_\uparrow^*/m < 1$. This is as calculated by BQ and others[10,11] for fully polarized ^3He. The more restrictive rôle played by the sum rule in $^3He\uparrow$ reflects the increased importance of the Pauli principle in determining the properties of a polarized Fermi liquid. Although it is not obvious that eq.(9) should truncate at $\ell = 1$, it is probably correct that the values of $f_\ell^{\uparrow\uparrow}$ for $\ell \geq 2$ are negligible. If one tried to fit the calculated compressibilty[13] and keep $f_1^{\uparrow\uparrow} > 0$, i.e. $m_\uparrow^*/m > 1$, then the higher moments would have to be much larger than in the unpolarized phase.

3. The Induced Interaction Model

The induced interaction model of Babu and Brown[1] has recently been applied to Pauli enhanced normal Fermi liquid[2]. More recently, this has been generalized to polarized Fermi liquids and to fully polarized ^{3}He[3]. In this model the quasiparticle interaction is separated into two pieces, $f_{\vec{p}\vec{p}'}^{\sigma\sigma'} = d_{\vec{p}\vec{p}'}^{\sigma\sigma'} + I_{\vec{p}\vec{p}'}^{\sigma\sigma'}$, where $d_{\vec{p}\vec{p}'}^{\sigma\sigma'}$ is the direct and $I_{\vec{p}\vec{p}'}^{\sigma\sigma'}$ is the induced interaction. The direct term, which has the same symmetry as the underlying potential, includes all diagrams that are not particle-hole reducible. The induced term corresponds to the exchange of phonons, density, spin-density, etc., between the quasiparticles. This is analogous to the separation of the electron-electron interactions in metals where the direct term is the screened Coulomb potential and the induced term is the piece mediated by lattice phonons. The essential difference here is that the direct and induced terms are dependent upon the interactions between the quasiparticles.

In principle, the quasiparticle interaction must be calculated self-consistently. In practice this is very complicated, so we have introduced a model approach. For the direct term we use the effective potential approach proposed by Bedell and Ainsworth[14]. The direct term is just the Fourier transform of an effective quasiparticle potential. From this we can construct the quantum mechanical scattering amplitude:

$$f_k(\phi) = -\frac{m^*}{4\pi}\int e^{i\vec{q}\cdot\vec{r}} V_{eff}(\vec{r};k)\,d^3r \quad , \tag{10}$$

where $q^2 = |\vec{k} - \vec{k}'|^2 = 2k^2(1-\cos\phi)$ with $2k^2 = k_F^2(1-\cos\theta)$ for states on the Fermi surface. Here $\vec{k}(\vec{k}')$ is the relative momentum of the incoming (outgoing) quasiparticles and ϕ is the angle between the scattering planes. In general, the effective potential is non-local. If we expand to order k^2 in the non-locality, which gives good results in ^{3}He,[14] then

$$V_{eff}(\vec{r};k) = U(r) + \tfrac{1}{3}k^2r^2W(r) \quad , \tag{11}$$

where $U(r)$ and $W(r)$ are local potentials.

In eq.(10) we keep terms of order q^2 and define the direct term as follows:

$$d^{\uparrow\downarrow}_{pp'} = -\frac{4\pi}{m^*} f_k(\phi = 0) = \frac{4\pi}{m}\left[a_s + \tfrac{3}{2}k_F^2 b_t - \tfrac{3}{2}k_F^2 b_t \cos\theta\right] \tag{12a}$$

and

$$d^{\uparrow\uparrow}_{pp'} = -\frac{4\pi}{m^*}\Big[f_k(\phi) - f_k(\phi+\pi)\Big]_{\phi=0}$$

$$= \frac{12\pi}{m} k_F^2 a_t (1 - \cos\theta) \quad , \tag{12b}$$

where $a_s = (m/m^*)a_0$, $a_t = (m/m^*)a_1$, and $b_t = (m/m^*)b_1$. Here the s-wave scattering length a_0 and the triplet scattering volume a_1 have the same form in terms of U(r) as in the Born approximation.[14] The non-local term b_1 is identical to a_1 with U(r) replaced by W(r).[14] For more details, consult ref.14. I will only discuss some of the qualitative aspects of this model.

If we ignore the finite range corrections introduced by a_t and b_t, the resulting direct interaction reduces to a contact interaction. This is analogous to the Hubbard Hamiltonian[15] in which the s-wave component corresponds to nearest neighbour interactions between particles of opposite spin. Finite range effects are introduced via a_t and b_t which take into account contributions from the other quasiparticles such as next nearest neighbours. If we increase the density the particles, on the average, will tend to sample more of the strong short-ranged repulsion. Thus, particles scattering in relative s-wave states, which sample the short-ranged correlations, will be more sensitive to changes in the density. On the other hand, particles scattering in relative p-wave states (a_t in the particle-particle and b_t in the particle-hole channels) should have, apart from the factor of k_F^2 in eqs.(12a) and (12b), only a weak density dependence. This follows since particles scattering via a_t or b_t will sample less of the short-ranged correlations.

In this model the direct term is a short-ranged potential; thus, only the first two moments are significant. The intermediate range correlations are generated by the induced interaction. The induced interaction is a model for treating the contributions to the quasiparticle interaction arising from the exchange of density, spin-density, etc., fluctuations in the crossed channel. The form of this interaction for an arbitrarily polarized system has been worked out.[3] Here we will concentrate on $f^{\uparrow\uparrow}_{\underset{pp'}{\leftrightarrow}}$, and keep only up to $\ell = 0$ in the crossed channel, then

$$f^{\uparrow\uparrow}_{\underset{pp'}{\leftrightarrow}} = d^{\uparrow\uparrow}_{\underset{pp'}{\leftrightarrow}} - \left\{\frac{f_0^{\uparrow\uparrow 2}\chi_\uparrow(q')\left[1 - f_0^{\downarrow\downarrow}\chi_\downarrow(q')\right] + f_0^{\uparrow\downarrow 2}\chi_\downarrow(q')\left[1 + f_0^{\uparrow\uparrow}\chi_\uparrow(q')\right]}{\left[1 - f_0^{\uparrow\uparrow}\chi_\uparrow(q')\right]\left[1 - f_0^{\downarrow\downarrow}\chi_\downarrow(q')\right] - f_0^{\uparrow\downarrow 2}\chi_\uparrow(q')\chi_\downarrow(q')}\right\} , \tag{13}$$

where $q'^2 = |\vec{p} - \vec{p}'|^2 = 2k_F^{\uparrow 2}(1 - \cos\theta)$ and $\chi_\sigma(q') = -N_\sigma(0)\chi_0(q')$, with $\chi_0(q')$ defined in ref.2. In general, the interactions $f_\ell^{\sigma\sigma'}$ will also depend on q';[16] however, we can ignore this in what follows.

Several general features of the quasiparticle interaction in a polarized Fermi liquid emerge from eq.(13). First we note that the driving term is a purely triplet interaction. In unpolarized ^{3}He it was found that[3] $|k_F^3 a_t| \ll |k_F a_s|$, where a_s is the dominant driving term for $f_{\vec{p}\vec{p}'}^{\uparrow\downarrow}$. However, it is known from experiment that $f_\ell^{\uparrow\uparrow} \simeq f_\ell^{\uparrow\downarrow}$, $\ell = 0,1$ in unpolarized ^{3}He. That $f_\ell^{\uparrow\uparrow}$ is so large comes partly from the explicit coupling to $f_0^{\uparrow\downarrow}$ in the induced interaction, see eq.(13). As the polarization is increased, the coupling to $f_0^{\uparrow\downarrow}$ is eventually frozen out, since $N_\downarrow(0) \xrightarrow[k_F^\downarrow \to 0]{} 0$, leaving $f_{\vec{p}\vec{p}'}^{\uparrow\uparrow}$ to be driven by a weak triplet interaction. This results in the rather dramatic change in the effective mass $m_\uparrow^*/m$ noted earlier. It should be noted that much of the change in $F_0^{\uparrow\uparrow} = N_\uparrow(0) f_0^{\uparrow\uparrow}$ also comes from the big change in $N_\uparrow(0)$.

The large change in $m_\uparrow^*/m$ arises from the combined effect of a reduction in the induced interaction and a sign change in a_t.[3] Actually, the sign change in a_t also arises from freezing out the spin-fluctuations, since the direct term contains contributions from two phonon, etc., exchange. Thus, a_t can have a strong polarization dependence. Clearly there will be two-phonon exchange (density fluctuations) even in the fully polarized system, but these will be much less important. The resulting direct interaction should then be dominated by the particle-particle T-matrix as calculated by Glyde.[11] His results for $m_\uparrow^*/m$ were very close to that found by BQ, suggesting that this may be a reasonable starting point for a microsopic calculation of the Landau parameters.

Another area where dramatic changes are expected to occur is in transport phenomena. In unpolarized ^{3}He, transport is dominated by a large singlet component in the scattering amplitude which is an order of magnitude larger than the triplet piece. With increasing polarization, this singlet component is 'frozen' out, thereby decreasing the scattering amplitude by an order of magnitude. This alone leads to two orders of magnitude increase in the transport coefficients η (viscosity) and κ (thermal conductivity). The additional order of magnitude increase found by BQ comes from the decrease in the effective mass and changes in the angular averages associated with η and κ.

Here I have touched upon only a few of the interesting features of polarized Fermi liquids. The points that I have emphasized were concerned with the rôle of spin-fluctuations and the Pauli principle in Pauli enhanced Fermi liquids. We have seen that polarizing the liquid leads to rather dramatic changes in some of the properties of these systems since polarizing tends to 'freeze' out both the spin-fluctuations and the singlet component in the scattering amplitude. This same feature should also make it easier to perform microscopic calculations in $^3He\uparrow$. With the singlet piece removed from the bare potential, we are left with a weaker triplet component of the bare potential from which to begin microscopic calculations. Althought the p-wave phase shift may not be small enough to permit perturbative calculations, it is not so large as to doom from the outset methods that perform partial summations of diagrams.[10,11]

It is clear that the polarized Fermi liquids provide us with rich systems on which to test both microscopic and phenomenological theories. Further study, both theoretical and experimental, is required to fully appreciate the richness of these systems.

I would like to thank T.L. Ainsworth, B.L. Friman and K.F. Quader for the contributions they have made in the development of the ideas discussed in this article. This work was supported in part by U.S. DOE Contract No. DEAC02-76ER13001.

References

1. S. Babu and G.E. Brown, Ann. Phys. 78(1973)1.
2. T.L. Ainsworth, K.S. Bedell, G.E. Brown and K.F. Quader, J. Low Temp. Phys. 50(1983)317; hereafter referred to as ABBQ.
3. K.S. Bedell and K.F. Quader, Phys. Lett. 96A(1983)91.
4. L.D. Landau, Zh. Eksp. Teor. Fiz. 30(1956)1058 [Sov. Phys. JETP 3 (1957)920].
5. A.A. Abrikosov and I.E. Dzyaloshinski, Zh. Eksp. Teor. Fiz. 35 (1959)771 [Sov. Phys. JETP 8(1959)535].
6. E.P. Bashkin and A.E. Meierovich, Usp. Fiz. Nauk 130(1980)279 [Sov. Phys. Usp. 23(1980)156].
7. K.S. Bedell and B.L. Friman, work in progress.
8. G. Baym and C.J. Pethick, in: The Physics of Liquid and Solid Helium, Part II, eds. K.H. Bennemann and J.B. Ketterson (Wiley, New York, 1978)1.
9. G. Baym and C.J. Pethick, in: The Physics of Liquid and Solid Helium, Part II, eds. K.H. Bennemann and J.B. Ketterson (Wiley, New York, 1978)123.
10. E. Krotscheck, J.W. Clark and A.D. Jackson, preprint 1983.
11. H.R. Glyde and S.I. Hernardi, invited talk presented at the Symposium on Quantum Fluids and Solids, Sanibel, Florida, AIP Conference Proceedings (1983).
12. G.R. Stewart, J.L. Smith and B.L. Brandt, Phys. Rev. B26(1982)3783.
13. C. Luhillier and D. Levesque, Phys. Rev. B23(1981)2203.
14. K.S. Bedell and T.L. Ainsworth, preprint 1982; T.L. Ainsworth, Ph.D. thesis, State University of New York at Stony Brook, 1983 (unpublished).
15. J. Hubbard, Proc. Roy. Soc. A276(1963)238.
16. T.L. Ainsworth, K.S. Bedell and Lin Yi, work in progress.

LINEAR AND NON LINEAR RESPONSE

A. Kallio, M. Puoskari, L. Lantto,
P. Pietiläinen and V. Halonen

University of Oulu
Department of Theoretical Physics
Linnanmaa, SF-90570 Oulu 57
Finland

I. Introduction

Suppose that we have a many-particle system in a weak external field $U(\vec{r},t)$ with hamiltonian

$$H = H_0 + \sum_{i=1}^{N} U(\vec{r},t) \tag{1}$$

If the approximate eigenstates of H are calculated in terms of the known eigenstates of H_0 by perturbation method keeping only terms linear in $U(\vec{r},t)$, when calculating the density fluctuation $\rho(\vec{k},\omega)$ around the equilibrium density ρ_0, the response function $\chi(\vec{k},\omega)$ is defined by

$$\rho(\vec{k},\omega) = \chi(\vec{k},\omega)\; U(\vec{k},\omega) \tag{2}$$

where $U(\vec{k},\omega)$ is the Fourier-transform of $U(\vec{r},t)$ multiplied by ρ_0. For static field $U(\vec{r})$ one simply drops the ω-dependence. A more detailed discussion of the general properties of $\chi(\vec{k})$ and $\chi(\vec{k},\omega)$ are to be found in references 1-2, where the conventional field theoretical approach is followed. In the instances where the correlations in H_0 are very strong the HNC-approach maybe also suitable. In the present paper we consider two systems, the electron gas and the liquid helium at $T = 0$. It is well known[2-6] that HNC theory works well for these systems. We have some preliminary results[7] also with non-homogeneous theory GHNC[8] applicable to nonuniform systems.

In this paper we will perform a calculation of the static response function $\chi(k)$. Clearly if we want to have a comparison with experiment in liquid helium or with previous field theoretical calculations in electron gas, we have to adopt the same phenomenological forms for the full response function $\chi(\vec{k},\omega)$ normally applied in these instances. For helium liquids exact and phenomenological response functions are defined by

$$\chi(\vec{k},\omega) = \frac{1}{N}\sum_n |(\rho_k)_{no}|^2 \frac{2\omega_{no}}{(\omega + i\eta)^2 - \omega_{no}^2} \tag{3a}$$

$$= \frac{\chi_0(\vec{k},\omega)}{1 - \left[f_s(k) + (\omega^2/\hbar^2 k^2)\, f_v(k)\right]\chi_0(\vec{k},\omega)} \tag{3b}$$

In formula (3a) ω_{no} is the excitation energy for the many-particle system with H_o and $\rho_k = \sum_i e^{i\vec{k}\vec{r}_i}$. In the phenomenological form (3b) used by Pines et al.[9-12] χ_o is the response function for non-interacting system

$$\chi_o^{-1} = \begin{cases} \chi_L^{-1} & \text{for Fermions} \\ m\omega^2/\hbar^2k^2 - \hbar^2k^2/4m, & \text{for Bosons} \end{cases} \tag{4}$$

The choice (4) is the simplest one but other choices are possible and have been used.[10-12] For the Coulomb systems one normally uses the form[2]

$$\chi(\vec{k},\omega) = \frac{\chi_o(\vec{k},\omega)}{1 - (4\pi e^2\rho_o/k^2)|1 - G(k)|\chi_o(\vec{k},\omega)} \tag{5}$$

Here G(k) is s.c. local field correction. From eq. (3b) one obtains the excitation spectrum

$$\omega(k) = \frac{\hbar^2k^2}{2m}\left[\frac{1 + 4mf_s/\hbar^2k^2}{1 - f_v/m}\right]^{1/2} = \frac{\hbar^2k^2}{2\tilde{m}^*(k)} \tag{6}$$

for liquid ^{4}He, which formally defines momentum dependent "effective mass" for excitations. In refs. 10-11, a momentum dependent effective mass is used already in $\chi_o(\vec{k},\omega)$ for the quasiparticles. It is determined from the backflow function $f_v(k)$ in the form

$$m^*(k) = m + \rho_o\, f_v(k) \tag{7}$$

This choice makes the comparison of their polarization potential f_s with our microscopic calculations very difficult. We therefore prefer to look for a possibility of using the definition (4) in a slightly different phenomenological analysis.

II. Response function from a mixture

In order to calculate the response function $\chi(k)$ we will apply the HNC-formalism for a mixture.[13-15] One can create external field by taking the limit of zero concentration and infinite mass for the second component. Then the cross component interaction $V_{\alpha\beta}$ becomes formally the external field U and the radial distribution function $g_{\alpha\beta}(r)$ gives the desired density fluctuation in the host system $\delta(r) = \rho(r) - \rho_o$

$$\rho_\alpha g_{\alpha\beta}(r) = \rho_o\, g_{\alpha\beta} = \rho(r) \tag{8}$$

The fully non-linear response to the impurity is obtained by solving the remaining Euler-Lagrange equation

$$-\frac{\hbar^2}{2\mu}\nabla^2\sqrt{\rho} + [U(r) + W(r)]\sqrt{\rho} = 0 \tag{9}$$

with

$$\begin{aligned} W(k) &= f_s(k)\left[1 - \frac{m-\mu}{\mu}\cdot S/S+1\right]\delta(k) \\ f_s(k) &= -(\hbar^2k^2/4m)(S^2-1/S^2) \end{aligned} \tag{10}$$

where S(k) is the liquid structure function of the host system and $\mu^{-1} = m_\alpha^{-1} + m_\beta^{-1}$. The linear response is obtained by taking the uniform limit solution of eq. (9) with $\mu = m$ (massive impurity) which is given by

$$\delta(k) = - U(k)/[\hbar k^2/4m + f_s(k)] \tag{11}$$

By eqs. (3b) and (4) we can identify $f_s(k)$ to be a microscopic expression for the polarization potential and the response function is simply given by

$$\chi(k) = - (4m/\hbar^2)\, S^2(k)/k^2 \tag{12}$$

For Coulomb Bose system one obtains by eqs. (5) and (12) the following expression to the local field correction[15] G(k)

$$\begin{aligned} G(k) &= \left[\chi^{-1} - \chi_0^{-1}\right]/v(k) \\ &= \frac{1}{v(k)}\left[-\frac{\hbar^2}{4m}\frac{k^2}{S^2} + v(k) + \frac{\hbar^2 k^2}{4m}\right] \end{aligned} \tag{13}$$

with

$$v(k) = \frac{4\pi e^2 \rho_0}{k^2}\ .$$

Equation (13) can also be solved for S(k) to obtain

$$S(k) = \frac{k}{\sqrt{k^2 - (4m/\hbar^2)\, v(k)(G - 1)}} \tag{14}$$

Here one notices that the normal uniform limit result for S(k) of Coulomb gas is obtained by setting G(k) = 0. If one applies the Lado approximation all above results apply to Fermion systems as well. Therefore in electron gas eq. (14) provides perhaps the simplest possible self-consistent relation between S(k) and G(k). The existence of such a relation has been discussed in connection with the STLS method[16] which in the lowest order gives a result very close to the one proposed by Hubbard.[17] The large k asymptotic limit of G(k) is 1 - g(0) in our case in agreement with VS-theory.[18] Our calculations show an overshoot G(k) with corresponding overshoot in S(k) whereas most of the other calculations mentioned above have none. Notably our results for G(k) are very similar to the ones obtained recently by Utsumi and Ichimaru.[19]

We close the discussion of electron gas by comparing our results for positron annihilation with the ones by Sjölander and Stott[28] and with experiment in Fig. 1. Especially at large values of r_s the linear response theory gives only a fraction of $\delta(r = 0)$ required by experiment.[15] In our case the singularity of ref. 28 near $r_s = 6$ appears at higher r_s-values and can be cured in a simple way.[20] Finally, we should mention that if one requires $\delta(r)$ to be exponential with cusp condition satisfied in addition to the screening condition[20] one essentially obtains our result in Fig. 1. In doing so one clearly ignores the Friedel oscillations whose amplitude we cannot as yet determine.

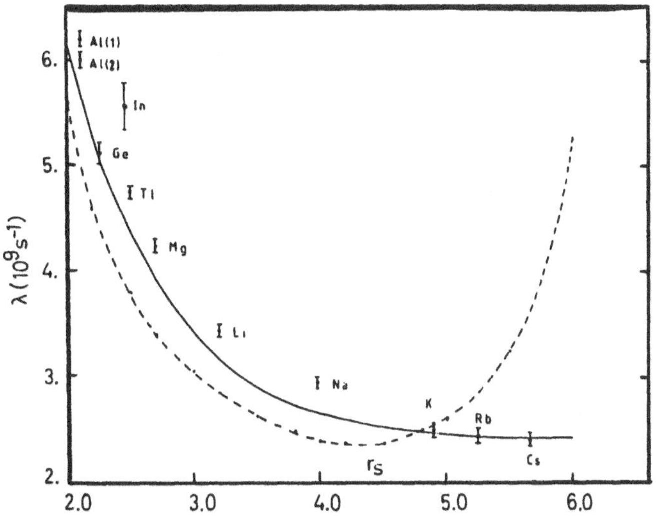

Fig. 1. Positron annihilation rate λ in metals. Solid curve gives the present result, dashed curve is from Ref. 28, and bars are the experimental results.

III. Backflow corrections in liquid heliums

If one wants to calculate $\chi(k)$ from the exact definition (3a) one needs the dispersion curve $\omega = \omega(k)$ and their corresponding wave functions as input. Suppose that ψ_o is the normalized groundstate wave function and we approximate the eigenstates ψ_k by the Feynman states

$$\psi_k = \rho_k \psi_o$$

we obtain from eq. (3a) by closure the expression

$$\chi(k) = -\frac{2}{N} \langle\psi_o|\rho_k^+ \frac{1}{H_o - E_o} \rho_k|\psi_o\rangle = \frac{-2S(k)}{\omega(k)} \tag{15}$$

which with Feynman dispersion law $\omega_F(k) = \hbar^2 k^2/2mS(k)$ reduces to our result in eq. (12). Keeping the ω -dependence this gives us dynamic structure factor

$$S(k,\omega) = S(k)\ \delta(\omega - \omega_F) \tag{16}$$

which satisfies all the sum rules.[1,5] At SVP $\chi(k)$ of eq. (12) or (15) behaves correctly both for small and large k. The small k limit is determined by the sound velocity or the compressibility

$$\chi(k = 0) = -\frac{1}{mc^2} = -\frac{1}{K} \tag{17}$$

As one might expect the agreement near the roton region is not very good. To improve things we have also done a calculation with backflow functions[21,5)]

$$\psi_k^{FC} = F_k \psi_o$$
$$F_k = \rho_k + \frac{A}{N} \sum_q 4\pi\rho_o \frac{\vec{k}\cdot\vec{q}}{q^2} \left[\rho_{k-q}\,\rho_q - \rho_k\right] \tag{18}$$

Using the notation of ref. 21 we now get from (3a)

$$\chi(k) = \frac{-2S(k)}{\omega(k)} \frac{[1 + .5\,Ak\,\xi_1(k)]^2}{[1 + Ak\,\xi_1(k) + (Ak)^2\xi_2(k)]} \tag{19}$$

with

$$\xi_1(k) = I_9(k)/S(k)$$
$$\xi_2(k) = I_{10}(k)/S(k)$$

Here I_9 and I_{10} are integrals defined in ref. 21, containing higher radial distribution functions. For the backflow parameter we use the classical value $A = -1/4\pi\rho$. The results obtained are compared with experiment in Fig. 2. It is seen that the backflow gives a dramatic improvement near the roton minimum. We have evaluated the expression (19) with experimental $\omega(k)$[23)] and $S(k)$[25)] and the χ_{exp} is from ref. 22. Since we can perhaps never obtain perfect fit to the experimental $\chi(k)$ by microscopic calculation, eq. (19) has motivated us to try a phenomenological model to push things further.

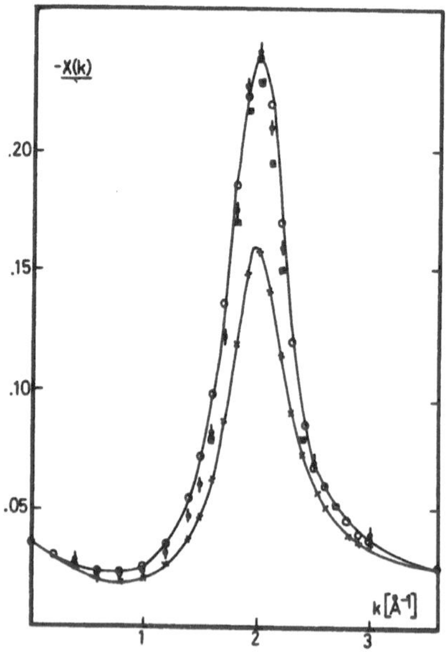

Fig. 2. The linear response function $\chi(k)$. Open circles give the experimental points from ref. 22, black circles with bar fit from eq.(20) and the crosses the lowest order result from eq. (12) calculated with S(k) of ref. 25, black squares from eq. (19).

It turns out to be rather easy to fit the experimental $\chi(k)$ in the form

$$\chi(k) = -4m/\hbar^2 \frac{S^2(k)}{k^2} \frac{1}{\gamma(k)} \tag{20}$$

which when the SVP data[25] for S(k) are used can be done fearly accurately by

$$\gamma(k) = 1 - \frac{S(k)}{1 + ak^2}$$
$$a = .73\ \text{Å}^2 \tag{21}$$

Since both large and small k-behaviour was correct, $\gamma(k)$ differs appreciably from 1 only in the middle. Requiring now that the full $\chi(k,\omega)$ of eq. (3b) has a pole at $\omega = \omega(k)$ and reduces to (20) when $\omega = 0$ we obtain the following suggestive dispersion law

$$\omega(k) = \frac{\hbar^2}{2m} \frac{k^2}{S(k)} \frac{\sqrt{\gamma(k)}}{\beta(k)} \tag{22}$$

with

$$\beta = (1 - f_v(k)/m)^{1/2}$$

We can also write it in the form of Feynman excitation with momentum dependent "effective mass"

$$\omega(k) = \frac{\hbar^2}{2m^*(k)} \frac{k^2}{S(k)} \tag{23}$$

Of course, our $f_v(k)$ or m* have nothing to do with the ones defined in refs. 10-12 and as we stressed this is all mean phenomenology except that the behaviour of $\beta(k)$ and $m^*(k)$ turns out to be rather striking, as is seen from Figs. 3 and 4.

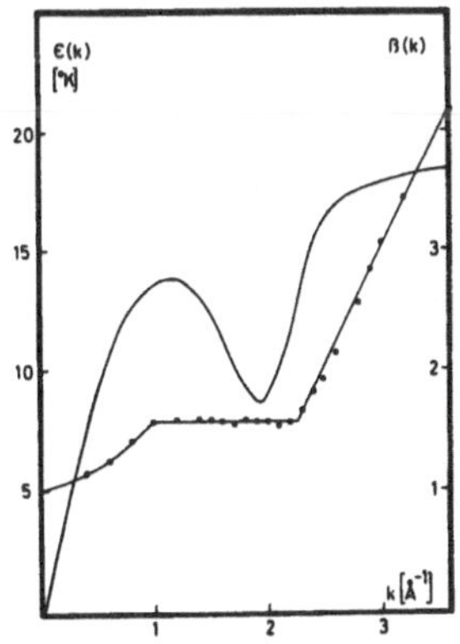

Fig. 3. The experimental excitation curve at SVP from ref. 23 and the effective mass function $\beta(k)$ (right scale).

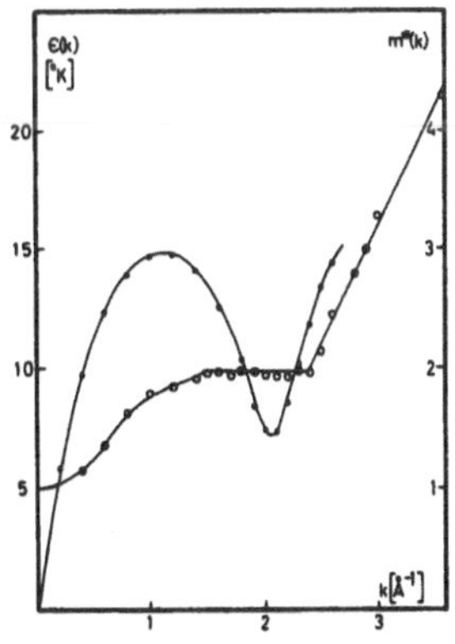

Fig. 4. The experimental excitation curve at 25.3 atm from ref.3 and the effective mass $m^*(k)$ at SVP (right scale).

We find a plateau for $m^*(k)$ and for $\beta(k)$ at wavevectors from 1.2 $Å^{-1}$ to 2.6 $Å^{-1}$ and from 1.0 $Å^{-1}$ to 2.3 $Å^{-1}$, respectively. This kind of behaviour for the effective mass was found already by Feynman and Cohen.[21] Miller, Pines and Nozieres[30] explained the constant effective mass at the roton region by regarding the Feynman excitations of wavevectors greater than 1 $Å^{-1}$ coupled to phonons in the same way as an impurity. This interpretation leads to regarding the roton as a He-atom moving through liquid surrounded by a phonon cloud that describes the backflow of the other atoms at long distances from the moving atom. The phonon cloud increases the effective mass of an impurity whose mass equals to that of ^{4}He-atom to the value[30] $m^* = 3m/2$, which is close to our effective backflow mass parameter $\beta = 1.6$ on the plateau. The endpoint of plateau for $m^*(k)$ at $k = 2.6$ $Å^{-1}$ might be attributed to the first momentum at which a decay of excitation becomes possible. Pitaevskii[31] has pointed out that an elementary excitation with the energy $\omega > 2\Delta$, where Δ is the energy of the roton minimum, can decay into two rotons. The excitation energy ω equals to 2Δ at $k = 2.68$ $Å^{-1}$ for the experimental dispersion curve and at $k = 2.4$ $Å^{-1}$ for the Feynman-Cohen curve, which agrees rather well with the endpoints of the plateaus for $m^*(k)$ and for $\beta(k)$.

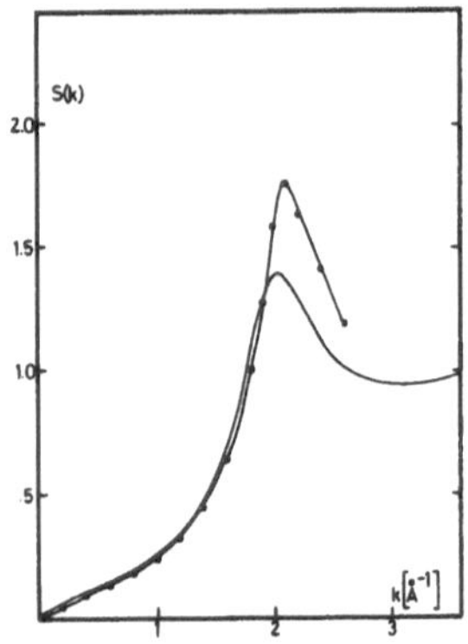

Fig. 5. The liquid structure function at SVP and at 25.3 atm (right circles).

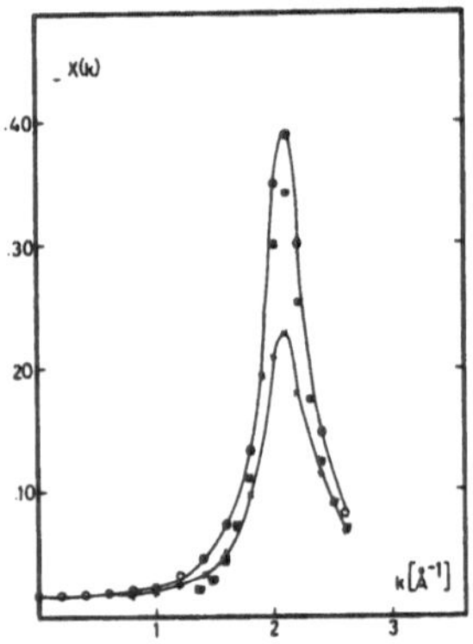

Fig.6. The linear response function at 25.3 atm predicted with corresponding S(k) of Fig. 5 and correction term of eq.(20)(open circles) and the lowest order result from eq. (12) (crosses), eq. (19) (black squares).

The main weakness of this model is that we have not studied the sum rules for $\chi(k,\omega)$ or $S(k,\omega)$ which would require a much deeper study of the analytical behaviours[6,26,27] of $\chi(k,\omega)$ than is possible here. Nevertheless, we can try to make a prediction to higher pressures, since this formalism is very simple. We have made a prediction of the unmeasured quantities $\chi(k)$ and $S(k)$ at the pressure 25.3 atm on the basis of the measured excitation curve which was read by eye from the ref. 12. With such a small amount of input data we had to assume that $\beta(k)$ is the same as in SVP and likewise the coefficient a in $\gamma(k)$ of eq. (21). The predictions are given in Figs. 5 and 6. Unfortunately, the excitation curve at 25.3 stops at $k = 2.2$ Å^{-1} so we cannot make prediction much beyond this. The nearest point in P.T-plane where $S(k)$ has been measured[24] (T = 1.67 P = 20 atm) agrees reasonably well with our result up to $k = 2.1$ Å^{-1}. Anyway the $S(k)$ obtained is not entirely reasonable, but any refinement would require more experimental data. A comparison of our microscopic expressions eqs. (12) and (20) for the polarization potential $f_s(k)$ with the one of Pines et al. is given in Fig. 7. The different usage of m* in χ_o shows up only at higher values of k. It is seen that already the lowest order gives nearly quantitative agreement.

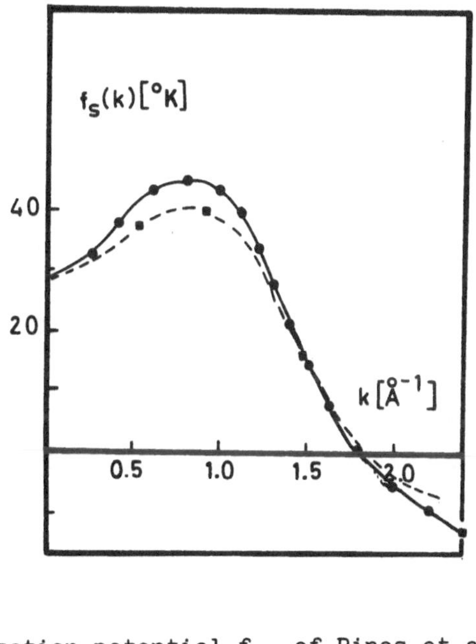

Fig. 7. Polarization potential f_s of Pines et al.[12] (dashed line) compared with microscopic expression of eq. (12) (solid line) and the fit with eq. (20) (squares).

Finally, we should mention that the inhomogeneous HNC-theory[29] also improves the $\chi(k)$ near roton minimum.

References

1. D. Pines and P. Nozieres, "The Theory of Quantum Fluids" Benjamin, New York, 1966.
2. K.S. Singwi and M.P. Tosi, "Correlations in Electron Liquid" Solid State Physics 36 (1981) p. 177.
3. J.G. Zabolitzky, Phys. Rev. B22 (1980) 2383.
4. L. Lantto and P.J. Siemens, Nucl. Phys. A317 (1979) 55.
5. C.E. Campbell, in Progess in Liquid Physics ed. by C.A. Croxton (Wiley, NY, 1978), L.R. Whitney, F.J. Pinski and C.E. Campbell, J. Low Temp. phys. 44 (1981) 367.
6. R.A. Smith, A. Kallio, M. Puoskari and P. Toropainen, Nucl. Phys. A238 (1979) 186.
7. M. Saarela, Univ. of Oulu, Dept. of Theor. Phys., Rep. 27 (To be published in Phys. Rev. B) (1983).
8. M. Saarela, P. Pietiläinen and A. Kallio, Phys. Rev. B27 (1983)231.
9. D. Pines and D. Bohm, Phys. Rev. 85 (1952) 338.
10. C.H. Aldrich and D.J. Pines, J. Low Temp. Phys. 25 (1979) 677.
11. C.H. Aldrich, C.J. Pethick and D.J. Pines, J. Low Temp. Phys. 25 (1976) 691.
12. D. Pines, Lecture Notes in Physics, vol. 142 (Springer-Verlag Berlin 1981) p. 202.
13. J.C. Owen, Phys. Rev. Lett.47 (1981) 586.
14. T. Chakraborty, Phys. Rev. B25 (1982) 3177.
15. P. Pietiläinen and A. Kallio, Phys. Rev. B27 (1983) 224.
16. K.S. Singwi, M.P. Tosi, R.H. Land and A. Sjölander, Phys. Rev. 176 (1968) 589.
17. J. Hubbard, Proc. R. Soc, London Ser. A 243 (1957) 336.
18. P. Vashista and K.S. Singwi, Phys. Rev. B6 (1972) 875.
19. K. Utsumi and S. Ichimaru, Phys. Rev. B22 (1980) 5203.
20. A. Kallio, P. Pietiläinen and L. Lantto, Physica Scripta 25 (1982) 943.
21. R.P. Feynman and M. Cohen, Phys. Rev. 102 (1956) 1189.
22. A.D.B. Woods and R.A. Cowley, Rep. Progr. Phys. 36 (1973) 1135.
23. R.J. Donelly, J.A. Donelly and R.N. Hills, J. Low Temp. Phys. 44 (1981) 471.
24. H.N. Robkoff and R.B. Hallock, Phys. Rev. B25 (1982) 1572.
25. V.F. Sears, E.C. Svensson, A.D.B. Woods and P. Martell, Atomic Energy of Canada Ltd. Report No. AECL-6779, 1979.
26. W. Götze and M. Lücke, Phys. Rev. B13 (1976) 3825.
27. H.W. Jackson, Phys. Rev. 185 (1969); A8 (1973) 1529.
28. A. Sjölander and M.J. Stott, Phys. Rev. B5 (1972) 2109.
29. M. Saarela, P. Pollari and J. Ylätalo, poster to the present conference.
30. A. Miller, D. Pines and P. Nozieres, Phys. Rev. 127 (1962) 1452.
31. L.P. Pitaevskii, Soviet Phys. JETP 9, (1959) 830; Soviet Phys. JETP 12 (1961) 155.

CORRELATIONS AND THE POSSIBILITY OF A CHARGE-DENSITY-WAVE (CDW) INSTABILITY IN QUANTUM ELECTRON LIQUIDS

K. S. Singwi
Department of Physics and Astronomy
Northwestern University, Evanston, IL 60201

Abstract

It is shown that the static local-field factor $G(\vec{q},o)$, as calculated from the quantum version of the STLS theory, has a structure that leads to the possibility of a charge-density-wave (CDW) instability in a 3D-homogeneous electron liquid beyond a certain r_s value.

Introduction

The problem of correlations in quantum electron liquids, although an old one, still holds some fascination for theoretical physicists for a number of reasons: (a) Electron liquid is relatively a simple many-body system and still not fully understood from a microscopic point of view[1]. (b) Recent "exact" results for the ground-state properties obtained by computer simulations[2] have given further stimulus to the theorist to refine the results of his approximate theories. (c) The discovery of the electron-hole liquid in semiconductors such as Germanium and Silicon has provided an ideal medium in which once again the many-body theories can be tested against laboratory experiments[3]. Perhaps, the main challenge now lies not so much in calculating the ground-state properties but the excitation spectrum of the electron liquid. It is somewhat unfortunate that more experimental work in this direction is not forthcoming. Let us hope that not before long quantum dynamics will be incorporated into computer simulation studies.

My objective in this talk is very limited. I wish to report to you on some new results[4] concerning the short-range aspect of the Coulomb and exchange correlations in an electron liquid within the framework of the quantum version of the theory of Singwi et al.[5] (STLS) as formulated by Hasegawa and Shimizu[6]. It turns out that the local field factor $G(\vec{q},o)$, which is a measure of short-range correlations, has a broad peak around momentum transfer $q \simeq 2q_F$. This behavior of $G(\vec{q},o)$, if indeed genuine, leads one to conclude that for a certain value of the density, r_s, the electron liquid will develop a charge-density-wave (CDW) instability. What other consequences this structure in $G(\vec{q},o)$ will have still need to be examined ?

II. Theoretical Considerations

Consider a homogeneous system of electrons of density n_o on a uniform, rigid positive background in the presence of a weak external potential $V_{ext}(\vec{r},t)$. The Fourier transforms of the induced density and the external potential are related by

$$\langle \delta n(\vec{q},\omega)\rangle = \chi(\vec{q},\omega)\left[V_{ext}(\vec{q},\omega)\right], \tag{1}$$

where χ is the density-density response function. Equation (1) is the defining equation for χ. In the presence of V_{ext}, the system will get polarized and this polarization will produce an added potential. Formally, one can write for the average induced density the expression[1]:

$$<\delta n(\vec{q},\omega)> = \chi_{eff}(\vec{q},\omega)\left[V_{ext}(\vec{q},\omega) + \Psi(\vec{q},\omega)<\delta n(\vec{q},\omega)>\right], \qquad (2)$$

where χ_{eff} is some effective response and $\Psi<\delta n>$ is the polarization potential. In eqn. (2), both χ_{eff} and Ψ are unknown functions. In the STLS scheme, one derives from microscopic considerations equation of the kind (2) where χ_{eff} is replaced by χ_o, the polarizability of a noninteracting electron gas, and Ψ is given by

$$\Psi(\vec{q},\omega) = v(\vec{q})\left[1-G(\vec{q})\right], \qquad (3)$$

where

$$v(\vec{q}) = 4\pi e^2/q^2$$

and

$$G(\vec{q}) = -\frac{1}{n_o}\int \frac{d\vec{q}\,'}{(2\pi)^3}\frac{\vec{q}\cdot\vec{q}\,'}{q'^2}\left[S(\vec{q}-\vec{q}\,')-1\right]. \qquad (4)$$

$S(q)$ is the structure factor (the F.T of the pair correlation function). From eqns. (1)-(3), it follows that

$$\chi(\vec{q},\omega) = \frac{\chi_o(\vec{q},\omega)}{1-v(\vec{q})\left[1-G(\vec{q})\right]\chi^o(\vec{q},\omega)} \qquad (5)$$

Knowing that the dielectric function is related to χ by

$$\frac{1}{\varepsilon(\vec{q},\omega)} - 1 = v(q)\chi(q,\omega), \qquad (6)$$

we have for the former the following expression,

$$\varepsilon(\vec{q},\omega) = 1 - \frac{v(q)\chi_o(\vec{q},\omega)}{1+G(\vec{q})v(\vec{q})\chi_o(\vec{q},\omega)} \qquad (7)$$

There are several points worth noting in the above set of equations. The so-called local-field factor $G(q)$ is ω-independent. When $G=o$, the above scheme reduces to the standard RPA. $v(q)G(q)$ represents the short-range part of the effective interaction between two electrons in the electron liquid. It can be viewed as a local t-matrix. Since it contains both exchange and Coulomb correlations in it, it is natural that it should be related to the exchange and Coulomb correlation hole around an electron. That it is indeed so in the STLS scheme is evident from eqn. (4). Since $S(q)$ is related to the $\mathrm{Im}\chi(q,\omega)$ through the fluctuation-dissipation theorem

$$S(\vec{q}) = \frac{\hbar}{\pi n_o}\int_o^\infty d\omega\ \mathrm{Im}\ \chi(\vec{q},\omega), \qquad (8)$$

the above scheme is self-consistent. Both the RPA and the well-known Hubbard approximation[7] are just special cases of this scheme; the latter is recovered[5] by substituting for $S(q)$ in eqn. (4) its Hartree-Fock value. The Hubbard local-field factor has the form

$$G_H(q) = \frac{1}{2}\frac{q^2}{q^2+k_F{}^2} \tag{9}$$

The above self-consistent scheme gives very good values for the ground-state energy and the pair-correlation function both in 3D- and 2D-electron liquids. A later modification of the theory by Vashishta and Singwi[8] yields good values for the compressibility. The theory is parameter free. Also, further refinements, in particular in the calculation of the excitation spectrum, are possible since the theory is based on microscopic equations of motion. One such refinement I shall be discussing here. But before I do that, let me make contact with the polarization potential approach recently extended by Iwamoto and Pines[9] to electron liquids from the earlier work of Aldrich and Pines[10] in liquid He^3. Their density-density response function is

$$\chi(\vec{q},\omega) = \frac{\chi_o(\vec{q},\omega)}{1-\left[v(q)+f^s(q)\right]\chi_o(\vec{q},\omega)} \tag{10}$$

On comparing eqns. (10) and (5), we have

$$f^s(q) = -v(q)G(q), \tag{11}$$

The function $f^s(q)$ is introduced here phenomenologically and is supposed to take care of sort-range correlations. Its functional form has to be guessed and the parameter (if only one) has to be fixed by some constraint such as the compressibility sum rule. The limitations of this approach are, therefore, obvious. To get some physical insight, let $V_q^{\uparrow\uparrow}$ and $V_q^{\downarrow\uparrow}$ represent, respectively, some effective interaction between electrons of parallel and antiparallel spin in the electron gas. Let

$$V_q^{\uparrow\uparrow} = v(q) + f^{\uparrow\uparrow}(q) \tag{12a}$$

and

$$V_q^{\uparrow\downarrow} = v(q) + f^{\uparrow\downarrow}(q), \tag{12b}$$

where the f's represent the short-range part of the interaction. One then has

$$V_q^s \equiv \frac{1}{2}(V_q^{\uparrow\uparrow} + V_q^{\uparrow\downarrow}) = v(q) + f^s(q), \tag{13a}$$

and

$$V_q^a \equiv \frac{1}{2}(V_q^{\uparrow\uparrow} - V_q^{\uparrow\downarrow}) = f^a(q), \tag{13b}$$

where

$$f^s(q) = \frac{1}{2}(f^{\uparrow\uparrow}(q) + f^{\uparrow\downarrow}(q)) \tag{14a}$$

and

$$f^a(q) = \frac{1}{2}(f^{\uparrow\uparrow}(q) - f^{\uparrow\downarrow}(q)) \tag{14b}$$

The spin symmetric $f^s(q)$ and the spin antisymmetric $f^a(q)$ functions have the significance of the usual f-functions occurring in the Landau Fermi-liquid theory. It is strightforward to show that they are related to the compressibility and the susceptibility of the electron gas, respectively. The relations are:

$$\lim_{q\to o} f^s(q) = \frac{2}{3}\frac{E_F^o}{n_o}\left[\frac{K_o}{K} - 1\right] \tag{15a}$$

and

$$\lim_{q\to o} f^a(q) = \frac{2}{3}\frac{E_F^o}{n_o}\left[\frac{\chi_p^o}{\chi_p} - 1\right], \tag{15b}$$

where E_F^o is the Fermi energy, $\frac{K_o}{K}$ and $\chi^o{}_p/\chi_p$ are, respectively, the ratios of free to interacting compressibility and paramagnetic susceptibility, n_o is the electron density. The right hand side of eqns. (15a) and (15b) is known through numerical simulation studies. This determines the value of the f'^s at one point i.e. q=o.

Iwamoto and Pines assume for the effective short-range potentials a Yukawa form i.e. they write

$$V^{\uparrow\uparrow}(\vec{r}) = \frac{e^2}{r} - \frac{e^2}{r} e^{-q_{\uparrow\uparrow} r}, \tag{16a}$$

and

$$V^{\uparrow\downarrow}(r) = \frac{e^2}{r} - \frac{e^2}{r} e^{-q_{\uparrow\downarrow} r} \tag{16b}$$

The parameters $q_{\uparrow\uparrow}$ and $q_{\uparrow\downarrow}$ are determined through the use of eqns. (13)-(15). The considerations which have probably guided these authors to assume a Yukawa form for the short-range potential are: (i) its simplicity and (ii) it gives them for G(q) (see eqn. (11)) a Hubbard-like form which is not too unreasonable. Besides, these "pseudo potentials" have to be attractive to cancel the singularity of the bare Coulomb potential at r=o, and hence the minus sign before the second term in (16). The form of G(q) in this approach is

$$G_{I-P}(q) = \frac{1}{2}\left[\frac{q^2}{q^2+q^2_{\uparrow\uparrow}} + \frac{q^2}{q^2+q^2_{\uparrow\downarrow}}\right] \tag{17}$$

$$\lim_{q\to o} G_{I-P}(q) = \frac{1}{2}\left[\frac{1}{q^2_{\uparrow\uparrow}} + \frac{1}{q^2_{\uparrow\downarrow}}\right] q^2$$

$$= (1-\frac{K_o}{K})\frac{q^2}{k^2_{TF}} \tag{18}$$

$k_{TF} = \left[\frac{6\pi n_o e^2}{E_F^o}\right]^{1/2}$ is the inverse of the Thomas-Fermi screening length.

$$\lim_{q\to\infty} G_{I-P}(q) = 1. \tag{19}$$

Note that the last relation is very different from that in the STLS theory and is independent of r_s. Niklasson has given the following exact result:

$$\lim_{q\to\infty} G(q) = \frac{2}{3}(1-g(o)), \tag{20}$$

where g(o) is the value of the pair correlation function at r=0.

This is in brief the polarization potential approach of Iwamoto and Pines. It has obviously the virtues and the faults of a phenomenological theory.

III $G(\vec{q})$ in the Quantum Version of the STLS Theory

The original STLS theory[5] is in a sense semiclassical since it starts with the equation of motion of one-particle classical phase-space distribution function. This equation involves in the interaction term two-particle phase-space distribution function. In the STLS approach this hierarchy of the BBGKY equations, in which phase-space distribution functions of higher order enter successively, is truncated right at the first equation of the hierarchy by expressing the two-particle distribution function as a product of one-particle distribution functions and the equilibrium pair-correlation function. This then leads[5] straightforwardly to a classical expression of G(q) given in eqn. (4). A quantum version of this approach was first given by Hasegawa and Shimizu[6] and later but independently by P.K. Aravind[11], who started with the equation of motion for one-particle Wigner distribution function. Their truncation procedure is exactly the same as that of STLS, i.e.

$$f^{(2)}_{\vec{k}\sigma,\vec{k}'\sigma'}(\vec{r},\vec{r}';t) = f^{(1)}_{\vec{k},\sigma}(\vec{r},t)\, f^{(1)}_{k',\sigma'}(r',t)g(|\vec{r}-\vec{r}'|),$$

where $f^{(2)}$ and $f^{(1)}$ are the two- and one-particle Wigner distribution functions, respectively; and g(r) is the equilibrium pair-correlation function. They then arrive at the following expressions for G:

$$G(\vec{q},\omega) = -\frac{1}{\chi_o(\vec{q},\omega)}\frac{1}{N}\sum_{\vec{q}'}\sum_{\vec{k},\sigma}\frac{n_{\vec{k}-\frac{1}{2}\vec{q}',\sigma} - n_{\vec{k}+\frac{1}{2}\vec{q}',\sigma}}{\hbar\omega - \frac{h^2}{m}\vec{k}\cdot\vec{q} + i\eta}\;.$$

$$\cdot\;\frac{q^2}{q'^2}\left[S(\vec{q}-\vec{q}')-1\right] \qquad (21)$$

Note that G is now frequency dependent and is a complex function. It is possible to show[11] that in the classical limit (21) reduces to (4). Also in the limit $\omega\to\infty$ $G(q,\omega)\to G_{STLS}(q) + O(\frac{1}{\omega^2})$. Hasegawa and Shimizu[6] have calculated the ground-state energy and the pair-correlation function for various r_s values for the electron liquid and have found that their results are of the same quality, if not slightly better, than those of the original STLS theory. My interest here is mainly in the static $G(\vec{q},o)$. We have recently[4] performed a fully self-consistent numerical calculation of $G(\vec{q},\omega)$ within the quantum version of the STLS theory for a number of r_s values. A typical result for $G_{QSTLS}(\vec{q},o)$ for two values of r_s=1 and 4 is shown in Fig. 1, where $G_{STLS}(q)$ is also displayed for comparison. The main difference between the two G's is that $G_{QSTLS}(\vec{q},o)$ has a broad peak in the neighborhood of $q=2.6k_F$, whereas $G_{STLS}(q)$ is a smooth monotonically increasing function which saturates to some fixed value

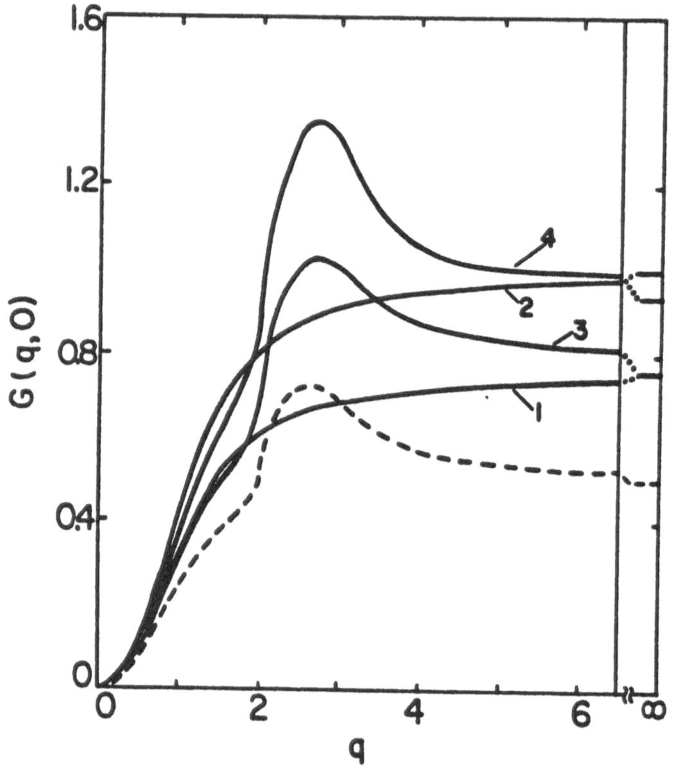

Fig. 1. Static part of the local field factor G(q,o) for two r_s values: ___ r_s=2 with S_{HF} as input in eqn. (21); (1) and (2) in STLS at r_s=1 and 4; (3) and (4) in QSTLS at r_s=1 and r_s=4.

for $q \to \infty$ (saturation values for the two G's are nearly the same). A more interesting point to note is that the peak value of $G_{QSTLS}(q,o)$ is greater than unity and which value increases as r_s increases. Our numerical results show that this value has the tendency to saturate around 1.4-1.5 for $r_s \geq 10$. In Fig. 2 we have compared $G_{QSTLS}(q,o)$ with the G(q)'s in other theories[12] for r_s=2. Note that the first-order theory[13] gives a large and a sharp peak at $q=2k_F$. The higher-order correlations, which are taken into account in the present theory, have considerably reduced the value of the peak in G(q,o) and have not only broadened but shifted it to somewhat larger value of q. The occurrence of a peak in G(q) in other theories, which are not of the Hartree-Fock kind, seems to indicate that it is genuine. The precise magnitude of the peak value is not at all certain because of the approximate nature of the theories.

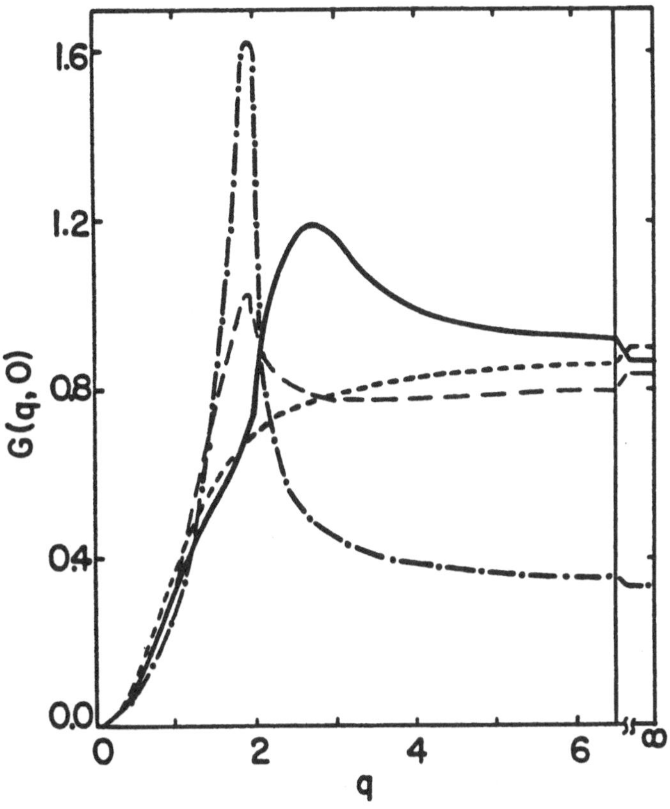

Fig. 2. Static part of the local-field factor G(q,o) at r_s=2: ____ in QSTLS Ref. 4, -.-.-.in the 1st-order theory (Ref. 13); _ _ in UI (Ref. 12), ___ in STLS (Ref. 5).

IV. Possibility of a CDW Instability

One of the interesting consequences of $G(\vec{q},o)$ being greater than unity is that the electron liquid can become unstable against the formation of a charge-density-wave[14] (CDW) at a certain r_s value. I am fully aware that I am here treading on a slippery path but nonetheless I would venture it. The quantity $1/\varepsilon(\omega,\vec{q})$ satisfies the Kramers-Kronig dispersion relation:

$$\mathrm{Re}\frac{1}{\varepsilon(\omega,\vec{q})} = 1 - \int_0^\infty \frac{dE^2\sigma(E,\vec{q})}{E^2-\omega^2-i\eta} \tag{22}$$

The spectral density

$$\sigma(\omega,\vec{q}) = -\frac{1}{\pi}\,\mathrm{Im}\left[\frac{1}{\varepsilon(\omega,\vec{q})}\right], \tag{23}$$

is a positive definite quantity since it is directly related to the experimentally observed dynamic form factor, which measures the probability of energy loss. From (22) it follows that

$$\frac{1}{\varepsilon(o,\vec{q})} = 1 - I, \tag{24}$$

where

$$I = 2\int_{o}^{\infty} \frac{dE}{E}\, \sigma(E,\vec{q}) \geq 0 \tag{25}$$

Therefore

$$\frac{1}{\varepsilon(o,\vec{q})} \leq 1 \tag{26}$$

On the other hand, $\varepsilon(o,\vec{q})$ should satisfy one of the inequalities

$$\varepsilon(o,\vec{q}) \geq 1, \quad \varepsilon(o,\vec{q}) \leq 0 \tag{27}$$

The first inequality corresponds to $I<1$, and the second to $I>1$.

Violation of the inequality (26) leads to the instability of the system and to the spontaneous breaking of the system's intrinsic symmetry. Let us now examine under what conditions this violation happens in the case of an electron liquid on a homogeneous, rigid, positive background. The inequality (26) is equivalent to saying that

$$\chi(o,\vec{q}) < 0, \tag{28}$$

because of eqn. (6). Since $\chi_o(\vec{q})$ is -ve, let us rewrite eqn. (5) as

$$\chi(\vec{q}) = -\frac{|\chi_o(\vec{q})|}{1+v(q)(1-G(\vec{q}))|\chi_o(\vec{q})|} \tag{29}$$

Inequality (28) implies that the demoninator of (29) is greater than zero. Instability will therefore set in when

$$1+v(\vec{q})\,(1-G(\vec{q}))|\chi_o(\vec{q})| = 0 \tag{30}$$

or

$$G(\vec{q}) = 1 + \frac{1}{v(\vec{q})\,|\chi_o(\vec{q})|} \tag{31}$$

Equation (31) can also be derived[14] (S. Ichimaru) using density-functional formalism. However, the above derivation is more satisfactory since it does not involve any approximation.

Obviously, if $G(q) \leq 1$ no instability occurs. This is so in the Hubbard and original STLS theories. In the quantum version of the latter this, however, is not the case since G(q) can become greater than unity. We can easily make a crude estimate of the r_s value at which equation (31) will be satisfied. Knowing $|\chi_o(q)|$, eqn. (31) can be written as

$$G(q) = 1 + \frac{\pi q^2}{4\alpha r_s} \frac{1}{g(q)} , \tag{32}$$

where the function g(q) has the following limiting behavior

$$\begin{aligned} g(q) &\simeq 1 + O(q^2) & q&<<1 \\ &\simeq \frac{1}{2} & q&=2 \\ &\sim \frac{4}{3} \frac{1}{q^2} & q&>>1 \end{aligned}$$

From the general behaviour of G(q) as we have seen in Fig. 1 and from an examination of the RHS of eqn. (32), one can easily convince oneself that this equation will be satisfied for some r_s value. Our numerical calculations indicate that G(q) changes very little, if any, beyond $r_s \simeq 10$. Numerical solution of eqn. (31) gives an r_s value of 65 and the wave vector at which the instability occurs is $q = 2.27k_F$.

V. Conclusion

We have shown that within the framework of the quantum version of the STLS theory, a homogeneous electron liquid on a uniform rigid positive background exhibits a CDW instability at the density $r_s \simeq 65$. This wave has a wave number $k \simeq 2k_F$.

Let me end with a note of caution. The above conclusion will be invalid if it turns out that in an exact theory the behaviour of G(q) is different and its value never exceeds unity. So far the STLS theory has been fairly trustworthy, but one can never tell.

Ceperley and Alder[2] do not report having seen this instability. It is not clear from their numerical work whether they actually looked for it. These authors also point out that (in the range of interest of r_s values here) "the energies of the three Fermion states (para, ferro and Wigner crystal) are sufficiently close that still more accurate calculations on larger systems would be desirable to confirm these results". In my view it would be very interesting to look specifically and carefully for the CDW instability.

Acknowledgements

I thank Ciovanni Vignale and S. Rahman for performing numerical calculations and for many discussions. I am also thankful to Professor Mario Tosi of the International Center for Theoretical Physics, Trieste, for critical remarks. This work was completed while the author was spending a summer month (1983) at the Center, whose hospitality is gratefully acknowledged, and was mainly supported by NSF Grant No. DMR-80-11934.

References

1. For a review, see K.S. Singwi and M.P. Tosi, in Solid State Physics, edited by H. Ehrenreich, F. Seitz and D. Turnbull (Academic, New York, 1981), Vol. 36.

2. D.M. Ceperley and B.J. Alder, Phys. Rev. Lett. 45, 566 (1980).

3. For a review, see P. Vashishta, R.K. Kalia and K.S. Singwi, in Electron-Hole Liquid, edited by L.V. Keldysh and D.C. Jeffries (North-Holland, Amsterdam, 1983).

4. S. Rahman, G. Vignale and K.S. Singwi (to be published). See also, S. Rahman, Ph.D. Thesis, 1983, Northwestern University, Evanston, Illinois 60201.

5. K.S. Singwi, M.P. Tosi, R.H. Land and A. Sjölander, Phys. Rev. 176, 589 (1968).

6. T. Hasegawa and M. Shimizu, J. Phys. Soc. Japan 38, 965 (1975). See also, G. Niklasson, Electron Correlations in Solids, Molecules and Atoms, edited by J.T. Devreese and F. Brosens (NATO Advanced Study Institutes Series, Volume 81, Plenum 1983).

7. J. Hubbard, Proc. Roy. Soc. (Lond.) A243, 336 (1957).

8. P. Vashishta and K.S. Singwi, Phys. Rev. B6, 875, 4883 (E) (1972).

9. Naoki Iwamoto and David Pines, preprint, 1983.

10. C.H. Aldrich, III and D. Pines, J. Low Temp. Phys. 32, 689 (1978).

11. P.K. Aravind, Ph.D. Thesis, 1979, Northwestern University, Evanston, Illinois 60201.

12. K. Utsumi and S. Ichimaru, Phys. Rev. B22, 1522, 5203 (1980); and L. Lantto, P. Pietilainen and A. Kallio, Phys. Rev. B26, 5568 (1982).

13. A. Holas, P.K. Aravind and K.S. Singwi, Phys. Rev. B20, 4912 (1979). See also, F. Brosens, J.T. Devreese and L.F. Lemens, Phys. Rev. B21, 1363 (1980).

14. See A.W. Overhauser in Electron Correlation in Solids, Molecules, and Atoms, Edited by J.T. Devreese and F. Brosens (Plenum Press, 1983) and references therein. Se also, S. Ichimaru, Rev. Mod. PHys., Vol. 54 (1982) for a discussion of instabilities in an electron liquid.

CBF THEORY OF METAL SURFACES: CHEMISORPTION

Chia-Wei Woo* and Mani Farjam
Department of Physics
University of California, San Diego
La Jolla, CA 92093, USA

This is a progress report on our correlated basis functions (CBF) work on the theory of metal surfaces and quantum adsorbates.

In a paper to be published in Phys. Rev. B,[1] we spelled out details of a CBF theory for the ground state of metal surfaces. Results were obtained for r_s in the range 2 to 6, and compared to both experimental data and theoretical results reported by other researchers. We shall present here a brief review of that work, and proceed to describe our approaches toward treating the chemisorption of hydrogen atoms on metal substrates.

I. Ground State of a Metal Surface

The conventional many-body method for dealing with metal surfaces is the density functional theory (DFT) for inhomogeneous systems, invented by Kohn and coworkers, and first applied to treating metal surfaces by Lang and Kohn.[2] In that method, the ground state energy is expressed as a functional of the density profile $n_M(\vec{r})$, and minimized with respect to the latter by solving an auxiliary self-consistent equation. One complication is that the exchange and correlation terms in the energy functional cannot be expressed in a closed form. In the lowest approximation, one employs a local density approximation and borrows from other independent work the exchange and correlation energies of homogeneous electron liquids calculated at various local densities. In higher orders, one introduces a density gradient expansion or effects a wave-vector interpolation scheme. Another complication centers on the description of the positive charge background. Low order perturbation corrections on the surface energy are sought when the jellium model is replaced by a more realistic discrete ion lattice. The convergence of such a perturbation expansion cannot be guaranteed.

In the CBF approach, one begins with a variational wave function which takes into account electron correlations. Suppressing spins, it takes the form:

$$\psi(1,2,\cdots,N) = \prod_{i<j} \exp\left[\tfrac{1}{2}u(r_{ij})\right] \cdot D[\varphi],$$

where $D[\varphi]$ denotes a determinant made up of single-particle elements $\{\varphi_{\vec{k}}(\vec{r})\}$. The expectation value of the full Hamiltonian is evaluated with this wave function, and minimized with respect to $u(r)$ and $\varphi_{\vec{k}}(\vec{r})$. The difficulty, as in all CBF calculations, resides in the evaluation of the expectation value. We do it by first calculating the density $n_o(\vec{r})$ for the uncorrelated wave function $D[\varphi]$, next solving an integral equation that relates $n_o(\vec{r})$ to the density $n_\psi(\vec{r})$ defined for the correlated wave function ψ, and finally evaluating the variational energy E_ψ from an expression involving $n_\psi(\vec{r})$ and two- and three-particle distribution functions. The latter functions have to be approximated by e.g. a convolution scheme, and normalized by a local sum rule.

Our progress was hampered by the appearance of a series of spurious divergences in the integral equation and the energy expectation value. They were removed one by one as we learned to take limits properly: the size of the system, the range of Coulomb correlations, and the distance between the cleaved surfaces have to be taken to their infinite limits in the proper order.

The surface energy per electron, σ, is obtained from subtracting the bulk energy exactly from the minimized variational energy. Throughout the metallic range ($2<r_s<6$), our calculated results turn out to be in good agreement with experiment and improved over the DFT. The work function, Φ, is in reasonably good agreement with experiment, though at small r_s poorer than the DFT results.

Some advantages of the CBF over the DFT are as follows.

1. The variational procedure is applied directly on the wave function, rather than the density; and the wave function is ours to design. In the present case, it has been constructed to account for single-particle and many-particle features in an explicitly balanced way. The method also lends itself to systematic improvements. Two options are available: introduce higher order correlating factors, or carry out perturbation calculations in the correlated representation.
2. The energy is expressed in terms of closed integrals. The theory is free of density gradient expansions or interpolation schemes. Nor is it necessary to

depend on a low order perturbation expansion when the ion lattice is introduced or varied. In fact, by adopting $\{\varphi_{\vec{k}}(\vec{r})\}$ from band theory calculations, one can retain all the successes of the latter and still account for many-body correlation effects in a straightforward manner.

3. The CBF theory gives us many-body wave functions. The knowledge of these wave functions, and not just the density, permits us to do microscopic, non-phenomenological theory for atomic and molecular adsorption. In particular, we can calculate directly the adsorption potentials, and the substrate-mediated interactions between a pair of adsorbed atoms/molecules. These form the basic ingredients of a self-contained theory of chemisorption. Applications include the determination of activation energies for dissociative chemisorption, the prediction of coverage-dependent sticking coefficients, configurational phase transformations in adsorbed layers, and orientational order-disorder transitions in adsorbates of elongated molecules.

II. Hydrogen Atom Chemisorbed on a Metal Surface

Let us consider the chemisorption of a hydrogen atom on a metal surface. The model consists of N+1 electrons in the field of a semi-infinite positive ion lattice and a stationary positive ion at a distance Z from the metal surface. If the substrate is described as a jellium, the lone model parameter Z suffices. If the ion lattice is realistically discrete, two other parameters are necessary, $\vec{R} \equiv (X,Y,Z)$, to locate the ion in a coordinate system from an origin fixed on some lattice point.

We wish to determine the adsorption energy at every surface-to-ion separation $\vec{R}$. If the linear dimensions of the substrate are measured by L, the bulk energy is of order L^3, as is the system energy. The surface energy, which is the difference between the two, is of order L^2. The adsorption energy is of order L^0. Our purpose is to identify this L^0 term.

We are pursuing this calculation in two directions. One may be referred to as a Heitler-London neutral adsorption model, and the other as a molecular-orbital positive-ion adsorption model.

In the Heitler-London case, we picture the substrate as a big molecule consisting of N positive ions and a like number of electrons. The hamiltonian of this subsystem is to be denoted by $H_M(1,2,\ldots,N+1|m)$: the same hamiltonian used in determining the properties of an isolated metal substrate containing N electrons (1 through N+1, minus the mth electron). We describe the adsorbed atom, the other subsystem, as for an isolated hydrogen atom, with a hamiltonian $H_A(m)$, where m labels the coordinates of the single electron. The combined system has a hamiltonian $H = H_M+H_A+H'$, where

$$H' = \sum_{\substack{i=1\\(\neq m)}}^{N+1} \frac{e^2}{|\vec{r}_i-\vec{r}_m|} - \int d\vec{r}\, \frac{e^2 n_+(\vec{r})}{|\vec{r}-\vec{r}_m|} - \sum_{\substack{i=1\\(\neq m)}}^{N+1} \frac{e^2}{|\vec{R}-\vec{r}_i|} + \int d\vec{r}\, \frac{e^2 n_+(\vec{r})}{|\vec{r}-\vec{R}|}$$

accounts for the interaction between constituents of one subsystem and those of the other subsystem, $n_+(\vec{r})$ representing the "known" positive charge distribution in the substrate. The thought underlying the Heitler-London approximation is the relative insignificance of H'. In that case, a good trial wave function for the combined system will be the antisymmetrized product of the ground state solutions for the two sybsystems. In other words,

$$\begin{aligned}\Psi(1,2,\cdots,N+1|m) &= \mathcal{A}\{\Phi_M(1,2,\cdots,N+1|m),\, \phi_A(m)\}\\ &= \sum_{i=1}^{N+1}(-1)^{i-1}\,\Phi_M(1,2,\cdots,N+1|i)\,\phi_A(i).\end{aligned}$$

The expectation value of H in Ψ gives an upper bound to the system energy. It consists of terms of orders L^3, L^2, and L^0, respectively. Since it is expressed in terms of the electronic density profile $n(\vec{r})$ and several-particle distribution functions, we once again need to call on the convolution approximation and a sum rule, and determine the new profile

$$n(\vec{r}_1) = (N+1)\,\frac{\int|\Psi|^2 d\vec{r}_2\cdots d\vec{r}_{N+1}}{\int|\Psi|^2 d\vec{r}_1\cdots d\vec{r}_{N+1}}$$

by solving an integral equation. The equation is much more complicated than in Ref. 1. The solution will hopefully reflect, to some degree, the existence of a chemical bond.

Much of the formalism has already been worked out.[3] In view of an argument put forth by Schrieffer[4], we will not expect to see bonding unless excited states of the substrate are taken into account to allow for spin flips and spin polarization on the substrate surface. In other words, a linear combination of Ψ's constructed from a "complete" set of Φ_M's will be required. Please note that Schrieffer employed a one-electron description for the substrate[4], while our Φ_M contains many-body correlations. Nevertheless, Schrieffer's reasoning appears to remain valid.

For the positive-ion adsorption model, we treat all N+1 electrons alike. The calculations proceed in a fashion which closely parallels that of Ref. 1.

A proton placed at $\vec{R}$ causes the electronic density profile $n_M(\vec{r})$ to shift to $n_{MA}(\vec{r})$, by an amount $\Delta n(\vec{r})$ which is as yet undetermined. This redistribution of electrons corresponds to adding a new charge distribution $\Delta n(\vec{r})$. Classically, it changes the surface potential seen by an electron by the amount

$$\Delta v(\vec{r}) = \frac{-e^2}{|\vec{r}-\vec{R}|} + \int d\vec{r}\,' \frac{e^2 \Delta n(\vec{r}\,')}{|\vec{r}-\vec{r}\,'|}$$

For the desorbed metal surface[1] we took $\{\varphi_{\vec{k}}(\vec{r})\}$ to be single-electron eigenfunctions of a variational model surface potential $v_M(\vec{r})$. Now we should, for consistency's sake, use single-electron eigenfunctions of the model potential $v_M(\vec{r}) + \Delta v(\vec{r})$. This produces a modified set of $\{\varphi_{\vec{k}}(\vec{r})\}$, subsequently a new $n_o(\vec{r})$, and finally -- through the solution of an integral equation -- a new $n_{MA}(\vec{r})$, or $\Delta n(\vec{r})$. Using this new $\Delta n(\vec{r})$, we repeat the process of calculating $\Delta v(\vec{r})$, etc., as described above. The variational calculation becomes complete when the iteration process converges.

In terms of physics, this approach closely resembles that of Lang and Williams.[5] The same comparison between our CBF approach[1] and Lang-Kohn's DFT[2] for the desorbed metal can, therefore, be made here for the metal-adsorbate system.

The main task that remains is the solution of the single-electron Schrödinger equation. We need the "self-consistent" density $n_o(\vec{r})$ and the eigenvalue spectrum. The latter should have an analytic form that permits us to turn its sum into an integral in the system energy calculation to follow. This requirement is harder to fulfill than one might offhand expect. We are approaching it by borrowing molecular-orbital techniques developed by Grimley[6] and Gunnarsson and Hjelmberg[7].

A natural extension of the present analysis leads to methods for determining substrated-mediated pairwise interactions between adsorbed atoms/molecules.

This work has been supported by the U. S. National Science Foundation through grant No. DMR81-20544.

References

*Present address: President, San Francisco State University, 1600 Holloway Avenue, San Francisco, CA 94132, USA.

1. X. Sun, M. Farjam, and C.-W. Woo, Phys. Rev. B, to be published.
2. N. D. Lang and W. Kohn, Phys. Rev. B1, 4555 (1970); N. D. Lang, Solid State Phys. 28, 225 (1973).
3. X. Sun, unpublished.
4. J. R. Schrieffer, in Varenna Summer School "Dynamics of Surface Physics" (1973).
5. N. D. Lang and A. R. Williams, Phys. Rev. B18, 616 (1978).
6. T. B. Grimley, in Varenna Summer School "Dynamics of Surface Physics" (1973).
7. O. Gunnarsson and H. Hjelmberg, Physica Scripta (Sweden) 11, 97 (1975).

MELTING OF ELECTRONS ON CORRUGATED SURFACES--STRUCTURAL AND DYNAMICAL PROPERTIES IN LIQUID AND SOLID PHASES*

P. Vashishta and R. K. Kalia
Argonne National Laboratory
Argonne, Illinois 60439 U.S.A.

and

J. J. Quinn
Physics Department
Brown University
Providence, R. I. 02912 U.S.A.

Abstract

Properties of a classical system of electrons on smooth and periodically corrugated surfaces are investigated using the molecular dynamics method. In the liquid phase, the electron-electron interaction, the total internal energy, the diffusivities parallel and perpendicular to the corrugation are all periodic functions of the wavevector of the corrugation. Thermodynamic and dynamical properties reveal a first-order melting transition for electrons on a smooth surface. Study of the melting transition on the commensurate corrugated surface of wavevector $\lambda = \frac{\sqrt{3}}{2} a$ (system S_2) shows a tricritical point separating a line of first-order transitions from a line of continuous transitions. On the commensurate corrugation of wavelength $\lambda = \frac{\sqrt{3}}{4} a$ (system S_4) melting transition is always first order. Topological analysis of particle coordinations shows bound dislocation pairs in the electron solid, a large number of defects and grain-boundary loops to be responsible for a first-order transition, and a small density of free dislocations to be the cause of continuous melting transition.

Introduction

In the past few years, a great deal of progress has been made in investigating the structural and dynamical properties of two-dimensional systems.[1] Experimental techniques such as LEED, neutron scattering, and synchrotron radiation have contributed a great deal toward understanding phase transitions in two dimensions.[2] In addition, computer-simulation methods have provided considerable insight into the nature of phase transitions and the role of defects in phase

*Work supported by the U. S. Department of Energy and National Science Foundation (J. J. Quinn) N.S.F. Grant DMR 8121069.

transitions.[3-5] However, experimental and computer simulation studies of melting in two dimensions disagree on the nature of the phase transition. While the results of most experiments are not inconsistent with the theoretical prediction[6] of Kosterlitz-Thouless, Halperin-Nelson, and Young in that the melting transition is continuous and induced by dislocations. Computer simulations show a grain-boundary mediated first-order transition.[3,4] One difference between computer simulations and experiments is the presence of a substrate in the latter.

In this paper, we shall discuss the effect of a periodically corrugated substrate on the properties of a two-dimensional system of electrons.[7,8] The Hamiltonian, H, of a system of electrons on a periodic corrugation of height U_0 and wave vector k is

$$H = \frac{1}{2} m \sum_i \dot{\vec{r}}_i^{\,2} + \sum_{i>j} \frac{e^2}{|\vec{r}_i - \vec{r}_j|} - U_b - U_0 \sum_{i=1}^{N} \sin k x_i \tag{1}$$

where U_b is the interaction between electrons and a rigid uniform, positive background.

Without the corrugation, the properties of the system can be described by the dimensionless variable $\Gamma = e^2(\pi\rho)^{1/2}/k_BT$, where ρ is the surface density of electrons. However, in the presence of a corrugation, the properties of the system depend not only on Γ but also on $\tilde{U}_0 = U_0/k_BT$ and the reduced wave vector $K = k/G_x$, G_x being the x-component of the smallest reciprocal vector of a triangular lattice of electrons. Figure 1 shows a triangular solid of electrons and its relationship with corrugations of different wave vectors K. In the ground state, S_2, S_4, ..S_{2n} form a triangular lattice independent of the height of the corrugation U_0. However, unlike the smooth surface, these systems are anisotropic. The ground state of electrons on S_1, S_3 ...depends on the value of U_0.

For K = 2, every particle resides in a corrugation minimum whereas for K = 4, the particles reside in alternate minima so that there is a two-fold valley degeneracy for every particle. A series of molecular dynamics (MD) calculations were performed in the liquid and solid phases of these systems. The calculations were carried out for N = 100, 256, and 576 electrons in a rectangle whose sides have a ratio of $\sqrt{3}/2$; the number density ρ was kept fixed at 1.477×10^8 cm^{-2} in all the calculations. Periodic boundary conditions were used and the long-range nature of 1/r interaction was properly taken into account by Ewald's summation. In section I, we shall describe the MD results for the electron liquid on various corrugated surfaces and in section II the melting transitions on smooth and corrugated surfaces. Section III deals with the nature of defects on smooth and corrugated surfaces and how they influence the melting transition.

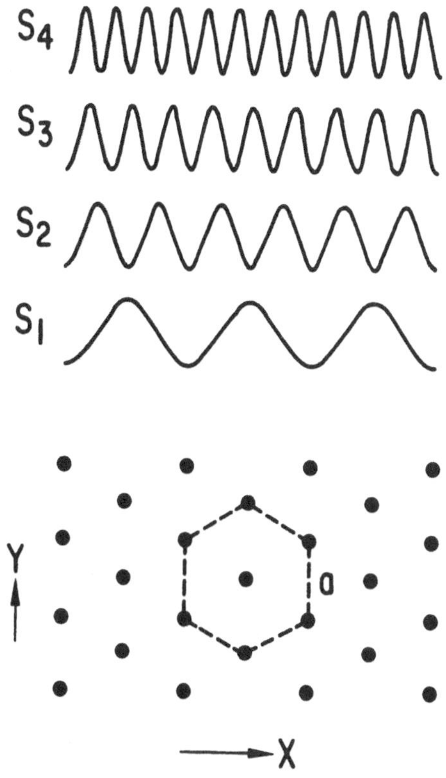

Fig. 1.

Electrons on periodic corrugations of reduced wave vectors K = 1, 2, 3, and 4, where K is measured in units of $G_x = 2\pi/\sqrt{3}\,a$, a being the lattice constant.

I. Properties of the Liquid

MD calculations were performed for several values of Γ, $\tilde{U}_0$, and K. Figure 2 shows the K dependence of the total internal energy and the self-diffusion constant at $\Gamma = 36$ and $\tilde{U}_0 = 1$. The internal energy of system S_1(K=1) is very close to that of the isotopic electron liquid on the smooth surface. Also, the constants of self-diffusion parallel and perpendicular to the corrugation have nearly the same values $\sim 5\ cm^2/sec$, which is almost equal to the self-diffusion on the smooth surface. The pair-correlation functions on smooth and corrugated surface K=1 are also the same. Thus, the electron liquid on S_1 is nearly isotropic and similar to the liquid on the smooth surface. For systems S_2, S_3, ... the electron-electron interaction, the internal energy, and the constant of self-diffusion parallel and perpendicular to the corrugation are nonmonotonic functions of K and have minimum values for the system S_2 (K=2).

The diffusivities parallel and perpendicular to the corrugations are quite different, the former being larger than the latter. Thus, the liquids on corrugated surfaces S_2, S_3, S_4 .... are anisotropic from structural and dynamical viewpoints.

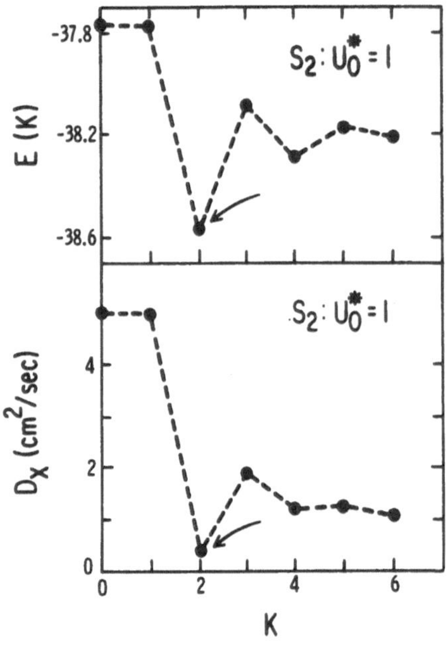

Fig. 2.

Wave vector dependence of the internal energy per particle, E, and the constant of self-diffusion along the corrugation, D_x. Reduced wave vector $K = \sqrt{3}\, \mathring{a}/\lambda$, where λ is the corrugation wavelength.

II. Melting on Smooth and Corrugated Surfaces

In this section we shall investigate the effects of height and wave vector of the corrugations on the nature of the melting transition. Using MD, we have studied melting on a smooth surface and on the corrugated surfaces S_2 and S_4. On the smooth surface, the temperature dependence of the internal energy and the constant of self-diffusion show hysteresis, supercooling, and a large specific heat in the transition region.[4] When the highest-temperature solid is heated, it melts by releasing latent heat and displays an abrupt jump in the constant of self-diffusion. The change in entropy on melting is 0.3 k_B per particle in agreement with the theory of freezing of Ramakrishnan.[9] Furthermore, the height of the first peak in the static structure factor agrees with the prediction of Ramakrishnan.[9] It has been shown by Weeks[10] that the change in area on melting is zero for potentials of the form $1/r^n$ if $n < d$. In this respect, the melting of electrons interacting via $1/r$ interaction is an interesting system because the first-order transition is not accompanied by a change in the area. This first-order melting transition occurs between Γ = 124-132, in excellent agreement with the experiments of Grimes and Adams and Mehrotra et al.[11]

For system S_2, the melting transitions are investigated for several values of the corrugation height U_0. For $U_0 > U_c$ ($\approx$ 1/16 K) the internal energy varies continuously with temperature and there is no hysteresis or latent heat of melting, as evident from Fig. 3. For $U_0 < U_c$ the transition is found to be first order. The critical value U_c where a line of first-order melting transition changes to a line of second-order transitions is a tricritical point, see Fig. 4. The present MD calculations provide the first evidence of a tricritical melting point in 2D.

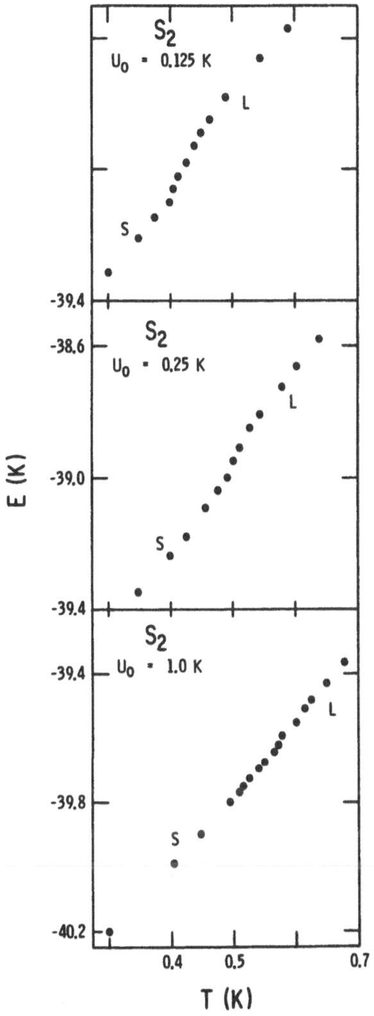

Fig. 3.

Temperature variation of the internal energy, E, for the system S_2 for three values of the height of corrugation. Electron density per unit area is 1.477×10^8 cm^{-2}. The symbols S and L denote solid and liquid.

For system S_4, Fig. 5 shows the temperature variation of the internal energy for the system heated monotonically from the ground state. The abrupt jump in the internal energy suggests the release of latent heat on melting and a first-order melting transition. The change in entrophy upon melting is again 0.3 k_B per particle. It should be pointed out that the melting transition on S_4 is first order for all values of the corrugation height.

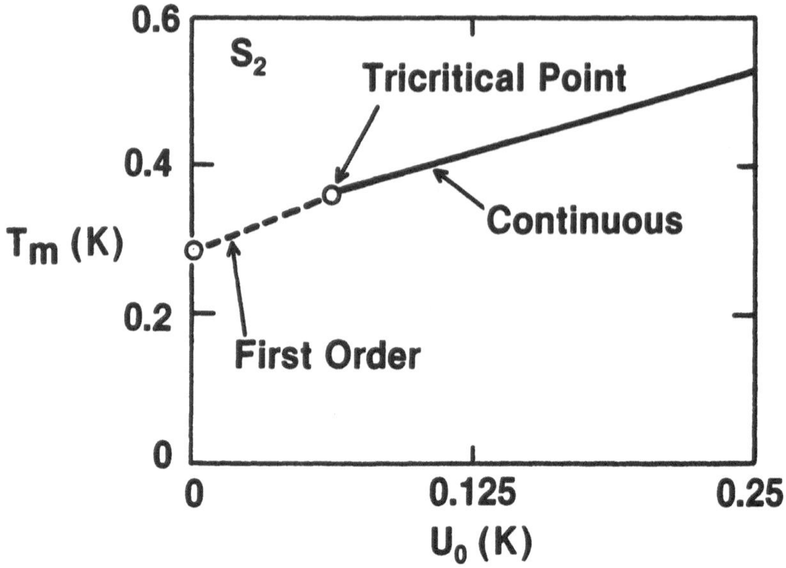

Fig. 4. Dependence of the melting temperature, T_m, on the corrugation height, U_o, for system, S_2. Dashed and continuous lines denote first-order and continuous transitions, respectively. The point where the two lines join is the tricritical point, denoted by the open circle.

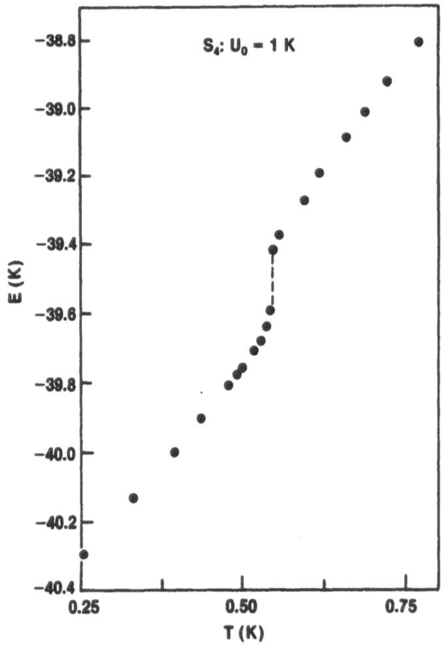

Fig. 5.

Temperature variation of the total internal energy per particle, E, for the system S_4. Solid circles denote monotonic heating from the ground state in which alternate valleys are occupied and electrons form a triangular lattice.

III. Analysis of Defects

Molecular dynamics calculations provide a microscopie understanding of defects since the positions of particles are known at all times. The topology of defects can be investigated by constructing around each particle a Wigner-Seitz cell; the number of faces of the cell gives the number of nearest neighbors. In an ideal triangular solid, each particle has six nearest neighbors (n.n.). However, near

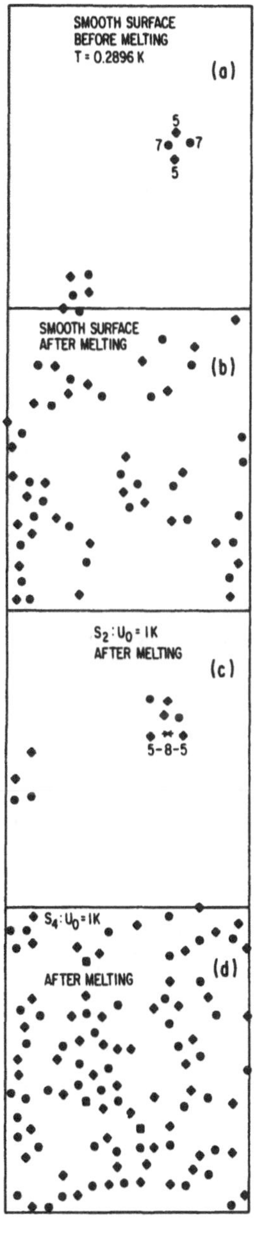

Fig. 6.

Snapshot of defects for the electron system on smooth (U_o = o) and corrugated (U_o = 1K) surfaces S_2 and S_4. Solid circles and diamonds denote particles with coordinations numbers 7 and 5, respectively. The star denotes a particle with coordination number 8.

Figure 6(a) shows two pairs of bound dislocations in an electron solid below the melting temperature. Note that each dislocation consists of a bound pair of 5- and 7-coordinated disclinations.

Figure 6(b) shows a high density of defects on a smooth surface above the melting transition. In addition to dislocation and disclinations, one finds a grain boundary loop.

Figure 6(c) shows the snapshots of defects on S_2, above melting. Low density of defects and the absence of grain-boundary loops are evident. A special defect configuration of 5-8-5 coordinated particles is found to be stable.

Figure 6(d) shows defects on S_4, above melting. A high density of defects is characteristic of first-order melting transition.

melting some particles have 5 or 7 n.n. and occasionally there are 4 or 8 n.n. In a triangular solid, a pair of adjacent particles with 5 and 7 n.n. constitute a free dislocation and a string of particles with 5 and 7 n.n. is a grain boundary. On the smooth surface and in S_4, we find a large number of defects and grain-boundary loops to be responsible for the melting transitions. In system S_2, the density of defects decreases with increasing U_0 and the grain-boundary loops disappear beyond the tricritical point. One of the defects found for $U_0 > U_c$ is a particle with 8 n.n. surrounded by two particles with 5 n.n. These MD results clearly suggest that first-order melting transitions are mediated by grain-boundaries and the continuous transitions by a relatively small density of dislocations.

Although defects have no concrete meaning in the liquid state, the present analysis is only meant to provide a way to examine the coordinations of particles in the solid and liquid phases. Figure 6 shows snapshots of particles with coordinations other than six on the smooth surface and on corrugations S_2 and S_4.

IV. Conclusions

Molecular dynamics study of electrons on a smooth surface reveals that melting is a first-order transition and it occurs between Γ_m = 124-132, in good agreement with experiments. The change in entropy per particle upon melting is found to be 0.3 k_B. The height of the first peak in the structure factor ($\simeq$5.5) before freezing and the change in entropy are in good agreement with the theory of Ramakrishnan.[9] Melting on smooth surface is found to be mediated by a high density of defects, predominantly grain boundaries which supports the theory of Chui.[12] There is no evidence for the two-step continuous melting transition and intervening hexatic phase as predicted by Halperin and Nelson.[6]

On the corrugated surface, S_2, melting is found to be a grain boundary mediated first-order transition if the height of the corrugation U_0 is less than the critical U_c ($\approx$ 1/16 K), and for $U_0 > U_c$ the melting transition is continuous and induced by a small density of defects. A tricritical point in the melting transition in 2-D is identified for the first time. From the MD results, it appears that corrugation on S_2 tends to increase the dislocation core energy, thereby reducing the density of defects and changing the nature of the melting transition from first-order to continuous. Recent Monte Carlo simulations of non-atomistic systems, interacting system of dislocations by Saito[13] and XY model by Swendsen,[14] support these conclusions.

Melting on S_4 is always a first-order transition, mediated by high density of defects. Defect energies as a function of the height of corrugation and temperature are different for S_2 and S_4 because alternate corrugation valleys are empty in the ground state of S_4 (degeneracy of two).

References

1. Ordering in Two Dimensions, edited by S. K. Sinha (North Holland 1980); Melting, Localization and Chaos, edited by R. K. Kalia and P. Vashishta (North Holland 1982); J. Vilain, in Ordering in Strongly Fluctuating Condensed Matter Systems, edited by T. Riste (Plenum 1979), p. 222; P. Bak, Reports on Progress in Physics 45, 587, (1982); D. Fisher, W. Brinkman, D. Moncton, Science 217, 693, (1982).
2. S. K. Sinha, in Nonlinear Phenomena at Phase Transitions and Instabilities, edited by T. Riste (Plenum 1981), p. 433
3. F. Abraham, in Melting, Localization and Chaos, edited by R. K. Kalia and P. Vashishta (North Holland 1982), p. 75
4. P. Vashishta and R. K. Kalia, in Melting, Localization and Chaos, edited by R. K. Kalia and P. Vashishta (North Holland 1982), p. 43
5. R. W. Hockney and T. R. Brown, J. Phys. C8, 1813 (1975); R. C. Gann, S. Chakravarty and G. V. Chester, Phys. Rev. B20, 326 (1979); R. H. Morf, Phys. Rev. Lett. 43, 931 (1979); R. H. Morf, in Physics of Intercalation Compounds, edited by L. Pietronero and E. Tosatti, Springer Series in Solid-State Science (Spring-Verlag 1982), Vol. 38, p. 252
6. J. M. Kosterlitz and D. J. Thouless, J. Phys. C6, 1181 (1973); B. I. Halperin and D. R. Nelson, Phys. Rev. Lett. 41, 121 (1978); A. P. Young, Phys. Rev. B19, 1855 (1979)
7. R. K. Kalia, P. Vashishta, S. D. Mahanti and J. J. Quinn, J. Phys. C16, L491 (1983)
8. P. Vashishta, R. K. Kalia and J. J. Quinn, J. Phys. C16, L405 (1983)
9. T. V. Ramakrishnan; Phys. Rev. Lett. 48, 541 (1982)
10. J. D. Weeks, Phys. Rev. B24, 1530 (1981)
11. C. C. Grimes and G. Adams, Phys. Rev. Lett. 42, 795 (1979); R. Mehrotra, B. M. Guenin and A. J. Dahm, Phys. Rev. Lett. 48, 641 (1982)
12. S-T Chui, in Melting, Localization and Chaos, edited by R. K. Kalia and P. Vashishta (North Holland 1982), p. 29
13. Y. Saito, Phys. Rev. Lett. 48, 1114 (1982)
14. R. H. Swendson, Phys. Rev. Lett. 49, 1302 (1982)

CORRELATIONS IN THE LAYERED ELECTRON-HOLE LIQUID

Tapash Chakraborty and C.E.Campbell
School of Physics and Astronomy
University of Minnesota,Minneapolis,MN 55455,USA.

Introduction.- At low temperatures and high carrier concentrations, the electrons and holes in semiconductors condense to form a high density metallic liquid.This electron-hole liquid(EHL)[1,2] is a collective state of electrons and holes,which is unique in many respects:It is the most metallic of metals,and the most quantum of fluids.The experimental results have indicated that the EHL has the characteristic properties of a liquid.Its volume is conserved,it has an equilibrium density etc. The EHL drops have a spherical shape with drop radius of 10^{-3}-10^{-4}cm, which is much larger than the exciton (bound electron-hole pairs) radius a_x=10^{-5}- 10^{-6}cm,which represents the characteristic interaction distance of the particles in the drop.Therefore,in calculating the bulk properties of the EHL,one can neglect the surface effects.Recently,a many-body variational approach[3,4] for two-component systems has been applied to study the ground-state properties of such a fascinating system.Some of the results obtained for the EHL are in very good agreement[5] with recent experiments.

The study of the EHL in two dimensions is also a very interesting problem for various reasons.Firstly,there are current experimental efforts[6] to obtain such systems.Secondly,all earlier works[7,8,10] were based on standard perturbative approaches[9].A major problem inherent in those methods,however,is the unphysical behavior of the short-range part of the correlation functions.Study of the inversion layer[9] indicated that the problem of negative correlation functions at small distances becomes accute in two-dimensions,which affects the ground-state properties significantly.

In the following,we shall investigate the two cases:a)the layered electron-hole liquid with variable interlayer separations,and b) systems with many-valley structure.

Layered electron-hole liquid.- We consider a model[11] where electrons and holes of finite two-dimensional density ρ,move in two different planes separated by a distance c.The tunneling of the particles between the planes in forbidden.This type of charge distribution is possible in two adjacent parts of a semiconductor by an electrical discharge. The unperturbed eigenstates in this model are

$$\Phi_{\vec{k},\sigma,m}(\vec{r},z) = A^{-\frac{1}{2}} a_\sigma e^{i\vec{k}.\vec{r}}\ \chi(z-m_i c) \qquad (1)$$

where A is the normalization area, $\vec{k}$ is a two-dimensional vector, a_σ is the spin eigenfunction, m_i=0 and 1 for electrons and holes respectively. We consider only the limit in which the one-dimensional eigenstate $\chi(z)$ is arbitarily highly localized, $\chi^*(z)\chi(z) = \delta(z)$. The nondynamical correlations for the electrons or holes are

$$g_f(r) = [1-\tfrac{1}{2}(2J_1(k_f r)/k_f r)^2] \qquad (2)$$

where k_f is the radius of a two-dimensional 'Fermi disk' and $J_1(x)$ is the Bessel function. The Fourier transform of (2) gives us the ideal gas structure function as,

$$S_f(k) = \begin{cases} \frac{2}{\pi}\sin^{-1}(k/2k_f) + \frac{k}{\pi k_f}[1-(k/2k_f)^2]^{\frac{1}{2}}, & k<2k_f \\ 1 & , k>2k_f \end{cases} \qquad (3)$$

For a finite separation between the layers ($c\neq 0$), the electron-hole interaction is, $v_{eh}=e^2/|\vec{c}+\vec{r}_i-\vec{r}_j|$, $\vec{c}$ = (0,0,c) and $\vec{r}$ is a two dimensional vector. Introducing the two-dimensional Coulomb units, $a_x=\hbar^2/2\mu e^2$, $E_x=2\mu e^4/\hbar^2$ where $\mu=m_e/(1+\sigma)$ is the reduced mass, $\sigma=m_e/m_h$, the dimensionless variables $x=r/\bar{r}$, $q=k\bar{r}$ and $c_s=c/a_x$, $\bar{r}=a_x r_s$, the mean interaction energy in the variational approach[3,4] is

$$\varepsilon_{int} = \frac{1}{r_s}\int_0^\infty [g_{ee}(x)+g_{hh}(x)-2x\{c_s^2/r_s^2+x^2\}^{-\frac{1}{2}} g_{eh}(x)]dx \qquad (4)$$

where $g_{\alpha\beta}(x)$ are the partial pair-correlation functions. Noticing that, $v_{eh}(q)=\frac{2\pi}{qr_s}\exp(-|qc_s/r_s|)$, ε_{int} is expressed in terms of the partial static-structure-functions as

$$\varepsilon_{int} = \frac{2c_s}{r_s^2} + \frac{1}{2r_s}\int_0^\infty [S_{ee}(q)+S_{hh}(q)-2S_{eh}(q)-2]dq + \frac{c_s}{2r_s^2}\int_0^\infty [2 - qc_s/r_s+\ldots]S_{eh}(q)qdq. \qquad (5)$$

The first term is, as expected, the energy contribution from the electrostatic interaction between the planes of opposite charges. The ground-state energy in the Hartree-Fock approximation is given by

$$\varepsilon_{HF} = \frac{1}{r_s^2} - \frac{8\sqrt{2}}{3\pi}\frac{1}{r_s}. \qquad (6)$$

Note the difference in the coefficient of the exchange energy, compared to Ref. 7, because of our different choice of the 2D-units.

The basic equation for the pair-correlation functions $g_{\alpha\beta}(x)$,

α,β =e,h in our approach is[3,4]

$$[-\nabla^2 + v_{\alpha\beta}(x) + W^b_{\alpha\beta}(x) + W^f_{\alpha\beta}(x)]g^{\frac{1}{2}}_{\alpha\beta}(x)=0 \quad (7)$$

where, $v_{\alpha\beta}(x)=[\eta_{\alpha\beta}r_s/2x-(1-\delta_{\alpha\beta})r_s/\sqrt{x^2+c_s^2/r_s^2}]$

$$\eta_{ee}=(1+\sigma),\ \eta_{hh}=(1+1/\sigma),\ \eta_{eh}=0.$$

The 'induced potentials' $W_{\alpha\beta}(x)$ are given in the earlier works[3,4],where the method of solving the above equations are also given.

Systems with many-valley structure.- So far,we have considered only the case of a single conduction band and a single valence band.In most cases,however,we have to take into account,the effect of the valley degeneracy on the ground-state results to be discussed below.Therefore,in the following,we will consider a case,where the conduction band has two minima and the valence band has a single maximum like in GaSe.The Hartree-Fock energy will now depend on the mass ratio as

$$\varepsilon_{HF}=\left(\frac{0.5+\sigma}{1+\sigma}\right)\frac{1}{r_s^2} - \frac{9.66}{3\pi}\frac{1}{r_s}. \quad (8)$$

Earlier results in this multi-valley case[7] have indicated that the ground-state energy depends more strongly on σ,compared to the single valley case and the effect is more pronounced in two-dimensions.This is mainly due to the difference in the density of states between the two- and three-dimensional systems.The correlational energy is affected,in our scheme,primarily through $W^f_{\alpha\beta}(x)$,since the fermi momentum now is $k_f^2=2\pi\rho/n_v$,where n_v=2 for the electrons and 1 for the holes.

Results and Discussions.- In Figs.1 and 2,we have plotted the electr-

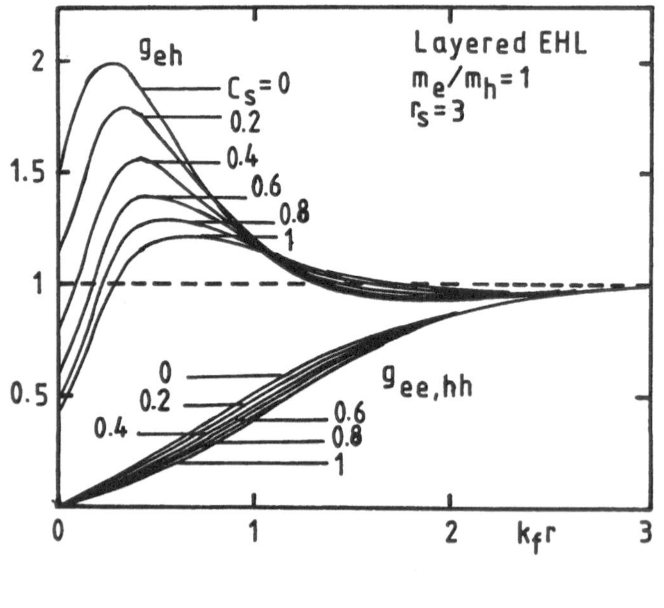

Fig.1: Pair-correlation functions $g_{ee}(r)$, $g_{hh}(r)$ and $g_{eh}(r)$ vs $k_f r$ for σ=1.

on-electron, hole-hole, and the electron-hole correlation functions as a function of $k_f r$, for various values of the dimensionless interlayer separation c_s, at $r_s=3$. For $\sigma=1$ (Fig.1), $g_{ee}(r)$ and $g_{hh}(r)$ are identical and they vary only slightly for different values of c_s. The effect of c_s is much stronger in $g_{eh}(r)$, which tends to show less structure as c_s is increased. For $m_h/m_e=10$ (Fig.2), there is more or less a similar pattern in

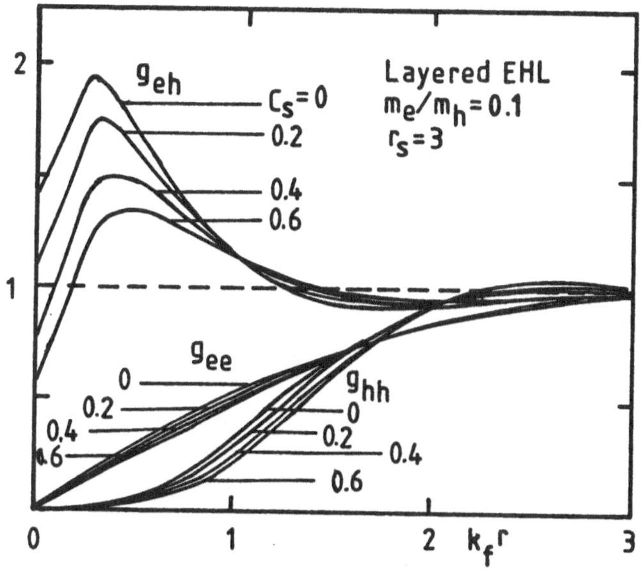

Fig.2: Same as in Fig.1, but for $\sigma=0.1$.

the distribution functions, except that the holes show stronger correlations among themselves. This behavior is reminiscent of the hole-hole distribution functions in three dimensions[3]. The noticeable difference in the e-h correlation functions, as compared to those in three dimensions is that the enhancement is much reduced in two dimensions. In calculating the enhanced density[5], it should however, be noted that the ratio of the enhanced density to the exciton density is,

$$\rho_{eh}/\rho_x = \begin{cases} 3g_{eh}(0)/4r_s^3 & ;3D \\ g_{eh}(0)/8r_s^2 & ;2D \end{cases} \qquad (9)$$

In Fig.3, we have plotted the partial-static-structure functions, and $D(k)=S_{ee}(k)S_{hh}(k)-S_{eh}^2(k)$, as a function of k/k_f, for $\sigma=1$ and $\sigma=0.1$. All the functions rapidly approach to zero for small k. These functions were used to obtain the collective modes in the electron-hole liquid. They are obtained by generalizing the Bijl-Feynman equation for the two-component systems[12]

$$\varepsilon_{1,2}/\varepsilon_f = \frac{q'^2}{2D}[(\sigma S_{ee}+S_{hh})\pm\{(\sigma S_{ee}+S_{hh})^2-4\sigma D\}^{\frac{1}{2}}] \tag{10}$$

where $q'=k/k_f$ and ε_f is the electron fermi energy. In Fig.4, we have plotted the two branches of the excitation energies ε_1 and ε_2 in units of electron fermi energy as a function of k/k_f, for $\sigma=1$ and $\sigma=0.1$. The dashed lines correspond to the threshold energy for the onset of Landau damping due to excitations of particle-hole pairs of the type 1

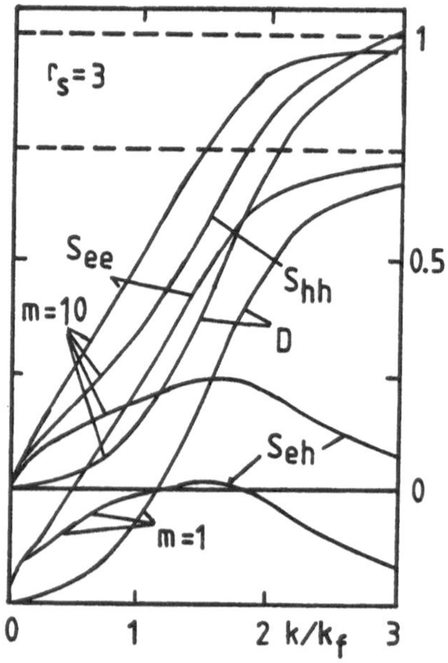

Fig.3: Static-structure functions $S_{ee}(k)$, $S_{hh}(k)$, $S_{eh}(k)$ and $D(k)$ as a function of k/k_f for $\sigma=1,0.1$.

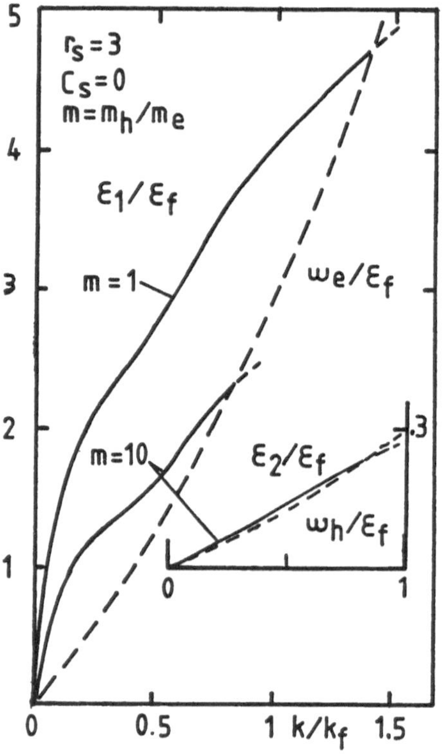

Fig.4: Dispersion relations of the two-dimensional plasmons in the EHL for m = 1,10, and the 'acoustic plasmon' mode for m=10, as inset.

$[\omega_e/\varepsilon_f=q'(q'+2)]$ and that of type 2 $[\omega_h/\varepsilon_f=q'(q'+2)\sigma]$. The plasmon mode rises sharply from zero with increasing k and has the characteristic plasma frequency[13] $\omega_p \simeq [2\pi e^2\rho k/\mu]^{\frac{1}{2}}$. The 'acoustic plasmon' mode (or ionic sound mode)[14] exists only in the case where the holes are much heavier than the electrons[12,15]. This mode for $\sigma=0.1$ is drawn as inset in Fig.4.

Finally, in Fig.5, we have plotted the ground-state energy minimum as a function of the interlayer separation c_s, for different values of the electron-hole mass ratio. In the single valley case, the effect of the heavy holes is clearly insignificant. However, in the multi-valley case,

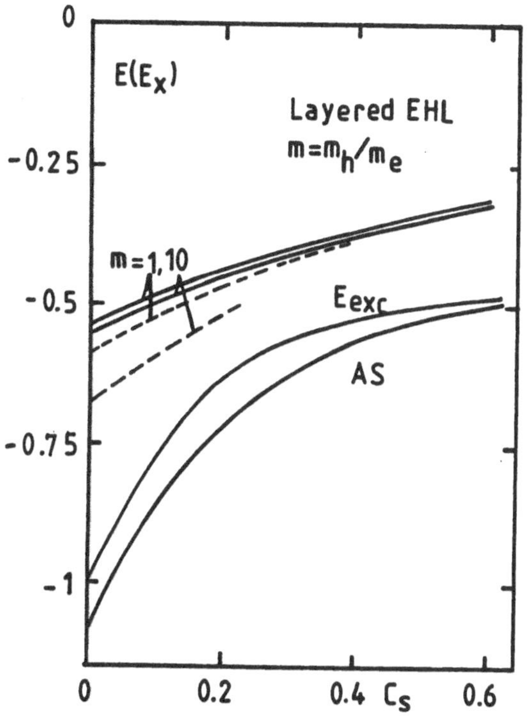

Fig.5: The ground-state energy minimum as a function of the dimensionless interlayer separation c_s. The solid lines are the single-valley case and the short-dashed lines are the many-valley results. The curve E_{exc} represents the exciton energy in the ground state and the curve 'AS' is the result of Ref.8.

we obtain a significant lowering of the energy. Comparing with the exciton energy in the ground state[8], we notice however, that the exciton state is clearly preferred. Nevertheless, a suitable choice of the electron-hole effective mass, the valley degeneracy and a finite seperation between the layers, might lead to the energy being lower than the exciton energy. Our result is apparently in contrast with the results of Ref.8 (curved marked AS), who obtained a lower energy for the EHL compared to the exciton energy E_{exc} for all values of c_s. However, the unphysical behavior of the correlation functions at small distances[8] and inadequate range in k, in their interpolation schemes, as pointed out in Ref.10, introduce large amount of uncertainty in those energy values. Qualitatively, results similar to Fig.5 were obtained in Refs.7 and 10, for c_s=0. More experimental efforts are undoubtedly needed to acieve a better understanding of this fascinating quantum liquid in two-dimensional systems.

This work was supported by a Grant No. DMR-7926447 from the National Science Foundation. One of us (T.C.) wishes to thank

L.J.Lantto for helpful discussions on the numerical computations.

References.-

1. T.M.Rice,Solid State Phys.32,1(1977);J.C.Hensel,T.G.Phillips,and G.A.Thomas,ibid.32,88(1977).
2. E.A.Andryushin and A.P.Silin,Sov.J.Low Temp.Phys.3,655(1977).
3. Tapash Chakraborty and P.Pietiläinen,Phys.Rev.Lett.49,1034(1982).
4. Tapash Chakraborty,Phys.Rev.B25,3177(1982);ibid.B26,6131(1982).
5. J.C.Culbertson and J.E.Furneaux,Phys.Rev.Lett.49,1528(1982); Tapash Chakraborty and C.E.Campbell,Solid State Commun.45,195(1983).
6. V.G.Litovchenko,D.V.Korbutyak and V.A.Zuev,Bull.USSR Acad.Sci.46, 1452(1982);see also,H.Kamimura,K.Nakao and Y.Nishina,Phys.Rev.Lett. 22,1379(1969).
7. Y.Kuramoto and H.Kamimura,J.Phys.Soc.(Japan)37,716(1974).
8. E.Andryushin,Sov.Phys.Solid State 18,1457(1976);E.A.Andryushin and A.P.Silin,Solid State Commun.20,453(1976).
9. A.K.Rajagopal and J.C.Kimball,Phys.Rev.B15,2819(1977);M.Jonson, J.Phys.C:Solid State Phys.9,3055(1976);S.M.Bose,Phys.Rev.B13,4192 (1976).
10. P.K.Isihara,Y.Nakane,and A,Isihara,J.Phys.C:Solid State Phys.15, 2929(1982).
11. P.B.Visscher and L.M.Falicov,Phys.Rev.B3,2541(1971).
12. C.E.Campbell,Ann.Phys.74,43(1972);C.E.Campbell and J.G.Zabolitzky, to be published.
13. F.Stern,Phys.Rev.Lett.18,546(1967).
14. P.M.Platzman and P.A.Wolff,Waves and Interactions in Solid State Plasmas,Ch.5 (Academic New York,1973).
15. Tapash Chakraborty,to be published.

DENSE COULOMB PLASMAS: QUANTUM STATISTICS AND ORDERING*

N. W. Ashcroft
Laboratory of Atomic and Solid State Physics
Cornell University, Ithaca, New York 14853

and

J. Oliva
Lawrence Livermore Laboratory
Livermore, California 94550

Abstract

The low temperature equilibrium and transport properties of dense liquid metallic phases of hydrogen and deuterium are examined. Both systems can be regarded as dense Coulomb plasmas, but are considered in temperature ranges where the statistical differences are strikingly evident.

Introduction

We are concerned in this paper with the physical properties of dense hydrogen and deuterium at densities where they exist in metallic form, i.e., for mean inter-electron spacings in the range $r_s \sim 1$ (where $\frac{4\pi}{3} r_s{}^3 a_o{}^3 = V/N$, for N electrons in volume V). Under these conditions, where the electrons are delocalized, the Hamiltonian is conveniently written as

$$\mathcal{H} = \mathcal{H}(\{r_e, R_i\})$$

where $\{r_e, R_i\}$ denotes the totality of electron and proton (i=p) or deuteron (i=d) coordinates. Since all interactions are Coulombic (with $v_c(k) = 4\pi e^2/k^2$), $\mathcal{H}$ takes the form

$$= \hat{T}_e + \frac{1}{2}(N/V) \sum_{k\neq 0} v_c(k)\{N^{-1}\hat{\rho}_e(\vec{k})\hat{\rho}_e(-\vec{k}) - 1\} \tag{1a}$$

$$+ \hat{T}_i + \frac{1}{2}(N/V) \sum_{k\neq 0} v_c(k)\{N^{-1}\hat{\rho}_i(\vec{k})\hat{\rho}_e(-\vec{k}) - 1\} \tag{1b}$$

$$- (N/V) \sum_{k\neq 0} v_c(k)\hat{\rho}_e(\vec{k})\hat{\rho}_i(-\vec{k}) \tag{1c}$$

in the thermodynamic limit. It will be noted that (1a) is the Hamiltonian appropriate to N electrons, of mass m, total kinetic energy $\hat{T}_e$, and with single-particle density $\hat{\rho}_e(\vec{k})$ in the presence of a uniform compensating background of mean charge

*Work supported in part by the National Science Foundation (Grant #DMR-80-20429, and in part by the National Aeronautics and Space Administration (Grant #NAG 2-159).

density eN/V. Term (1b), correspondingly, is the Hamiltonian of N protons (or N deuterons) of mass m_i, total kinetic energy $\hat{T}_i$, and with single-particle density $\hat{\rho}_i(\vec{k})$ in the presence of a uniform compensating background of mean charge density --eN/V. These two Hamiltonians are themselves connected by the Coulomb coupling represented by term (1c) whose average contribution, in states of fixed N, is removed by the overall requirements of charge neutrality. In the charged Fermi liquid problem discussed below we shall assume that $v_c(k=0) = 0$, as is often convenient.

The questions of primary physical interest center on the nature of the states resulting from (1), particularly the low temperature states where the crucial quantum statistical differences between proton and deuteron subsystems can become quite apparent for liquid phases. That such phases are possible near T = 0 was first suggested by Brovman et al. [1] and follow from noting [2] that a typical zero-point energy per ion is $\sim 6 \left(\frac{m}{m_i}\right)^{1/2} \frac{1}{r_s^2}$ Ry, or about 0.14 Ry for protons at $r_s \sim 1$. These energies, and more particularly their differences in different crystal structures, can considerably exceed the energy differences per ion (typically millirydbergs) associated with the electronic and electrostatic energies of the same structures [3]. Accordingly, at some densities there is sufficient zero-point energy to cause an apparent continuous arrangement of structures, including liquid-like structures, as can be verified by direct simulation methods [3]. The latter show that the existence of ground state or near ground state liquids depends very much on the choice of density. For both liquid and crystalline phases, however, a range of possible electronic and ionic orderings is likely, and our major purpose here is to discuss the physical properties, particularly the low temperature properties, of the ensuing phases. In states of liquid symmetry, liquid metallic hydrogen (LMH) is a two-component Fermi liquid characterized, as noted, by long-range interactions and possessing a large component mass ratio (m_p/m_e). On the other hand, liquid metallic deuterium (LMD) is a Boson-Fermion fluid, characterized again by long-range interactions and a large mass ratio. As we shall see, the presence of non-zero spin actually leads to a new non-magnetic excitation in Bose condensed phases. We will also see that the statistical differences between LMH and LMD lead to physical properties that are dramatically distinct.

Dense Hydrogen as a Two-Component Fermi Liquid

(a) Normal States

We begin by assuming that the system under discussion is normal, recognizing, of course, that the possibilities of ordering are manifold and include ferromagnetism in either, or both, electrons and protons, electronic superconductivity, pairing of the protons, charge or spin density waves, molecular pairing, and even liquid-crystal formation. As shown by Akhiezer and Chudnovsky [5], and Oliva and

Ashcroft [6], the phenomenological theory of Landau for one-component Fermi liquids [7] is readily generalized to uniform two-component spin 1/2 systems. In hydrogen, the bare masses of the two components are vastly different, though for neutral systems the Fermi momenta p_{F_i} are identical. For protons, at $r_s \sim 1$, we expect fully degenerate behavior to be quite evident at temperatures of order $\sim$ 10K and possibly higher. The equilibrium properties of the system are described in terms of quasi-particle distribution functions $n^0_{i\vec{p},\sigma}$ (σ the spin projection), and the quasi-particle energies $\varepsilon_{i\vec{p},\sigma}$ which are functions of the $n_{i\vec{p},\sigma}$. As in the one-component case, effective interaction energies between quasi-particles are defined by (in this case four) second variational derivatives of the total internal energy with respect to the distribution functions. These "f-functions" in turn can be expanded in the usual Legendre series for both the (spin) symmetric and antisymmetric components, ${}^s f_\ell^{ij}$ and ${}^a f_\ell^{ij}$. Thermodynamic stability of the system then requires that

$$1 + {}^a f_\ell^{ii} \nu_i/(2\ell+1) > 0 \tag{2a}$$

$$1 + {}^s f_\ell^{ii} \nu_i/(2\ell+1) > 0 \tag{2b}$$

$$(1 + {}^s f_\ell^{11} \nu_1/(2\ell+1))(1 + {}^a f_\ell^{22} \nu_2/(2\ell+1)) - ({}^s f_\ell^{12})\nu_1\nu_2/(2\ell+1)^2 > 0 \tag{2c}$$

where ν_i are the component densities of states at the Fermi energy. These inequalities embody in part the physical requirement that if the system is to remain normal, then there can be neither too much interspecies repulsion (2c) nor too much inter- or intra-species attraction (2c, or 2a and 2b). The standard argument [6] of Galilean invariance relating the quasi-particle effective masses to the f's also apply here [6]. Thus

$$\frac{m_i^*}{m_i} = \frac{1}{1 - (V p_F^2/3\pi^2\hbar^3)({}^s f_1^{ii} m_i + {}^s f_1^{ij} m_j)} \qquad (i,j = e,p) \tag{3}$$

These appear directly in the specific heat of the system which, for temperatures less than the proton Fermi temperature, is given by

$$C_V = \frac{p_F^2 k_B^2}{3\hbar^3}(m_e^* + m_p^*)T = C_{V_e} + C_{V_p}. \tag{4}$$

To appreciate the content of (4), we compare

$$C_V = C_{V_e}\left(1 + \frac{m_p^*}{m_e^*}\right) \tag{5}$$

with the corresponding specific heat C_V^s expected for a crystalline form of hydrogen for which

$$C_V^s = C_{V_e}^s(1 + AT^2) \tag{6}$$

where A is determined by the phonon spectrum of the assumed crystalline phase. The major point, however, is that in the range of temperature of interest here the linear term in the specific heat (normally gauged by C_{V_e}) acquires an enormous enhancement ($\sim m_p^*/m_e$) that is not present in a crystalline counterpart. This difference reflects the density of states and phase-space available to a proton fluid that are not accessible to the corresponding crystal. An equally dramatic realization of this effect is apparent in the coefficient of thermal expansion which, for normal liquid metallic hydrogen, is shown by Oliva and Ashcroft to be

$$\alpha = K_T C_V\left(\frac{2}{3} - \frac{\partial(\ell n\ m_p^*)}{\partial(\ell n\ n)}\right) \tag{7}$$

where n is the particle density and where K_T is the compressibility. Since $C_V \sim (\frac{m_p^*}{m_e^*})C_{V_e}$ it follows that the expansion coefficient in the liquid can be enormously greater than that of the solid. Further, α can in principle be negative (as is the case in normal ^3He) and indicates a tendency for ordering in momentum space at the expense of becoming more free particle-like in real space [8].

(b) Transport

Both of these equilibrium properties clearly distinguish a liquid phase of metallic hydrogen from a solid. However, it is more likely that any experimental distinction will be made through the measurement of a transport property, such as the conductivity. To calculate such a quantity, the Landau-Silin-Boltzmann equations for the two-component system can be treated by a procedure which is a straightforward generalization of the methods introduced by Abrikosov and Khalatnikov [9] for reducing the kinetic equations. In fact, Oliva and Ashcroft [10] have shown that the resistivity, for example, is given by

$$\rho \sim \frac{m_p^{*2}}{p_F^5}\, T^2 \tag{8}$$

in marked contrast to the phonon-scattering result for pure crystals ($\rho \sim CT^5$). As is well known, electron-electron scattering effects in crystals, when Umklapp mediated, also give a T^2 behavior. However, because of the higher density of states of the scatterers (protons) the T^2 term represented by (8) completely swamps this contribution, as it does the phonon contribution.

In references [6] and [10], other transport and equilibrium properties of LMH have been discussed in detail. It is worth noting that the collective modes of the system have also been analyzed within Landau theory. As expected there occur both optical plasmons and acoustic plasmons, the components oscillating out of phase in the latter.

(c) Electron Pairing: Superconductivity

From the existence of superconductivity in glassy or amorphous metals we know that disorder in the ions of a metal is not inimical to the phenomenon of electron pairing and superconductivity. A superconducting metallic liquid is, in principle, possible, a concept that applies here to the case of metallic hydrogen. In fact, to calculate the superconducting transition temperature of such a system [11] we need only to solve the Eliashberg equations for the gap function and to find the temperature at which this gap is suppressed by the presence of a vanishingly small pair-breaking field. The starting information required for this method is the Eliashberg function $\alpha^2F(\omega)$ which for solids is generally obtained from the spectral weight of the phonon Green's function [12]. In highly excited quantum liquids, well defined phonons do not exist; accordingly a more general description is needed which in Ref. [12] is approached from a knowledge of the density-density response function of the proton fluid. An expression for $\alpha^2F(\omega)$, similar to those derived for amorphous metals [13], is readily obtained. One important physical difference, however, is that while phonon-like excitations of longitudinal character are still present in LMH, the corresponding transverse modes are absent. In fact, they are replaced qualitatively by an interesting electron pairing mechanism involving particle-hole excitation of the proton Fermi fluid.

From the numerical analysis of $\alpha^2F(\omega)$, transition temperatures comparably high to those of the crystalline solid are found [12]. Even more interesting, however, are the magnetic properties of LMH in a possible superconducting phase, for a reason closely connected with the behavior, with temperature, of the normal state conductivity, as discussed above. Thus at temperatures near T_c, the scattering rate for normal transport is very high. However, as T is lowered, we would find that this rate, were the system actually normal, decreases very dramatically. As a consequence, it has been found by Jaffe and Ashcroft [14] that as temperature is lowered from above, superconductivity first develops as Type II. But because of the remarkable temperature dependence of the transport relaxation time, a quite unusual change takes place with decreasing temperature in which the system passes from Type II to Type I behavior. During this process the upper critical field progresses through a maximum, an effect that will be present also in liquid deuterium provided, however, that the deuterons are not themselves ordered.

Liquid Metallic Deuterium

As noted above, a quite noticeable aspect of liquid metallic deuterium (LMD) is the presence of non-zero nuclear spin in the ionic component. To understand the implications of this we regard LMD as a spin 1 Boson fluid with electronically screened interactions U that are effectively short-ranged. We then recast (1) in a manner that will bring out the essential physics, i.e.,

$$\mathcal{H} = E_0 + \sum_{\vec{k}} \sum_{\lambda=1} \frac{h^2k^2}{2\overline{m}_d} a^+_{\lambda k} a_{\lambda k}$$

$$+ \frac{1}{2V} \sum_{\vec{k}_1 \vec{k}_2} \sum_{\lambda,\lambda'=1}^{3} U(|\vec{k}_1-\vec{k}_3|)\delta(\vec{k}_1+\vec{k}_2-\vec{k}_3-\vec{k}_4) a^+_{\lambda k_3} a^+_{\lambda' k_4} a_{\lambda' \vec{k}_1} a_{\lambda \vec{k}_2} \tag{9}$$

where $\overline{m}_d$ is a renormalized deuteron effective mass and E_0 is a constant. In (9) $a^+_{\lambda\vec{k}}$ and $a_{\lambda\vec{k}}$ are Boson creation and destruction operators for spin λ and momentum $\vec{k}$.

The first observation to be made is that for an ideal spin 1 Boson system the Bose condensation temperature at $r_s = 1.6$ is $T_B = 43K$. The interacting system will probably condense at a temperature close to this and therefore experimentally quite accessible. Next, we consider the character of the quasi-particle branches for the interacting system. These are best understood by analyzing the weak coupling problem via the non-zero spin generalization of the Bogolyubov method [15,16]. We assume the most general form of Bose condensation and write

$$N_{o\lambda} = N_o u_\lambda^{\ 2} \tag{10a}$$

$$\sum u_\lambda^{\ 2} = 1 \tag{10b}$$

where $N_{o\lambda}$ is the number of condensed Bosons in the λth spin sub-level. Here N_o is the total condensate number and u_λ are components of a real arbitrary unit vector. As in the usual spin-zero Bogolyubov procedure we treat the zero momentum operator as a c-number and retain in H only terms of order N_o and $N_o^{\ 2}$. The resulting approximation for $\mathcal{H}$ is quadratic in the a's and can be diagonalized in a two-step procedure to give

$$\mathcal{H} = \sum_k \varepsilon^B_k \alpha_k + \alpha_k + \sum_{i=1}^{3} \sum_{\vec{k}} \frac{\hbar k^2}{2\overline{m}_d} \beta^+_{i\vec{k}} \beta_{i\vec{k}} \tag{11}$$

where the operators α take the usual spin-<u>zero</u> Bogolyubov form, but in terms of the linear combinations

$$\tilde{a}_k \equiv \sum_{\lambda=1} u_\lambda a_{\lambda\vec{k}}. \tag{12}$$

Accordingly

$$\alpha_{\vec{k}} = (1 - A_k^{\ 2})^{-1/2}(\tilde{a}_k - A_k \tilde{a}_{-k}^{\ +}) \tag{13}$$

where

$$A_k = (V/U(k)N)\left[\frac{-\hbar^2k^2}{2\overline{m}_d} - U(k)\,\frac{N}{V} + \left\{\left(\frac{\hbar^2k^2}{2\overline{m}_d} + U(k)\,\frac{N}{V}\right)^2 - \left(U(k)\,\frac{N}{V}\right)^2\right\}^{1/2}\right]. \qquad (14)$$

The physical meaning of (12) is this: the first term describes a phonon branch with $\varepsilon_k^B \sim ck$ as $k \to 0$ (c being the sound speed). The second set of terms is entirely a consequence of the non-zero spin. They are impurity-like modes, with quadratic dispersion. (The β's are straightforward to determine in terms of the u_λ and $a_{\lambda k}$ [16].) As has been observed previously [17], such modes are Goldstone or broken symmetry modes, and arise when the system condenses into a particular direction in "u-space", thereby breaking a continuous symmetry of the Hamiltonian. The new modes have an energy and associated damping that both vanish as $k \to 0$. Though they can be thought of as magnons in u-space, it is crucial to note that they arise without the presence of any explicit magnetic or spin-flip interactions in the Hamiltonian [18].

It is particularly interesting to examine the dynamic structure factor which in the Bogolyubov approximation takes the form [16]

$$S(k,\omega) = N_0^{1/2}\left(\frac{1 + A_k}{1 - A_k}\right)^{1/2} \delta(\omega - \varepsilon_k^B) \qquad (15)$$

Thus a longitudinal probe will see only a "Bogolon" branch as would be expected for a spin-zero Boson system. The impuriton branches do not contribute to the structure factor.

In addition to these two distinct sets of Boson branches, the usual Fermi quasi-particle and quasi-hole branches will also exist in LMD, assuming that the electrons are normal. The corresponding dispersions will also be fully renormalized by all the interactions in the system. It follows that at finite temperatures we may describe dense liquid metallic deuterium in terms of "gases" of three fundamentally different types of elementary excitations, each distributed according to quantum ideal gas distribution functions appropriate to the corresponding statistics. (In particular it should be noted that the number of "impuriton" excitations is temperature dependent.) This mixture of elementary excitation gases leads to particularly interesting thermodynamic functions. A distinct and quite novel feature of the specific heat, for example, is that the impuriton contribution, because of the large deuteron/electron mass ratio, dominates for all but extremely low temperatures [19], and yields a characteristic $T^{3/2}$ dependence. As is the case with liquid and solid metallic hydrogen, the specific heat difference between solid and liquid metallic deuterium is striking. The thermal expansion coefficient is also dominated by a $T^{3/2}$ dependence, once again a consequence of the excitation of impuritons [16], and once again quite different from the behavior of the solid.

We conclude by noting that so far as electrical transport in Bose-condensed but otherwise normal states of LMD is concerned, the scattering of electrons at low temperatures will be dominated by the impuriton modes, barring possible impuriton drag effects. A variational estimate [20] yields a remarkable $T^{7/2}$ behavior in the low T resistivity. This arises as a consequence of both the $T^{3/2}$ variation in the number of impuriton targets, and the final state restriction on the electrons imposed by Fermi statistics. With respect to electron pairing and superconductivity in a Bose condensed phase, the usual phonon mechanism will be present and will also be supplemented by pairing arising from the excitation of impuritons. The latter will replace to some extent the transverse-phonon coupling expected in a solid phase of the system.

References and Footnotes

[1] E. G. Brovman, Yu Kagan and A. K. Kholas, Sov. Phys. JETP 34, 1300 (1972).

[2] D. M. Straus and N. W. Ashcroft, Phys. Rev. Lett. 38, 415 (1977).

[3] J. Hammerberg and N. W. Ashcroft, Phys. Rev. B 9, 5025 (1974).

[4] K. K. Mon, N. W. Ashcroft, and G. V. Chester, Phys. Rev. B 21, 2641 (1980).

[5] L. A. Akhiezer and E. M. Chudnovsky, Sov. Phys. JETP 39, 1135 (1974).

[6] J. Oliva and N. W. Ashcroft, Phys. Rev. B 23, 6399 (1981).

[7] See, for example, D. Pines and P. Nozieres, The Theory of Quantum Liquids (Benjamin, N.Y., 1966).

[8] K. A. Brueckner and K. R. Atkins, Phys. Rev. Lett. 1, 315 (1959).

[9] A. A. Abrikosov and I. M. Khalatnikov, Rep. Prog. Phys. 22, 329 (1959).

[10] J. Oliva and N. W. Ashcroft, Phys. Rev. B 25, 223 (1982).

[11] J. Jaffe and N. W. Ashcroft, Phys. Rev. B 23, 6176 (1981).

[12] D. J. Scalapino, Superconductivity, Ed. R. D. Parks (Marcel Dekker, N.Y., 1969), p. 488.

[13] J. Jäckle and K. Fröbose, J. Phys. F 10, 471 (1980).

[14] J. Jaffe and N. W. Ashcroft, Phys. Rev. B 27, 5852 (1983).

[15] W. H. Bassichis, Phys. Rev. A 134, 543 (1964).

[16] J. Oliva and N. W. Ashcroft, Phys. Rev. B (to be published). Many of the results of this section are discussed more fully in this paper.

[17] B. I. Halperin, Phys. Rev. B 11, 178 (1975).

[18] Analysis of the problem beyond the Bogolyubov approximation suggests that the "Bogolon" and "impuriton" dispersions are both renormalized but otherwise retain their weak coupling k-dependences in the $k \to 0$ limit. (Y.A. Nepomnyaschi, Zh. Eksp. Teor. Fiz. 70, 1070 (1976) [Sov. Phys. JETP 43, 559 (1976)]).

[19] The crossover to electron dominance occurs at $T \sim 10^{-6}/r_s^2$ °K. The phonon contribution ($\sim T^3$) is also very small.

[20] J. Oliva and N. W. Ashcroft (to be published).

A CONSERVING DYNAMIC THEORY FOR THE ELECTRON GAS IN METALLIC SYSTEMS

F. Green, D. Neilson and J. Szymanski

School of Physics,
University of New South Wales
Kensington, Sydney 2033 Australia.

INTRODUCTION

We report here on a new microscopic theory for the dynamical behaviour of the interacting electron gas at metallic densities. A number of significant properties are built into this theory. Firstly, our model includes both the linearised screening effects of the R.P.A. and also the leading correction terms to the R.P.A. at small momentum transfers arising from electron-hole scattering[1]. Secondly the model reproduces the strong short-range Coulomb correlations between pairs of electrons which are known to dominate correlations at large momentum transfer[2]. Finally, the theory strictly conserves particle number, momentum and energy. In particular the dynamic structure factor exactly satisfies the f-sum and conductivity sum rules.

In a preliminary application of this theory we examined the dynamical structure factor at $r_s = 2$, and concluded that the multiple peak structure observed at large momentum transfer in Be and Al[3] can be quantitatively associated with dynamic Coulomb correlations between pairs of electrons[4]. Here we apply the theory at $r_s = 3$ and draw similar conclusions.

THEORY

We employ a method developed by Baym and Kadanoff[5] to generate a strictly conserving approximation for the irreducible polarisation function $\chi^{sc}(\vec{q},\omega)$.

We start by approximating the ground state energy $\Phi[G]$, which is a functional of the fully renormalised one-body propagator G, by the set of terms shown in Fig. 1. The solid lines represent G in the Feynman representation and the dashed horizontal lines the bare Coulomb interaction.

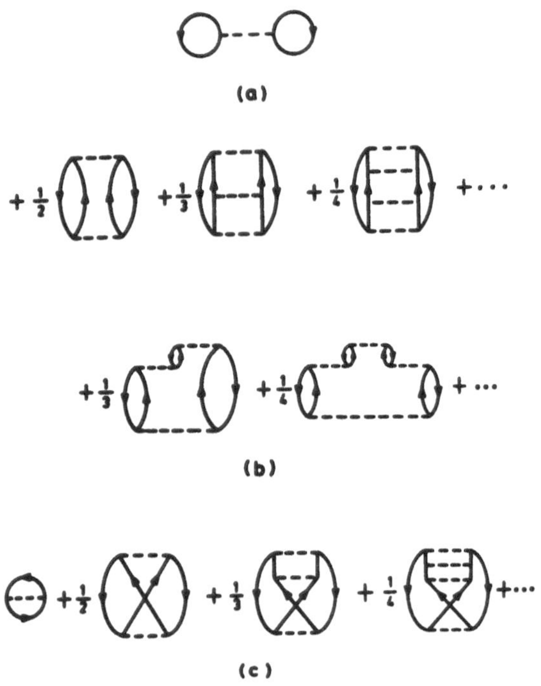

Figure 1
Contributions to the ground state energy functional $\Phi[G]$.

The fractions are weighting factors, with an overall implicit factor of ½. By construction, within each diagram of this set every propagator G is equivalent to every other propagator G, which ensures that the conservation laws are strictly satisfied. The Hartree - Fock diagrams together with the ring diagrams account for the overall collective polarisation of the electron gas - including the leading corrections to the R.P.A. . Once collective screening plus corrections to it have been taken care of, the most important corrections to the single - particle excitation terms at all momentum transfers will be the binary collisions between single - particle excitations[1,2]. This is the reason for the presence of the T-matrix ladder terms in $\Phi[G]$.

The second variation of $\Phi[G]$ with respect to G generates the irreducible electron-hole propagator $\Lambda^{sc}[G]$:

$$\Lambda^{sc}[G] = -iGG\,(1 + \Xi^{sc}[G]\,\Lambda^{sc}[G]\,) \tag{1}$$

$$\Xi^{sc}[G] = i^2\,\frac{\delta^2\,\Phi^{sc}[G]}{\delta G\,\delta G} \quad . \tag{2}$$

The trace of $\Lambda^{sc}[G]$ over the hole states (i.e. integral over momentum and energy) gives the irreducible polarisation function $\chi^{sc}(\vec{q},\omega) = \mathrm{tr}\,\Lambda^{sc}[G]$.

The interaction functional $\Xi^{sc}[G]$ can be separated into a part $\Xi^{loc}[G]$ which is approximately local (i.e. it can be well represented by the standard mean field or local approximation), and a nonlocal part $\Xi^{nl}[G]$. Equation (1) can be then resummed,

$$\Lambda^{sc}[G] = \Lambda^{loc}[G] + \Lambda^{loc}[G]\,\Xi^{nl}[G]\,\Lambda^{loc}[G]$$

$$+ \{\text{terms of higher order in } \Xi^{nl}[G]\} \quad , \tag{3}$$

where $\Lambda^{loc}[G]$ satisfies Eq. (1) with $\Xi^{loc}[G]$ replacing $\Xi^{sc}[G]$. The functional $\Lambda^{loc}[G]$ can be obtained in a closed form if $\Xi^{loc}[G]$ is approximated by the mean field construction, because $\Xi^{loc}[G]$ will then completely decouple from $\Lambda^{loc}[G]$ in Eq. (1).

A systematic examination of all the terms contributing to $\Lambda^{sc}[G]$ for large momentum transfers shows that the only terms which cannot be approximated by a local construction, and hence must be put into the $\Xi^{nl}[G]$ part, are certain of the electron-electron and hole-hole multiple scattering terms. We display these terms by showing in Fig. 2

$$\chi^{(1)}\,(\vec{q},\omega) \equiv \mathrm{tr}\,\{(-iGG)\,\Xi^{nl}[G]\,(-iGG)\} \quad . \tag{4}$$

The two parallel horizontal lines schematically stand for a dynamic interaction consisting of the T-matrix ladder sum of bare Coulomb interactions or a pair of R.P.A. dynamically screened interactions. Of all the terms contributing to $\chi^{sc}(\vec{q},\omega)$ for large $\vec{q}$, it is only these terms which are characterised by having both large momentum transfer passing across the correlation vertex, and also having ω - dependent energy denominators associated with the vertex. It is this property which makes these terms both highly non-local and strongly ω-dependent.

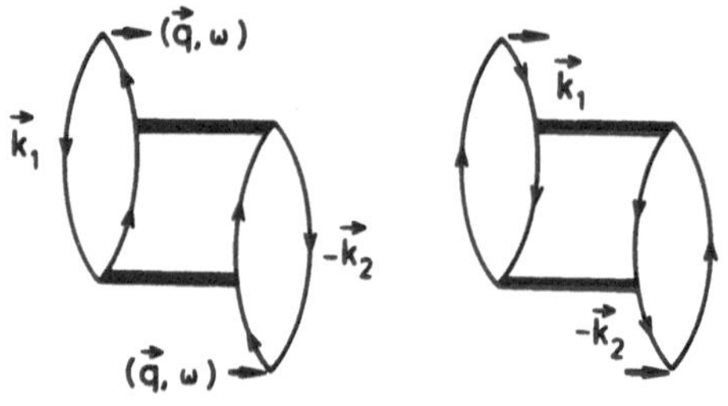

Fig. 2. Non-local contributions to the irreducible polarisation function.

In contrast, all the remaining contributions to $\chi^{sc}(\vec{q},\omega)$ at large $\vec{q}$ can be well approximated by a local construction and are consequently included in $\Lambda^{loc}[G]$. Figure 3 shows two typical diagrams of this type. In Fig. (3a) the energy denominators are ω-dependent in the correlation region, but the momentum transfer remains small even for large $\vec{q}$. In Fig. (3b) the momentum transfer is large but there is no ω-dependence in the energy denominators for the correlation region. We have found that these and all other terms which are placed with $\Lambda^{loc}[G]$, exhibit only a weak functional dependence on ω and consequently result in only an overall relaxation of $\chi^{sc}(\vec{q},\omega)$, with no fine structure as a function of ω.

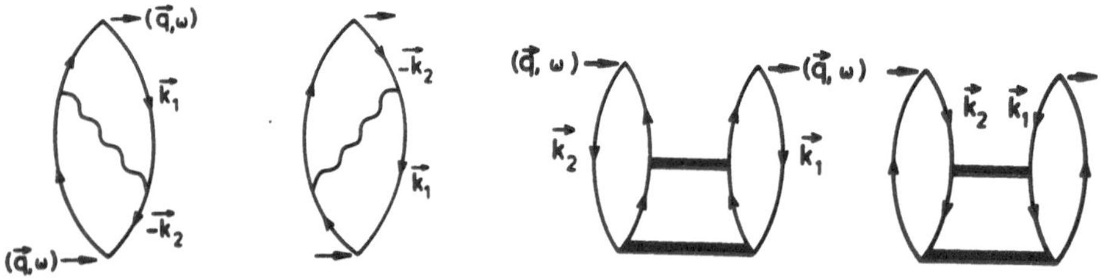

(a) (b)

Fig. 3. Approximately local contributions to the irreducible polarisation function.

RESULTS

A numerical examination of Eq. (3) in the metallic density range $2 \lesssim r_s \lesssim 4$, and for momentum transfers $q/k_F > 1.5$ shows that the second term on the right hand side of this equation is small compared with the first for the complete range of ω. Hence, the expansion can be truncated after the first two terms. We find that the strong functional dependence on ω of the non-local terms carries through to affect the ω-dependence of the structure factor $S(\vec{q},\omega)$. In Fig. 4 we show $S(\vec{q},\omega)$ calculated for $r_s = 3.2$ together with experimental data for Li^6. For this calculation we neglect the overall relaxation effects caused by the local contributions $\Xi^{loc}[G]$, so that we replace $\Lambda^{loc}[G]$ by the unrelaxed electron-hole polarisation (-iGG). This

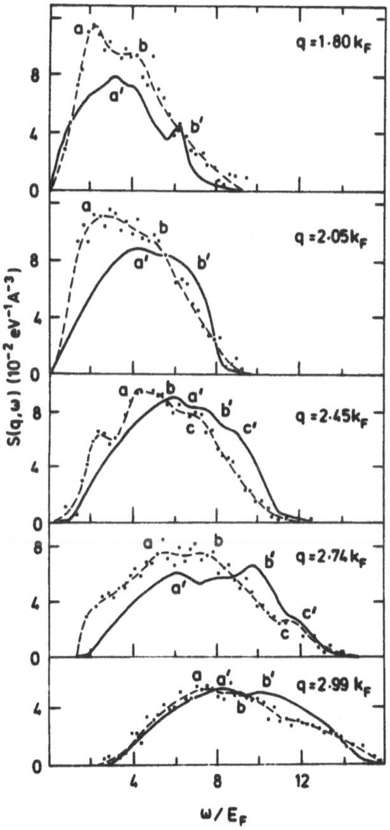

Fig. 4. Structure factor for Li.

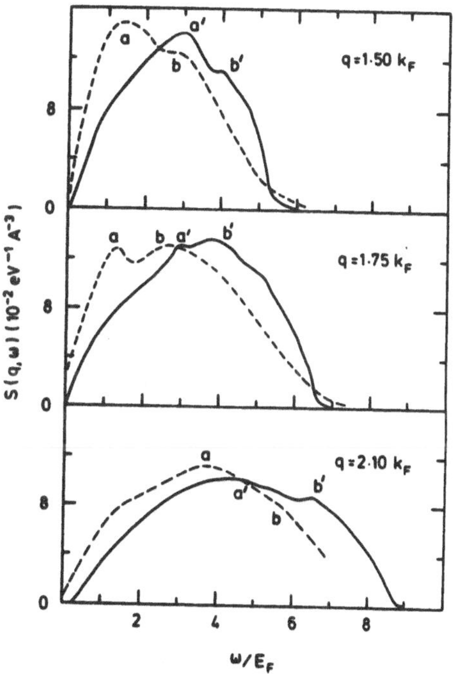

Fig. 5. Structure factor for Be.

means that our overall curve is centred on the R.P.A. structure factor rather than its fully relaxed counterpart. In Fig. 4 the experimental data points are shown together with a dashed guide-line, also taken from Ref. 6. The solid line is our calculated $S(\vec{q},\omega)$. The labels (a) - (a'), etc., are our matchup of the fine-structure peaks. We note the quantitative agreement between the position of the peaks at all reported momentum transfers. In the experimental data there is an unaccounted-for low-energy peak around $\omega/\varepsilon_F \simeq 2.5$ at momentum transfers $q/k_F = 2.45$ and 2.74. A careful examination of all our many-body contributions to $S(\vec{q},\omega)$ reveals no such peak, and we conclude that it is almost certainly related to some band-structure effect. We speculate that a possible cause for this peak would be a plasmon excitation accompanied by a single-electron interband excitation which has undergone an Umklapp. The kinematics for this process would predict a peak at this energy over the restricted momentum transfer range $2.2 \lesssim q/k_F \lesssim 2.9$, in agreement with the data.

Placing these results alongside the similar quantitative agreement we previously reported[4] for Be ($r_s = 2$) (Fig. 5) leads us to conclude that the dynamic electron-electron correlations in binary collisions are the primary cause of the striking multi-peak behaviour observed in the large momentum transfer dynamic structure factor.

ACKNOWLEDGEMENTS

We thank G.E. Brown, K. Golden, D. Pines, and R. Taylor for helpful and stimulating discussions.

REFERENCES

1. D.F. DuBois and M.G. Kivelson, Phys. Rev. 186, 409 (1969).
2. D.N. Lowy and G.E. Brown, Phys. Rev. B12, 2138 (1975).
3. P.M. Platzman and P.E. Eisenberger, Phys. Rev. Lett. 33, 152 (1974).
4. F. Green, D.N. Lowy and J. Szymanski, Phys. Rev. Lett. 48, 638 (1982).
5. G. Baym and L.P. Kadanoff, Phys. Rev. 124, 287 (1961).
   G. Baym, Phys. Rev. 127, 1391 (1962).
6. G.D. Priftis, J. Boviatsis and A. Vradis, Phys. Lett. A68, 482 (1978).

What present theory of superconductivity needs from many-body physicists.

D. Rainer
Physikalisches Institut der Universität Bayreuth

We discuss the microscopic meaning of the various input parameters for the theory of superconductivity.

The microscopic theory of superconductivity is probably the most successful theory in condensed matter physics. It is very accurate, reliable in its prognoses, valid in a wide range of temperature, frequencies, etc., and needs a surprisingly small calculational effort. The theory owes this success to the existence of various very good expansion parameters like ω_{Debye}/E_F, T_c/E_F, $1/k_F\ell$, etc., and to many-body physicists of the early sixties who taught us how to benefit from these parameters. Their key message is that superconductivity is a low-energy, long-wavelength phenomenon if compared with the typical electronic scales E_F and k_F^{-1}; it is not strong enough to break up short range electron correlations. This allows us to start from the correlated electron system, and to calculate all superconducting phenomena at fixed short range correlations. The simplifications are considerable: Short range correlations are among the most intricate theoretical problems in condensed matter physics. Because it is free of this burden, the theory of superconductivity has become one of the few simple and practicable high precision theories. This does not mean that short range correlations are neglected in the theory of superconductivity. They enter the theory via its input parameters.

Unfortunately, we have to pay a prize for the simplicity. There are important questions which are undecidable within the theory of superconductivity. A prominent one is: 'Is there a material with a superconducting T_c above 25 K (30 K, 100 K, room temperature, ...)?'. The existence of an '25 K superconductor' can neither be proved nor disproved by this theory. The theory of superconductivity puts no (relevant) limit on T_c. (Early limits derived from $T_c \simeq \omega_{Debye} \exp(-\ldots)$, i.e. $T_c \lesssim \omega_{Debye}$, are false). Suitable changes of the input parameters can produce any desired high T_c. Nature, however, limits T_c by putting constraints on the input parameters. A theoretical understanding of these constraints must come, and is expected to come from outside the theory of superconductivity, especially from many-body theorists. The theory of superconductivity can only help somewhat by calculating (within a few seconds CPU-time) the effect of the input constraints on superconducting data like the critical temperature, critical fields, tunneling spectra, etc. This paper presents a list and a brief discussion of the input parameters required by the theory of superconductivity. Until now, no substantial progress has been achieved in calculating these parameters from first principles. Probably, what is needed are novel techniques and ideas, and some help from many-body physicists.

The theory of superconductivity is most conveniently formulated in terms of Feynman diagrams for time ordered propagators /1/. This technique provides also the justification for the high accuracy of the theory of superconductivity. It allows to prove the correctness in leading order in the small expansion parameters ω_{Debye}/E_F, T_c/E_F, etc. /2,3/. The proof is constructive and yields quite naturally, as a by-product, a precise definition of the input parameters for the theory /4/. Unfortunately, the method of time ordered propagators seems to be not well suited for 'constructing' explicitely the input parameters on a computer. The various numerically more handy many-body methods /5/, on the other hand, could not yet be linked neatly to the standard theory of superconductivity.

The theory of superconductivity needs as input only zero temperature properties of metals with a frozen-in lattice. It can handle fast and perfectly any corrections due to finite temperatures and dynamical lattice degrees of freedom. Furthermore, one only needs properties of the low-lying fundamental excitations up to energies of the order ω_{Debye}. For most metals these are the <u>electronic quasiparticles</u> near the Fermi surface and the <u>phonons</u>. The required properties are:

1. spectrum and normal modes of phonons
2. spectrum (Fermi surface and Fermi velocity) of electronic quasiparticles in the frozen-in lattice
3. coupling matrix elements between electronic quasiparticles and phonons

4. quasiparticle-quasiparticle interaction parameters (Landau parameters, Coulomb pseudopotential)

At first sight, all seems easy to get from a reasonably equipped bandstructure calculator, perhaps with the exception of point 4. However, the problem is more subtle, as is indicated by my pedantic use of the term 'electronic quasiparticle' instead of 'electron'. The quasiparticle differs from the electron by its dressing with a local correlation cloud which might change its physical properties appreciably. The quasiparticle-phonon interaction consists of the contribution of the original electron plus the interaction of its correlation cloud with the phonon. Of course, the indistinguishability of the original electron from those in the correlation cloud must be taken into account, and leads to additional complications. What is usually (with some understatement) simply called electron-phonon interaction in the theory of superconductivity is, in fact, an involved many-body problem. A future microscopic theory of material parameters in superconductors needs a reasonable solution of the correlation problem in the electron-phonon interaction. At present, this problem is bypassed by the hypothesis that many-body effects cancel to a large amount and can be ignored.

In the following I sketch the microscopic definitions of the four classes of input quantities.

on 1:

The Born-Oppenheimer approximation holds for metals in leading order in ω_{Debye}/E_F. It reduces a first principle calculation of phonons to the problem of calculating the ground state energy of electrons in the static potential of the nuclei (+ core electrons). This is a widely examined and successfully attacked many-body problem, and need not be discussed here.

on 2:

The theory of superconductivity needs as input the Fermi surface,(it shall be parametrized by the 2-D variable s) the partial density of states $N(s,E_F)$ at point s on the Fermi surface, and the Fermi velocities $\vec{v}_F(s)$. These quantities are normal state quasiparticle properties of the electron liquid in the presence of the frozen-in lattice. They are most easily defined within the theory of time ordered propagators

(I know of no alternative and equally correct definition, although it is desirable to get hold of these quantities in numerically more efficient schemes). Knowing the selfenergy $\Sigma(x,x',\varepsilon)$ and its energy derivative at the Fermi energy (ε=0), one obtains the needed data, e.g., by solving the effective single particle Schrödingers equation

$$\left[\varepsilon(\delta(x-x')-\frac{\partial\Sigma}{\partial\varepsilon}(x,x';0))-(T+V+\Sigma(x,x';0)-E_F\delta(x-x'))\right]\psi(x') = 0\ . \tag{1}$$

It is convenient to introduce the wave-function renormalization factor:

$$a^{1/2}(x,x') = (\delta(x-x')-\frac{\partial\Sigma}{\partial\varepsilon}(x,x';0))^{-1/2}\ , \tag{2}$$

and the single (quasi-) particle Hamiltonian

$$H_O(x,x') = a^{1/2}(x,x_1)(T+V+\Sigma(x_1,x_2;0)-E_F\delta(x_1-x_2))a^{1/2}(x_2,x') \tag{3}$$

whose spectrum near ε = 0 determines $N(s,E_F)$, the density of states $N(E_F) = \int d^2sN(s,E_F)$, and $\vec{v}_F(s)$. Modern bandstructure programs are ready to calculate the spectrum; many-body theory is expected to supply Σ and $a^{1/2}$.

on 3:

To leading order in the fundamental expansion parameters of the theory of superconductivity the electron-phonon coupling (more precise: quasiparticle-phonon coupling) is determined by $H_O(x,x')$ of eq. (3). H_O depends implicitely on the positions R_i of the nuclei. The linear terms in δR_i (δR_i = deviation from the equilibrium position) give the coupling operator between quasiparticles and phonons:

$$H_{el-ph}(x,x';\delta R) = \sum_j \partial_{R_j} H_O(x,x')\cdot\delta R_j \tag{4}$$

The theory of superconductivity needs the matrix elements $\langle s,p|H_{el-ph}|s',p'\rangle$ for electrons on the Fermi surface. $\langle s,p|$ are eigenfunctions of H_O supplemented by the phonon hamiltonian. (p characterizes the phonon states). All many-body effects on the electron-phonon coupling are comprised in $\partial_R\Sigma(x,x',0)$ and the renormalization factors $a^{1/2}(x,x')$; ($\partial_R\ a^{1/2}(x,x')$ can be omitted!). At present, we have no clue on how important these effects are. Any insight is welcome.

on 4:

The theory of superconductivity needs two types of electronic interaction parameters, the electronic Landau interactions A(s,s') (quasiparticle interaction at zero momentum in the particle-hole channel), and the electronic pairing interactions $\mu^*(s,s')$ (quasiparticle interaction at zero momentum in the particle-particle channel). They are related to the general interaction vertex $\Gamma(x_1,x_2,x_3,x_4)$ =

$$= a^{1/2}(x_1,x_1')a^{1/2}(x_2,x_2')a^{1/2}(x_3,x_3')a^{1/2}(x_4,x_4')\Gamma_0(x_1',x_2',x_3',x_4') \qquad (5)$$

where Γ_0 is the sum of all connected 4-point vertex diagrams with no low energy propagators. A(s,s') and $\mu^*(s,s')$ are obtained from the matrix elements of $N(E_F)\Gamma(x_1,x_2,x_3,x_4)$ with the $\varepsilon = 0$ eigenfunctions of H_0 by taking the appropriate zero momentum limits. $\mu^*(s,s')$ is usually called Coulomb pseudopotential. It depends in general on s and s', the positions of the initial and final states on the Fermi surface. A(s,s') influences various quantities like Hc_2 in high field superconductors but drops out of T_c in zero magnetic field. $\mu^*(s,s')$ is a relevant interaction parameter for T_c because this (usually) repulsive interaction cancels off part of the attraction from the electron-phonon mechanism. $\mu^*(s,s')$ (but not A(s,s')) depends on an artificial cutoff ω_c (ω_c is introduced to separate 'low energies' ($\lesssim O(\omega_{Debye})$) from 'high energies'). The conventional choice is $\omega_c = 5\ \omega_{Debye}$, and most values of μ^* in the literature refer to this cutoff. Although theory has given a precise and constructive definition of $\mu^*(s,s')$ no reliable calculation from first principles exists. Measurements of μ^* by tunneling spectroscopy are possible, in principle, but have very limited accuracy. A popular alchemistic approach is to set $\mu^*(s,s')$ constant and equal to 0.13, or to take μ^* from the Bennemann-Garland formula /6/. Many-body physics is challenged to produce deeper insight.

I conclude by repeating that the theory of superconductivity needs an amazingly small number of input parameters compared to the large variety of effects it can calculate. The parameters are Fermi surface data, the electron-phonon interaction, the Coulomb pseudopotential, Landau parameters, and a few scattering lifetimes in the case of dirty metals. Our present information on these parameters comes exclusively from analyzing experiments, often with strong help of the theory of superconductivity. This procedure allows to understand the superconducting properties of experimentally well established materials. Any theoretical prognoses on the possibilities and limitations for future materials need first principle calculations of the input parameters, and hence the solution of several presently still open many-body problems.

References

/1/ A recent review on the microscopic theory of superconductivity is: P.B. Allen and B. Mitrović, Solid State Physics (H. Ehrenreich, F. Seitz, D. Turnbull, ed.), vol. 37, Academic Press, New York (1982).

/2/ A.I. Larkin, A.B. Migdal, Sov. Phys. JETP 17, 1146 (1963).

/3/ P. Morel, P. Nozières, Phys. Rev. 126, 1909 (1962).

/4/ J. Serene, D. Rainer, 'The quasiclassical approach to superfluid ^{3}He', to be published in Physics Reports (1983). The arguments given in this review can be immediately transferred to superconductors.

/5/ A collection of applied many-body methods can be found in: Recent Progress in Many Body Theories (J.G. Zabolitzky, M. de Llano, M. Fortes, J.W. Clark, ed.), Springer, Berlin (1981).

/6/ K.H. Bennemann and J.W. Garland, AIP Conf. Proc. 4, 103 (1972).

$$K = H - \mu N_{op} \tag{1}$$

$$H = T_{el} + V_{Coul} + V_{pseudo}^{bare} + H_{ion-ion}^{bare} \tag{2}$$

This Hamiltonian, written in terms of normal coordinates Q_{qj} and normal momenta P_{qj} of the ions is to be exposed to an unitary transformation given by [4]

$$Q_{qj} = (2\omega_j(q))^{-1/2}\,(b^+_{-qj} + b_{qj}) \qquad P_{qj} = i\left(\omega_j(q)/2\right)^{1/2}(b^+_{qj} - b_{-qj}) \tag{3}$$

This yields the usual electron-phonon Hamiltonian

$$H = H_{Bloch} + V_{Coul} + V_{el-phon} + H_{ion-ion}^{bare} \tag{4}$$

with

$$H_{Bloch} = \sum_{K} \frac{K^2}{2m} c^+_K c_K + \sum_{xx'l} V_{pseudo}^{bare}(xx', X_l^0)\, c^+(x)\, c(x') \tag{5}$$

$$V_{el-phon} = \sum_{qj} \sum_{KK'} c^+_K c_{K'} (b^+_{-qj} + b_{qj})(2\omega_j(q))^{-1/2} \sum_{x} (n M_x)^{-1/2}$$
$$\langle K+q | \sum_l e^{iqX_l^0} \vec{e}(x|qj) \cdot \vec{\nabla} V_{pseudo}^{bare} | K \rangle \tag{6}$$

$$H_{ion-ion}^{bare} = \sum_{qj} \sum_{nm} f_{nm} \left(b^+_{q_1 j_1} \ldots b^+_{q_n j_n} ; b_{q'_1 j'_1} \ldots b_{q'_m j'_m} \right) \tag{7a}$$

$$+ \sum_{qj} \omega_j(q) \left(b^+_{qj} b_{qj} + \tfrac{1}{2} \right) \tag{7b}$$

Here x x', denote the electron coordinates, X_l^0 sites, n the number density of unit cells [4], M_x the mass of the ion x, $\vec{e}(x|qj)$ the phonon-polarisation operator. We can be brief here because equations 1-6 are standard and may be found in textbooks [5] and monographs [6]. The expression 7a is - contrary to 7b - a complicated sum of operators and in general not well known. However it always occurs together with other terms in the CC-equations, so that the sum of these terms always yields contributions of relative order $(m/M)^{1/2}$ to the ground state energy, the gap Δ etc. The ground state wave function $|\Psi_0\rangle$ and excited state wave functions $|\Psi_e\rangle$ are given by the usual coupled cluster ansatz [1,2,3,7]

COUPLED CLUSTER EQUATIONS FOR SUPERCONDUCTING SYSTEMS [+)]

K. Emrich
Institut für theoretische Physik, Ruhr-Universität Bochum
4630 Bochum 1, Universitätsstraße 150

J.G. Zabolitzky
Institut für theoretische Physik, Universität zu Köln
5000 Köln 41, Zülpicher Straße 77

Abstract

The coupled cluster formalism is combined with the Bogoljubov transformation to describe superconducting systems. One obtains the well known BCS-formula

$$\Delta_0(k) = \int d^3k' \; v_{eff}(k,k') \; \Delta_0(k') / E(k')$$

as an exact result, valid for both weak and strong coupling superconductors. The effective interaction can be calculated as a sum over some clusters.

1. Introduction

We want to calculate the gap of a superconductor as a function of some "input parameters". These are given by the parameters of the bare electron-ion pseudopotential V^{bare}_{pseudo}, the number density of electrons, the ion mass, the phonon dispersion relation $\omega_j(q)$ (and phonon polarisation operator $\vec{e}(x\,|\,\vec{q}\,j)$), the lattice type and lattice constants. The parameters shall be taken from other work, if possible from first principle calculations. Here we intend to calculate the gaps of uniform systems. Calculations may later be extended to the gaps of real metals (crystals). Furthermore we hope to get answers to the questions "what makes the gap of a superconductor large" and "how large is the upper limit of the gap".

2. The Hamiltonian and the general form of the wave functions

The coupled cluster (CC)-theory is a method for the calculation of eigenvalues [1] or differences of eigenvalues [2, 3] of the Schrödinger-equation. We have therefore to start from a Hamiltonian H or, for superconductors, from a thermodynamical potential K. These operators are given by

$$|\Psi_0\rangle = \exp(S)|\Phi\rangle \quad , \qquad |\Psi_e\rangle = R\exp(S)|\Phi\rangle \tag{8}$$

with

$$S = \sum_n \left(S_n^{el} + S_n^{el\text{-}phon} + S_n^{phon}\right) , \quad R = \sum_n \left(R_n^{el} + R_n^{el\text{-}phon} + R_n^{phon}\right)$$

The phonon-clusters S_n^{phon} are zero in the harmonic approximation. For $|\Phi\rangle$ we use the ansatz [4,8,9]

$$|\Phi\rangle = |\Phi_{BCS}\rangle\, |0_{Phon}\rangle \tag{9}$$

Here $|0_{Phon}\rangle$ is the phonon vacuum, defined by [4]

$$b_{qj}\,|0_{Phon}\rangle = 0 \tag{10}$$

and $|\Phi_{BCS}\rangle$ the BCS determinant [8,9]

$$|\Phi_{BCS}\rangle = \prod_K \alpha_K \beta_{-K}\; |vac\rangle \tag{11}$$

where α_K and β_K are Bogoljubov-quasiparticle operators defined by

$$\alpha_K = u_K\, c_{K\uparrow} - v_K\, c^+_{-K\downarrow} \qquad \beta_{-K} = u_K\, c_{-K\downarrow} + v_K\, c^+_{K\uparrow}$$
$$u_K^2 + v_K^2 = 1 \tag{12}$$

Of course the S-operators have to be expressed in terms of these quasiparticle operators. Details may be found in ref. 10. We only mention here that the BCS-functions u_k and v_k have to be determined by the lowest order CC-ground state equation:

$$\langle\Phi|\,\alpha_K\beta_{-K}\; e^{-S}\, H\, e^{S}\,|\Phi\rangle = 0 \tag{13}$$

3. The equation for the zero temperature gap

We define the one particle excitation spectrum as usual by

$$E(K) = \begin{cases} \varepsilon^{part}(K) - \mu & K > K_F \\ \mu - \varepsilon^{hole}(K) & K \leq K_F \end{cases} \tag{14}$$

where

$$\varepsilon^{part}(K) = \tilde{E}_K(N+1) - \tilde{E}_0(N)$$
$$\varepsilon^{hole}(K) = \tilde{E}_0(N) - \tilde{E}_K(N-1) \tag{15}$$

$$\mu = \tilde{E}_0(N+1) - \tilde{E}_0(N) = \tilde{E}_0(N) - \tilde{E}_0(N-1) + O(1/N) \tag{16}$$

The energies $\tilde{E}_K$ denote exact eigenvalues of the Schrödinger-equation and are therefore real. For normal systems E(k) defined by eq. (14) is a continuous function, which is zero at k_F and has a discontinuous derivative at this point. For a superconducting system $E(k_F)$ is different from zero.

We define the temperature zero gap $\Delta_0(k_F)$ by the equation

$$E(k_F) = \Delta_0(k_F) \tag{17}$$

Eq. (14) defines the gap, as usual, by energy differences of states with different particle numbers. If the excitation energies of the N-particle system, i.e.

$$E_{exc} = \tilde{E}_K(N) - \tilde{E}_0(N) \tag{18}$$

are considered one has to distinguish between collective and non collective states. The minimum of the latter ones is just given by $2\,\Delta_0(k_F)$ as shown in the Appendix. Of course, this corresponds to the break-up of a BCS-pair.

In the following we use the definition eq. (17). It would be very inconvenient and, in practice, very inaccurate to calculate the Schrödinger eigenvalues occuring in eqs. (14) and (17) and substract them afterwards. Actually, the difference is calculated by the formalism of Offermann, Kümmel and Ey[2]. A graphical representation of their exact result is given by (compare Fig. 5a, f of Ref. 7)

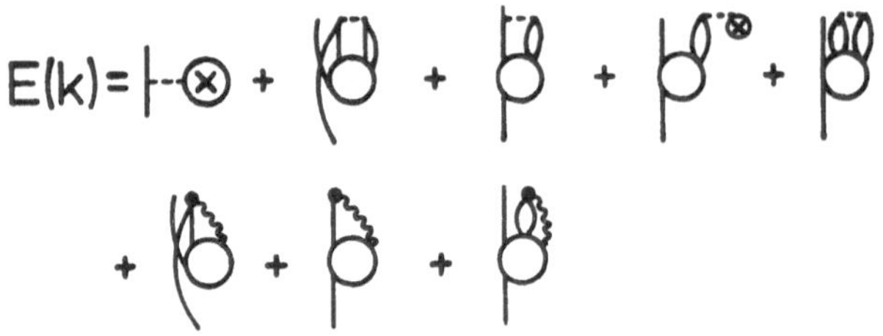

(19)

In the following we give some exact results deduced from eq. (19). From eq. (19) it follows without any restriction that $E(K)$ can be written in the following way

$$E(K) = a_K\,(\varepsilon(K) - \mu) + b_K\,\Delta_0(K) \tag{20}$$

where: $a_K = u_K^2 - v_K^2\,,\qquad b_K = 2u_K v_K$ (21)

$$\varepsilon(K) = \int d^3K'\, V_{eff}(K,K')\, a_{K'} + \begin{cases} T_K & \text{for uniform systems} \\ \langle K | H_{Bloch} | K \rangle & \text{for crystals} \end{cases} \tag{22}$$

$$\Delta_0(K) = -\int d^3K'\, V_{eff}(K,K')\, b_{K'} \tag{23}$$

There is a one-to-one-correspondence between $\mathcal{V}_{eff}(kk')$ and the sum given by the right hand side of eq. (19). Note also that eq. (17) is fulfilled by eq. (2o). One still needs eq. (13) in order to determine u_k and v_k and therewith a_k and b_k. One obtains, again as a strict result

$$b_k \left(\varepsilon(k) - \mu \right) = a_k \Delta_0(k) \tag{24}$$

From eqs. (2o), (21) and (24) it follows

$$E(k) = \left[(\varepsilon(k) - \mu)^2 + \Delta_0^2(k) \right]^{1/2} \tag{25}$$

$$b_k = \frac{\Delta_0(k)}{E(k)} \qquad a_k = \frac{\varepsilon(k)}{E(k)} \tag{26}$$

This yields, together with (23)

$$\Delta_0(k) = -\int d^3k' \, \mathcal{V}_{eff}(k,k') \, \Delta_0(k') / E(k') \tag{27}$$

The proof of eqs. (2o-24) is too lengthy to be given here and will be published elsewhere. Some remarks are in order, however:

i) $\mathcal{V}_{eff}(kk')$ is independent on Δ_0 up to terms $\mathcal{O}(\Delta_0/E_F)$

ii) Eq. (26) does not depend on a weak coupling assumption

iii) We are well aware of the fact (and see the preceding article by D. Rainer) that one can describe a large amount of experimental facts in superconducting systems, using some few parameters describing the Landau quasiparticle-quasiparticle interaction, the quasiparticle-phonon-interaction and the Landau renormalization factors. We admire this theory but unfortunately we are completely unable to calculate these wonderful and ingenious theorists inventions! We are only capable to calculate eigenvalues and differences of eigenvalues of the Schrödinger equation. With some additional effort also expectation values of other observables than energy can be calculated by us.

iV) For details of the calculation of clusters we refer to ref. (1o). We only mention here that the singularity occuring in the Fock-part of $\mathcal{V}_{eff}(kk')$ is just cancelled by singularities of the correlation part of $\mathcal{V}_{eff}$. The Fermi velocity $(d\varepsilon/dk)_{k=k_F}$ which is needed to solve eq. (27) has to be taken from eq. (22). Of course in this expression $a_{k'}$ can be replaced by +1(-1) for $k > k_F$ $(k \leq k_F)$, i.e. the Fermi velocity is given by the normal state Fermi velocity if (Δ_0/E_F)-terms in $\mathcal{V}_{eff}$ are neglected.

4. Numerical results

Numerical solution of eqs. (2o-27) requires as input the two-electron and the electron-phonon correlation amplitudes implicit in the effective interactions. Numerical solution of this coupled set of non-linear integral equations is presently under study. In order to test our numerical approximation schemes we studied the one-component problem, the electron fluid with homogeneous background. The problem in our formalism is a truncated version of the coupled electron-phonon problem. Without approximations, we would have to solve for functions of six continuous variables which is not feasible. The study of approximation schemes is therefore of paramount importance in order to solve eqs. (2o-27).

The general recipe of our approximation scheme is rather simple: correlation amplitudes are averaged over single-particle momenta to yield functions of only one continuous variable, the momentum transfer q. This is in line with the approximations introduced by Bishop and Lührmann, ref. (11). In contrast to their calculation, no further approximation is introduced by us i.e. all purely geometrical integrals over products of step functions are carried out exactly. Details may be found in ref. (1o).

Results for the correlation energy per electron are presented in table 1. In order to describe accurately the one-electron spectrum, eq. (22), it was found necessary to include parts of up to four-electron correlations. Therefore, our complete calculation is denoted SUB4-approximation. It is seen that there is excellent agreement with the exact Monte-Carlo data of Ceperley and Alder, ref. (12). As a remarkable result we find that just summing the ring diagrams, i.e. doing nothing but standard RPA, with kinetic energy denominators replaced by Hartree-Fock energy denominators, however, almost the same accuracy is achieved (RPA($T+U_{HF}$) column). This observation allows us to introduce significant simplifications into the algorithm for solving the coupled set of equations when including the phonons.

r_s	SUB4	GFMC	GFMC interpolated	CCref 7	RPA (T+U)
2.0	90.4	90.2	89.6	91.7	96.2
3.0	73.8		73.8	75.1	76.7
4.0	63.4		63.6	64.4	64.3
5.0	56.0	56.3	56.3	56.8	55.7
6.0	50.5		50.7		49.3
10.0	37.0	37.22	37.1		34.3

Tab. 1 The ground state correlation energy of the electron fluid, given in Millirydberg. The results of columns 2-6 are obtained by our CC(SUB4)-approximation, the Green's function Monte Carlo method [12], the latter results interpolated by a Padé approximant [13], the Coupled Cluster method of ref. 7 and our most simple "a posteriori approximation".

Appendix

The gap $\Delta_0(\kappa_F)$ is defined by eq. (17), i.e. by states with different particle numbers. Here we will prove that $2\,\Delta_0(\kappa_F)$ is just the minimum energy of the noncollective particle-hole states.

An exact equation describing such states (and collective ones, too) is given by eq. (3.1) and Fig. 4 of ref. 7. This equation can be written in shorthand notation (compare fig. 4 and fig. 5 of ref. 7) in the following way ($q = \kappa - \kappa'$, diagram $= \langle \kappa | R_1^{d} | \kappa' \rangle$)

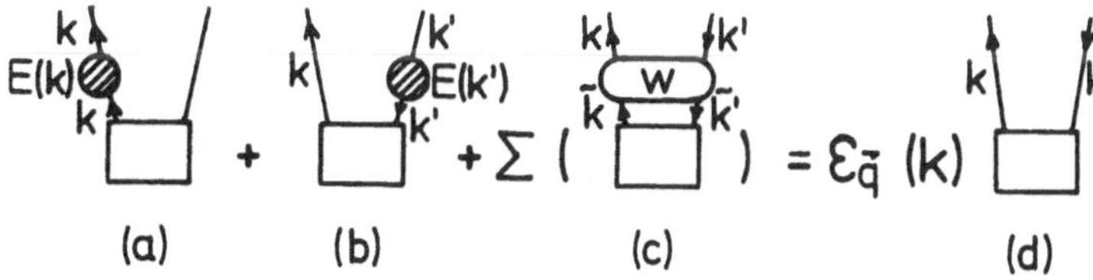

(A1)

We have written the equation in the basis diagonalizing the energies $E(\kappa)$ eq. (19). The block-diagram W connects the κ- and κ'-lines. We now consider a strictly non collective state [7]

$$|\Psi_e\rangle = \langle \kappa | R_1^{d} | \kappa' \rangle \, \alpha^+_{\kappa} \beta^+_{-\kappa'} \, |\Psi_0\rangle \qquad (\underline{\text{not}}\ |\Phi\rangle\ !) \tag{A2}$$

Evidently this is an eigenfunction of (A1) in the limit $N \to \infty$ because the integral over $\tilde{\kappa}, \tilde{\kappa}'$ in A1c only contains one single term and is therefore zero. The eigenvalue is given by

$$\mathcal{E}_q(\kappa) = E(\kappa) + E(\kappa') \qquad \text{(A3)}$$

From eq. (A3) the assertion follows immediately.
Of course eq. (A1) contains collective solutions, too. Because the leading terms of A1c are attractive, some of them will appear below the minimum of the continuous spectrum of noncollective states.

References

+) Work supported in part by the Deutsche Forschungsgemeinschaft

1) H. Kümmel, K.H. Lührmann, J.G. Zabolitzky, Phys.Rep.36C(1978)1 and refs given here

2) R. Offermann, H. Kümmel and W. Ey, Nucl.Phys.A273(1976)349

3) K. Emrich, Nucl.Phys.A351(1981)379

4) A.A. Maraduddin, E.W. Montroll, G.H. Weiss and I.P. Ipatova, Solid State phys.Suppl. 3(1971)

5) C. Kittel, Quantum Theory of Solids, J. Wiley, New York 1963

6) J.R. Schrieffer, Theory of Superconductivity, Benjamin, N.Y.(1964) V.L. Ginzbury and D.A. Kirzhnits (Ed.), High Temperature Superconductivity, 1982, Consultants Bureau, N.Y.

7) K. Emrich, Nucl.Phys.A351(1981)397

8) J. Bardeen, L.N. Cooper, J.R. Schrieffer, Phys.Rev.1o8(1957)1175

9) N.N. Bogoliubov, Sov.Phys.JETP 7(1958)41

1o) K. Emrich and J.G. Zabolitzky, submitted to Phys.Rev.B

11) R.F. Bishop and K.H. Lührmann, Phys.Rev.B26(1982)5523

12) D.M. Ceperley, B.J. Alder, Phys.Rev.Lett.45(198o)566

13) S.H. Vosko, L. Wilk, M. Nusair, Can.J.Phys.58(198o)12oo

COUPLED CLUSTER APPROACH WITH EXPLICITLY CORRELATED CLUSTER FUNCTIONS

Bogumil Jeziorski,[a,b] Hendrik J. Monkhorst,[a] Krzysztof Szalewicz[a,b,c]
and John G. Zabolitzky[a,c]

[a]Quantum Theory Project, Department of Physics, University of Florida,
Gainesville, Florida 32611

[b]Department of Chemistry, University of Warsaw, Pasteura 1,
02-093 Warsaw, Poland

[c]Institute for Theoretical Physics, University of Cologne,
5000 Cologne 41, West Germany

1. Introduction

In this paper we present a new approach to the electron correlation problem in atoms and molecules. Generally speaking, this approach consists in using explicitly correlated functions, i.e. functions depending explicitly on the interelectronic distance r_{12}, to expand the pair and higher cluster functions appearing in the coupled cluster (CC) theory of Coester and Kümmel [1,2]. The linear and nonlinear parameters in the expansion are determined from the CC equations using a combination of iteration and variation techniques. The strong orthogonality (SO) of the cluster functions is a consequence of the variational treatment and is approached gradually as the correlated basis set becomes complete. For an incomplete basis set both the CC equations and the SO condition are fulfilled only approximately. This treatment of the strong orthogonality eliminates most of many-electron integrals and enables a practical implementation of the method. Considering only two-electron clusters [3] and using the basis set of explicitly correlated Gaussian geminals [4] we recovered [5] 98.5 and 97.9 percent of the experimental correlation energies for the Be atom and for the LiH molecule, respectively. For Be our result is identical with that obtained by solving CC equations numerically [6], while for LiH it significantly surpasses in accuracy the results of all previous calculations of the correlation energy. It is expected that the remaining 1.5-2.0 percent of the correlation energy can be obtained by including the one- and three-electron clusters.

2. General definitions, spin elimination and coupled-pair equations

The starting point of the closed-shell CC theory is the equation

$$\Psi = e^{T}\Phi \tag{1}$$

relating the exact singlet wave function Ψ to a closed-shell determinant Φ providing a zeroth-order description of a many-electron system. We shall always assume that Φ is the Hartree-Fock determinant. Eq. (1) defines T uniquely (for a given Φ and Ψ) if we assume that T belongs to the operator algebra generated by all operators of the form $a_r^\dagger a_\alpha$, where a_α annihilates a spinorbital occupied in Φ and $a_r^\dagger$ creates a particle in a state orthogonal to all occupied spinorbitals. The operator T can be determined from the Schrödinger equation [1]

$$Q\, e^{-T} H e^{T} \Phi = 0, \tag{2}$$

where H is the Hamiltonian and $Q = 1 - |\Phi\rangle\langle\Phi|$. When T is known the correlation energy can be obtained from the formula

$$E_{corr} = \langle\Phi|HT\Phi\rangle + \frac{1}{2}\langle\Phi|HT^2\Phi\rangle, \tag{3}$$

where we have assumed that H contains only two-particle interactions. To permit practical calculations one has to constrain the cluster structure of T. In atomic and molecular applications the interaction potential is of a soft-core type and one may hope to obtain good results by expanding T as a linear combination of "double-excitation" operators $a_r^\dagger a_\alpha a_s^\dagger a_\beta$ only. The approximate T, denoted by T_2, can then be practically obtained from Eq. (2) in which Q is replaced by the operator Q_2 projecting on the space of all doubly excited determinants. Such a method of calculating E_{corr} has been introduced by Čížek [3] and will be referred to as the complete coupled pair (CCP) method. The CCP method can be extended in a natural way to include single and triple excitation operators T_1 and T_3 [7].

It can be shown [5] that if H is a spin scalar then both the exact T of Eq. (1) and the approximate cluster operators T_2, $T_1 + T_2$, or $T_1 + T_2 + T_3$ (defined by Eq. (2)) are spin free operators. In particular the T_2 operator can be represented in the form [5]

$$T_2 = \sum_{i<j}^{N} t(ij), \tag{4}$$

$$t = \sum_{\alpha,\beta=1}^{N/2} |\tau_{\alpha\beta}\rangle\langle\phi_\alpha\phi_\beta|, \tag{5}$$

where N is the number of electrons, ϕ_α and ϕ_β are occupied orbitals and $\tau_{\alpha\beta}$ are spinless pair functions to be determined. These pair-functions must satisfy the SO condition

$$q(12)\tau_{\alpha\beta}(12) = \tau_{\alpha\beta}(12), \tag{6}$$

where q(12) is the strong-orthogonality projector $q(12) = [1-p(1)][1-p(2)]$ and

$$p = \sum_{\alpha=1}^{N/2} |\phi_\alpha\rangle\langle\phi_\alpha|. \tag{7}$$

In practice it is convenient to work with the symmetry adapted pair functions

$\tau^1_{\alpha\beta} = \tau_{\alpha\beta} + \tau_{\beta\alpha}$ and $\tau^3_{\alpha\beta} = \tau_{\alpha\beta} - \tau_{\beta\alpha}$. The number of these functions is $N^2/4$. For obvious reasons $\tau^1_{\alpha\beta}$ and $\tau^3_{\alpha\beta}$ are referred to as the "singlet" and "triplet" pair functions. When they are known the correlation energy can be easily evaluated from

$$E_{corr} = \frac{1}{2} \sum_{\alpha,\beta=1}^{N/2} \langle \phi_\alpha \phi_\beta | r_{12}^{-1} | \tau^1_{\alpha\beta} + 3\tau^3_{\alpha\beta} \rangle. \tag{8}$$

The integro-differential equations for $\tau^s_{\alpha\beta}(12)$, $s = 1,3$ are easily obtained [5] by inserting (4) and (5) into the projected Schrödinger equation $Q_2 e^{-T} H e^{T} \Phi = 0$ and making use of the elementary reduced density matrix theory. The explicit form of these equations is [5]

$$[f(1) + f(2) - e_\alpha - e_\beta]\tau^s_{\alpha\beta}(12) = -q(12) r_{12}^{-1} \phi^s_{\alpha\beta}(12) + L^s_{\alpha\beta}(12) + F^s_{\alpha\beta}(12) + Q^s_{\alpha\beta}(12), \tag{9}$$

where f is the Fock operator, e_α, e_β are the orbital energies such that $f\phi_\alpha = e_\alpha \phi_\alpha$, $\phi^s_{\alpha\beta}(12) = \phi_\alpha(1)\phi_\beta(2) + (2-s)\phi_\beta(1)\phi_\alpha(2)$, and

$$\begin{aligned} L^s_{\alpha\beta}(12) = &-q(12) r_{12}^{-1} \tau^s_{\alpha\beta}(12) - \frac{1}{4} \sum_{\gamma,\delta=1}^{N/2} \langle \phi^s_{\gamma\delta} | r_{12}^{-1} \phi^s_{\alpha\beta} \rangle \tau^s_{\gamma\delta}(12) \\ &+ \frac{1}{2} A^s_{12}\, q(1) \sum_{\gamma=1}^{N/2} \int \phi^*_\gamma(3) r_{13}^{-1} [2\tau^s_{\alpha\gamma}(12)\phi_\beta(3) + 2\tau^s_{\gamma\beta}(12)\phi_\alpha(3) \\ &- \phi_\alpha(1)\Theta_{\beta\gamma}(23) - (2-s)\phi_\beta(1)\Theta_{\alpha\gamma}(23)] d3, \end{aligned} \tag{10}$$

$$\begin{aligned} F^s_{\alpha\beta}(12) = &- \frac{1}{4} \sum_{\gamma,\delta=1}^{N/2} \langle \phi^s_{\gamma\delta} | r_{12}^{-1} \tau^s_{\alpha\beta} \rangle \tau^s_{\gamma\delta}(12) + \\ &+ \frac{1}{2} \sum_{\gamma,\delta=1}^{N/2} [\langle \phi_\gamma \phi_\delta | r_{12}^{-1} \Theta_{\beta\delta} \rangle \tau^s_{\alpha\gamma}(12) + \langle \phi_\gamma \phi_\delta | r_{12}^{-1} \Theta_{\alpha\delta} \rangle \tau^s_{\gamma\beta}(12)], \end{aligned} \tag{11}$$

$$\begin{aligned} Q^s_{\alpha\beta}(12) = &\frac{1}{2} A^s_{12} \sum_{\gamma,\delta=1}^{N/2} \iint \phi^*_\gamma(3)\phi^*_\delta(4) r_{34}^{-1} \tau^s_{\alpha\beta}(13)\Theta_{\gamma\delta}(24) d3d4 \\ &- \frac{1}{4} A^s_{12} \sum_{\gamma,\delta=1}^{N/2} \iint \phi^*_\gamma(3)\phi^*_\delta(4) r_{34}^{-1} [\Theta_{\alpha\gamma}(13)\Theta_{\beta\gamma}(24) - \tau^1_{\alpha\delta}(13)\Theta_{\beta\gamma}(24) \\ &- \tau^3_{\alpha\delta}(13)\Theta_{\beta\gamma}(24) + 2\tau^1_{\alpha\delta}(13)\tau^s_{\beta\gamma}(24) - 2\tau^3_{\alpha\delta}(13)\tau^s_{\beta\gamma}(24)] d3d4. \end{aligned} \tag{12}$$

In Eqs. (10)-(12) $\Theta_{\mu\nu}$ is a short hand notation for $\tau^1_{\mu\nu} + 3\tau^3_{\mu\nu}$, $q = 1 - p$, and $A^s_{12} = 1 + (2-s)P_{12}$, where P_{12} interchanges the spatial coordinates of the electrons.

It should be emphasized that Eq. (9) may have solutions which do not satisfy the SO condition (6). Since such solutions do not have a physical meaning, Eq. (9) must be supplemented by Eq. (6) to obtain a complete set of integro-differential equations defining physically acceptable pair functions $\tau^s_{\alpha\beta}(12)$.

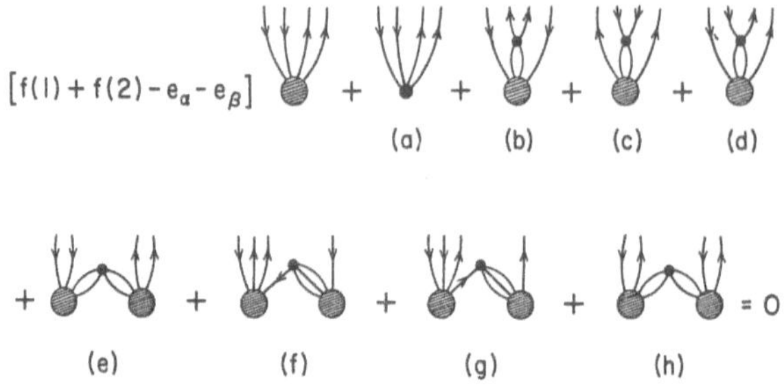

Fig. 1. Diagrammatic interpretation of the CCP equation (9). The equation of the FCP method is obtained by deleting diagrams (g) and (h).

In Fig. 1 we give the diagrammatic interpretation (in the Hugenholtz form) of the CCP equation (9). Diagram (a) corresponds to the inhomogenity in Eq. (9). Diagrams (b), (c) and (d) correspond to the three successive terms in Eq. (10) while diagrams (e) and (f) correspond to the first and second double summation in Eq. (11), respectively. Diagrams (g) and (h) correspond to the two double integrals composing $Q^s_{\alpha\beta}$ of Eq. (12). Since in $F^s_{\alpha\beta}(12)$ the pair-functions are explicitly factored out (while in $Q^s_{\alpha\beta}$ they are coupled via the double integration), the diagrams (e) and (f) are much easier to evaluate than the diagrams (g) and (h). Moreover, if the variational procedure of the next section is applied, the latter diagrams generate a great number of four-electron integrals. It has been suggested [6,8] that the unfactorizable diagrams (g) and (h) might be neglected without loosing much of accuracy. In Section 4 we shall show that for Be and LiH this is really the case and the so-called factorizable coupled pair (FCP) method, defined by neglecting $Q^s_{\alpha\beta}$ provides an excellent approximation to the CCP results. If the diagrams (e) and (f) are also neglected we obtain the linear coupled pair (LCP) method introduced already by Čížek [3]. This method is expected to be considerably less accurate than the FCP or CCP procedures.

3. Iteration schemes, variational principles and the strong orthogonality problem

The CCP equations (9) and (6) form a system of coupled nonlinear integro-differential equations for $N^2/4$ functions of six variables. Since solving them analytically is entirely hopeless, one has to take recourse to some kind of an iterative or variational procedure. When the nonlinear terms in (9) are neglected the pair functions and the resulting correlation energy E_{LCP} can be obtained by minimizing the functional

$$E_{LCP} \leq J_{LCP}[\tilde{\tau}] = \frac{1}{4} \sum_{\alpha,\beta=1}^{N/2} \sum_{s=1,3} s[\langle\tilde{\tau}^s_{\alpha\beta}|f(1) + f(2) - e_\alpha - e_\beta|\tilde{\tau}^s_{\alpha\beta}\rangle - \langle\tilde{\tau}^s_{\alpha\beta}|L^s_{\alpha\beta}\{\tilde{\tau}\}\rangle + 2\mathrm{Re}\langle\tilde{\tau}^s_{\alpha\beta}|r^{-1}_{12}\phi^s_{\alpha\beta}\rangle], \qquad (13)$$

where $\tilde{\tau}$ denotes the set of trial pair functions $\tilde{\tau}^s_{\alpha\beta}$ and $L^s_{\alpha\beta}\{\tilde{\tau}\}$ is a linear function of all $\tilde{\tau}^s_{\alpha\beta}$ defined by the right-hand side of Eq. (10). For the complete coupled pair method we could not find a similar functional providing an upper bound to E_{CCP}. Therefore we iterated the pair equations and applied a variational procedure in each iteration step. In fact many iteration schemes are possible, the simplest one consists in generating the following sequence of approximate pair functions $\tau^{[n]s}_{\alpha\beta}$, n = 1,2,...:

$$[f(1) + f(2) - e_\alpha - e_\beta]\tau^{[n]s}_{\alpha\beta} = R^{[n-1]s}_{\alpha\beta}, \tag{14}$$

where $R^{[n-1]s}_{\alpha\beta}$ is the r.h.s. of Eq. (9) calculated with the pair functions of (n-1)th iteration. When one sets $\tau^{[0]s}_{\alpha\beta} = 0$ then $\tau^{[1]s}_{\alpha\beta}$ are equal to the first-order pair functions $\tau^{(1)s}_{\alpha\beta}$ and, generally, the whole iterative process (14) is closely related to the order by order many-body perturbation theory expansion for T_2 and for the correlation energy E_{corr}.

The strongly orthogonal solution to Eq. (14) can be found by unconstrained minimization of the following functional [5]

$$J[\tilde{\tau}] = \langle\tilde{\tau}|h(1) + h(2) - e_\alpha - e_\beta|\tilde{\tau}\rangle - 2\mathrm{Re}\langle\tilde{\tau}|R^{[n-1]s}_{\alpha\beta}\rangle, \tag{15}$$

where $h = f + \Delta_{\alpha\beta}p$ and $\Delta_{\alpha\beta} = (e_\alpha + e_\beta)/2 - e_1 + \eta$. e_1 is here the lowest orbital energy and η is an arbitrary positive parameter having the dimension of energy. The functional (15) represents a simple generalization of the one used by us previously for the calculation of the second-order correlation energies [9].

Unfortunately there is a price we have to pay for the formal simplicity of the straightforward iteration process (14). As shown in Table I, its convergence is slow and in unfavorable situations, like the Be atom, a converged result may be hard to obtain. To remedy this we modified the straightforward iteration procedure by shifting to the l.h.s. of Eq. (9) all terms from $L^s_{\alpha\beta}$ and $F^s_{\alpha\beta}$ involving $\tau^s_{\alpha\beta}$. This modified iteration process generates the following sequence of approximate pair functions

$$[f(1) + q(1)v(1) + f(2) + q(2)v(2) + q(12)r^{-1}_{12} - e_\alpha - e_\beta - e^{[n-1]s}_{\alpha\beta}]\tau^{[n]s}_{\alpha\beta} = \bar{R}^{[n-1]s}_{\alpha\beta}, \tag{16}$$

where $v(i) = \frac{1}{2}sK_\alpha(i) + \frac{1}{2}sK_\beta(i) - J_\alpha(i) - J_\beta(i)$, J_α and K_α are the components of the usual Coulomb and exchange operators corresponding to orbital ϕ_α,

$$e^{[n]s}_{\alpha\beta} = -\frac{1}{1+\delta_{\alpha\beta}}\langle\phi_\alpha\phi_\beta|r^{-1}_{12}|\phi^s_{\alpha\beta} + \tau^{[n]s}_{\alpha\beta}\rangle + \frac{1}{2}\sum_{\gamma=1}^{N/2}[\langle\phi_\alpha\phi_\gamma|r^{-1}_{12}\theta^{[n]}_{\alpha\gamma}\rangle + \langle\phi_\beta\phi_\gamma|r^{-1}_{12}\theta^{[n]}_{\beta\gamma}\rangle] \tag{17}$$

and

$$\bar{R}^{[n]s}_{\alpha\beta} = R^{[n]s}_{\alpha\beta} + q(12)r^{-1}_{12}\tau^{[n]s}_{\alpha\beta} + A^s_{12}q(1)v(1)\tau^{[n]s}_{\alpha\beta} - e^{[n]s}_{\alpha\beta}\tau^{[n]s}_{\alpha\beta}\ . \tag{18}$$

One can show that the strongly orthogonal solution to Eq. (16) can be obtained by unconstrained minimization of the following functional

$$I[\tilde{\tau}] = \langle\tilde{\tau}|g(1) + g(2) + q(12)r_{12}^{-1}q(12) - e_\alpha - e_\beta - e_{\alpha\beta}^{[n-1]s}|\tilde{\tau}\rangle - 2\mathrm{Re}\langle\tilde{\tau}|\bar{R}_{\alpha\beta}^{[n-1]s}\rangle, \qquad (19)$$

where $g = f + qvq + \Delta_{\alpha\beta}p$. Unfortunately, even for $n = 1$ the functional (19) is much more time-consuming to evaluate than the functional (15). When only linear parameters in $\tau_{\alpha\beta}^{[n]s}$ are to be optimized one can replace (19) by a much simpler bilinear functional [5]

$$B[\tilde{\rho},\tilde{\tau}] = \langle\tilde{\rho}|h(1) + q(1)v(1) + h(2) + q(2)v(2) + q(12)r_{12}^{-1} - e_\alpha - e_\beta - e_{\alpha\beta}^{[n-1]s}|\tilde{\tau}\rangle$$

$$- \langle\tilde{\rho}|\bar{R}_{\alpha\beta}^{[n-1]s}\rangle - \langle\bar{R}_{\alpha\beta}^{[n-1]s}|\tilde{\tau}\rangle. \qquad (20)$$

This is possible since $B[\tilde{\rho},\tilde{\tau}]$ is stationary only if $\tilde{\tau}$ satisfies both Eq. (16) and the SO condition (6). In fact we optimized the nonlinear parameters only for $n = 1$

Table I. Convergence of the straightforward and modified iteration procedures for solving the CCP equations. The unit of energy is 1 mhartree = 27.2 meV. The equilibrium internuclear separation of 3.015 bohr was assumed for LiH. All numbers are quoted from Ref. [5] .

number of iterations	Be SIP[a]	Be MIP[b]	LiH SIP[a]	LiH MIP[b]
1	-76.35	-76.35	-72.17	-72.17
2	-84.08	-94.43	-78.43	-81.70
3	-88.57	-92.64	-80.69	-81.47
4	-90.42	-92.89	-81.20	-81.48
5	-91.52	-92.86	-81.39	
6	-92.10		-81.44	
7	-92.44		-81.47	
8	-92.62		-81.47	
9	-92.73		-81.48	
10	-92.78			
11	-92.82			
12	-92.83			
13	-92.86			

[a] straightforward iteration procedure. [b] modified iteration procedure.

using the functional (15). For $n > 1$ only linear parameters were optimized using either $J[\tilde{\tau}]$ or $B[\tilde{\rho},\tilde{\tau}]$. As shown in Table I the convergence of the modified iteration process, based on Eqs. (16) and (20), is excellent even for Be where the convergence of the straightforward procedure is poor. This result is very gratifying since $B[\tilde{\rho},\tilde{\tau}]$ is not more time consuming to calculate than $J[\tilde{\tau}]$.

One should emphasize that explicitly correlated basis functions (geminals) used to represent the trial functions $\tilde{\tau}$ do not have to satisfy the SO condition. The optimal fulfillment of this condition is guaranteed by the variational determination of adjustable parameters in $\tilde{\tau}$. For a complete geminal basis set the SO condition will be satisfied exactly. When the geminal basis set is incomplete both the SO condition (6) and the CCP equations (9) can be fulfilled only approximately. It is always possible, however, to improve the strong orthogonality of $\tau^s_{\alpha\beta}$ (at the expense of less precise fulfillment of Eq. (9)) by increasing the value of the parameter η. As a criterion of deviation from the strong orthogonality we used the quantity $S = \langle\tau|p(1)|\tau\rangle/\langle\tau|\tau\rangle$. In our calculation S was always of the order 10^{-3} - 10^{-6} and did not increase in the successive iterations.

4. Results and prospects

To minimize the functional (15) or to find the stationary point of the functional (20) the trial functions $\tilde{\tau}$ and $\tilde{\rho}$ were represented by explicitly correlated Gaussian type geminal expansions of the form

$$\sum_{k=1}^{K} a_k A^s_{12} \exp[-\alpha_k(\mathbf{r}_1-\mathbf{A}_k)^2 - \beta_k(\mathbf{r}_2-\mathbf{B}_k)^2 - \gamma_k(\mathbf{r}_1-\mathbf{r}_2)^2], \quad (21)$$

where a_k are linear and α_k, $\mathbf{A}_k$, β_k, $\mathbf{B}_k$, γ_k nonlinear parameters and K is the number of geminals. In our most accurate calculations [5] we set K = 60 for Be and K = 40 for LiH, while the occupied orbitals ϕ_α were represented by 20 spherical Gaussian type orbitals with optimized centers and exponents. We found that setting $\eta = 1$ hartree guarantees sufficiently accurate fulfillment of the SO condition. Obviously, explicitly correlated functions with linear or exponential $\exp(-\gamma r_{12})$ dependence on r_{12} would be more appropriate to expand $\tilde{\tau}$. The expansion (21) has however the great advantage that all many-electron integrals can be expressed in closed form.

In Table II we compare the results obtained using the present method with the results of previous CCP calculations and with other most accurate correlation energies for Be and LiH. The results of Refs. 10 and 11 were obtained using the original Čížek formulation of the CCP method. This amounts to expanding $\tau^s_{\alpha\beta}$ in terms of products of a finite number of excited Hartree-Fock orbitals. Such an expansion converges very slowly [6,9]. In particular orbitals with high angular momentum are needed to obtain accurate results. Since for molecules only s,p,d and f orbitals are practical to use, the virtual orbital expansion cannot provide us with a high accuracy in this case.

It is gratifying to observe that our CCP result for Be agrees very well with the result of the numerical integration of the CCP equations performed by Lindgren and Salomonson [6]. The small difference between their result and ours may be caused to some extent by the fact that these authors did not iterate the unfactorizable diagrams until self consistency.

For LiH our CCP result surpasses in accuracy the results of all previous correlation energy calculations, including a fixed node Monte-Carlo treatment [16]. It should be emphasized that the fact that our CCP results lie above the experimental correlation energy is not accidental. One can show that the leading perturbation theory correction to the CCP energy (appearing in the fourth-order of perturbation

Table II. Comparison of correlation energies for the Be atom and for the LiH molecule. The unit of energy is 1 mhartree = 27.2 meV. The equilibrium internuclear separation of 3.015 bohr was assumed for LiH.

	Be	LiH
literature CCP results	-91.37[a]	-27.2[b]
	-92.96[c]	
other literature results	-92.85[d]	-78.2[e]
	-93.88[f]	-79.1[g]
		-80±2[h]
present method[i]		
LCP	-98.34	-82.7
FCP	-92.89	-81.5
CCP	-92.86	-81.5
"experiment"[j]	-94.31±0.03	-83.2±0.1

[a]virtual orbital expansion with 96 exponential (Slater) type orbitals of s,p,d,f and g symmetry, Ref. 10.
[b]virtual orbital expansion with Gaussian type orbitals, Ref. 11.
[c]partial wave expansion and numerical treatment of two-dimensional radial equations, Ref. 6.
[d]numerical multiconfigurational Hartree-Fock procedure, Ref. 12.
[e]best CI result, Ref. 13.
[f]best CI result, Ref. 14.
[g]Møller-Plesset perturbation expansion through third order, Ref. 15.
[h]fixed node Monte Carlo result, Ref. 16.
[i]Ref. 5.
[j]Ref. 14 for Be and Ref. 15 for LiH.

theory) must be negative. Including all fourth-order terms is necessary if one wants to recover the missing 1.5 and 2.1 percent of E_{corr} for Be and LiH, respectively. This can be done by extending our method to include the T_1 and T_3 cluster operators. These operators can be represented in the following spin-free form

$$T_1 = \sum_{i=1}^{N} t_1(i), \tag{22}$$

$$t_1 = \sum_{\alpha=1}^{N/2} |\tau_\alpha\rangle\langle\phi_\alpha|, \tag{23}$$

$$T_3 = \sum_{i<j<k}^{N} t_3(ijk), \tag{24}$$

$$t_3 = \sum_{\alpha,\beta,\gamma}^{N/2} |\tau_{\alpha\beta\gamma}\rangle\langle\phi_\alpha\phi_\beta\phi_\gamma|, \tag{25}$$

where $\tau_\alpha(1)$ and $\tau_{\alpha\beta\gamma}(123)$ are the one- and three-particle spinless cluster functions. The missing fourth-order contribution to the correlation energy can be obtained by minimizing the following functional

$$E^{(4)}(T_n) \leq \langle\tilde{T}_n\Phi|[H_0,\tilde{T}_n]\Phi\rangle + 2\mathrm{Re}\langle\tilde{T}_n\Phi|[H,T_2^{(1)}]\Phi\rangle, \tag{26}$$

where $\tilde{T}_n$, n = 1,3 are trial cluster operators of the form (22) - (25), $H_0 = f(1) + \ldots + f(N)$ is the sum of the Fock operators and $T_2^{(1)}$ is the first-order cluster operator defined through the first-order pair functions $\tau_{\alpha\beta}^{(1)s}$. The functional (26) breaks down into a set of independent functionals for the second-order cluster functions $\tau_\alpha^{(2)}(1)$ and $\tau_{\alpha\beta\gamma}^{(2)}(123)$. These cluster functions, together with $\tau_{\alpha\beta}^{(2)}(12)$, can be subsequently used to evaluate all correlation energy diagrams through the fifth-order of perturbation theory. Some small basis set full CI and perturbation calculations suggest [17] that for well closed shells this may give us more than 99.9% of E_{corr}. We are currently engaged in calculating $E^{(4)}(T_n)$, n = 1,3, by minimizing the functional (26) with explicitly correlated expansions for $\tau_{\alpha\beta}^{(1)s}(12)$ and for $\tilde{\tau}_{\alpha\beta\gamma}(123)$. Our present work can also be extended to allow for an open-shell description of the breakage of the chemical bond. Since the open-shell CC theories [18] assuming a unitary invariance of the model space cannot be employed in this case, we have developed a very general CC formalism applicable for both complete and incomplete model spaces [19]. We are now working on specializing this formalism to a chemical bond problem and on implementing it using explicitly correlated Gaussian geminals. Another possibility, which is worth mentioning, is to use these geminals to expand the frequency dependent pair functions appearing in the time-dependent CC theory [20]. This should lead to more accurate values of dynamic polarizabilities and other response functions.

Acknowledgement

This work was supported by grants from: NSF CHE-79006129, PAN MR.I.9, Deutche Forschungsgemeinshaft and Heinrich-Hertz Foundation.

References

[1] F. Coester, Nucl. Phys. 7, 421 (1958); F. Coester and H. Kümmel, ibid. 17, 477 (1960).
[2] H. Kümmel, K. H. Lührmann, and J. G. Zabolitzky, Phys. Rep. 36C, 1 (1978).
[3] J. Čížek, J. Chem. Phys. 45, 4256 (1966); Adv. Chem. Phys. 14, 35 (1969).
[4] S. F. Boys, Proc. R. Soc. London A258, 402 (1960); K. Singer, ibid. A258, 412 (1960).
[5] B. Jeziorski, H. J. Monkhorst, K. Szalewicz and J. G. Zabolitzky, J. Chem. Phys., (submitted).
[6] I. Lindgren and S. Salomonson, Phys. Scr. 21, 335 (1980).
[7] J. Paldus, J. Čížek and I. Shavitt, Phys. Rev. A5, 50 (1972).
[8] K. Jankowski and J. Paldus, Int. J. Quantum Chem. 18, 1243 (1980).
[9] K. Szalewicz, B. Jeziorski, H. J. Monkhorst and J. G. Zabolitzky, Chem. Phys. Lett. 91, 169 (1982); J. Chem. Phys. 78, 1420 (1983).
[10] B. G. Adams, K. Jankowski and J. Paldus, Phys. Rev. A24, 2316 (1981).
[11] V. Kvasnička, Chem. Phys. Lett, 78, 98 (1981).
[12] C. Froese-Fisher and K. M. S. Saxena, Phys. Rev. A9, 1498 (1974).
[13] D. M. Bishop and L. M. Cheung, J. Chem. Phys. 78, 1396 (1983).
[14] C. F. Bunge, Phys. Rev. A14, 1965 (1976).
[15] K. Szalewicz, B. Jeziorski, H. J. Monkhorst and J. G. Zabolitzky, J. Chem. Phys. (in press).
[16] P. J. Reynolds, D. M. Ceperley, B. J. Alder and W. A. Lester, Jr., J. Chem. Phys. 77, 5593 (1982).
[17] P. Saxe, H. F. Schaefer III and N. C. Handy, Chem. Phys. Lett. 79, 202 (1981); G. D. Purvis III and R. J. Bartlett, J. Chem. Phys. 76, 1910 (1982); S. Wilson and M. F. Guest, Chem. Phys. Lett. 73, 607 (1980).
[18] R. Offermann, W. Ey and H. Kümmel, Nucl. Phys. A273, 349 (1976); R. Offermann, ibid. A273, 368 (1976); W. Ey, ibid. A296, 189 (1978); I. Lindgren, Int. J. Quantum Chem. Symp. 12, 33 (1978).
[19] B. Jeziorski and H. J. Monkhorst, Phys. Rev. A24, 1668 (1981).
[20] H. J. Monkhorst, Int. J. Quantum Chem. Symp. 11, 421 (1977).

PERTURBATION THEORY IN A CORRELATED BASIS +)

S.Fantoni
Dipartimento di Fisica dell'Università, Pisa, Italy
Istituto Nazionale di Fisica Nucleare, Pisa, Italy

B.L.Friman and V.R.Pandharipande
Department of Physics, University of Illinois at Urbana-Champaign
1110 W.Green Street, Urbana, Illinois 61801

Abstract

We give a brief outline of correlated basis functions (CBF) theory focusing the discussion on the linked cluster property of the theory, the diagrammatical rules of the perturbative series and its similarities with standard perturbation theory. Next, we discuss the choice of the correlation operator and give results obtained for the binding energy, the optical potential and the momentum distribution of nuclear matter by using variational theory plus second order CBF perturbation theory. When a state dependent pair correlation is used for the correlation operator, the second order perturbative corrections are generally small, nevertheless they are essential to reproduce important effects like the enhancement of the effective mass and the logarithmic slope of the momentum distribution at $k=k_F$. The results obtained are compared with other theoretical estimates and with the available experimental data. The agreement with the empirical data is all over fairly good. Most of the discrepancies are attributed to the coupling of low lying states to surface vibrations, not present in infinite nuclear matter.

1.-Correlated basis perturbation theory.

Correlated basis functions Ψ_i are generated by a many-body correlation operator acting on Fermi gas functions Φ_i

$$\Psi_i = G\Phi_i \quad . \tag{1.1}$$

The correlation operator G is generally not unitary, thus the basis functions Ψ_i are not necessarily orthogonal. A perturbative theory based on CBF is easily set up [1]. The eigenvalue E_n relative to the eigenstate $|n\rangle$ of the hamiltonian H is given by

$$E_n = H_{nn} + \sum_{p\neq n} \frac{(H_{np}-E_nN_{np})(H_{pn}-E_nN_{pn})}{E_n-H_{pp}} + \sum_{\substack{p,q\neq n\\ p\neq q}} \frac{(H_{np}-E_nN_{np})(H_{pq}-E_nN_{pq})(H_{qn}-E_nN_{qn})}{(E_n-H_{pp})(E_n-H_{qq})} + \ldots \quad , \tag{1.2}$$

where H_{ij} and N_{ij} are the normalized matrix elements of the hamiltonian and the identity operator respectively. The various orders $\Delta E_n^{(i)}=E_n^{(i)}-H_{nn}$ of the perturbative series can be obtained by expanding eq.(1.2) around $E_n=E_n^V=H_{nn}$ and then inserting ΔE_n back on the r.h.s. repeatedly. The second order correction is given by

$$E_n^{(2)} = \sum_{p\neq n} \frac{(H_{np}-H_{nn}N_{np})(H_{pn}-H_{nn}N_{pn})}{H_{nn}-H_{pp}} \tag{1.3}$$

An awkard feature of the series obtained by the above procedure is that each order $\Delta E_n^{(i)}$ is divergent in the large A-limit. For instance, the terms with p differing from n in more than three orbitals ($d_{np}>3$) in eq.(1.3) have a bad A-behavior. However a proper sum of the terms of the series shows that the perturbative series is linked. Such a property has been assumed to hold after the original structural studies[2] of the low orders of the series and the later developments of the correlated coupled-cluster formalism[3] and the correlated RPA theory[4]. Recently, a rigorous proof of the linked cluster property of the CBF perturbation theory has been given[5] by following the diagrammatical technique devised by Brandow[6] to prove the Goldstone theorem. In CBF theory one has to consider two types of "interactions", namely the overlapping matrix N_{ij} and the interaction matrix $V_{ij}= H_{ij}-H_{ii}N_{ij}$. Each term of the perturbative series is a product of N and V factors and energy denominators and can be represented by a Goldstone-like diagram. Unlinked diagrams correspond to terms having a bad A-behavior. A second main difference between standard perturbation theory and CBF theory is that the interactions V and N are matrix elements of sums of many-body operators and contain unlinked cluster contributions if i and j differ in more than three orbitals. Such contributions lead to unlinked Goldstone diagrams which are then cancelled by higher order diagrams in virtue of the following factorization property[&]:

$$V(d=d_{ij}) = \{ V^{(L)}(d=d_{ij})+ \sum V^{(L)}(d=d_{ij}-2)N(d=2)+\ldots$$
$$\ldots+ \sum V(d=2)N(d=2)N(d=2)\ldots \} \quad (1+O(1/A)) \quad , \tag{1.4}$$

where d is the number of entering and outgoing particle-hole lines and $V^{(L)}$ and $N^{(L)}$ mean linked parts of V and N respectively. If d<4 both V and N are linked[1].

A detailed diagrammatical analysis leads[5] to the result that the perturbative correction ΔE_o to the ground state energy is given by the sum of all the linked diagrams which, in addition to the diagrammatical rules of standard Goldstone diagrams, obey the following ones:

(i) the interaction lines may have any number d>1 of entering and outgoing particle-hole lines; (ii) the topmost line of a diagram must be a V-line;(iii) each N-line carries a minus sign and no energy denominator is present between a N-line and the interaction line immediately over it; (iv) a N-line not directly connected by particle-hole lines to the interaction line above it cannot occur; (v) a broken orizonthal line made up of two or more interaction lines, corresponding to some unlinked parts of V or N interaction is not allowed.

A few examples of allowed and not allowed diagrams contributing to ΔE_o are given in Fig.1

The convergence of the CBF perturbative series crucially depends on the choice of the correlation operator G. When G=1 the CBF perturbation theory reduces to Bethe-Bru-

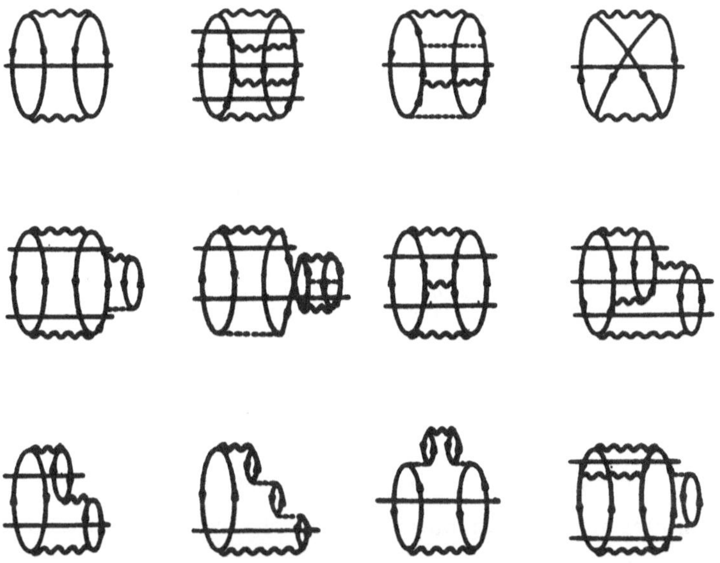

Fig.1: Linked Goldstone-like diagrams occurring in CBF theory for ΔE_o. The wavy and dashed lines represent the interactions V and N respectively, the orizonthal bars correspond to energy denominators and the up- and down-going lines are the particle-hole lines. The last two diagrams are not allowed.

eckner-Goldstone theory which is not convergent order-by-order. A symmetrized product of pair correlation functions

$$G = \mathcal{S} \prod_{i<j=1}^{A} f(ij) \tag{1.5}$$

has been proved to be suitable in microscopic calculations of nuclear matter[7]. Three body correlations have to be taken into account when more dense systems, like liquid helium, are treated. If f(ij) is state independent the symmetrizer $\mathcal{S}$ is not effective and the corresponding correlation operator is of the Jastrow type.

The strategy generally followed in CBF calculations is, first to choose the type of pair correlation f(ij), next to compute the f(ij) which gives a minimum for the binding energy of the system and then to use the calculated f(ij) in the perturbative calculations. In nuclear matter calculations a state dependent f(ij) is essential to get variational estimates of the energies which are reasonably close to the true values. For instance, for the Reid-v_6 model of nuclear matter a state dependent f(ij) gives[8] -18. MeV of binding, whereas in Jastrow approximation nuclear matter is not even bound and one gets $E_o^V/A=+15.8$MeV. However a Jastrow choice for f(ij) greatly simplify the calculations of the matrix elements entering in CBF theory, thus one hopes to calculate higher order diagrams, like third-order digrams or do ladder summations[5,9] etc...

In what follows we present and discuss a few results obtained in nuclear matter calculations by using a state dependent f(ij) and second order CBF perturbation theory.

2.-Ground state calculations.

A realistic hamiltonian is given by

$$H = -\frac{\hbar^2}{2m}\sum_i \nabla_i^2 + \sum_{i<j} v(ij) + \sum_{i<j<k} (v^{2\pi}(ijk)+v^R(ijk)) \quad , \tag{2.1}$$

where v(ij) is the Urbana-v_{14} interaction[10], $v^{2\pi}(ijk)$ is the Fujita-Miyazawa two-pion exchange three-nucleon interaction and $v^R(ijk)$ is a phenomenological representation of multipion exchange three-nucleon interaction, both taken from model V of ref.11 . The three-nucleon interaction seems to be necessary to obtain reasonable binding energies for ^{3}H and ^{4}He nuclei and a satisfactory value of the saturation density of nuclear matter[11]. Most of the calculations discussed in this section are done with an approximate hamiltonian in which $v^{2\pi}$ is neglected and v^R is approximated by a density dependent two-nucleon interaction[12].

A state dependent pair correlation f(ij) of the form

$$f(ij) = \sum_{p=1,8} f^p(r_{ij})\, O^p(ij) \quad , \tag{2.2}$$

allows for variational estimates of the energy eigenvalues which are fairly close to the true values. The eight operators involved in eq.(2.2) are: 1, $\vec{\sigma}\cdot\vec{\sigma}$, S, $\vec{L}\cdot\vec{S}$, plus the same operators multiplied by $\vec{\tau}\cdot\vec{\tau}$. The $f^p(r)$ are taken as solutions of approximate "Euler-Lagrange" equations with the constraints $f^p(r<d_p) = \delta_{p1}$. The 0th-order (or variational estimates) E_i^V of the energy eigenvalues E_i have been computed by using FHNC/SOC approximation[7]. Table I gives E_o^V at $k_F=1.33\text{fm}^{-1}$ calculated as a function of the range d_t of the tensor correlations.

The quantities to be computed in second order CBF perturbation theory are the matrix elements of (H-E) between states having only two different orbitals, like for instance $\langle 0\,|H-E\,|h_1h_2p_1p_2\rangle$. We have computed[8] these quantities at the level of two-body plus three-body separable diagrams (2+3S) of the cluster expansion. Such an approximation is generally not successful to variationally determine the correlation operator G, but it may not be a bad approximation to compute the matrix elements with the "optimal" G. The cluster expansion obtained with the correlation operator of ref.12 converges reasonably well for the ground state energy. The FHNC/SOC energy at k_F=1.33fm^{-1} is -16.MeV (as reported in ref.12), that obtained in the two-body cluster approximation is -21.MeV and a (2+3S) calculation gives -14.MeV[8]. Moreover, the (2+3S) results[8] for the imaginary part of the optical potential of Reid-v_6 model of nuclear matter agree within few per cent with the corresponding results in FHNC approximation[13].

$d_t/r_\circ$	$E_\circ^V/A$	$\Delta E_\circ^{(2)}/A$	$\kappa^{(2)}$
2.5	-9.2	-3.8	0.032
3.0	-9.9	-1.8	0.015
3.5	-9.3	-1.7	0.011
4.0	-7.9	-4.0	0.025

Table I: $E_\circ^V$, $\Delta E_\circ^{(2)}$ and $\kappa^{(2)}$ in nuclear matter at k_F=1.33fm^{-1} obtained with the approximate hamiltonian. The energies are in MeV and $r_\circ = (3/4\pi\rho)^{1/3}$.

The second order perturbative correction to the variational ground state energy is given by

$$\Delta E_\circ^{(2)}/A = \frac{1}{4}\sum \frac{|\langle 0\,|H-E_\circ^V\,|h_1h_2p_1p_2\rangle|^2}{e^V(h_1)+e^V(h_2)-e^V(p_1)-e^V(p_2)} , \tag{2.3}$$

where $e^V(p)=E_p^V-E_\circ^V$ and $e^V(h)=E_\circ^V-E_h^V$ are the variational single particle and single hole excitation energies. $\Delta E_\circ^{(2)}$ calculated at the experimental saturation density is given in Table I as a function of d_t, together with the quantity

$$\kappa^{(2)} = \frac{1}{4}\,\frac{1}{A}\sum \frac{|\langle 0\,|H-E_\circ^V\,|h_1h_2p_1p_2\rangle|^2}{(e^V(h_1)+e^V(h_2)-e^V(p_1)-e^V(p_2))^2} , \tag{2.4}$$

which measures the additional depletion of the Fermi sea due to the CBF two particle-two hole admixtures in the ground state. The minima of $E_\circ^V/A$ and of $\kappa^{(2)}$ are obtained for different values of d_t, which indicates that the better way to choose the correlation operator G may be to minimize $\kappa^{(2)}$ and not $E_\circ^V$. In fact $E_\circ^V$ is not sensitive to large admixtures of low lying states corresponding to $E_i \sim E_\circ$. In Fig.2 the variational energies $E_\circ^V(d_{tV})$ obtained with the full hamiltonian (2.1)[11] are compared with $E_\circ^V(d_{tV})$

$+\Delta E_2$, where $\Delta E_2 = E_o^V(d_{tP}) + \Delta E_o^{(2)}(d_{tP}) - E_o^V(d_{tV})$ is taken from ref.14 where the approximate hamiltonian has been used, and with the empirical saturation curve (d_{tV} and d_{tP} denote the values of d_t at which E_o^V and $\kappa^{(2)}$ have their minima).

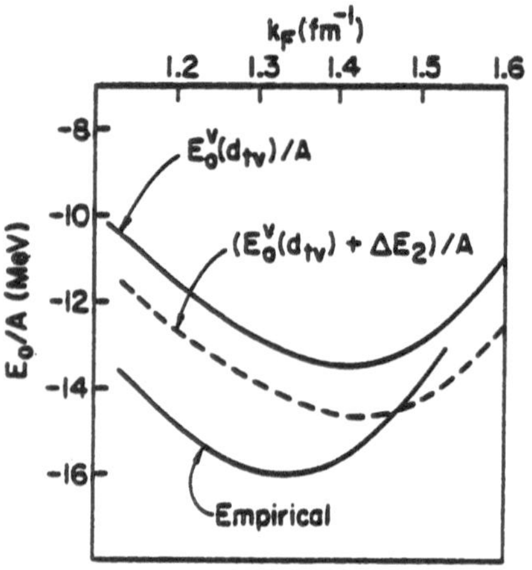

Fig.2: Saturation curves of nuclear matter

The real part $U(\rho,e)$ of the optical potential is defined by the following equation

$$e(k) = \frac{h^2}{2m} k^2 + U(\rho,e) \quad , \tag{2.5}$$

where ρ is the density of the system and k can be either $p(>k_F)$ or $h(<k_F)$. The 0th-order U_V of U is calculated from the variational E_o^V and E_k^V . The $U_2(\rho,e)$ is calculated with e(k) correct up to the second order of CBF perturbation theory (except for rearrangement terms) and given by $e^V(k)+\Delta e^{(2)}(k)$, where

$$\Delta e^{(2)}(p) = \tfrac{1}{2}\sum \frac{|\langle p\ |H-E_p^V\ |p_1p_2h_1\rangle|^2}{e^V(p)+e^V(h_1)-e^V(p_1)-e^V(p_2)} - \tfrac{1}{2}\sum \frac{|\langle 0\ |H-E_o^V\ |pp_1h_1h_2\rangle|^2}{e^V(h_1)+e^V(h_2)-e^V(p)-e^V(p_1)} \quad , \tag{2.6}$$

$$\Delta e^{(2)}(h) = \tfrac{1}{2}\sum \frac{|\langle 0\ |H-E_o^V\ |hh_1p_1p_2\rangle|^2}{e^V(h)+e^V(h_1)-e^V(p_1)-e^V(p_2)} - \tfrac{1}{2}\sum \frac{|\langle h\ |H-E_h^V\ |p_1h_1h_2\rangle|^2}{e^V(h_1)+e^V(h_2)-e^V(h)-e^V(p_1)} \quad . \tag{2.7}$$

The $U_V(\rho,e)$ and $U_2(\rho,e)$ calculated with the approximate hamiltonian are shown in Fig.3. The difference between U_V and U_2 is rather small, although U_2 is in a slightly better agreement with the empirical well depths in the energy range from -60 to +60 MeV.

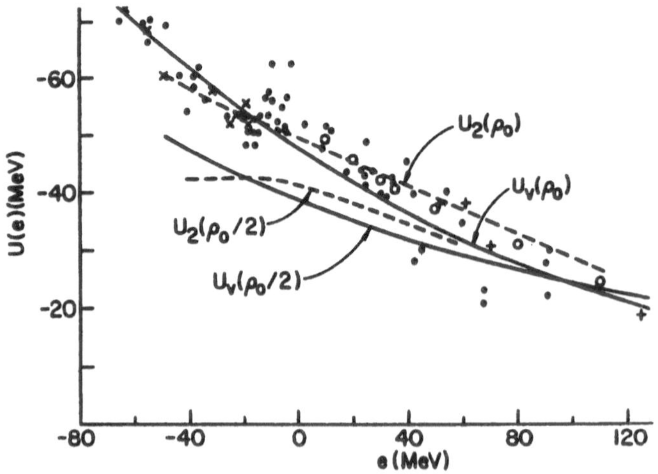

Fig.3: U_V and U_2 calculated at the experimental equilibrium density ρ_o and at $\rho_o/2$ are compared with the empirical data from various compilations given in ref.15 .

The effective mass, which is related to $U(\rho,e)$ by the equation

$$m^*(\rho,e)/m = 1-\partial U(\rho,e)/\partial e \quad , \tag{2.8}$$

determines the velocity of nucleons and the density of states in nuclear matter. The curves labelled $m^*_V(e)$ and $m^*_2(e)$ in Fig.4 are obtained from U_V and U_2 respectively. m^*_2 has a peak in the region of $e \sim e_F$. Such a behavior of m^* was first suggested by Brown et al.[16] and later analyzed and discussed extensively by Mahaux and collaborators[17,18].

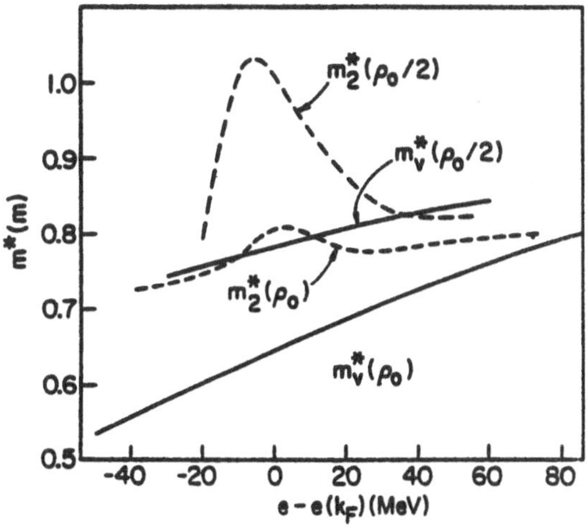

Fig.4: The effective mass at densities ρ_0 and $\rho_0/2$. The enhancement effect is bigger at lower density indicating surface absorption.

The imaginary part $W_0(e)$ of the optical potential, which is used to analyze elastic scattering, is related to the lifetime $\tau(p)$ of the CBF state $|p\rangle$ through the following equation[19,20]

$$W_0(e)=2\hbar\,\tau(p)\,(m^*/m) \quad . \tag{2.9}$$

The lifetime $\tau(p)$ is computed from the total transition rate of the state $|p\rangle$ to decay into two-particle one-hole states $|p_1p_2h_1\rangle$. Fig.5 displays the calculated $W_0(e)$ at ρ_0, together with the results obtained by using impulse approximation[21], corrected by a factor $(m^*/m)^2$, and Brueckner theory[17], corrected by (m^*/m) in accordance with eq.(2.9). The various theoretical estimates agree quite well amongst themselves and are a little larger than the strengths of Wood-

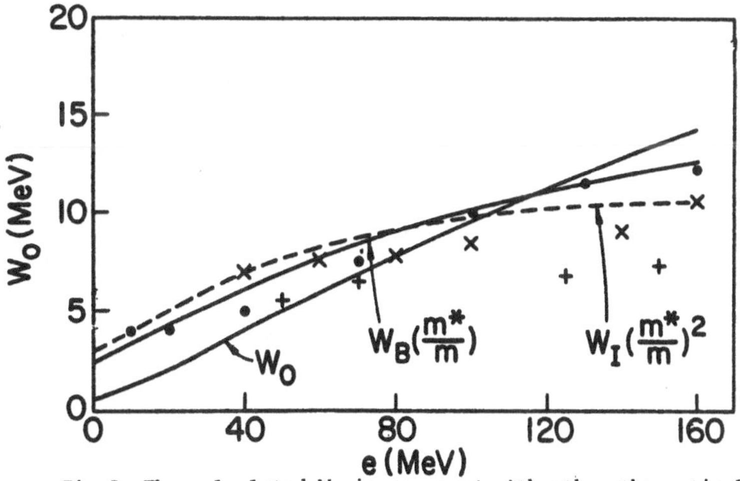

Fig.5: The calculated W_0 is compared with other theoretical estimates obtained in impulse approximation $W_I(m^*/m)^2$ and Brueckner theory $W_B(m^*/m)$ and the empirical data (see ref.20)

Saxon potential used in the standard optical model. The $W_o(e)$ obtained in Jastrow approximation for the Reid-v_6 model of nuclear matter is almost the double of that given by state dependent correlation operator[8], indicating that second order perturbation theory may not be sufficient when Jastrow approximation is used.

$k(fm^{-1})$	$n_V(k)$	$n_V^J(k)$
0.07	0.90	0.97
1.26	0.90	0.97
1.40	.0081	.0023
4.06	.0006	.0003

Table II: The variational occupation probability with (n_V) and without (n_V^J) spin-isospin correlations.

Another interesting quantity is the occupation probability n(k) of states with momentum k, which can be measured by deep inelastic scattering experiments. The deviation of n(k) from $\Theta(k_F-k)$ is indicative of the strength of correlations. The 0th order evaluation $n_V(k)$ of n(k) is given by the expectation value of $a_k^\dagger a_k$ on the variational wavefunction (1.1). FHNC/SOC approximation has been used in the calculations[22] and the results at $k_F=1.33fm^{-1}$ are given in Table II and compared with those obtained by setting all the $f^{p>1}$(spin-isospin correlations) in eq.(2.2) equal to zero. The comparison confirms that the spin-isospin correlations are extremely important in nuclear matter. The normalization and the kinetic energy sum rules are satisfied within .3% and 2% respectively, indicating that a FHNC/SOC calculation of $n_V(k)$ is quite accurate.

A variational calculation of n(k) is not adequate in the region $k\sim k_F$, since the correlation operator is too simple to realistically represent the correlations of particles close to the Fermi surface. Second order perturbative calculations are obtained from the expectation value of $a_k^\dagger a_k$ on the state $|\Omega_2\rangle = |0\rangle+\frac{1}{4}\sum\alpha(h_1h_2p_1p_2)|h_1h_2p_1p_2\rangle$, where

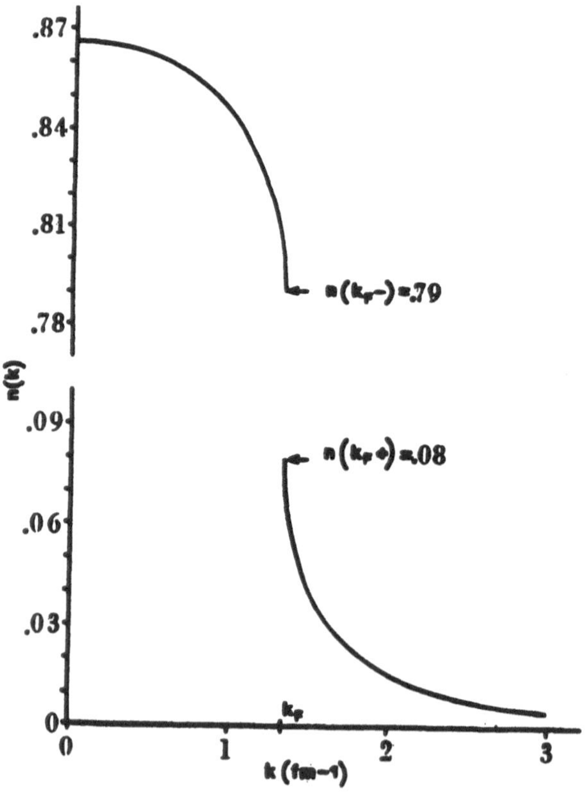

Fig.6: The calculated momentum distribution as a function of k

$$\alpha(h_1h_2p_1p_2)=\frac{\langle h_1h_2p_1p_2|\ H-E_o^V\ |0\rangle}{e^V(h_1)+e^V(h_2)-e^V(p_1)-e^V(p_2)}\quad, \tag{2.10}$$

and keeping only terms of order α^2 in its evaluation. The full n(k) is plotted in Fig. 6. Both n(h) and n(p) have infinite slopes as h or $p\to k_F$. This singularity is well known[18,23] and, in the present treatment comes from one of the perturbative correction terms, namely $\delta n_2(h)=\frac{1}{2}\sum\alpha^2(hh_1p_1p_2)$ or $\delta n_2(p)=\sum\alpha^2(h_1h_2pp_1)$. It results that

$$n(k\sim k_F) = n(k_F\pm) \pm A\ln\frac{|k-k_F|}{k_F}\quad, \tag{2.11}$$

where the plus(minus) sign is for particle (hole) state. The coeficient A in eq.(2.11) is proportional to the strength W of the quadratic term in $(e-e_F)$ of the imaginary part of the self-energy at $e\sim e_F$. In our calculations $W\sim 0.0025\text{MeV}^{-1}$ and $A\sim 0.2$. In nuclei the low lying single particle states have large widths due to their coupling with surface. These correspond[24] to a value of W of $\sim 0.02\text{MeV}^{-1}$, thus one expects a smoother behavior of n(e) at $e\sim e_F$ in nuclear matter than in nuclei.

The amount of discontinuity of n(k) at $k=k_F$ gives[25] the renormalization constant Z of the Green's function, which in turn is related to the E-mass of Mahaux and coworkers[17,26]. In our calculations the E-mass of nuclear matter at $k=k_F$ results to be 1.43. If we express the total effective mass $m^*(k_F)$, found to be 0.81m, as $m^*(k_F)$=(bare mass) $\times$(k-mass)/Z , the k-mass comes to be 0.57, in fair agreement with the results found by the Liège groupe[26], and quite different from the variational effective mass m_V=0.65m.

The calculated Z ($\sim$0.7) is also in reasonable agreement with the recent experimental data[27] on the difference between charge densities of ^{206}Pb and ^{205}Te, which seem to indicate a Z of about 0.6 .

References

+) Work supported in part by NSF grant PHY81-21399 and NATO grant 0453/82
1. J.W.Clark: contribution in this volume.
2. C.W.Woo: Phys.Rev.151(1966) 138.
3. E.Krotscheck, H.Kümmel and J.G.Zabolitzky: Phys.Rev. A22(1980)1243.
4. E.Krotscheck: Phys.Rev. A26(1982)3536.
5. S.Fantoni:Phys.Rev.B to be published.

&) The factorization holds only up to the order 1/A, which implies that also the so called "diagonal correction terms"(see refs. 3 and 5) must be taken into account. In the following discussion we consider them included into $V^{(L)}$ or $N^{(L)}$.

6. B.H.Brandow: Rev.Mod.Phys. 39(1967)771.
7. V.R.Pandharipande and R.B.Wiringa: Rev.Mod.Phys. 51(1979)821; S.Rosati:Proceedings of the International school of Physics Enrico Fermi, course LXXIX, ed. A.Molinari(North-Holland,Amsterdam 1982), pag.73.

8. S.Fantoni,B.L.Friman and V.R.Pandharipande: Nucl.Phys.A386(1982)1.
9. E.Krotscheck: contribution in this volume.
10. I.E.Lagaris and V.R.Pandharipande: Nucl.Phys. A359(1981)331.
11. J.Carlson, V.R.Pandharipande and R.B.Wiringa: Nucl.Phys.A in press
12. I.E.Lagaris and V.R.Pandharipande: Nucl.Phys. A359(1981)349
13. S.Fantoni, B.L.Friman and V.R.Pandharipande: Nucl.Phys. A399(1983)51.
14. E.Krotscheck, R.A.Smith and A.D.Jackson: Phys.Lett. 104B(1981)421.
15. B.Friedman and V.R.Pandharipande: Phys.Lett. 100B(1981)205.
16. G.E.Brown, J.H.Gunn and P.Gould: Nucl.Phys. 46(1963)598.
17. J.P.Jeukenne,A.Lejeune and C.Mahaux: Phys.Reports 25C(1976)83.
18. R.Sartor and C.Mahaux: Phys.Rev. C21(1980) 2613.
19. J.W.Negele and K.Yazaki: Phys.Rev.Lett. 47(1981)71.
20. S.Fantoni, B.L.Friman and V.R.Pandharipande: Phys.Lett. 104B(1981)89.
21. J.Dabrowski and A.Sobiczewski: Phys.Lett. 5(1963)87.
22. S.Fantoni and V.R.Pandharipande: preprint (1983) .
23. V.A.Belyakov: Sov.Phys. JETP 13(1961)850.
24. G.F.Bertsch, P.F.Bortignon and R.A.Broglia: Rev.Mod.Phys. 55(1983)287.
25. A.B.Migdal: Sov.Phys. JETP 5(1957)333.
26. C.Mahaux: contribution in this volume.
27. J.M.Cavedon et al. : Phys.Rev.Lett. 49(1982)978.

RECENT DEVELOPMENTS IN A CORRELATED THEORY OF LINEAR RESPONSE

David G. Sandler
Department of Physics
Loomis Laboratory of Physics
University of Illinois at Urbana-Champaign
Urbana, Illinois 61801, U.S.A.

N.-H. Kwong
Max-Planck Institut für Kernphysik
D-6900 Heidelberg 1, West Germany

I. INTRODUCTION

This contribution gives a brief account of some recent advances in our application of a generalization of the random-phase approximation (RPA) to the elementary-excitation spectra of Fermi liquids at zero temperature. We first sketch the essentials of linear-response theory formulated in terms of a specific choice of free-space interaction between fermions and a corresponding variational description of the correlated ground state. We then apply the theory to i) an initial, exploratory study of normal liquid ^{3}He, and ii) a semi-realistic model of neutron matter in which the neutrons experience a hard-core repulsion at small interparticle separation. In the course of our analysis and discussion, we shall pay particular attention to comparative aspects with respect to standard mean-field and phenomenological theories of linear response.

II. CORRELATED RANDOM-PHASE APPROXIMATION

In References 1-4, derivations of a theory of correlated linear response (CLR) are presented, extending the usual analysis[5] based on time-dependent Hartree-Fock (TDHF) theory to a situation where strong many-body correlations are explicitly included in the trial ground state, which is to be subjected to a weak, external one-body perturbation $\hat{\rho}(t) = \hat{\rho}(\omega)[e^{i\omega t} + e^{-i\omega t}]$. This variational ground state, $|\Psi_0\rangle = F|\Phi_0\rangle I_{00}^{-1/2}$, with energy $H_{00} = \langle\Psi_0|H|\Psi_0\rangle$, is the lowest configuration in the complete set of correlated basis functions[2,6] (CBF) $\{|\Psi_m\rangle = F|\Phi_m\rangle I_{mm}^{-1/2}$, $I_{mm} \equiv \langle\Phi_m|F^\dagger F|\Phi_m\rangle\}$. The least-action principle

$$\delta\int_{t_1}^{t_2}\langle\Psi(t)|H - i\,\frac{\partial}{\partial t} + \hat{\rho}(t)|\Psi(t)\rangle dt = 0 \qquad \text{(II.1)}$$

is applied to the space of time-dependent states

$$|\Psi(t)\rangle = F|\Phi(t)\rangle/\langle\Phi(t)|F^\dagger F|\Phi(t)\rangle^{1/2} \qquad , \qquad \text{(II.2)}$$

where

$$|\Phi(t)\rangle = e^{-iH_{00}t}\, e^{\sum_{ph} C_{ph}(t)a_p^\dagger a_h}\,|\Phi_0\rangle \quad . \qquad \text{(II.3)}$$

The correlation operator F is constrained to be static, so that the variation is carried out with respect to the complex particle(p)-hole(h) amplitudes $C_{ph}(t)$ and $C^*_{ph}(t)$, which are assumed small in magnitude. Decomposing each amplitude into both positive- and negative-frequency components, and requiring the (uncorrelated) single-particle basis be compatible with a "correlated Brillouin condition",[1-4] the equations of correlated linear response (CLR) follow:

$$\begin{pmatrix} A & B \\ B^* & A^* \end{pmatrix}\begin{pmatrix} X \\ Y \end{pmatrix} = \omega \begin{pmatrix} M & 0 \\ 0 & -M^* \end{pmatrix}\begin{pmatrix} X \\ Y \end{pmatrix} - \begin{pmatrix} P \\ P^* \end{pmatrix} \tag{II.4}$$

In this supermatrix equation, the X's and Y's are column vectors with elements X_{ph}, Y_{ph}, and the other members have elements

$$A_{ph,p'h'} = z_{ph} z_{p'h'} \left(H_{ph,p'h'} - H_{oo} N_{ph,p'h'}\right) \quad ,$$

$$B_{ph,p'h'} = z_{php'h'} \left(H_{php'h',o} - H_{oo} N_{php'h',o}\right) \quad ,$$

$$M_{ph,p'h'} = z_{ph} z_{p'h'} \left(N_{ph,p'h'} - N_{ph,o} N_{o,p'h'}\right) \quad ,$$

$$P_{ph} = z_{ph} \left(\hat{p}_{ph,o} - \hat{p}_{oo} N_{ph,o}\right) \quad . \tag{II.5}$$

(A notation for CBF matrix elements is employed convenient when two orbitals Φ_m, Φ_n are either Φ_o, or ph, php'h' states.) In (II.5), N stands for the identity operator, and $z_m = (I_{mm}/I_{oo})^{\frac{1}{2}}$. Letting $\hat{p} \to 0$, (II.4) becomes the "correlated RPA" (CRPA) eigenvalue problem, the non-trivial metric explicitly reflecting the non-orthogonality of the basis. General considerations based on (II.2)-(II.3), expanded to first order in the perturbation, in conjunction with (II.4), lead to the polarization propagator[8,9] (or density-density response function) $\chi^{CRPA}_{ph;p'h'}(\omega)$ in terms of the CRPA eigenvectors and eigenfrequencies.[1,4,7] As expected, the collective energy levels of the given Fermi system correspond to the poles of this two-particle Green's function. Also, the presence of negative-frequency components in $\hat{p}(t)$ implies that the CRPA propagator contains information about the "physical" vacuum not contained in $|\Psi_0\rangle$ alone. This is taken up in detail in Ref. 4, following closely the approach of Ref. 10.

For the remainder of this report, we shall be concerned with uniform, extended Fermi systems, and further specialize to the case of state-independent Jastrow correlations; i.e., $F = \Pi_{i<j} f(r_{ij})$. Then the first of equations (II.4) may be written, for $\hat{p} = 0$,

$$(e_p - e_h) X^{(n)}_{ph} + \sum_{p'h'} \langle h'p|V(12)|p'h\rangle_a X^{(n)}_{p'h'} + \sum_{p'h'} \langle pp'|V(12)|hh'\rangle_a Y^{(n)}_{p'h'}$$

$$= \omega_n \sum_{p'h'} \langle h'p|M(12)|p'h'\rangle_a X^{(n)}_{p'h'} \tag{II.6}$$

The single-particle energies, $e_k = \delta H_{oo}/\delta n_k$, k being p or h, correlation operator $M(12) = 1 + N(12)$, and particle-hole interaction are all generic CBF quantities defined, for example, in Refs. 2 and 6, and evaluated using Fermi-hypernetted-chain (FHNC) techniques in Ref. 11. In general, we have $e_k = k^2/2m + u_v(k)$, thus isolating a variational, mean-field contribution to the static self-energy. The matrix elements of V(12) depend on an effective interaction W(12), as well as the e's and the compact operator N(12). For example,

$$\langle pp'|V(12)|hh'\rangle_a = \langle pp'|W(12)|hh'\rangle_a + \frac{1}{2}[e_p+e_{p'}-e_h-e_{h'}]\langle pp'|N(12)|hh'\rangle_a \quad ; \quad \text{(II.7)}$$

structurally similar forms hold for the other 2p2h channels. If the quantity $f^{(v)}_{kk'} = \delta^2 H_{oo}/\delta n_k \delta n_{k'}$ (suppressing spin indices), where δn_k is again taken independent of $\mathfrak{z}(r)$, is identified as the variational estimate to the quasiparticle interaction,[6,12,13] one has the important formal relation

$$f^{(v)}_{kk'} = \lim_{q\to 0} \langle k+q,k'-q|V(12)|kk'\rangle_a \equiv \langle kk'|W(12)|kk'\rangle_a \quad . \quad \text{(II.8)}$$

At the level of practical evaluation, however, the two prescriptions for calculating $f_{kk'}$ may disagree somewhat due to the inclusion of differing classes of FHNC diagrams. The relevance of (II.8) becomes apparent upon noticing that, at least for a short-range (i.e., non-optimal) Jastrow function, in the Landau limit the CRPA equations collapse to a collisionless transport equation,[4,14] containing the effective mass $m^*_v = k_F(de_k/dk)^{-1}_{k_F}$. As is the case with generalized-RPA[3-5,7,8,14] (GRPA) (which obtains from CRPA by letting $F \to 1$, replacing the bare two-body potential by a weak HF interaction), the approximant to the full particle-hole interaction in CRPA contains no <u>explicit</u> frequency-dependence, and the overall renormalization depends in part on how closely the lowest-order spectrum approximates the <u>true</u> effective mass. In most Fermi systems (an exception being spin-polarized liquid ^{3}He),[15] one expects the self-energy to acquire perturbatively a marked frequency-dependence in the vicinity of the Fermi surface.[13,16,19] Regardless, then, of the value of m^*_v, we believe that the ultimate CRPA description of a given Fermi liquid should encompass <u>using</u> $\chi^{CRPA}(\omega)$ to achieve the dominant renormalization of the correlated mean-field description developed so far.[4] This entails, at the very least, calculating self-energies, and corrections to the ground-state energy and quasiparticle interaction.[4,13] Substantial breakthroughs have already been made along these lines within the "CBF-3" interpretation[18,19] of CRPA. Carrying through the type of program suggested in Ref. 4 will indeed require concerted effort, such as we have witnessed in the development of Monte-Carlo methods for calculating ground-state properties.[31] It seems reasonable that elementary excitations receive a similar amount of attention.

III. THE DYNAMIC FORM FACTOR AND SUM RULES IN CRPA

For a uniform medium, a CRPA excitation can be characterized by quantum numbers $\{\vec{q} = \vec{p} - \vec{h}, \omega_n; \tilde{S}(\tilde{T}) M_J\}$, where J is its total angular momentum in the coupled p-h space, and $\tilde{S}(\tilde{T})$ its spin (isospin). For the sake of economy, we ignore spin-dependence in this and the following section. (More general expressions may be found in Refs. 4,7,20,21.) Then the CRPA response function[4,7] is specified by the excitation spectrum $\{\omega_n\}$ and transition amplitudes for the density-fluctuation operator via the familiar formula[8,9]

$$\chi(q,\omega) = \frac{1}{\Omega} \sum_n \left[\frac{|\langle n|\rho_{\vec{q}}|0\rangle|^2}{\omega - \omega_n + i\eta} - \frac{|\langle n|\rho_{-\vec{q}}|0\rangle|^2}{\omega + \omega_n + i\eta} \right] , \tag{III.1}$$

where Ω is the normalization volume and η a positive infinitesimal. The required matrix elements are given by

$$\langle n|\rho_{\vec{q}}|0\rangle = \sum_{\omega_n > 0} \sum_{\vec{h}} \left[\bar{X}^{(n)*}_{\vec{h}+\vec{q},\vec{h}} + \bar{Y}^{(n)*}_{\vec{h}-\vec{q},\vec{h}} \right] . \tag{III.2}$$

The bar on the eigenvectors stands for contraction with the M-matrix:

$$\bar{X}^{(n)}_{\vec{h}+\vec{q},\vec{h}} = \sum_{\vec{h}} M_{\vec{h}+\vec{q},\vec{h};\vec{h}'+\vec{q},\vec{h}'} \, X^{(n)}_{\vec{h}'+\vec{q},\vec{h}'} . \tag{III.3}$$

It is not surprising that the amplitude (III.2) depends on $\mathfrak{f}(r)$ via the metric; in fact, we may expect the one-particle propagator implicit in the CRPA equations to exhibit a corresponding dependence since the one-body density matrix is not diagonal in the correlated 1p1h basis.[4] A quantity more easily calculated, which determines χ uniquely,[8] is the dynamic form factor $S(q,\omega) = -\mathrm{Im}\,\chi(q,\omega)/\pi$. We point out that the maximum correlated single-pair energy, $\omega_m^{s.p} = e_{k_F+q} - e_{k_F}$, is modified from its free-spectrum value due to the static field $u_v(k)$, thus reflecting the level of renormalization and affecting the nature of Landau damping. The compressibility sum-rule[8] reads $S_0(q) \equiv \int S(q,\omega)d\omega = S_v(q) + \Delta S(q)$, $S_v(q)$ being the static structure function for the correlated state $|\Psi_0\rangle$, which formally coincides with the correlated-Tamm-Dancoff (CTD) value (i.e., $Y_{ph} = 0$, all ph). The quantity $\Delta S(q)$ is then the correction arising from having probed the ground state, in the usual RPA manner. The energy-weighted sum $S_1(q) \equiv \int S(q,\omega)\omega\, d\omega$ can be evaluated exactly using closure,[4,7] with the result $S_1(q) = q^2/2m(1+\kappa)$, the factor κ accounting for an enhancement of the f-sum rule when spin-(isospin-) exchange forces are present in the potential. The CRPA contains no effects of multipair excitations (c.f. comment at the end of Ref. 22).

IV. METHOD OF SOLUTION

It is clear that the appearance of $u_v(k)$, exchange matrix elements of V(12) (which is non-local), and N(12) in our eigenvalue problem prohibit analytic solution à la ring summation. Since exact solution is required for the f-sum rule to be satisfied, we must keep all nonzero spherical components in expanding the exchange terms $V(q;\vec{h},\vec{h}')$. In so doing, we include all the effects of current-current correlations, or repeated ph scattering during (de)excitation of the medium. Numerical solution is achieved by expressing X_{ph} as $X_q(h,\alpha_h)$, $\alpha_h = \cos\theta_{\vec{h}\vec{q}}$, leading, at a given q, to paired two-dimensional integral equations. These are solved by matrix diagonalization on a two-fold mesh; the dynamic structure factor is calculated from the transition matrix elements (III.2), at each discrete frequency ω_n. A collective mode, if present, appears as obvious to the eye, being removed from the continuum at frequency $\omega_c > \omega_m^{s.p.}$, with strength s_c. The continuum is then smoothed using a procedure similar to that of Ref. 23.

This general approach is discussed at some length, within the context of GRPA calculations for symmetrical nuclear matter, in Refs. 7 and 21. The fact that $M \neq 1$ in CRPA necessitates an additional diagonalization; presumably the similarity transformation[3,4] employed here numerically accomplishes the same cancellations discussed by Clark[18] and Krotscheck.[19]

As pointed out by D. Pines,[24] it would be instructive to compare $S^{CRPA}(q,\omega)$ with the corresponding Landau-theory result, especially with regard to sum rules and damping. To do so consistently, one would solve the transport equation, eliminating (h,α_n) in favor of $\lambda = \omega/qv_F$, being careful to include all non-zero Landau parameters--calculated from (II.8). (See, e.g., Refs. 10 and 25.) The compressibility sum-rule could then be used to identify the value of q where departures from Landau theory set in.[4,7]

V. EXAMPLES OF $S^{CRPA}(q,\omega)$: LIQUID ^{3}He AND NEUTRON MATTER

Until recently, our applications of CLR theory have focused on symmetrical nuclear matter, at small (but finite)[20] momentum transfer and moderate-to-large q,[26] employing the hard-core OMY-6 potential.[2,6] We are currently moving on to soft-core two-nucleon potentials which contain tensor components (e.g., Reid v_6,[13] pi- and rho-meson exchange forces[7]). In this section, we report on our first skirmish with normal liquid ^{3}He, and summarize results for pure neutron matter obtained using the OMY-4 potential. The studies of superfluid neutron matter by Yang and Clark[27] suggested this model force as a logical and potentially fruitful choice for our first investigations of the linear response of this system. A recent application of Landau theory to neutron matter is described in Ref. 28. The results of neutron-matter calculations will be presented in greater detail in Ref. 26.

A. Liquid ^{3}He

Figures 1a and 1b display the incoherent dynamic form factor, $S^{\tilde{s}=1}(q,\omega)$, for $q = 0.5k_F$ and $q = k_F$, respectively, at equilibrium density $\rho = 0.0166$ A^{-3} (k_F=0.789A^{-1}).

We used the Lennard-Jones potential and a Schiff-Verlet Jastrow function;[29] CBF quantities were evaluated at FHNC/C level, thus including "elementary" FHNC diagrams in the approximate manner of Ref. 29. In a), results are also shown for CTD and the CRPA "bubble" (obtained by setting $V \equiv 0$, and retaining the M-matrix)--we consider it very relevant to witness the usual RPA mechanism at work for this correlated generalization. One sees that the CBF particle-hole force is sufficiently repulsive in this channel to produce a pronounced spin-fluctuation peak, with associated strong enhancement of the static structure function over its one-bubble value. We find $m_V^* = 0.92$ m, so it is tempting to consider the excitations in this microscopic model as "correlated paramagnons";[17] however, we find $F_0^a \simeq -0.5$, so the static spin susceptibility is off by a factor $\sim$4.5. At $q \simeq 1.25\ k_F$, a local ph instability occurs for incoherent density fluctuations. Since our microscopic treatment is far from being fully-renormalized,[17,30] we are not predicting an experimental instability. However, CRPA instabilities do pertain to the relevance of calculating CBF perturbation corrections using a specified $f(r)$ and scheme for evaluating matrix elements.

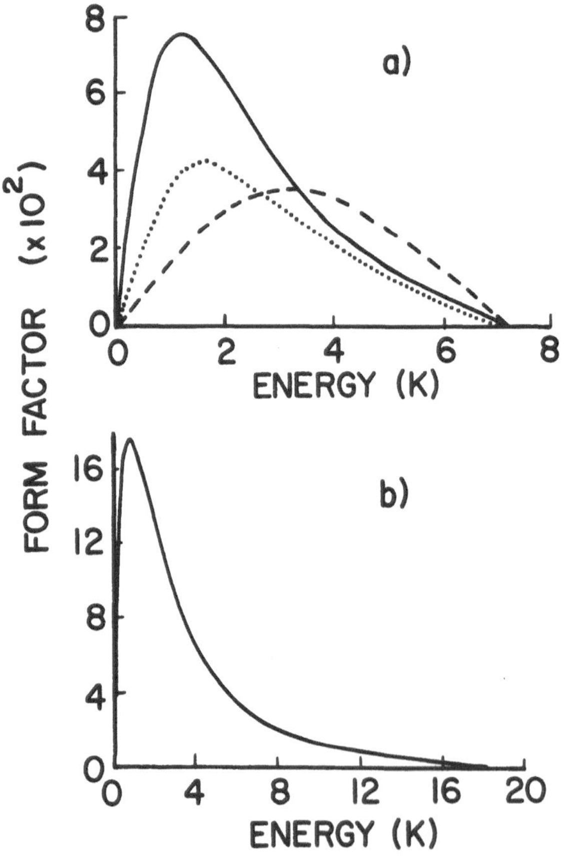

Figure 1
a) $S^I(q = 0.5\ k_F, \omega)$ for liquid ^3He. Solid: CRPA; dotted: CTD; dashed: one bubble.
b) $S^I(q = k_F, \omega)$ for liquid ^3He.

Moving on to the density channel ($\tilde{s} = 0$), the value $F_0^s \simeq -0.4$ ensuing from (II.8) hardly makes coherent excitations worth considering in this model. Nonetheless, it is interesting that neglect of elementary diagrams causes the metric-matrix to have negative eigenvalues for $q \gtrsim 0.25\ k_F$. We believe that, at the very least, for a CRPA treatment as we carry it out to be viable, it would be necessary to include effective three-body correlations in F, and to evaluate more accurately elementary-diagram contributions to the particle-hole interaction. Even then, considering the coherent collective behavior (damped or not) in ^{3}He up to $q \sim 1.2\ k_F$,[30] in the final analysis it will most likely be necessary to make F time-dependent, thereby allowing the excitations to experience different correlations than do ^{3}He atoms in the ground state.

B. Neutron Matter

We shall focus on three values of k_F, covering a wide range of neutron number densities $\rho = k_F^3/3\pi^2$. Table 1 lists quantities relating to the Landau limit of our model, although we do not mean to imply higher-order Landau parameters are negligible. Numerical inconsistencies between m_V^* and F_1^s have been alluded to in Sec. II; their consequences for the f-sum rule are described in Ref. 26.

Table 1

k_F (fm^{-1})	m_V^*/m	F_0^s	F_1^s	F_0^a
1.0	0.89	-0.57	-0.34	0.68
2.0	0.72	-0.11	-0.57	0.92
2.5	0.62	0.32	-0.52	1.01

Results for $S^{S=1}(q,\omega)$ at $k_F = 1.0\ fm^{-1}$ are shown in Figs. 2a ($q = 0.1\ fm^{-1}$) and 2b ($q = 0.25\ fm^{-1}$). An undamped spin-zero-sound mode is present up to $q \simeq 0.2\ fm^{-1}$. When a collective mode is present, its height is set at s_c/ω_c, and the fractions of the compressibility- and f-sum rules are indicated in parentheses. Dynamic form factors for the spin-channel at $k_F = 2.0\ fm^{-1}$ are plotted in Figs. 3a ($q = 0.1\ fm^{-1}$) and 2b ($q = 0.3\ fm^{-1}$). At this density, distinct collectivity persists until $q \simeq 0.5\ fm^{-1}$. At the still greater value of $k_F = 2.5$, we find no evidence for a spontaneous instability of the correlated ground state against spin-density fluctuations, which would be a possible precursor to neutral pion condensation.[32] On the contrary, undamped spin-zero-sound is present for $q \lesssim 0.8\ fm^{-1}$, as illustrated by the dispersion curve of Fig. 4. If coupling to (spin-)density-fluctuations should increase m* to nearly the bare mass, from Fig. 4 we might expect the threshold to increase considerably.

Although F_0^s turns positive at $k_F \simeq 2.2\ fm^{-1}$, we find no evidence for zero sound in the density channel up to $k_F = 2.5\ fm^{-1}$ due to the increase in $\omega_m^{s.p.}$ accompanying the drop in m_V^*.

When CRPA results seem reasonable in light of experimental information and/or other theoretical results, it is both informative and very useful to relate a CRPA approach to a finite-q Landau description, at the same level of renormalization.

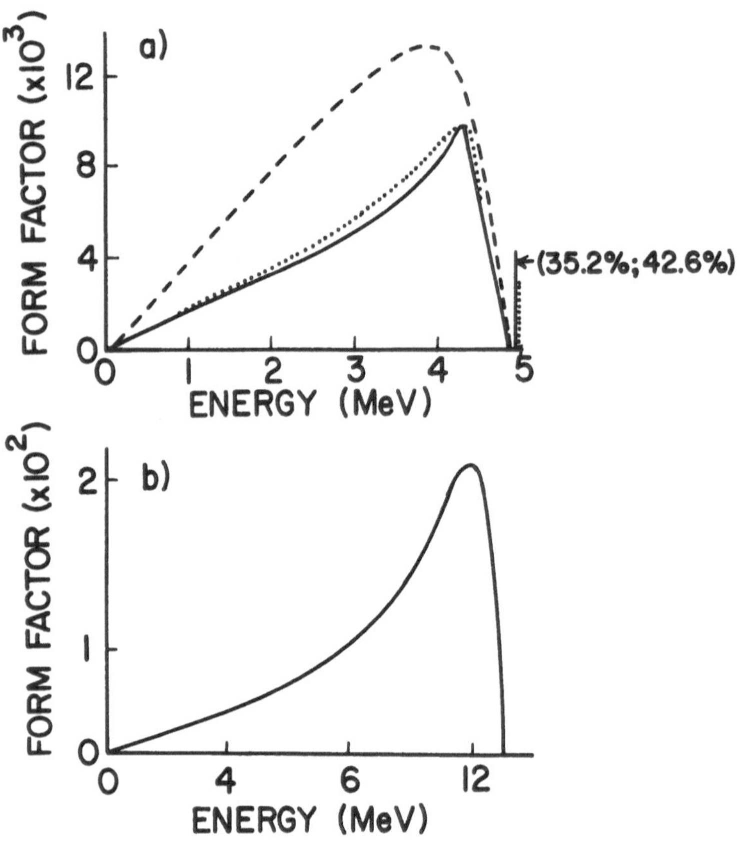

Figure 2

Toward this end, we have explored the Aldrich-Pines[30] forms

$$\chi^{\tilde{s}=0,1}(q,\omega) = \frac{\chi_0^\star(q,\omega)}{1 - [f^{s,a}(q) + \frac{\omega^2}{q^2} g^{s,a}(q)]\chi_0^\star(q,\omega)} \tag{V.1}$$

with $m^\star(q)$ appearing in the Lindhard function defined through $\omega_m^{s.p.} \equiv qk_F/m^\star(q) + q^2/2m^\star(q)$. The scalar functions $f^{s,a}(q)$ are calculated from the $\ell = 0$ component of

$$f_{\vec{k}\vec{k}'}(q) = \langle\vec{k} + \frac{\vec{q}}{2},\vec{k}' - \frac{\vec{q}}{2}\ |W(12)|\,kk'\rangle_a \quad , \quad |\vec{k}| = |\vec{k}'| = k_F \quad , \tag{V.2}$$

thus preserving quasiparticle-quasihole symmetry while moving off the Fermi surface, and retaining exchange in an angle-averaged manner. The quantities $g^{s,a}(q)$ are determined by matching to the CRPA results for the f-sum rules. Figure 5 shows the scalar polarization fields deriving from (V.2), for $k_F = 2.0\ \mathrm{fm}^{-1}$. The form factors $S^{\tilde{s}=0,1}(q,\omega)$ obtained from (V.1), in the comparisons we have made to date for neutron matter, agree encouragingly well with their CRPA counterparts. We conclude by mentioning that choices for effective interaction other than W, in particular the vertex U of Refs. 18 and 19, should be investigated.

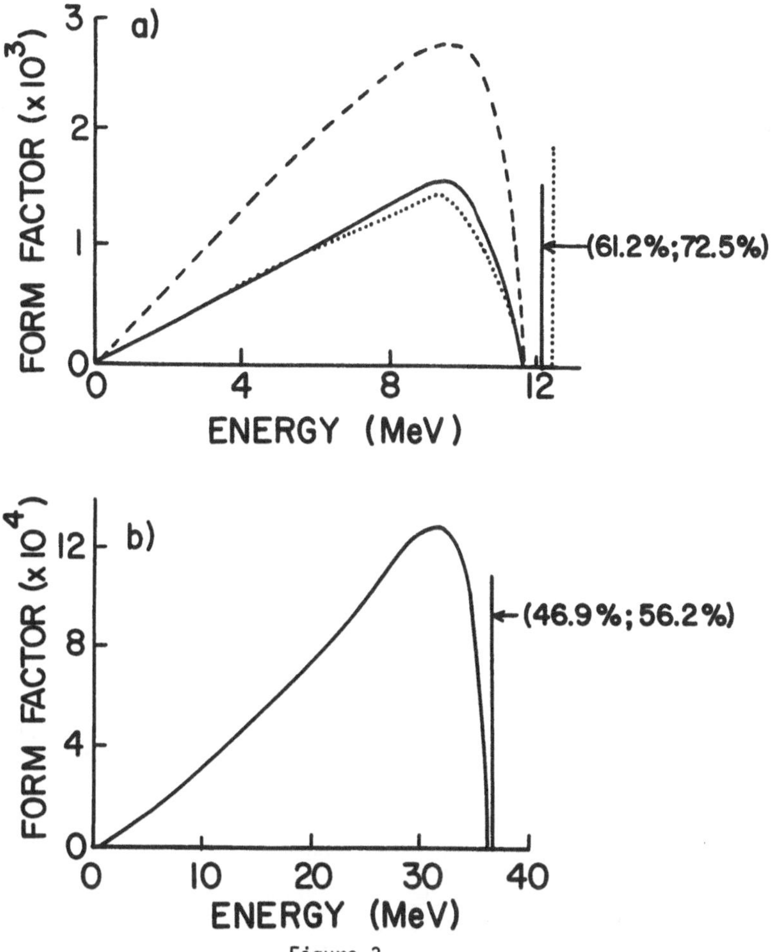

Figure 3

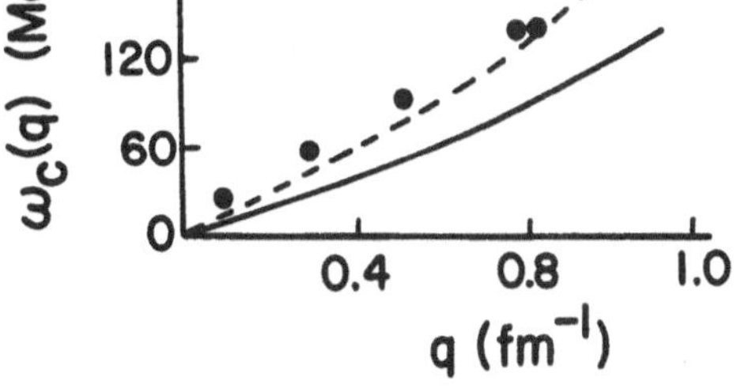

Figure 4

Spin-zero sound dispersion for neutron matter for $k_F = 2.5\ \text{fm}^{-1}$.
Dashed: $\omega_m^{s.p.}$ as in CRPA; solid: $\omega_m^{s.p.}$ for free spectrum.

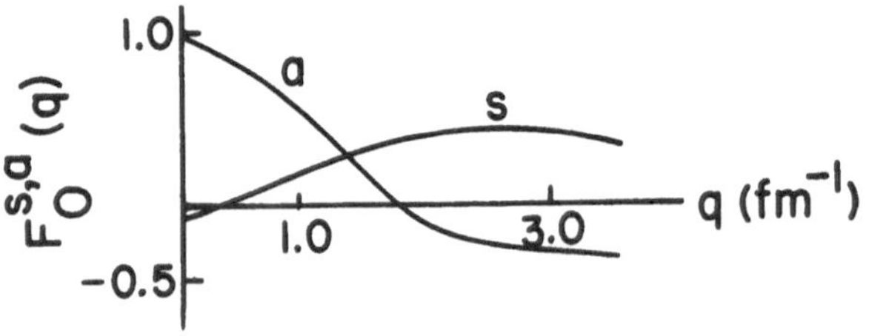

Figure 5

Scalar polarization potentials for neutron matter, at $k_F = 2.0\ fm^{-1}$.

D. G. S. wishes to thank D. Pines and C. J. Pethick for support under grant NSF DMR 82-15128, and for valuable exposure to Landau theory. Also, we are extremely grateful to Mrs. P. Suntharothok-Priesmeyer for preparing the typescript and figures.

References

1. J. M. C. Chen, J. W. Clark, and D. G. Sandler, Z. Physik A 305, 223 (1982).
2. J. W. Clark, Lecture Notes in Physics 138, 184 (1981).
3. D. G. Sandler, N.-H. Kwong, J. W. Clark, and E. Krotscheck, Lecture Notes in Physics, 142, 228 (1981).
4. D. G. Sandler, N.-H. Kwong, and J. W. Clark, to be published.
5. D. J. Thouless, The Quantum Mechanics of Many-Particle Systems (Academic, N. Y., 1972).
6. E. Feenberg, Theory of Quantum Fluids (Academic, N.Y., 1969); J. W. Clark, in Progress in Nuclear and Particle Physics, ed. D. H. Wilkenson (Pergamon, Oxford, 1979), Vol. 2, p. 89.
7. N.-H. Kwong, Ph.D. thesis, California Institute of Technology (unpublished).
8. D. Pines and P. Nozières, Theory of Quantum Liquids (Benjamin, N.Y., 1966).
9. A. L. Fetter and J. D. Walecka, Quantum Theory of Many-Particle Systems (Mc-Graw Hill, N.Y., 1971).
10. J.-P. Blaizot, Ph.D. thesis, University of Paris (unpublished); Phys. Rep. 64, 172 (1980).
11. E. Krotscheck and J. W. Clark, Nucl. Phys. A328, 73 (1979).
12. E. Krotscheck, R. A. Smith, and J. W. Clark, Lecture Notes in Physics 142, 270 (1981).
13. A. D. Jackson, E. Krotscheck, D. Meltzer, and R. A. Smith, Nucl. Phys. A386, 125 (1982).
14. G. E. Brown, Many Body Problems (North-Holland, Amsterdam, 1972).
15. K. Bedell, these Proceedings.
16. C. Mahaux, these Proceedings.
17. G. E. Brown, C. J. Pethick, and A. Zaringhalam, J. Low Temp. Phys. 48, 349 (1982).
18. J. W. Clark, these Proceedings.
19. E. Krotscheck, Phys. Rev. A 26, 3536 (1982), and these Proceedings.
20. N.-H. Kwong and D. G. Sandler, submitted to Phys. Lett.
21. N.-H. Kwong and D. G. Sandler, to be published.

22. D. G. Sandler, in Proceedings of the VIth Pan-American Workshop on Condensed Matter Theories, eds. J. M. Chen, J. W. Clark and P. Suntharothok-Priesmeyer (Washington University, St. Louis, 1983).

23. G. Bertsch and S. F. Tsai, Phys. Rep. 18C, 125 (1975).

24. D. Pines, private communication.

25. D. Gogny and R. Padjen, Nucl. Phys. A293, 365 (1977).

26. D. G. Sandler, N.-H. Kwong and J. W. Clark, to be published.

27. C.-H. Yang and J. W. Clark, Nucl. Phys. A174, 49 (1971); C.-H. Yang, Ph.D. thesis, Washington University (unpublished).

28. N. Iwamoto and C. J. Pethick, Phys. Rev. D 25, 313 (1982).

29. E. Krotscheck, R. A. Smith, J. W. Clark, and R. M. Panoff, Phys. Rev. B 24, 6383 (1981).

30. C. H. Aldrich, III and D. Pines, J. Low Temp. Phys. 32, 689 (1978).

31. M. H. Kalos, Lecture Notes in Physics 142, 252 (1981).

32. D. G. Sandler and J. W. Clark, Phys. Lett. B100, 213 (1981); O. Benhar, these Proceedings; R. Takatsuka, these Proceedings.

SUM RULES AND A COUPLED CLUSTER FORMULATION OF LINEAR RESPONSE THEORY

R.F. Bishop
Department of Mathematics
University of Manchester Institute of Science and Technology
P.O. Box 88, Manchester M60 1QD, England

1. INTRODUCTION

It is my intention here to describe the recent developments of the coupled-cluster formulation of quantum many-body theory in which we have succeeded in imbedding the well-known theory of linear response within this formalism, and have shown how a new hierarchy of very useful sum rules thereby emerges. It will transpire in so doing that the new formalism also provides a very convenient bridge between the previously somewhat separate (although, of course, related) coupled-cluster formalisms for the ground and excited states respectively of the many-body system under consideration.

2. COUPLED-CLUSTER DECOMPOSITION OF THE SCHRÖDINGER EQUATION

A very brief outline is first presented of such of the main elements of the coupled-cluster formalism as are needed here.

2.1 Ground-state formalism

The usual starting-point for the ground-state (g.s.) coupled-cluster formalism (CCF) is an exact re-expression of the many-body g.s. Schrödinger equation in terms of a set of non-linear coupled equations for the so-called correlation amplitudes.

Purely for ease of present exposition the discussion is given here wholly in terms of infinite, homogenous systems of bosons, for which the coupled-cluster ansatz for the exact g.s. wavefunction $|\Psi\rangle$ is given as,

$$|\Psi\rangle = e^{S}|\Phi\rangle \quad ; \quad S = \sum_{n=1}^{N} S_n \quad , \tag{1}$$

in terms of an N-body model or reference g.s. $|\Phi\rangle$ which is taken to be a single-state (usually zero-momentum) condensate,

$$|\Phi\rangle = (N!)^{-\frac{1}{2}}(b_0^{\dagger})^{N}|0\rangle \quad , \tag{2}$$

where $|0\rangle$ is the vacuum state. In terms of a complete set of boson creation operators $b_\alpha^\dagger$, which create the complete orthonormal single-particle (s.p.) basis $|\alpha\rangle$ when acting on the vacuum, the correlation operators S_n, which excite n particle-hole pairs from this condensate, may be written as

$$S_n = (n!)^{-1} \sum_{\rho_1 \cdots \rho_n} b_{\rho_1}^{\dagger} \cdots b_{\rho_n}^{\dagger} (N^{-\frac{1}{2}} b_0)^{n} \, s_n(\rho_1 \cdots \rho_n) \tag{3}$$

where the labels $\rho_1 \cdots \rho_n$ indicate non-condensate s.p. states, thereby displaying the linked-cluster aspect of the original e^S ansatz of Eq. (1). The derivation of the g.s. coupled-cluster equations is now formally performed in two simple steps. The g.s. Schrödinger equation, with energy eigenvalue E is first pre-multiplied by the operator e^{-S},

$$e^{-S} H e^{S} |\Phi\rangle = E|\Phi\rangle \quad , \tag{4}$$

which may be considered as a purely formal step to eliminate some "unlinked" terms from the outset that otherwise need to be eliminated later. The scalar product is then finally taken of Eq. (4) either with the model state $|\Phi\rangle$ or with the states

$$b^{\dagger}_{\rho_1} \cdots b^{\dagger}_{\rho_n} (N^{-\frac{1}{2}} b_0)^n |\Phi\rangle \ . \tag{5}$$

Clearly, when $\rho_1 \cdots \rho_n$ run over all (non-condensate) s.p. states of the complete s.p. basis, and when n runs from 1 to N, the vectors $|\Phi\rangle$ and (5) span the entire N-body Hilbert space. Thus the set of equations

$$\begin{aligned} \langle\Phi| e^{-S} H e^{S} |\Phi\rangle &= E \quad , \\ \langle\Phi| (N^{-\frac{1}{2}} b_0^{\dagger})^n b_{\rho_n} \cdots b_{\rho_1} e^{-S} H e^{S} |\Phi\rangle &= 0 \quad , \end{aligned} \tag{6}$$

which are the g.s. coupled-cluster equations, are hence fully equivalent to the N-body Schrödinger equation. They are a coupled set of nonlinear equations for the matrix elements (or subsystem amplitudes) $S_n(\rho_1 \cdots \rho_n)$ of the correlation operators S_n. In order to be useful in practice one has to truncate this hierarchy and, for example, the "natural" truncation of the so-called SUBn scheme, wherein each of the amplitudes S_i is set to zero for $i > n$ and the remaining n coupled equations are solved for the amplitudes S_i with $i \leq n$, has by now been thoroughly investigated. Thus it is by now well known that the numerical solution of appropriately truncated subsets of the equations (6) has lead to excellent quantitative g.s. results for systems as diverse as closed-shell atomic nuclei, the one-component Coulomb plasma, and even quite complex systems from the realm of quantum chemistry.

For further details of a formulation of the g.s. formalism that perhaps best stresses its physical content, the interested reader is referred to the article by Lührmann.[1] A full review has also been given[2] in the context of applications to nuclear physics, and hence where interest is particularly focussed on short-range correlations. For the particular problems inherent to the cases of long-range interactions and long-range correlations, the reader is also directed to the essentially self-contained articles by Lührmann and the present author,[3,4] which deal with the one-component Coulomb plasma (or 'jellium') in SUB2 approximation.

2.2 Excited-state formalism

The g.s. formalism already described presumably may be employed not only for the g.s. but also for those states (with the same imposed symmetry as the g.s.) that have non-zero overlap with the model state $|\Phi\rangle$. (We note that Eq. (1) automatically implies a normalisation $\langle\Phi|\Psi\rangle = 1$.) Thus restricting ourselves to excited states $|\Psi_\ell\rangle$ which are orthogonal to both $|\Phi\rangle$ and $|\Psi\rangle$, Emrich[5] has recently shown that an appropriate choice of CCF wavefunction is,

$$|\Psi_\ell\rangle = S^{(\ell)}e^S|\Phi\rangle \quad , \quad S^{(\ell)} = \sum_{n=1}^{N} S_n^{(\ell)} \quad ,$$
$$S_n^{(\ell)} = (n!)^{-1} \sum_{\rho_1\cdots\rho_n} b^\dagger_{\rho_1}\cdots b^\dagger_{\rho_n}(N^{-\frac{1}{2}}b_0)^n S_n^{(\ell)}(\rho_1\cdots\rho_n) \quad . \tag{7}$$

Each non-zero vector $S_n^{(\ell)}|\Phi\rangle$ is assumed to have a non-zero overlap with $|\Psi_\ell\rangle$. If the excited state is a momentum eigenstate with eigenvalue $\vec{q}$ and we choose to work in a plane-wave s.p. basis, this implies that the s.p. momenta $\rho_1\cdots\rho_n$ in Eq. (7) must add to $\vec{q}$, whereas in the g.s. Eq. (3) they must add to zero.

The formal derivation of the excited-state (e.s.) coupled-cluster equations is now again easily performed. The e.s. Schrödinger equation, with energy eigenvalue $E_\ell \equiv E + \omega$ (i.e. with excitation energy, ω), is first combined with its g.s. counterpart, to give

$$[H,S^{(\ell)}]|\Psi\rangle = \omega S^{(\ell)}|\Psi\rangle \quad . \tag{8}$$

A similar procedure as in the g.s. case above then leads to the e.s. counterpart of Eq. (6) as

$$\langle\Phi|(N^{-\frac{1}{2}}b_0^\dagger)^n b_{\rho_n}\cdots b_{\rho_1} e^{-S}[H,S^{(\ell)}]e^S|\Phi\rangle = \omega S_n^{(\ell)}(\rho_1\cdots\rho_n) \quad . \tag{9}$$

Equations (9) are thus the linked e.s. coupled-cluster equations, and we note that they take the form of a coupled set of linear eigenvalue equations for the e.s. subsystem amplitudes, with the same (excitation energy) eigenvalue ω in each equation. In each equation, the g.s. solution is assumed already known so that the g.s. correlation amplitudes are input to Eqs. (9).

Just as in the g.s. case, the e.s. Eqs. (9) also have to be truncated to be useful in practice. As an obvious extension of the g.s. SUBn scheme, for example, we mention the SUB(m,n) scheme where the n lowest equations of Eq. (6) and the m lowest equations of Eq. (9) are solved in the approximation that the operators $S_{m+k}^{(\ell)}$ and S_{n+k} are set to zero for all $k \geq 1$. However, one obvious point that arises immediately is the choice of "compatible" (m,n) pairs. For example, one would like to know *a priori* whether for a given n, higher values of m in the SUB(m,n) scheme necessarily lead to a "better" approximation. Such questions are difficult to answer without further information, and it is in this sense that the theory of linear response to be presented, provides a bridge between the otherwise essentially disparate g.s. and e.s. formalisms already described.

3. LINEAR RESPONSE AND GENERAL SUM RULES

Let us now consider the response of the system to the addition of a small perturbation λv to the hamiltonian H, by expanding the g.s. energy and wavefunction in powers of the coupling parameter λ,

$$H' = H + \lambda v \; ,$$
$$E' = E + \lambda E^{(1)} + \lambda^2 E^{(2)} + \cdots \; , \qquad (10)$$
$$|\Psi'\rangle = |\Psi\rangle + \lambda|\Psi^{(1)}\rangle + \lambda^2|\Psi^{(2)}\rangle + \cdots \; .$$

One possible means of progressing within the CCF at this point is to define $e^{S'}|\Phi\rangle$ as the perturbed g.s. $|\Psi'\rangle$, and to use Eqs. (10) to determine the perturbed correlation operator S', as has recently been discussed by Arponen.[6] As an alternative however, and guided by the usual derivation of sum rules, we make contact at this point with (at least part of) the excitation spectrum $|\Psi_\ell\rangle$ of the unperturbed hamiltonian H by expanding the first-order correction to the g.s. wavefunction as

$$|\Psi^{(1)}\rangle = \sum_\ell g_\ell |\Psi_\ell\rangle \; ; \quad H|\Psi_\ell\rangle = E_\ell |\Psi_\ell\rangle \; , \qquad (11)$$

where the coefficients g_ℓ are as yet unknown. We now restrict ourselves to first-order changes in the g.s. wavefunction only (linear response theory), and also impose as further restrictions from the outset that the excited states $|\Psi_\ell\rangle$ entering the expansion (11) are orthogonal to the model condensate state $|\Phi\rangle$ (as in Sect. 2.2). Further, we restrict the ensuing discussion to perturbations v such that the inner products of the vector $v|\Psi\rangle$ with both $|\Phi\rangle$ and $|\Psi\rangle$ are zero. The standard analysis of linear response then readily shows both that the first-order energy change $E^{(1)}$ vanishes, and the further results,

$$\sum_\ell \omega_\ell g_\ell |\Psi_\ell\rangle = -v|\Psi\rangle \; ; \qquad (12)$$

$$g_\ell = -\frac{1}{\omega_\ell}\,\frac{\langle\Psi_\ell|v|\Psi\rangle}{\langle\Psi_\ell|\Psi_\ell\rangle} \; ; \qquad (13)$$

$$E^{(2)} = \sum_\ell g_\ell \,\frac{\langle\Psi|v|\Psi_\ell\rangle}{\langle\Psi|\Psi\rangle} \, . \qquad (14)$$

For future purposes it is also convenient to consider the "$m^{\underline{th}}$ power" of the perturbed Schrödinger equation, namely

$$(H+\lambda v)^m |\Psi'\rangle = E'^m |\Psi'\rangle \; . \qquad (15)$$

It is then straightforward to use Eqs. (10) and (11) in expanding Eq. (15) to first order in λ, to show that for any integral m,

$$\sum_\ell \omega_\ell{}^m g_\ell |\Psi_\ell\rangle = -v_{(m)}|\Psi\rangle \; , \qquad (16)$$

where $v_{(m)}$ is a nested commutator, defined iteratively as

$$v_{(1)} = v \; ; \quad v_{(m)} = [H, v_{(m-1)}] \; , \; m > 1 \; . \qquad (17)$$

Equations (12) and (16) now constitute the basis for our general hierarchies of sum rules. Thus, by taking their inner products with the states given in Eq. (5), after a prior pre-multiplication by e^{-S}, gives the sum rules

$$\sum_\ell \omega_\ell{}^m g_\ell S_n^{(\ell)}(\rho_1 \cdots \rho_n) = -\langle\Phi|\,(N^{-\frac{1}{2}}b_0^\dagger)^n\, b_{\rho_n} \cdots b_{\rho_1}\, e^{-S} v_{(m)} e^{S}\,|\Phi\rangle \equiv -F_{mn}(\rho_1 \cdots \rho_n) \; . \qquad (18)$$

We note in particular that the sum rules (18) relate the excitation energies and correlation amplitudes on the one hand with the ground-state correlations on the other. Equation (14) [which together with Eq. (13) is just second-order perturbation theory for the g.s. energy] may also be regarded as a kind of zeroth order sum rule.

4. RELATION TO SUM RULES FOR THE STRUCTURE FUNCTION

A particularly important application of the above analysis, motivated by the restrictions discussed below Eq. (11), now follows from the choice,

$$v = \tfrac{1}{2}(\rho_{\vec{q}} + \rho_{\vec{q}}^{\dagger}) \equiv v^{\dagger} \quad ; \quad (q \neq 0) \quad , \tag{19}$$

$$\rho_{\vec{q}} \equiv N^{-\frac{1}{2}} \sum_{\vec{k}} b_{\vec{k}}^{\dagger} b_{\vec{k}+\vec{q}} \equiv \rho_{-\vec{q}}^{\dagger} \quad . \tag{20}$$

The operator $\rho_{\vec{q}}^{\dagger}$ creates a density fluctuation with momentum $\vec{q}$, and the perturbation v thus destroys the translational invariance of the original hamiltonian. Working again in a momentum-eigenstate (i.e., plane-wave) s.p. basis, with $|\Phi\rangle$ the zero-momentum condensate, it is clear from Eq. (13) that the only excited states of interest, namely those that carry non-zero weight g_ℓ in Eq. (11), are momentum eigenstates with eigenvalue $\vec{q}$ or $-\vec{q}$. Hence to obtain non-trivial results from Eq. (18), the momenta $\rho_1 \cdots \rho_n$ must also add either to $\vec{q}$ or $-\vec{q}$, and in the following we assume they sum to $\vec{q}$. In the simplest case, $m = 1$, the right-hand side of Eq. (18) can now be evaluated with the perturbation of Eq. (19), to give,

$$F_{11}(\vec{q}) = \tfrac{1}{2}[1 + S_2(q)] \quad ; \quad S_2(q) \equiv S_2(\vec{q}, -\vec{q}) \quad , \tag{21}$$

and the other functions F_{1n} with $n > 1$ can be similarly evaluated in terms of the g.s. subsystem amplitudes S_n. (Note that in the plane-wave s.p. basis, $S_1 \equiv 0$ by momentum conservation.) In this way, one can show for example that the lowest order ($n = 1$) sum rules derived from Eq. (18) with $m = 1,2$ are respectively given as

$$\sum_\ell \omega_\ell g_\ell S_1^{(\ell)}(\vec{q}) = -F_{11}(\vec{q}) \quad , \tag{22}$$

$$\sum_\ell \omega_\ell^2 g_\ell S_1^{(\ell)}(\vec{q}) = -\frac{\hbar^2 q^2}{2m} \{1 - F_{11}(\vec{q})\} \quad , \tag{23}$$

and the other (m,n) sum rules can be similarly evaluated.

In the limit of vanishing momentum transfer, the energy shift due to the perturbation of Eq. (19) can also be calculated macroscopically in the usual well-known fashion, to give the "compressibility sum rule" for the dynamic structure function $T(q,\omega)$,

$$T(q,\omega) \equiv \sum_\ell \frac{\langle\Psi|\rho_{\vec{q}}|\Psi_\ell\rangle\langle\Psi_\ell|\rho_{\vec{q}}^{\dagger}|\Psi\rangle}{\langle\Psi|\Psi\rangle\langle\Psi_\ell|\Psi_\ell\rangle} \delta(\omega - \omega_\ell) \quad , \tag{24}$$

which can be expressed in our CCF language and in the usual way as

$$\lim_{q\to 0} 2 \sum_\ell g_\ell \frac{\langle\Psi|\rho_{\vec{q}}|\Psi_\ell\rangle}{\langle\Psi|\Psi\rangle} = -\frac{1}{2mc^2} \iff \lim_{q\to 0} \int_0^\infty d\omega\, \omega^{-1}\, T(q,\omega) = \frac{1}{2mc^2} \quad , \tag{25}$$

in terms of the first-sound velocity, c. Two other well-known sum rules for $T(q,\omega)$, namely the "static sum rule" for the static function $T(q)$ and the "f-sum rule", can also easily be derived by taking the inner product with the vector $v|\Psi\rangle$ of Eq. (16) with $n = 1$ and $m = 1,2$ respectively:

$$2\sum_{\ell}\omega_{\ell}g_{\ell}\frac{\langle\Psi|\rho_{\vec{q}}|\Psi_{\ell}\rangle}{\langle\Psi|\Psi\rangle} = -T(q) \iff \int_0^{\infty} d\omega\, T(q,\omega) = T(q) \; ; \qquad (26)$$

$$2\sum_{\ell}\omega_{\ell}^2 g_{\ell}\frac{\langle\Psi|\rho_{\vec{q}}|\Psi_{\ell}\rangle}{\langle\Psi|\Psi\rangle} = -\frac{\hbar^2q^2}{2m} \iff \int_0^{\infty} d\omega\,\omega\, T(q,\omega) = \frac{\hbar^2q^2}{2m} \; . \qquad (27)$$

In order to compare Eqs. (26), (27) with Eqs. (22), (23) it is most useful to rewrite the term $\langle\Psi|\rho_{\vec{q}}|\Psi_{\ell}\rangle$ in Eqs. (26), (27) as $\langle\Psi|e^{S}\mathbb{1}e^{-S}\rho_{\vec{q}}|\Psi_{\ell}\rangle$, and then to insert the following identity in the N-particle Hilbert space for the unit operator so indicated,

$$\mathbb{1} \equiv |\Phi\rangle\langle\Phi| + \sum_n \frac{1}{n!}\sum_{\rho_1\cdots\rho_n} b^{\dagger}_{\rho_1}\cdots b^{\dagger}_{\rho_n}(N^{-\frac{1}{2}}b_0)^n|\Phi\rangle\langle\Phi|(N^{-\frac{1}{2}}b_0^{\dagger})^n b_{\rho_n}\cdots b_{\rho_1} \; . \qquad (28)$$

In this way it is easy to see that the first term on the right-hand side of Eq. (28) when inserted as indicated in the left-hand sides of Eqs. (26) and (27) yields just exactly the left-hand sides of Eqs. (22) and (23) respectively. From the remaining term on the right-hand side of Eq. (28) we get for larger n a very complex and nonlinear dependence on the g.s. correlation amplitudes S_n but with each term linear in the e.s. amplitudes $S_n^{(\ell)}$. In this way one shows that the sets of sum rules (18) with $m = 1,2$ respectively (and in each case for all n) correspond to the sum rules of Eqs. (26) and (27) for $T(q,\omega)$. The same is clearly also true for the (perhaps less familiar) higher ($m > 2$) sum rules which represent the higher moments of the structure function $T(q,\omega)$. In each case, for a given moment (or index m), our system of sum rules constitutes a cluster decomposition (in index n) of the corresponding sum rule for the dynamic structure function, into sub-sum-rules.

4.1 One-state approximation

As a preliminary indication of the usefulness of the new sum rules, we now assume that a single excited state exhausts the sum rules -- the so-called one-state approximation. In this case, division of Eq. (23) by Eq. (22) yields the relation,

$$\omega \to \omega(q) = \frac{\hbar^2q^2}{2m}\,\frac{1-S_2(q)}{1+S_2(q)} \; , \qquad (29)$$

where use has also been made of Eq. (21). In order to make use of Eq. (25) to make further progress we now need to evaluate the left-hand side of Eq. (25) by the same procedure as already indicated between Eqs. (27) and (28). To make the calculation tractable we also work in the random-phase approximation (RPA) as the obvious approximation scheme, and which is characterised by neglecting all contributions which depend on momenta other than $\pm\vec{q}$, and by setting $S_i = 0$ for $i \geq 3$. In this approximation, the second term on the right-hand side of Eq. (28) gains a non-zero contribution only from the n = 2 term in the sum. The remaining factors can now be evaluated using similar techniques of inserting appropriate unit operators. Space considerations preclude a detailed analysis, and we quote only the final result (true only in the RPA):

$$\langle\Psi|\rho_{\vec{q}}|\Psi_\ell\rangle / \langle\Psi|\Psi\rangle \simeq S_1^{(\ell)}(q)[1 - S_2(q)]^{-1} . \tag{30}$$

Inserting Eq. (30) into Eq. (25), and combining the resulting equation with Eqs. (21) - (23), yields the further results in the one-state approximation,

$$\lim_{q\to0} S_2(q) = -1 + \frac{\hbar q}{mc} \quad ; \quad \lim_{q\to0} \omega(q) = \hbar cq . \tag{31}$$

Equation (31) shows that this simple approximation leads to the universal existence of a low-momentum phonon branch of the excitation spectrum. Furthermore, the Bose equation for $S_2(q)$ from the coupled-cluster equations (6) is given in the RPA by[4]

$$\frac{\hbar^2 q^2}{m} S_2(q) + NV(q)[1 + S_2(q)]^2 = 0 , \tag{32}$$

in terms of the two-body potential V. Use of Eq. (31) in Eq. (32) then also yields the well-known relation for the sound velocity,

$$c = [NV(0)/m]^{\frac{1}{2}} . \tag{33}$$

The static structure function $T(q)$ can also be evaluated in the RPA from Eq. (26) by use of Eq. (30). Comparing the resulting equation with Eq. (22), evaluating both in the one-state approximation, and using also Eq. (21), then yields the result valid in the RPA,

$$T(q) \simeq [1 + S_2(q)] / [1 - S_2(q)] . \tag{34}$$

A comparison of Eqs. (29) and (34) finally yields the well-known Bijl-Feynman relation between the static structure function and the excitation spectrum,

$$T(q)\omega(q) \simeq \frac{\hbar^2 q^2}{2m} . \tag{35}$$

It is of some interest to note that the only approximation involved in deriving Eq. (29) is the one-state approximation, whereas Eqs. (34) and (35) have also involved the RPA. We note however that making the one-state approximation also in Eq. (27), which we have so far not used, and comparing with Eq. (26), yields Eq. (35) directly but now <u>without</u> using the RPA. It must be carefully noted that an evaluation of $T(q)$ in a <u>different</u> approximation to RPA will therefore <u>not</u> be wholly compatible with the one-state approximation expressed by Eq. (35). We finally note that although the one-state approximation is compatible with the RPA at the level of Eqs. (22), (23), (26) and (27), it is certainly not exact. When higher sum rules (18) (<u>e.g.</u> m = 3, n = 1) are also simultaneously considered, inconsistencies start to arise.

5. FINAL REMARKS

We have seen how to derive a set of exact sum rules in the CCF which connect properties of the excitation spectrum characterised by $|\Psi_\ell\rangle$ and $S_n^{(\ell)}$, with ground-state properties characterised by $|\Psi\rangle$ and S_n. Our new double manifold (in indices m,n) of sub-sum-rules (18) represents a cluster decomposition of the usual single manifold of sum rules for the energy-weighted moments of the dynamic structure function, and thereby is able to provide much more detailed information about many-body systems than the latter. The new formalism should also be useful in applications to systems as liquid ^4He where the first step would probably be to use it to determine the g.s. correlation function $S_2(q)$ from the experimental spectrum $\omega(q)$. In this way one could investigate the use of excited-state properties to determine the correlations present in the g.s., which have otherwise proven notoriously difficult to unravel. It is also possible that our new sub-sum-rules will enable the CCF and the alternative many-body moment methods[7] mutually to illuminate one another.

We have seen how even the simplest approximation for these new sum rules of assuming that the excitation spectrum consists of a single state (which therefore exhausts the sum rules), leads to a phonon spectrum in the long-wavelength limit, and also to the Bijl-Feynman relation when the RPA is further assumed. We note that these results, which can also be derived from the structure function sum rules in the same one-state approximation, were derived using the (m,n) sub-sum-rules (18) for the case n = 1 only. This is not surprising since phonons are just (collective) coherent superpositions of one-particle/one-hole excitations, and which are therefore fully described by $S_1^{(\ell)}$ and which hence correspond to the case $n = 1$ only. Extensions of the use of the sub-sum-rules (18) to the case $n > 1$ is likely also to be of some interest. How to improve upon the one-state approximation, perhaps with the aim of making arbitrarily many of the (m,n) sub-sum-rules (18) compatible with each other (and with the corresponding sum rules for the structure function), also remains an open question of considerable importance. In this same spirit, the difficulty of really proving that phonons become <u>exact</u> eigenstates in the low-q limit has also been emphasised in other contexts by Feenberg.[8]

Finally, we hope that the sum rule formalism developed here may open the door to a wider range of applications of the CCF, which has hitherto largely been used only to solve (approximately) the g.s. and e.s. Schrödinger equation with given many-body hamiltonian. In particular, we believe that the sum rules are likely to prove useful in investigating the compatibility between otherwise disparate approximations in the essentially distinct g.s. and e.s. CCF methods. More generally, it seems clear that in principle, <u>all</u> systems to which the CCF has already very successfully been applied, are worth studying anew with these newly-developed tools.

References

[1] K.H. Lührmann, Ann.Phys.(NY) 103 (1977) 253.

[2] H. Kümmel, K.H. Lührmann and J.G. Zabolitzky, Phys. Reports 36C (1978) 1.

[3] R.F. Bishop and K.H. Lührmann, Phys.Rev. B 17 (1978) 3757.

[4] R.F. Bishop and K.H. Lührmann, Phys.Rev. B 26 (1982) 5523.

[5] K. Emrich, Nucl.Phys. A351 (1981) 379, 397.

[6] J. Arponen, University of Helsinki preprint HU-TFT-81-41 (1981).

[7] B.J. Dalton, S.M. Grimes, J.P. Vary and S.A. Williams (Eds.), "Moment Methods in Many-Fermion Systems," Plenum, New York (1980).

[8] E. Feenberg, "Theory of Quantum Liquids," Academic Press, New York, (1969), p.72.

VARIATIONAL EXP S METHODS

J. Arponen and E. Pajanne

Research Institute for
Theoretical Physics
University of Helsinki
Siltavuorenpenger 20 C
00170 Helsinki 17, Finland

I. Principles

The exp S method of Coester[1], Kümmel and others[2-3] originates from the observation of Hubbard[4] that the ground-state wave function of a many-fermion system can be expressed in the form

$$|\psi\rangle = e^{S}|\phi\rangle , \tag{1}$$

where $|\phi\rangle$ is the non-perturbed ground state (a Slater determinant), and S is an operator which creates particle and hole excitations,

$$S = \sum_{m\geq 1} \frac{1}{m!m!} \sum_{(\rho\nu)} \langle \rho_1 \ldots \rho_m | S_m | \nu_1 \ldots \nu_m \rangle a^{\dagger}_{\rho_1} \ldots a^{\dagger}_{\rho_m} a_{\nu_m} \ldots a_{\nu_1} . \tag{2}$$

Here the indices ρ denote unoccupied and ν occupied one-particle states. The amplitudes S_m are sums of definite open-ended linked Goldstone diagrams, and describe m-body correlations. The conventional way to proceed is to write down the energy eigenvalue equation in the form[2-3]

$$e^{-S}He^{S}|\phi\rangle = E|\phi\rangle , \tag{3}$$

and to observe that the similarity-transformed operator

$$e^{-S}He^{S} = H + [H,S] + \ldots + \frac{1}{4!} [[[[H,S],S],S],S] \tag{4}$$

can be explicitely constructed as a finite-order expression in S. If the Hamiltonian contains 2-body interactions, the series truncates at fourth order, because each commuting S erases at least one destruction operator from H, and there are at most 4 destruction operators to be erased.

To obtain a practical working scheme the index m in the sum (2) has to be restricted by $m \leq n$, which defines a SUBn approximation for S. For a general $S \in$ SUBn one has

$$e^{-S}He^{S}|\phi\rangle = \sum_{m=0}^{N(n)} |\psi_m\rangle, \tag{5}$$

where $|\psi_m\rangle$ is an m-particle-m-hole state, and the maximum number of particle-hole pairs is $N(n) = 4n-2$. In order to find the proper amplitudes S_m $(m \leq n)$, conditions (3) clearly cannot be completely satisfied because they would impose too many restrictions on the unknown quantities. One obvious way to generate a proper number of conditions on S is to require (3) to be valid only in the projected subspace spanned by states containing at most n particle-hole pairs:

$$P_{SUBn}e^{-S}He^{S}|\phi\rangle = E|\phi\rangle. \tag{6}$$

Approximation (6) has been called the "SUBn approximation" or the "complete n-body subsystem approximation"[5], and has been shown to be applicable to systems with sufficiently soft-core potentials[3,5]. By restricting further the set of Godstone diagrams that are summed by approximation (6) one can apply the method also to hard-core cases[3].

Conditions (6) can be satisfied by introducing a variational principle[6] (or a stationary principle) to the functional

$$\mathcal{J}_H\left[S,\Omega\right] = \langle\phi|\Omega e^{-S}He^{S}|\phi\rangle, \tag{7}$$

where $\Omega = 1+\sum_1^n \Omega_m$, and the Ω_m are amplitudes that destroy m particle-hole pairs. The requirement of stationariness with respect to variations of Ω_m yields the SUBn equations (6), while the same requirement with respect to S_m allows one to solve for Ω_m. The procedure generates a density matrix which is diagrammatically fully consistent with the energy evaluation scheme (in the Feynman-Hellman sense), and any average values can be calculated as $\langle A\rangle = \mathcal{J}_A\left[S,\Omega\right]$.

A closer inspection of the diagrams summed by the "normal SUBn" procedure described above reveals certain undesirable features in the amplitudes Ω_m, but offers a clue to a more satisfactory and complete diagram summation scheme which is a derivative of the well-known Rayleigh-Ritz principle. Namely, without doing any approximations, the R-R average-value functional

$$J_H = \frac{\langle\psi|H|\psi\rangle}{\langle\psi|\psi\rangle} = \frac{\langle\phi|e^{S^\dagger}He^{S}|\phi\rangle}{\langle\phi|e^{S^\dagger}e^{S}|\phi\rangle} \tag{8}$$

can be put into form (7) by choosing

$$\langle\phi|\Omega = \frac{\langle\phi|e^{S^\dagger}e^{S}}{\langle\phi|e^{S^\dagger}e^{S}|\phi\rangle}\ . \tag{9}$$

The bra state (9) can be expressed in the form

$$\frac{\langle\phi|e^{S^\dagger}e^{S}}{\langle\phi|e^{S^\dagger}e^{S}|\phi\rangle} = \langle\phi|e^{S''}, \tag{10}$$

where S" is represented by a series similar to eq. (2), but with absorption operators. The exact amplitudes S''_m turn out to be linked-cluster quantities and they are represented by the diagrams in figure 1. The

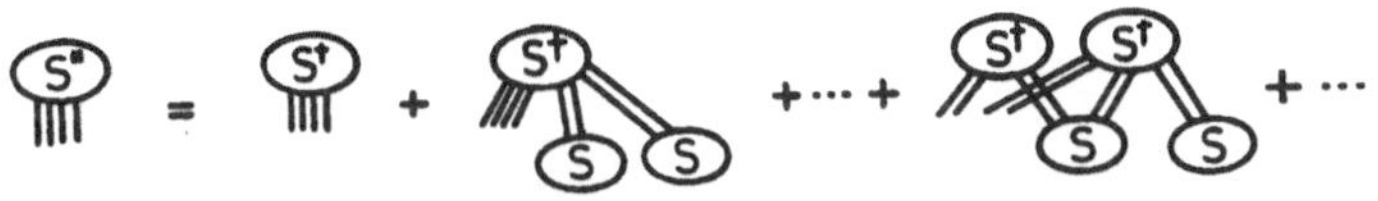

Figure 1.

functional (8), if written in the form

$$J_H\left[S, S''\right] = \langle\phi|e^{S''}e^{-S}He^{S}|\phi\rangle, \tag{11}$$

then immediately suggests a new truncation scheme, the "extended SUBn approximation", in which both S_m and S''_m are restricted to the SUBn class ($m \leqq n$), and considered to be freely variable. The classes of Goldstone diagrams that are summed by the new scheme at the stationary point are analyzed in more detail in ref. 7, and the result can only be sketched here. The diagrams are generalized trees, figures 2a-c, with links represented by wavy lines and composed of $m \leqq n$ particle-hole line

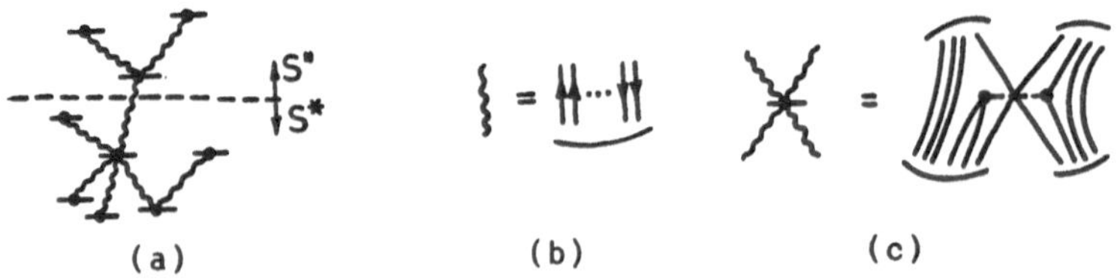

Figure 2.

pairs, and with vertices that are related to interactions at which the lines of the links meet and are grouped together according to specific rules. If a tree is cut into two pieces, the upper part is a possible contribution to S''_m, and the lower part is a contribution to the quantity

$$S^*_i = \frac{\delta}{\delta S_i''} \langle \phi | e^{S''} S | \phi \rangle \, . \tag{12}$$

In this scheme the average values are again given by the rule $\langle A \rangle = J_A[S, S'']$.

A very important property of a variationally formulated theory is the possibility to introduce dynamics into the system. Consider the functional[7)]

$$I = \int dt \, \langle \phi | e^{S''(t)} e^{-S(t)} \left[i \frac{\partial}{\partial t} - H(t) \right] e^{S(t)} | \phi \rangle . \tag{13}$$

The conditions for I being stationary,

$$\frac{\delta I}{\delta S_i''(t)} = \frac{\delta I}{\delta S_i(t)} = 0 \, , \tag{14}$$

lead to the equations of motion for the linked-cluster amplitudes,

$$i \frac{d}{dt} S^*_i = \frac{\delta J^*_H}{\delta S_i''} \, , \tag{15a}$$

$$i \frac{d}{dt} S''_i = - \frac{\delta J^*_H}{\delta S^*_i} \, , \tag{15b}$$

where J^*_H is the same as in (11) but expressed in terms of variables S^*, S''. When linearized around a stable stationary point, equations (15) can be used to find the linear response of the system to any perturbations. The second derivatives of J^*_H at the stationary point define the dynamical matrix, the eigenvalues of which give the energies of the excited states relative to the ground-state energy.

We wish to make a few comments on the typical features of the new methods.

1) The expS methods "solve" the topological diagram classification problem: it is not necessary to consider diagrams at all, the algebraic equations in each SUBn are written down in a straightforward, automatic way (at least in principle). The extended SUBn schemes represent a very high degree of renormalization, and are able to sum much more diagrams than the normal SUBn or coupled-cluster methods.

2) The extended SUBn approximation is formally a mean-field theory for "fields" that are something like m-point functions. The SUB1 approximation for bosons is the classical mean-field theory for one-point fields, and sums the conventional tree diagrams[8)]. The fermion SUB1 approximation can be shown to give exactly the Hartree-Fock result[7)]. Addition of cluster amplitudes with $m>1$ allows the treatment of inter-particle correlations and can only be expected to improve upon the mean-field result. The method works well even for large perturbations and is insensitive to the choice of the model state $|\phi\rangle$. In fact, the extended expS method can even properly treat certain ground-state phase transitions (see below).

3) Also the extended SUBn scheme can be modified to become applicable to hard-core problems.

II. Applications

We can only briefly comment on the numerical applications of the variational schemes described above. In references 9-10 we studied the homogeneous electron gas in an approximation which is roughly the normal SUB2 scheme. However, we introduced as an intermediate step the RPA or Sawada bosons to describe particle-hole excitations, which leads to a somewhat simplified treatment of exchange diagrams in comparison with the full fermionic normal SUB2. The results were found quite satisfactory in the metallic density regime, although our correlation energies are in general not quite as good as in the expS study of Bishop et al[11)]. On the other hand, we were able to calculate easily other physical observables such as momentum density and pair correlation function, using the prescription below eq. (7). Out of the numerous results of the work we choose - rather randomly - to reproduce in fig. 3 the value g(0) of the pair-correlation function at zero separation, which quantity is

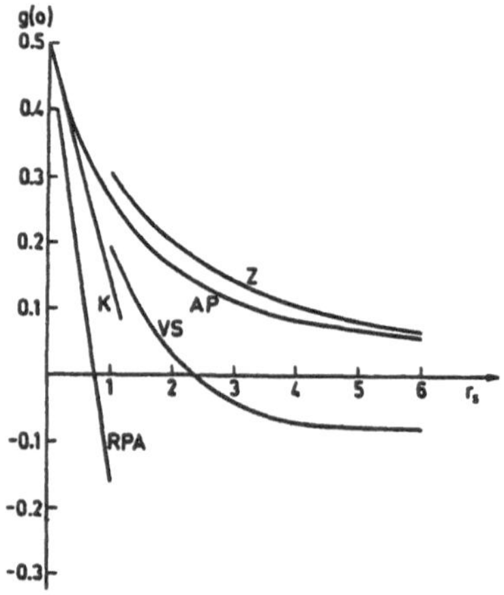

Figure 3. Pair-correlation function at zero distance for homogeneous electron gas as function of the density parameter r_s. AP: our result[10], Z: Zabolitzky[13], VS: Vashishta and Singwi[14], K: Kimball's asymptotically exact formula[15], RPA: result of random-phase approximation.

perhaps the most sensitive characteristic of the electron gas and rarely well described.

In another work[12] the well-known Lipkin-Meshkov-Glick (LMG) model was studied in the extended SUB1-SUB4 approximations. The LMG model is a particularly illuminating example where all the required expS expressions can be readily written down and programmed to a computer. In the limit of large particle number $N \to \infty$ the LMG model displays a phase transition at a critical coupling strength $g_c = NV_c/\varepsilon = 1$, above which the ground state is doubly degenerate. There exists a symmetry-breaking order parameter, which is the y-component of the pseudospin operator. The extended expS results are highly interesting expecially around the critical coupling strength, and for another notable reason: the obtained ground-state energy is always an upper limit to the true ground-state energy. These features, taken together with the possibility to calculate accurate estimates for the lowest excitation energies, make the extended SUBn approximations far superior to the normal expS method.

References

1) F. Coester, Nucl.Phys. 7 (1958), 421.
2) F. Coester and H. Kümmel, Nucl.Phys. 17 (1960), 477.
3) H. Kümmel, K.H. Lührmann and J.G. Zabolitzky, Phys. Reports 36 (1978) 1.
4) J. Hubbard, Proc.Roy.Soc. A240 (1957), 539.
5) K.H. Lührmann, Ann.Phys. 103 (1977), 253.
6) J. Arponen, Helsinki University preprint HU-TFT-81-41 (1981).
7) J. Arponen, Ann.Phys. (to be published).
8) D.J. Amit, "Field Theory, the Renormalization Group, and Critical

Phenomena", McGraw-Hill, New York (1978).

9) J. Arponen and E. Pajanne, J.Phys. C15 (1982), 2665.

10) E. Pajanne and J. Arponen, J.Phys. C15 (1982),2683.

11) R.F. Bishop and K.H. Lührmann, Phys.Rev. B26 (1982), 5523.

12) J. Arponen and J. Rantakivi, Nucl.Phys. A407 (1983), 141.

13) J.G. Zabolitzky, Phys.Rev. B22 (1980), 2353.

14) P. Vashishta and K.S. Singwi, Phys.Rev. B6 (1972), 875; ibid. 4883.

15) J.C. Kimball, Phys.Rev. B14 (1976), 2371.

COMPUTATIONAL QUANTUM MECHANICS AND THE BASIS SET PROBLEM

C.L. Davis, H.-J. Aa. Jensen
and H.J. Monkhorst
Quantum Theory Project
University of Florida
Gainesville, FL 32611 U.S.A.

Abstract

It is advocated to use singularity-matching basis functions in wavefunction expansions. Atomic and molecular Hartree-Fock calculation strongly support this approach. Nonlinear optimization is then totally avoided.

Setting the Stage

Few problems in quantum mechanics have simple solutions. Such happy cases usually belong to the class of so-called "exactly solvable" problems. Other problems can be greatly simplified when a full separation of variables is possible, leading to equations in one variable only. These equations are solved, to any degree of accuracy desired, with a mixture of well-tested analytic and numerical methods. The overwhelming majority of problems, however, cannot be mapped onto exactly solvable or fully separable ones. We are then left to our own devices how to proceed to numerically and physically acceptable solutions. At this point, many argue, quantum science blends with artistry, and an infinity of methods, approaches and approximations springs eternally. This Conference is a beautiful reflection of this phenomenon. It guarantees lively discussions which are only quieted when entirely different methods lead to the same results.

For many years, expansion of the wavefunctions in basis sets has been the standard answer. Mathematically, such sets must be infinite in size, by necessity; in practice these are truncated, because of computational limitations. This irreconcilable conflict has led to great inventiveness in choosing the "best" basis sets, somehow simultaneously mimicking completeness requirements and emphasizing salient features of the wavefunctions to be computed. These aspects are often known from general or model considerations. Particular potentials (microscopic or effective) and qualitative insights in the problem at hand often suggest usable basis sets. Examples are the Gaussian home base functions in solid helium, harmonic oscillator functions for nuclei, Slater functions (SF's) in atoms and (to a lesser extent) molecules. By and large, such choices have been successful in practice. However, considerable criticism can be leveled against the mathematical behavior of most basis functions.

Basis Sets in Quantum Chemistry--A Case History

The present situation in computational quantum chemistry epitomizes the basis set problem and its pragmatic solutions with their troubles because of unsatisfactory systematics and mathematics.

In the early days of atomic physics it was realized that atomic orbitals can be best approximated by linear exponential functions in the electron-nucleus distance. Slater[1] devised a set of rules, the Slater rules, that relate the exponents to the atomic nuclear charge, the filled shells closer to the nucleus than the orbital in question and the number of other electrons in the shell to which the orbital belongs. These rules had considerable qualitative usefulness, particularly at a time when quantitative aspects of the many-body problem were not taken seriously. With the advent of computers, starting around 1950, shell model calculations on atoms began in earnest, particularly by Roothaan[2] and his group at Chicago. Accurate analytic HF calculations were performed, using Slater functions (SF's) of the form

$$\chi_n^s(r) = N_n \, r^{n-1} \, e^{-\zeta r} \tag{1}$$

The ζ values according to Slater's rules were quantitatively incapable of giving satisfactory solutions to the HF equations. In fact, it became obvious that for better than 10^{-2} hartree (or 0.5 eV) accuracy of the total energy, at least two SF's with different exponents are required. Thus was born the irrepressible urge of quantum chemists to energy-optimize non-linear parameters that characterize basis sets. The tables of atomic HF wave functions by Clementi and Roetti[3], obtained by extensively optimizing SF exponents, are a classic example of a numerical high-technology in quantum chemistry. But the connection with the physically motivated Slater rules is totally lost; the optimal exponents bear no relationship to shielding constants, ionization stages, etc.

The status of molecular electronic calculations is yet different. Although the situation is less serious for diatomic molecules, analytic computations with SF's optimized for atoms have caused essentially insurmountable numerical difficulties. There are mainly two reasons. Most troublesome, notwithstanding many intense efforts, no satisfactory (i.e., numerically reliable and fast) evaluation methods for most of the molecular integrals in such calculations have been found. Furthermore, high accuracy results demand the reoptimization of the SF's used in the linear combination of atomic orbital (LCAO) representation of the molecular orbitals (MO's). Although, inner-shell AO's are often adequate for inner-shell MO's (because of their strong atomic localization) valence shell orbitals are greatly altered upon molecule formation. Failure to optimize can lead to errors of several millihartrees per electron. However, reoptimization with respect to a greatly increased number of nonlinear parameters is painful and fraught with numerical uncertainties. Another, albeit less clearcut problem is the intrinsic inability of the LCAO representation to ever satisfy the HF equation pointwise. This question will be addressed in the next section.

As early as 1950, Boys proposed the use of Gaussian functions (GF's) for AO and MO basis sets[4]. In 1960, Boys, Singer and Longstaff[5] proposed the use of explicit electron correlation through factors like $\exp(-\gamma r_{12}^2)$ in GF's. These functions were introduced for one reason only: their integrals are easy. The fact that GF's, with or without correlation, are quite inadequate at short and large distances between Coulomb particles, was not considered serious. It was reasoned that linear combinations of many, sometimes contracted GF's with properly chosen exponents and origins of localization (not necessarily coinciding with nuclear positions) give adequate accuracies. Extensive optimizations of atomic HF energies and wavefunctions using various GF basis set sizes have appeared since 1960. The most complete and systematic work is that by Ruedenberg and Schmidt[6], using even-tempered exponents to reduce the number of non-linear parameters. It is found that from two to three times as many GF's than SF's are needed for comparable accuracies. Gaussian orbital technology now permeates all of quantum chemistry from HF to all presently practical methods dealing with correlation effects (configuration interaction, many-body perturbation theory, coupled-cluster approach, etc.). Reoptimization is totally ignored although quite necessary[6]. Starting from atomic basis sets, quite arbitrary supplements are made such as polarization and diffuse GF's, using positions and exponents obtained with artistry.

In the present drive to achieve "chemical" accuracy of one millihartree, and beyond, for atomic and molecular energy levels, the basis set problem re-emerges in full force. MBPT and coupled cluster calculations, starting with HF zeroth order states, have clearly shown that at least fourth-order corrections are needed to reach such accuracies. But these results have also exposed that standard Gaussian basis sets cause errors about an order of magnitude greater than fourth-order corrections. It is imperative that the quality of basis sets is improved before calculating these corrections.

This situation presents a considerable problem. Increasing the number of functions is barely feasible, because of the inherent slow convergence of expansions in these functions; it would greatly increase the computational effort, resulting in only slight improvements. The alternative is an extensive non-linear optimization. But this is no real solution because of the labor involved, already substantial for small molecules, and becoming worse for larger systems. Moreover, occurrence of multiple local minima is a numerical plague which no non-linear optimization algorithm can avoid.

During the past three years one of us (HJM) has been involved with a project to very accurately compute atomic and molecular correlation energies using explicitly correlated Gaussian geminal expansions[7]. With a novel functional upperbounding the second order correlation energy, both linear and non-linear parameters for first-order pair-functions can be obtained with greater ease than previously possible. The particularly efficient enforcement of near-strong orthogonality avoids the need for calculating four-electron integrals, and greatly reduces the number of integrals needed to evaluate this functional. Larger Gaussian orbital and geminal expansions could therefore be used. This has led to second and third-order correlation energies for He, Be, H_2 and LiH with accuracies higher than ever obtained before. A review of these and full coupled-pair results has been presented at this conference[8].

Problems with non-linear optimizations did arise, but they were relatively minor for those small systems. However, on contemplating larger systems and geometry variations for potential energy surface calculations, it became clear that serious problems will arise. Too many non-linear parameters will have to be optimized, and too many repetitions of that procedure will be needed. Lack of systematics and multiple local minima in multi-dimensional parameter space will haunt us. An entirely new approach will be needed.

Back to Basics

All Schrödinger (or Fock, perturbation, coupled-cluster, etc.) equations for wavefunctions (or their components) are mathematically well-defined when the potentials are specified. Self-consistent field equations, although non-linear, can also be regarded as well-defined, but questions about existence and multiplicity of solutions are hard to answer[9]. All equations are differential (or integro-differential) equations in position space representation, and integral (or non-linear integral) equations in momentum representation.

Having reminded ourselves of those well-known facts, it is surprising that so little attention has been paid to the relationship between singularities in the potentials of these equations, the singularities in the solutions and the appropriateness of adopted basis functions for their expansions. For too long the attitude has been that "any complete set of functions will do." But this overlooks the practical necessity of basis truncation, and then the rapidity of convergence is very relevant. Some complete sets are more equal than others. The observation above about the need for a larger number of GF's than SF's for comparable accuracies is a point in case. We should turn from computational expediency considerations to an analysis of the mathematical desiderata for basis sets. This can then be followed by a study of how to best evaluate the integrals needed to reduce the problem to an algebraic one, solvable with standard methods.

A Working Hypothesis

We propose that the best convergence results when each basis function has singularities matching, both in position and in character, all of the singularities of the wavefunction (or components, such as cluster functions) which it helps to describe.

Such singularities can be identified, since they are related to those of the potential. In other than one-dimensional cases (e.g., non-separable potentials) the relationship between potential and wavefunction singularities is not immediately obvious; work on this question is in progress[10]. But recent work on HF systems, mainly atoms, has shown that exponent optimization can be eliminated entirely in HF calculations, without sacrificing accuracy or significantly increasing basis set sizes[11]. This is achieved by exploiting a simple relationship between SF exponents and occupied orbital energies. The results are so convincing to us, and the principle so general rather than particular for HF systems, that we are confident to entirely eliminate exponent optimization from quantum chemical calculations. Here we wish to illustrate this approach with some results published in great detail elsewhere[11]. In later publications we will return to the motivation for this work: the elimination of a non-linear parameter search from calculations of explicitly correlated cluster functions.

An Example: Singularities in Atomic and Molecular Hartree-Fock Theory

The closed-shell Fock equations are of the form

$$(F-\varepsilon_i)\Phi_i = 0 \tag{2}$$

where, in atomic units,

$$F = -\tfrac{1}{2}\nabla^2 - \sum_\alpha Z_\alpha/r_\alpha + 2J-K. \tag{3}$$

r_α are the electron distances to the α-th nucleus, with charge Z_α, and J and K are the usual Coulomb and exchange operators, respectively. To find the singularities in $\Phi_i(\underset{\sim}{r})$ we have to identify those of the potential terms in Eq. (3).

(i) Singularities at $\underset{\sim}{r}_\alpha = 0$. Singularities for finite r values are for $\underset{\sim}{r}_\alpha = 0$, i.e., for electron-nucleus coalescence. These are known as cusp singularities, and have been studied quite extensively[12]. The combination of the Cartesian-separable Laplace operator and locally spherical potential causes a branchpoint-like behavior (in the Cartesian coordinates) of Φ_i near $\underset{\sim}{r}_\alpha = 0$. For atoms, where the J and K operators are spherically symmetric, Φ_i factorizes with spherical coordinates, and an accurate description of its r-dependent factor is easy. In fact, as explained in Ref. [11], any linear combination of SF's can describe that factor quite well for $r < \infty$, regardless of the exponent values. GF's are totally incapable of a point-wise description: any finite sum of such functions will be analytic for all $r < \infty$, and therefore unable to cancel singular terms in Eq. (2). This is the main reason for the slow energy-convergence with GF expansions pointed out above.

For molecules, having no spherically symmetric potential, no coordinate system exists to obtain full separability of Φ_i. A systematic description is then not clear; we have yet to find the best basis set capable of that. But any linear combination of atomic basis functions, having at best only singularities at their nuclear origins, is clearly inadequate. If χ_A is an atomic basis function centered on nucleus A, then

$(F-\varepsilon_i)\chi_A$ will be infinite at another nucleus B: the $(-\frac{1}{2}\nabla^2)$ operator acting on χ_A does not produce an infinity at B to cancel the infinite potential there. Consequently, each member of a basis $\{\chi_\mu\}$ used to expand Φ_i must have singularities at each nuclear position. Elliptic functions (EF's) for diatomic molecular orbital expansions satisfy this requirement, but our calculations suggest that the singularities have the wrong angular character[11].

(ii) Singularities at $r = \infty$. All the potential terms in Eq. (3) have essential singularities at $r = \infty$. Since in that point these terms behave similarly for atoms and molecules, the associated singularities in Φ_i also must be similar. Therefore we can mainly study atoms, and expect the results to be equally valid for atomic and molecular orbital calculations. This is fortunate since atomic HF calculations are easier, more accurate and can be compared with high-accuracy numerical calculations. We also can examine the effects of the singularity at $r = \infty$ in isolation from nuclear and correlation cusp effects.

Since we are interested in the behavior of Φ_i at infinity, we should study its asymptotic limits. A detailed derivation can be found in Ref. 11. Briefly, the asymptotic solution of Eq. (2) can be obtained by expanding $\Phi_i(\underset{\sim}{r})$ in spherical harmonics $Y_p^q(\theta,\phi)$ and radial functions $f_{ipq}(r)$. For $r \to \infty$, the combined nuclear, Coulomb and exchange potential behaves as $(-1/r)$. Proposing a trial solution valid as $r \to \infty$, of the form

$$f_{ipq}(r) \sim \sum_{k=1}^{N} e^{-\alpha_k r} r^{\beta_k} \sum_{n=o} a_n^{kpq}(i) r^{-n} \tag{4}$$

with N the number of doubly occupied orbitals, and substituting the leading terms of the potential, Eq. (2) produces an expression which must vanish identically. By equating coefficients of different powers of r to zero, equations for α_k and β_k are obtained, and recursion relations for $a_n^{kpq}(i)$. The exponents α_k, here termed "primary asymptotic exponents," are related to orbital energies by

$$\alpha_k = \sqrt{-2\varepsilon_k} \tag{5}$$

Normally, each orbital involves all N primary exponents, due to the exchange operator, except when all orbitals have s symmetry. We showed that

$$\beta_k = 1/\alpha_k - 1 \tag{6}$$

Except for hydrogen-like atoms, these powers will be irrational numbers, indicating logarithmic branch-point-like singularities at $r = \infty$.

Eq. (4) gives only part of the asymptotic expansion of f_{ipq}. More rapidly decreasing terms can be obtained by an iterative procedure. The next step involves recalculating the Coulomb and exchange potentials with Eq. (4) for the orbitals. In addition to the inverse powers of r, the result contains exponentially decreasing terms with exponents $\alpha_k + \alpha_\ell$, for k, = 1,...,N. These terms in the Fock operator generate additional terms in the asymptotic expansion of f_{ipq} of the form Eq. (4), but now with exponents

$$\alpha_{klm} = \alpha_k + \alpha_l + \alpha_m, k,l,m = 1...N \tag{7}$$

and irrational powers

$$\beta_{klm} = \beta_k + \beta_l + \beta_m; \tag{8}$$

we refer to these quantities as "secondary asymptotic exponents and powers." The iterations can be continued indefinitely to obtain "tertiary" and higher asymptotic exponents and powers, which are sums of five, seven, nine, etc., primary exponents and powers. Each term in the asymptotic expansion represents a distinct essential singularity at $r = \infty$.

Numerical Calculations

As a result of above analysis and other mathematical observations explained in Ref. 11, we set out to perform HF calculations on several closed-shell systems. The orbitals of several closed-shell systems were expanded with a basis of SF's for atoms and a basis of EF's for diatomic molecules, choosing the exponents from a list of primary, secondary, etc., asymptotic values. Some, but little, trial and error was required to arrive at the best combinations of exponents and powers in Eq. (1). Our calculations support an earlier conjecture that primary exponents are the most important, followed in order by secondary, tertiary, etc., exponents[13]. Some of the latter exponents must be omitted to avoid linear dependence problems. Orbital energy estimates are required before asymptotic exponents can be evaluated. This is no problem, since it is found that any reasonable guesses for orbital energies can be used in a small basis set calculation to obtain very much more accurate values[11]. Few systematic and unique iterations are required for convergence to the desired accuracies, with computing time considerably less than that needed to optimize exponents.

Rather than presenting reams of numbers to support the correctness of our working hypothesis we summarize the numerical evidence in the following points:

(i) For atomic basis sets with less than five or six SF's per orbital it is energetically favorable to optimize the exponents. At the associated accuracy level, contributions to the energy integrals from the asymptotic tail is insignificant. Nevertheless, SF's making the largest contributions have optimized exponents close to the asymptotic values.

(ii) Atomic basis sets with more than five or six SF's per orbital exponent optimization around asymptotic values gives only slight energy improvements. In fact, optimization with larger sets tends to bring optimal exponents ever closer to asymptotic values.

(iii) Calculations of the orbital moments $<r^n>$ show that only AO moments with $N \leq M$, the number of SF's, can be evaluated with accuracy, regardless of optimization of exponents. Roughly M SF's per AO give $< r^n >$ to (M - n) significant figures with $n \leq M$. This "moment effect" can be attributed to the inability for any finite SF basis to correctly describe the irrational power in the asymptotic behavior of Φ_i. The energy optimization emphasizes a good fit in the small-r region, at the expense of a balance between (r^k) factors (in the case of asymptotic-exponent SF's) or various exponential tails (in optimized-exponent SF's) to mimick the irrational power over some significant large-r region. The "fit" gradually deteriorates as r increases, leading to a loss of accuracy of $< r^n >$ when n increases. This surprising finding exposes an intrinsic inadequacy of SF's. Only an infinite number of them can give an accurate large-r irrational power. But term-by-term energy integration is then not allowed because of a lack of uniform convergence in intervals including $r = \infty$. Functions correctly matching singularities at $r = 0$ and $r = \infty$ do exist. These will be studied in a future publication.

(iv) The results for the diatomic molecules H_2 and LiH suggest that basis functions other than EF's are needed which more adequately describe the nuclear cusp singularities to meet the accuracy of numerical HF calculations. The results for H_2 are a striking demonstration of this problem. Whereas total HF energies could be computed to seven significant figures (with about ten EF's), the kinetic energy has only one significant figure accuracy.

(v) Finally, in the Figures below, we present typical examples of energy variations upon changes in Ne basis sizes and various exponents around asymptotic values (see legend). The curves show minima nearer asymptotic exponent values for the larger basis sets.

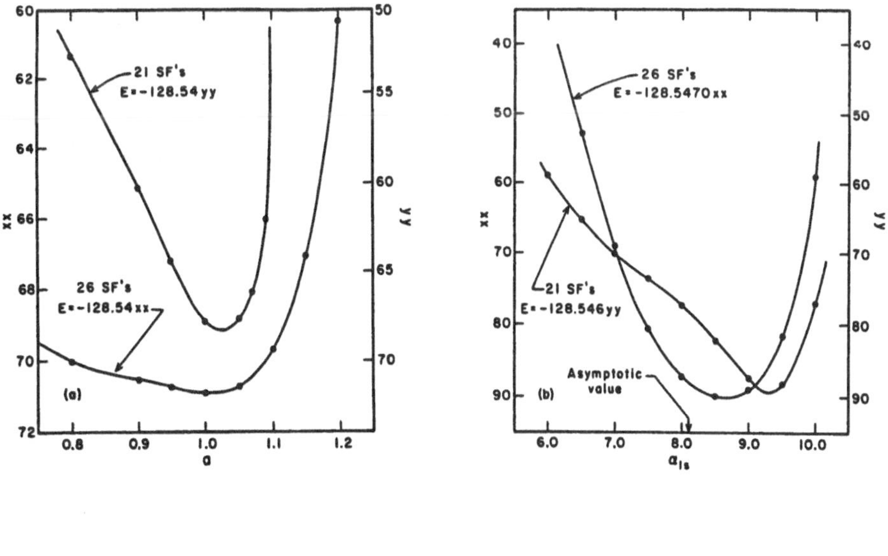

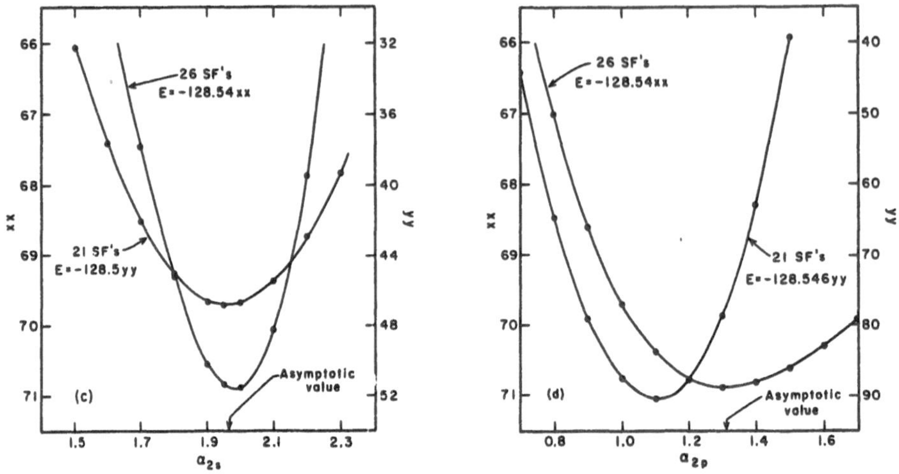

Figure 1. Variation in groundstate energy of Ne, calculated for 21 and 26 SF basis sets, when (a) all exponents are simultaneously varied around their asymptotic values by uniform scale factor a; (b) α_{1s} is varied with asymptotic α_{2s} and α_{2p}; (c) α_{2s} is varied with asymptotic α_{1s} and α_{2p}; (d) α_{2p} is varied with asymptotic α_{1s} and α_{2s}.

Concluding Remarks

The results suggest that singularity-matching HF basis functions give the best HF convergence. We did not use anything special from the HF approximation other than its independent-particle aspect enabling an easy identification of singularities, a mathematical-technical virtue, but not a simplification in principle.

Integrals over optimally singularity-matching basis functions can be evaluated by introducing proper integral transforms, preferably of Gaussian type. Interchanging integrations can result in accurate evaluations. Initial experiments are encouraging, and many untried possibilities are left[10].

References

1. J.C. Slater, Phys. Rev. 36, 57 (1930).
2. C.C.J. Roothaan and P. Bagus, Methods in Computational Physics, Vol. 2, Acad. Press, New York (1963).
3. E. Clementi and C. Roetti, At. Data and Nucl. Data Tables 14, 177 (1974).
4. S.F. Boys, Proc. Roy. Soc. A200, 542 (1950); ibid. A206, 489 (1951).
5. S.F. Boys, Proc. Roy. Soc. A258, 402 (1960); K. Singer, ibid. 412 (1960); J.V.H. Longstaff and K. Singer, ibid. 421 (1960).
6. M.W. Schmidt and K. Ruedenberg, J. Chem. Phys. 71, 3951 (1979).
7. K. Szalewicz, B. Jeziorski, H.J. Monkhorst and J.G. Zabolitzky, J. Chem. Phys. 78, 1420 (1983) and unpublished work.
8. B. Jeziorski, H.J. Monkhorst, K. Szalewicz and J.G. Zabolitzky, this issue.
9. T.P. Zivkovic and H.J. Monkhorst, J. Math. Phys. 19, 1007 (1978).
10. C.L. Davis and H.J. Monkhorst, to be published.
11. C.L. Davis, H.-J. Aa. Jensen and H.J. Monkhorst, J. Chem. Phys. (1984) in print.
12. T. Kato, Commun. Pure Appl. Math. 10, 151 (1957); W.A. Bingel, Z. Naturforsch. Teil A, 18, 1249 (1963); H. Conroy, J. Chem. Phys. 41, 1327 (1964).
13. W.M. Huo and E.N. Lassettre, J. Chem. Phys. 72, 2374 (1980).

PARQUET PERTURBED

Alexander Lande

Instituut voor Theoretische Natuurkunde
Groningen University
Groningen, The Netherlands

I. INTRODUCTION

This talk will describe work carried out in collaboration with A. D. Jackson and R. A. Smith. Our approach to the description of zero-temperature quantum liquids is based on the approximate summation of parquet diagrams [1]. Following a brief review of the formalism I shall show how it may be perturbatively improved upon and present results of calculations for ^{4}He.

The formalism constitutes essentially a self-consistent summation of particle - particle ladders of Brueckner theory and particle-hole ladder diagrams of the RPA which lead to the description of phonons in the system. The latter do not converge, which led us to believe that the parquet diagrams represent the minimum set of diagrams which must be summed (at least approximately) for a reliable description of quantum liquids. This rewriting of many-body theory has a number of virtues.

(a) The parquet approach is interpretative. HNC variational approaches to the many-body problem are essentially identical in formal content to an approximate parquet summation. Remaining differences were shown to be small in a number of physically interesting numerical cases. Thus the parquet diagrams provide an approximate diagrammatic interpretation of HNC calculations and suggest which features of the HNC approach are physically important.

(b) The parquet equations offer a clean and well defined approach to quantitative calculations of quantum fluids. In the simplest approximation it provides a relatively reliable reproduction of full GFMC in liquid ^{4}He. We shall demonstrate here an obvious and systematic way to improve on our earlier approximate results.

(c) The parquet approach permits analogies to surprisingly different fields. Planar diagrams play a dominant rôle in QCD in the large color limit. To the extent that the parquet subset (completely two-particle reducible planar diagrams) captures the essence of the planar diagrams, it may prove possible to establish a detailed microscopic connection between familiar quantum fluids and QCD.

In Section II we summarize the parquet equations for Bose systems and the propagator approximations which render them soluble. (Although we confine our attention to Bose systems, the extension to Fermi systems has been performed [2] . In Section III we shall consider perturbative improvements and present results for liquid ^{4}He.

II. THE PARQUET EQUATIONS

We summarize here those features of the parquet formulation essential for the present perturbative extension. In constructing pp-ladders, it is clear that any pp-irreducible diagram may be included in the driving term. For example, any diagram which is ph-reducible is pp-irreducible. One is immediately led to the parquet equations for the two-particle vertex, Γ, in terms of the pp-reducible ladder diagrams, L, the ph-reducible chain diagrams, C, and the bare potential, V.

$$\begin{aligned} L &= (V+C)\, G_{pp}\, (V+C) + (V+C)\, G_{pp}\, L \\ C &= (V+L)\, G_{ph}\, (V+L) + (V+L)\, G_{ph}\, C \\ \Gamma &= V + L + C \end{aligned} \tag{1}$$

In order to use eqn.(1) we must specify the pp- and ph-propagators and provide a complete set of momentum variables for the quantities L, C, and Γ. We use oriented Feynman propagators for individual particles along with the convention that the momentum of individual particle lines be determined by their orientation and by Kirchhoff's law.

The full propagator for a single-particle of four-momentum $(\underset{\sim}{k},\omega)$ is

$$G^{\alpha\beta}(\underset{\sim}{k},\omega) = \delta_{\alpha\beta}\left[\frac{\theta(k-k_F)}{\omega-\omega_k-\Sigma^*(\underset{\sim}{k},\omega)+i\varepsilon} + \frac{\theta(k_F-k)}{\omega-\omega_k-\Sigma^*(\underset{\sim}{k},\omega)-i\varepsilon}\right] . \tag{2}$$

Although we are concerned with a boson problem, it is convenient to adopt a fermion formalism to facilitate the treatment of the zero-momentum condensate. We ultimately take limits of k_F going to zero and the fermion degeneracy, υ, going to infinity in such a way that the density (proportional to υk_F^3) is constant. The kinetic energy, ω_k, is defined as $k^2/2$. The proper self-energy, $\Sigma^*(\underset{\sim}{k},\omega)$, must ultimately be determined (in a self-consistent fashion) from the two-particle vertex, Γ. The free single-particle propagator, $G_0^{\alpha\beta}(\underset{\sim}{k},\omega)$ is obtained by setting $\Sigma^*(\underset{\sim}{k},\omega)$ equal to zero.

The two-particle propagators in eqn.(1) are simply products of two suitably chosen single-particle propagators. Thus we can define the joining of two arbitrary diagrams, X and Y, with the pp-ladder operation as (see also the schematic figure below),

$$\left[X G_{pp} Y\right]_{adp;bcp'} = i\int\frac{d^4z}{(2\pi)^4}\, X(azp;bz'p')\, G(z)G(z')\, Y(zdp;z'cp') \tag{3}$$

where e.g. the indices a,d and p, represent the left final momentum, initial momentum, and its orientation (± 1). Momentum conservation imposes the constraint, $z'=b+pp'(a-z)$. Similar rules apply for joining two diagrams through the ph-ladder operation. (Here, $z' = z + pp''(a-d)$.)

$$\begin{aligned}\left[X G_{ph} Y\right]_{adp;bcp'} = -i\int\frac{d^4z}{(2\pi)^4}\sum_{p''} X(adp;zz'p'')\, G(z)\, G(z') \\ \times\ \frac{1}{2}\left[Y(zz',-p'';bcp') + Y(z'zp'';bcp')\right] .\end{aligned} \tag{4}$$

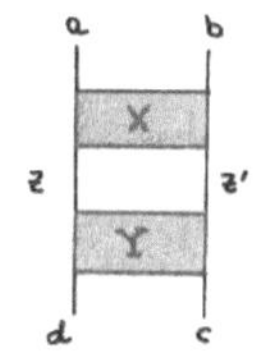

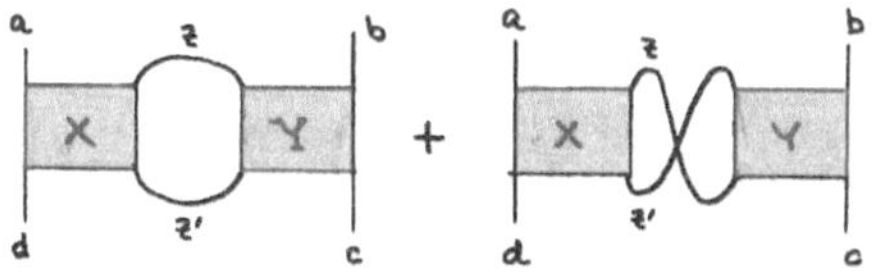

The two-particle vertex is a sum of parquet diagrams. The simultaneous description of pp- and ph-ladders requires the self-consistent solution of the non-linear equations (1). They are not sufficient to determine Γ, since the full G(z) involves Σ^*. It is clear that Γ could be obtained from Σ^* by a single functional differentiation with respect to particle number; each term in Σ^* of order-n in the interaction giving rise to n distinct contributions to Γ. The inverse problem of working from Γ to Σ^* would be equally simple if the various contributions to Γ could be regrouped in precisely such n-element sets. The parquet contributions to Γ do not admit to such treatment: the fact that one contribution to Γ arising from a given self-energy diagram is of the parquet class does not guarantee that all contributions are parquet. Thus in [1] we studied the structure of all contributions to Σ^* which could arise from closing one single-particle line in the parquet approximation to Γ. This analysis led to the observation that the correctly counted contributions to Σ^* could be obtained from the pp-ladder contributions to Γ which have a top rung of V. This led us to define the class of diagrams,

$$\bar{L} = \frac{1}{2} V G_{pp} (\Gamma + L) \tag{5}$$

in terms of which

$$\Sigma^*(k) = \rho\tilde{V}(0) - i\upsilon \sum_{p'} \int \frac{d^4k'}{(2\pi)^4} \bar{L}(kk,p=+1;k'k'p')\, G(k') \tag{6}$$

This completes the set of equations to be solved in a grand self-consistency scheme for L, C, and Γ. In practice we have neglected this final element of self-consistency and have used instead G_o, the free propagator, in eqns. (3)-(6).

Even when full self-consistency is not sought, eqn.(6) still represents an important intermediate step on the way to the calculation of the total energy of the system. Furthermore, the parquet contributions to Σ^* do posses the simplifying property that was absent in relating Γ to Σ^*: All Σ^* diagrams generated by the functional differentiation of a given energy diagram are generated by the parquet procedure is any one of them is. This important result leads to considerable simplification in determining the energy of the system. We can write

$$E/A = -\frac{i\upsilon}{2\rho} \int \frac{d^4k}{(2\pi)^4} G_o(k) \left[\omega + \omega_k\right] \Sigma^*(k)\, G(k) \tag{7}$$

Exploiting the completeness of our self-energy diagrams, we can also obtain the energy from the Feynman-Hellmann theorem

$$E/A = \frac{1}{2} \int_0^1 d\alpha \int \frac{d^4k}{(2\pi)^4} V(k) \left[S^\alpha(k) - S^o(k)\right] + \frac{1}{2}\rho\,\tilde{V}(0) \tag{8}$$

where $S^{\alpha}(k)$ is the dynamic structure function obtained by replacing the potential V by αV. Eqn.(8) enables us to proceed from E/A to S(k) [1].

The full parquet equations present a formidable numerical challenge. We make the following approximations. Throughout, G is replaced by G_o. The two-particle vertex is a function of six scalar variables. The success of HNC calculations suggest that adequate results can be obtained by regarding Γ as a function of three-momentum transfer only. This reduces the ph-ladder equation to an algebraic equation and, exploiting the spherical symmetry of V, renders the pp-ladder equation to a function of one variable. The spirit of the approximations is that we wish to preserve exact results for E/A and S(k) in the two cases when either only pure pp-ladders or pure ph-ladders are retained. Specifically, we make the local approximations for the pp- and ph-propagators,

$$G_o(k-p_1)\;G_o(k'+p_1) \sim \begin{cases} \delta(\omega_1)/2\omega_{p_1} & (\uparrow\uparrow) \\ 0 & (\uparrow\downarrow) \end{cases} \tag{9}$$

$$G_o(q_1)\;G_o(q_1+p) \sim \delta(\omega_1)\left[\frac{\theta(k_F-q_1)}{\bar{\omega}_p-\omega_p} - \frac{\theta(k_F-|q_1+p|)}{\bar{\omega}_p+\omega_p}\right] \tag{10}$$

The specific choice of $\bar{\omega}_p$, along with the form of all the equations needed in this approximate calculation is given in the Table below.

A summary of the approximate parquet equations

(i) $$L(k) = \int \frac{d^3p}{(2\pi)^3}\,[V(p+k)+C(p+k)]\,\frac{1}{(-2\omega_p)}\,[V(p)+C(p)+L(p)]$$

(ii) $$C(k) = [V(k)+L(k)]\,\frac{2\rho\omega_k}{\bar{\omega}_k^2-\omega_k^2}\,[V(k)+L(k)+C(k)]$$

$$\bar{\omega}_k^2 = \epsilon_k^2 - (\epsilon_k+\omega_k)^3/(\epsilon_k+3\omega_k)$$

$$\epsilon_k^2 = \omega_k^2 + 2\rho\omega_k[V(k)+L(k)]$$

(ii)′ $$C(k) = \frac{-\rho[V(k)+L(k)]\,[V(k)+L(k)+C(k)]}{\omega_k} + \frac{\rho^2[V(k)+L(k)]\,[V(k)+L(k)+C(k)]^2}{4\omega_k^2}$$

(iii) $$E/A = \tfrac{1}{2}\rho\tilde{V}(0) + \tfrac{1}{2}\rho\int\frac{d^3p}{(2\pi)^3}\,\frac{V(p)[V(p)+L(p)+C(p)]}{(-2\omega_p)}$$

(iv) $$S(k) = 1 - \frac{\rho}{k^2}\frac{d}{d\alpha}\left[\alpha(\alpha V+L^{\alpha}+C^{\alpha})\right]\Big|_{\alpha=1}$$

In liquid ^{4}He the approximate parquet equations lead (for the Lennard-Jones potential) to a binding energy of 5.28 K per particle at an equilibrium density of 0.018 Å^{-3}. The optimized HNC yields 5.43 K at 0.0185 Å^{-3}. This small difference was used in [1] as grounds for regarding the parquet sums as a diagrammatic interpretation of the HNC variational calculation. Essentially exact GFMC calculations [3] give 6.85 K at ρ = 0.0222 Å^{-3}. It is this remaining discrepency that we shall now attempt to account for by essentially perturbative techniques.

III. PERTURBATION THEORY

Our aim is to provide a more reliable approximation to Γ , and hence to the energy and S(k). In so far as is possible, we shall cast our pertubative improvements in terms Γ_o, the local approximation obtained by solving the equations in the Table. It is intended that Γ retain all of the dynamical content of the underlying local potential while exploiting all of the cancellations inherent in the parquet sums: (a) Γ contains the pp-ladder sums and is well behaved at small-r where V(r) is badly divergent, (b) it respects the cancellations attendant to the ph-ladder sums so that $\Gamma(k)$ is indeed zero in the limit of small k.

To see how pertubative results can be expressed solely in terms of $\Gamma(k)$ it is useful to eliminate the ladders and chains appearing in eqn.(1). This leads to the manifestly non-linear equation relating Γ and V,

$$\Gamma = V + \sum_{i=pp,ph} \Gamma\,(1+G_i\Gamma)^{-1} G_i \Gamma \tag{11}$$

In [1] this relation was largely of formal interest as dramatic illustration of the crossing symmetry. Here it provides the vehicle for systematic improvement of approximate parquet results.

Our perturbation scheme has a number of related goals: (a) to improve the evaluation of parquet diagrams, (b) to include self-energy diagrams neglected in our approximate parquet sums, (c) to include effects of planar (non-parquet) diagrams and, more generally, non-planar diagrams. Non-planar diagrams are properly included as modifications to the driving term V. Non-parquet planar diagrams contribute to E/A in fifth order (in V) and make fourth order contributions to Γ. We shall ignore these effects in favor of those mentioned in (a) and (b). These modify E/A in fourth order and Γ in third-order. Although our results will arise as a correction to Γ, our primary concern here is with E/A and S(k).

Starting from eqn.(11) we can write the contributions to chains and ladders as

$$L = (\Gamma G_{pp}\Gamma)\left[1 + G_{pp}\Gamma\right]^{-1} \tag{12}$$

$$C = (\Gamma+L)\, G_{ph}(\Gamma+L)\,\left[1 + G_{ph}(\Gamma+L)\right]^{-1} \tag{13}$$

The approximate parquet equations can be written for the local quantities L_o and C_o in a precisely equivalent form in terms of the local approximate propagators g_{pp} and g_{ph}. Thus,

$$L = L_o + (\Gamma G_{pp}\Gamma)\left[1+G_{pp}\Gamma\right]^{-1} - (\Gamma_o g_{pp}\Gamma_o)\left[1+g_{pp}\Gamma_o\right]^{-1} \tag{14}$$

$$C = C_o + (\Gamma+L)G_{ph}(\Gamma+L)\left[1+G_{ph}(\Gamma+L)\right]^{-1} - (\Gamma_o+L_o)g_{ph}(\Gamma_o+L_o)\left[1+g_{ph}(\Gamma_o+L_o)\right]^{-1} \tag{15}$$

We wish to solve these equation pertubatively. This involves propagator corrections such as $(G_{pp}-g_{pp})$ and the local quantities L_o and Γ_o. Note that in our approximation g_{pp} is non-zero only for parallel orientations. Also,our definitions of G_{ph} and g_{ph} render the chain diagrams independent of orientation.(This is somewhat different than in [1] and rather more economical.

The lowest order corrections to L_o and C_o are simply,

$$\delta L_1 = L^{(1)} - L_o = \Gamma_o (G_{pp} - g_{pp}) \Gamma_o \quad (16)$$

$$\delta C_1 = C^{(1)} - C_o = (\Gamma_o + L_o)(G_{ph} - g_{ph})(\Gamma_o + L_o) \quad (17)$$

giving

$$\Gamma^{(1)} = \Gamma_o + \delta L_1 + \delta C_1 \quad (18)$$

Both terms represent non-trivial corrections to the two-particle vertex which can be evaluated. They are of second-order in Γ_o and L_o and, thus of second (and higher) order in V. Following the route from Γ to $\bar{L}$ to Σ^* to E/A these corrections should make third-order corrections to E/A. However all third-order energy diagrams are either pure rings or pure ladders. In constructing our approximate propagators we tailored them to ensure that the sum of pure rings and pure ladders in every order are correctly evaluated. We anticipate therefore substantial cancellations by the time (16) and (17) have been used in evaluating corrections to E/A and S(k). It is easy to verify that δL_1 does not contribute to either E/A or S(k). δC_1 does yield a finite correction,

$$E_1/A = \rho^2 \int \frac{d^3p}{(2\pi)^3} V(p)\Gamma_o^2(p) \left[\frac{1}{(2\omega_p)^2} + \frac{1}{2(\omega_p^2 - \bar{\omega}_p^2)} \right] \quad (19)$$

A number of comments are in order. Eqn.(19) involves only $\Gamma_o(k)$ and, of course, the common V(p) arising from eqn.(5) even though δC_1 involved both Γ_o and L_o. This is a consequence of sums over orientations and is a general result. The reason that eqn.(19) is non-zero is our definition of $\bar{\omega}_p$ which was chosen to reproduce ring diagram results in spite of our local approximation. We selected an approximate g_{ph} which, while it did not yield the correct ring contribution to Γ , was constrained to yield exact ring contributions to E/A and S(k). Any attempts to improve Γ perturbatively will lead to incorrect results for E/A and S(k).

To circumvent this difficulty, we rewrite eqn.(19) as though it were a modification to the underlying (local) potential. Specifically,

$$E/A = \frac{\rho}{2} \int \frac{d^3p}{(2\pi)^3} V(p) \frac{1}{(-2\omega_p)} \Delta V(p)$$

which allows the immediate identification $\Delta V_{II}(p) = -\rho^2 \Gamma_o^3(p)/k^4$.

One obvious concern regarding eqn.(20) is that ΔV does not contain short-range correlations essential if the integral is to avoid (unphysical) domination by the short-range repulsion in the bare V. This may be avoided by restoring these correlations and defining, in general,

$$\Delta V^* = \Delta V \left[1 + g_{pp}\Gamma_o\right] \quad (21)$$

The general approach is now obvious. Having constructed the perturbative improvements to L and C via eqns.(14) and (15), we construct the related correction to the energy. We then write this correction as if it were a local correction to the potential ΔV, as in eqn.(20), and solve the approximate parquet equations for V+ ΔV to determine a new, local Γ and in turn, E/A and S(k). This procedure is systematic in providing perturbative improvements to E/A and S(k) and concurrently preserves the correct limiting behaviours of S(k) and g(r) in all orders.

Implementing this scheme in lowest order, with ΔV_{II} added to the driving term, we expect very small effects. To the extent that they are non-zero, they are the results of the necessary but awkward suppression of the dependence of the ph-ladder diagrams on the fourth component of momentum transfer in favor of a local form. The results are small. At $\rho = 0.020\ \text{Å}^{-3}$ the binding energy per particle is decreased by 0.20 K for ΔV_{II}, and by 0.14 K for ΔV^{*}_{II}. This 0.06 K difference between the two shifts appears also in the third-order calculation (described below) where use of the correlated ΔV^{*}_{III} in place of ΔV_{III} reduces the overall effect by 0.06 K. It seems fair to regard, say 0.15 K, as a measure of the uncertainty in the precise way in which perturbation theory is realized.

We turn now to the third-order corrections to ΔV and the related corrections to the energy of the system. These fall into three groups. First there are the corrections to the parquet diagrams due to differences between exact and approximate propagators. The related fourth-order contributions to the energy involve the first terms which are neither pure rings nor pure ladders. These should be significant. Second there are the parquet corrections due to differences between exact and free propagators, ignored in our approximate equations. They make their first contribution to the energy in fourth-order. Finally there are corrections aimed at reinstating missing non-parquet (in this order, non-planar) diagrams. There is one such term to fourth-order in the energy. Using eqns. (14)-(15) we can determine the δL_2 and δC_2 appearing in

$$\Gamma^{(2)} = \Gamma + \delta L_1 + \delta L_2 + \delta C_2 \tag{22}$$

The fourth order corrections to the energy $\Gamma^{(2)}$ generates have the general structure of the parquet diagrams (of fourth order) in an ordinary perturbation theory. They are shown in figs. (a)-(f). Again, the pure ladder diagram will not affect E/A. The associated ΔV_{III} will be given in Ref. 4 .

(a)

(b)

(c)

(d)

(e)

(f)

The self-energy insertions, dropped in our replacement of G by G_0, may be reinstated perturbbatively. The self energy diagram appearing in (g) will, for local interactions, represent a common displacement of all single-particle energies and therefore not affect the binding energy. The first non-trivial effects arise in fourth order as shown in (h) and (i). The corresponding ΔV's do not vanish in the p=0 limit, but their sum does.

(g)

(h)

(i)

(j)

Finally, we turn to (j) the only diagram of fourth-order in the energy that is not of the parquet class. The associated ΔV also vanishes in the zero-p limit. In some sense (h) - (j) are reminiscent of a Ward identity, with (j) having the appearance of a vertex correction related to the self-energy processes (h) and (i). Taken together they display a strong tendency towards cancellation. At $\rho = 0.020$ Å^{-3} the self-energy processes (h) + (i) reduce the binding energy of ^{4}He by 1.3 K per particle. When combined with the non-parquet term (j) one finds an increase in binding energy by a modest 0.27 K. This suggests that it may be important to maintain a symmetric description of self-energy terms and non-parquet diagrams.

All calculations reported here are for the usual Lennard-Jones potential. As shown in the figure below, the approximate parquet equations ($\Delta V = 0$) lead to substantial underbinding of ^{4}He (1.6 K per particle) and an underestimate of the equilibrium density (0.0175 Å^{-3} instead of 0.02185 Å^{-3}). As discussed above, we neither expect, nor find, significant effects from the lowest order improvement on these results. Third order offers more richness and we consider their effects separately. (We emphasize that since each calculation represents a separate solution of the parquet equations, the various changes in E/A are not strictly additive). Third-order modifications to the parquet diagrams, taken alone, lead to significant improvement in both E/A and the equilibrium density. The curve $\Delta V_{III\text{-}FULL}$ represents the results obtained upon adding the contributions from the self-energy and non-parquet diagrams. The binding energy of 6.73 K per particle is now in remarkable agreement with GFMC results. For the present purpose, these exact results, based on the same potential, may be regarded as better than data. The calculated equilibrium density of 0.0203 Å^{-3} is a considerable improvement over the initial parquet results.

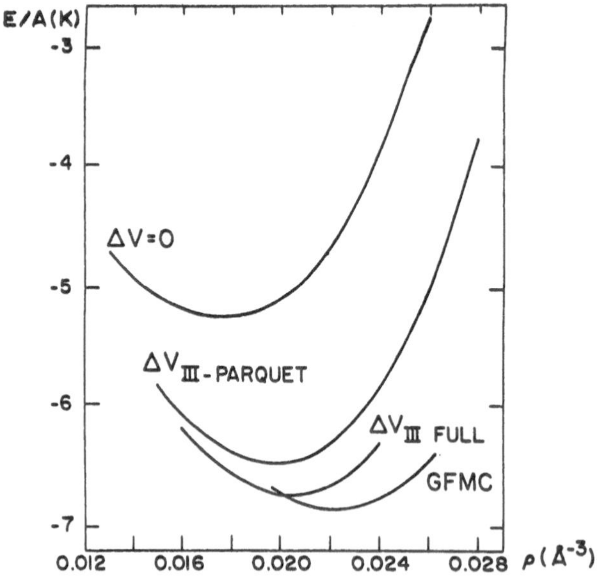

References

1. A.D. Jackson, A. Lande and R.A. Smith, Phys. Rep. 86 (1982) 55.
2. A. Lande and R.A. Smith, Phys. Lett. B, in press.
3. M.H. Kalos, D. Levesque and L. Verlet, Phys. Rev. A9 (1974) 2178.
4. A.D. Jackson, A. Lande and R.A. Smith, to be published.

CROSSING SYMMETRIC RINGS, LADDERS, AND EXCHANGES

R. A. Smith
Department of Physics
Texas A&M University
College Station, TX 77843

Alexander Lande
Instituut voor Theoretische Natuurkunde
Groningen University
Groningen, The Netherlands

I. INTRODUCTION

Diagrammatic techniques are one of the cornerstones of many-body theory. The Ursell-Mayer diagrams [1,2] for classical statistical mechanics, the Feynman-Goldstone[3,4] perturbation theory, the Fermion hypernetted-chain diagrams for central[5,6] and single-operator chain[7] state-dependent correlations, the correlated basis function (CBF) diagrams[8], and the exp(S) diagrams[9] all provide very useful ways of analyzing the behavior of many-particle systems. It is interesting to note strong similarities among these different diagrammatic schemes, particularly in regions where several methods are reasonably reliable.

We have described in some detail the relationship between the optimized hypernetted chain variational approach for bosons at zero-temperature and the sum of parquet diagrams for the two-body vertex in perturbation theory[10]. This comparison could be made relatively unambiguously because of the relatively simple structure of the boson hypernetted chain theory and the absence of a Fermi sea. It was shown that through fourth order, the optimized hypernetted chain energy is equivalent to that computed from a partial sum of a proper subset of the parquet diagrams; the summation is partial because not all parquet diagrams were generated with the correct numerical factor. This characteristic is expected to persist to all orders, although the difficulty of deriving the incorrect factors increases rapidly with the order in perturbation theory. At that time, the importance of using crossing symmetry to extend the technique to generate the exchange diagrams necessary for fermions and finite temperatures was clearly seen. More recently, this construction has been carried out explicitly[11]. I would like to discuss some aspects of the construction and look briefly at the classical limit.

II. Diagrammatic summations

We will be working in a constructive way with the two-body vertex, Γ. This is the two-body Green's function with the legs removed. It is often useful to draw it with legs attached to make it easy to locate the external points and to indicate the orientations which the external directed legs must have. The lines are Feynman one-body propagators appropriate for zero or finite temperature theory. On the first pass, these propagators may be thought of as bare propagators, but ultimately they will be promoted to the level of a bare propagator fully dressed by a self-consistently calculated self-energy. The bare interaction is denoted by a wiggly line. For a given labelling of the external coordinates, there are six possible orientations for the external legs. Half of these are shown in Fig. 1.

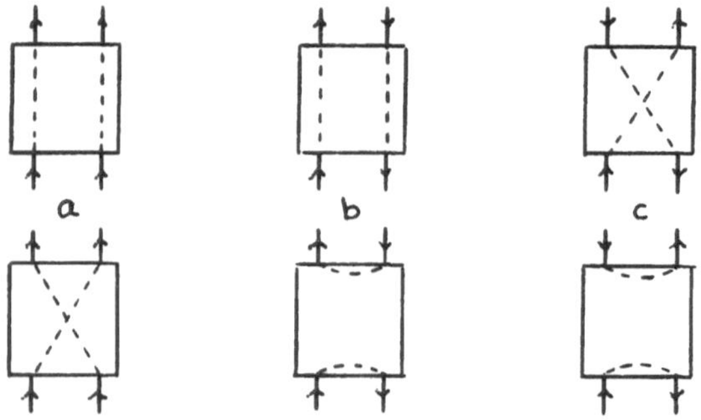

Figure 1. Orientations for the full vertex

The other half are obtained by simply reversing the directions of all lines. In addition, internal lines may connect the "in" and "out" points in two distinct ways. Examples are illustrated in Fig. 1.

Diagrams which may be broken into disconnected "top" and "bottom" parts by cutting two internal propagators are called s-channel reducible; these are the familiar ladder diagrams. Diagrams which may be broken into disconnected "left" and "right" parts are t-channel reducible ring diagrams. The other way in which a diagram can be divided into two disconnected parts by cutting two internal propagators gives the u-channel reducible diagrams. In each channel, the four cut directed ends can be reassembled in two distinct ways; these will be discussed as separate channels. The various channels are illustrated in Fig. 2.

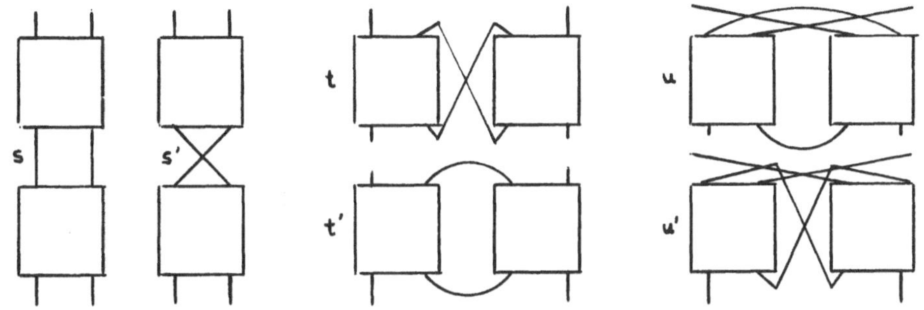

Figure 2. The s,s',t,t',u,u' channels

We denote by S, T, and U, the sum of diagrams which are reducible in the (s,s'), (t,t'), or (u,u') channels. All other diagrams are irreducible. The lowest-order irreducible diagram is the bare potential, which may have its external legs drawn in various ways. This set of diagrams is divided into direct and exchange diagrams, I and I^*.

The crossing-symmetric equations are schematically of the form [12]

$$\begin{aligned} \Gamma &= I \pm I^* + S + T + U \\ S &= (I \pm I^* + T + U) G_s \Gamma \\ T &= (I \pm I^* + S + U) G_t \Gamma \\ U &= (I \pm I^* + S + T) G_u \Gamma , \end{aligned} \tag{1}$$

where the G's denote a pair of single-particle propagators connecting the subdiagrams as appropriate for both the primed and unprimed channels; the G includes an overall factor of 1/4 to avoid overcounting of direct and exchange diagrams. The S, T, and U are (anti)symmetric in their respective channels; the Γ is completely crossing symmetric. The plus signs pertain to boson systems; the minus signs are for fermions.

By iterating these equations, it is possible to generate these diagrams to all orders. Upon calculating the second-order diagrams, one finds that the U diagrams calculated that way are just the s-channel exchange of the T diagrams. This feature persists to all orders. More generally, any of the S, T, and U sets can be generated as an exchange of another set in the remaining channel. An example of this is illustrated in Fig. 3.

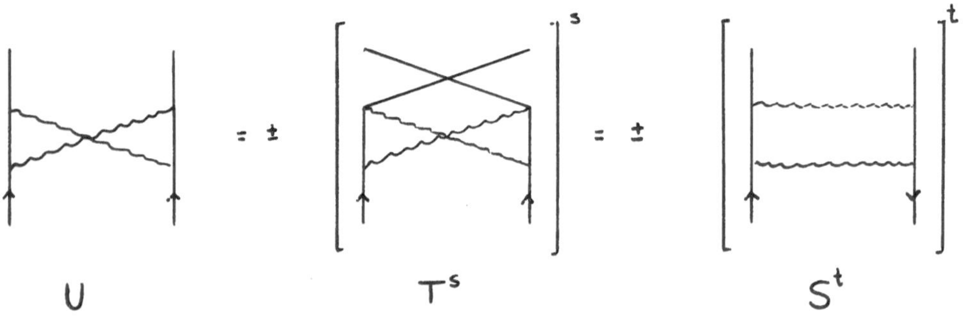

Figure 3. Diagrammatic rearrangements

These relationships may be used to eliminate U or even T and U from the crossing-symmetric equations of Eq.(1) in favor of exchanges of the remaining channel(s). The set of diagrams for Γ constructed with these equations in second and third order is shown in Fig. 4.

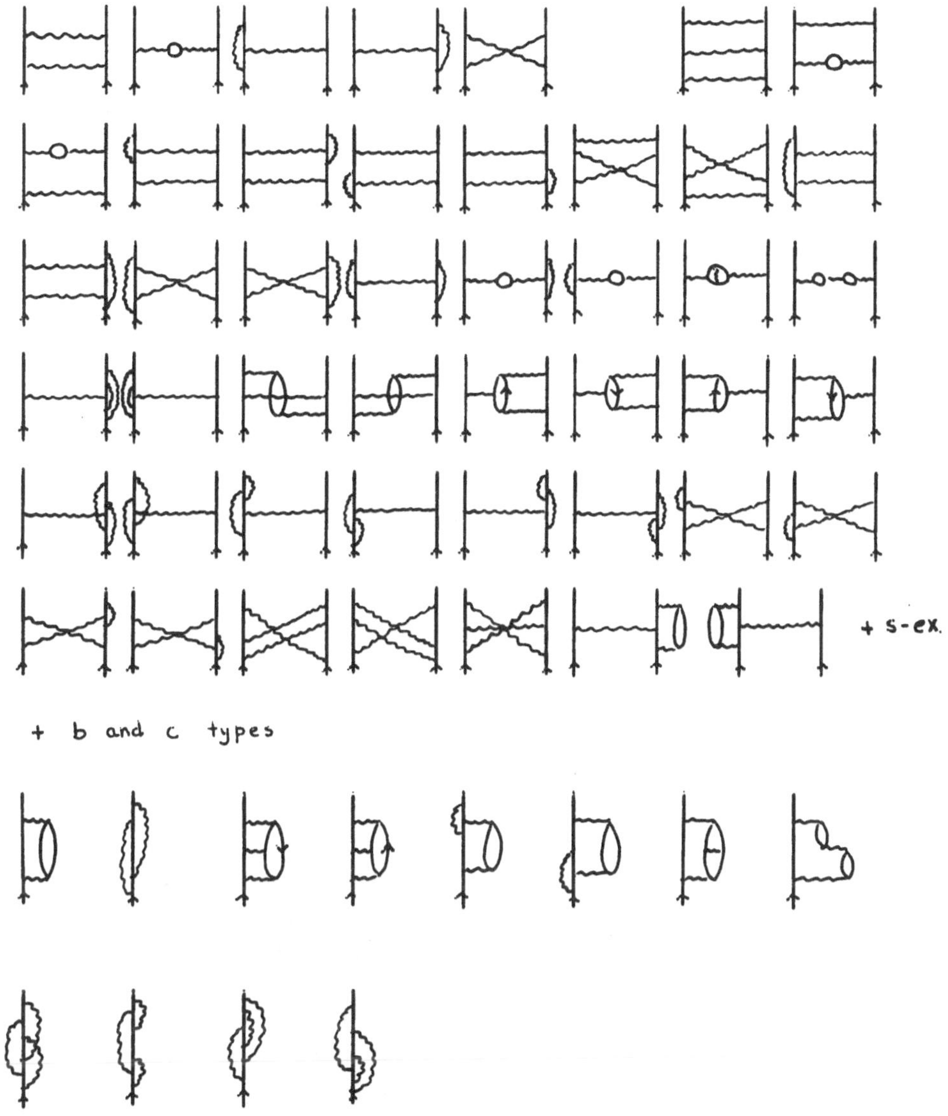

Figure 4. Second-order and third-order vertex and self-energy diagrams

The energy of the system at zero temperature or the thermodynamic potential at finite temperature may be computed by integrating the product of the vertex as a function of coupling constant with the potential over both coupling constant and the coordinates of the endpoints. This properly counts diagrams for an arbitrary initial choice of the set of irreducible diagrams I.

For the case when the irreducible diagrams are taken to be just the bare potential, the self-energy is computed by closing two of the four external points of the vertex with a propagator. To avoid double counting, one can simply use the S, I, and I^* diagrams with external legs oriented as shown in the first part of Fig.1. When a larger class of irreducible diagrams is considered, the situation is complicated, since a self-energy diagram can be generated by both reducible and irreducible diagrams. In other words, while the self-energy is computed from the vertex directly, functional differentiation of the self-energy would give back a larger set of vertex diagrams. On the other hand, as in the boson case, functional differentiation of the energy, starting with just the bare potential as I, gives the self-energy.

It is possible to generate larger classes of irreducible diagrams by substituting for each interaction line in a set of "basic" irreducible diagrams the full vertex. This is analogous to similar operations in the hypernetted-chain variational theories.

III. THE CLASSICAL LIMIT

In the classical limit, the structure of perturbation is greatly simplified. To begin, all exchange diagrams may be neglected. Then, each particle loop may be contracted to a point. The two-body vertex becomes, essentially, the pair distribution function. The topological structure of the S and T equations is then the same as that of the classical hypernetted-chain equations. However, the numerical factors associated with the diagrams are not those that would be obtained in the classical limit. As an example, consider the parquet diagrams illustrated in Fig. 5 which contribute to the indicated Ursell-Mayer diagram.

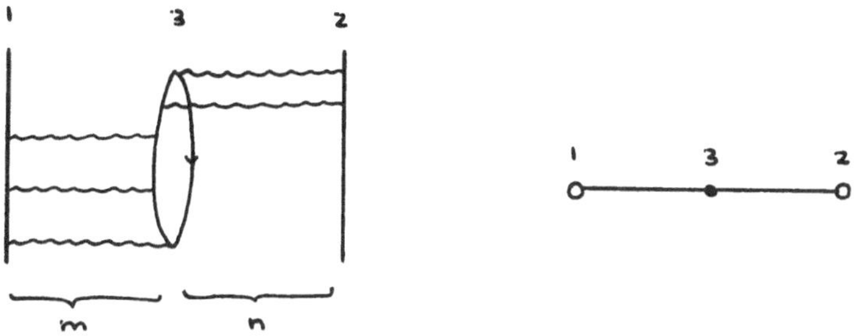

Figure 5. Taking the classical limit

If there are m lines connecting the central loop to the left and n lines connecting it to the right, then the total number of ways in which this can be done without interleaving the two sets is (m+n). With interleaving this number becomes (m+n)!/(m!n!). Summing over all m,n>1 with all interleavings allowed would give a contribution to g(r) of

$$\rho\int dx_3 \sum_{m,n=1}^{\infty} \frac{[-V(x_1 - x_3)]^m[-V(x_2 - x_3)]^n}{(m+n)!}\frac{(m+n)!}{m!n!}$$

$$= \int dx_3\ \rho\ [\exp(-V(x_{13}))-1][\exp(-V(x_{23}))-1]$$

the classical limit and the result of the classical hypernetted chain. The corresponding parquet limit, obtained without interleaving, is a sum which cannot be written as the product of two sums, each depending on just one interparticle separation.

$$\rho\int dx_3 \sum_{m,n=1}^{\infty} \frac{[-V(x_1 - x_3)]^m[-V(x_2 - x_3)]^n}{(m+n)!}\ (m+n)$$

The analysis of more complicated diagrams is similar but more tedious. Further study of this interesting case is in order. Empirically, the classical hypernetted chain works best for long-ranged forces, where the ring diagrams are particularly important. The contributions coming from diagrams in which either m or n equal to 1 are identical for the classical hypernetted chain and the parquet diagrams.

IV. DISCUSSION

In this talk, I have discussed some of the structure of the parquet equations which are suitable for fermion and boson systems at finite temperature. There are still a number of interesting things to be done. High on the priority list is finding an approximation scheme to reduce the complete equations to a set which can be handled conveniently numerically. At the simplest level, this will mean ignoring some of the non-local structure of the vertex completely and energies occuring in propagator denominators by energies averaged over the Fermi sea. The simple structure of the equations should make it clear, at least in principle, how to include corrections perturbatively. Calculations at both zero and finite temperatures would be very interesting. I hope that some efforts in these directions can be discussed at the next of these conferences.

This work was supported in part by the National Science Foundation under Grant PHY-8206325 and the research program of FOM supported by ZWO. R. A. S. appreciates the hospitality of the KVI where some of this work was carried out.

REFERENCES

1. H. D. Ursell, Proc. Royal Soc. London A106(1924)463.
2. J. E. Mayer and M. G. Mayer, Statistical Mechanics (John Wiley & Sons, 1940).
3. R. P. Feynman, Phys. Rev. 76(1949)749,769.
4. J. Goldstone, Proc. Royal Soc. London A239(1957)267.
5. S. Fantoni and S. Rosati, Nuovo Cim. 20A(1974)179.
6. E. Krotscheck, Phys. Lett. 54A(1975)123.
7. V. R. Pandharipande and R. B. Wiringa, Rev. Mod. Phys. 51(1979)821.
8. E. Krotscheck and J. W. Clark, Nucl. Phys. A328(1979)73.
9. F. Coester and H. Kümmel, Nucl. Phys. 17(1960)477.
10. A. D. Jackson, A. Lande and R. A. Smith, Phys. Rep. 86(1982)55.
11. A. Lande and R. A. Smith, Phys. Lett B, in press.
12. R. W. Haymaker and R. Blankenbecler, Phys. Rev. 171(1968)1581.
    I. T. Diatlov, V. V. Sudakov and K. A. Ter-Martirosian,JETP 5(1957)631.

NEW PERTURBATION SCHEME FOR QUANTUM FLUIDS BASED ON LOW-DENSITY EXPANSIONS

George A. Baker, Jr.
Theoretical Division, Los Alamos National Laboratory,
Los Alamos, NM 87545

and

Mauricio Fortes & Manuel de Llano
Instituto de Física, Universidad Nacional Autónoma de México
01000 México, D.F.

Abstract. We address the question of how close one can approach the ground energy of the Schrödinger equation for a many-particle system interacting via simple, but not necessarily unrealistic, two-body potentials. For this, existing low-density expansions are rearranged so as to correspond to a perturbation scheme not about the ideal gas but about a fluid with properly defined repulsions. The resulting density- and attractive-coupling- expansion is then treated with Padé and similar techniques to extend their validity to physical regions. Results for hard sphere fluids, homework and hard core square well fermions and bosons are presented.

1. INTRODUCTION

Computer simulations for both classical and quantum fluids have shown that a surprising similarity exists between the radial distribution function of, say, a Lennard-Jones fluid and that of a hard-sphere fluid, whereas both differ markedly from that of an ideal gas. This has revived an old suggestion by van der Waals of basing perturbative schemes for classical fluids[1)] on as good an equation of state as can be obtained for the corresponding hard- (or soft-) sphere fluid---instead of on the ideal gas system. What is perhaps the best attainable microscopic description of a classical liquid has apparently thus emerged.

Extensive investigations are being carried out for the case of quantum fluids, both fermion and boson, starting from the well-known low-density expansions. These are in general non-power series in the density and contain various low-energy scattering parameters which can be expanded in powers of the attractive well coupling strength. Modern extrapolation techniques (e.g., Padé and generalizations thereof) are then applied to the ensuing double series to predict the equation of state at moderate (i.e., physical) density and coupling. Here, we report results on i) hard sphere fluids for bosons and for 2- and 4-species fermions ii) V_0 (homework) neutrons and iii) hard core square well fermion and boson matter as compared with empirical ^{3}He and ^{4}He data.

2. PADE APPROXIMANTS

We consider here a very simple example. Suppose a function $f(x)$ is represented by a power series

$$f(x) = 1 + a_1 x + a_2 x^2 + O(x^3)$$

but only the first two (nontrivial) coefficients, a_1 and a_2, are known. The [1/1] Padé approximant to this truncated series is by definition

$$[1/1](x) \equiv \frac{1+p_1 x}{1+q_1 x} \underset{x \ll 1}{\simeq} (1+p_1 x)(1 - q_1 x + q_1^2 x^2 - \ldots)$$

$$\simeq 1 + (p_1 - q_1) + q_1(q_1 - p_1)x^2 + O(x^3)$$

We require the first two coefficients to be a_1 and a_2, respectively. Hence, our approximant is just

$$[1/1](x) = \frac{1 + (a_1 - a_2/a_1)x}{1 + (a_2/a_1)x},$$

and is now in a form which can have poles, in addition to zeros.

3. BOSONS

It is well known that in the Rayleigh-Schrödinger perturbation series (about an ideal boson gas) for the ground state energy of a fully-interacting Bose system divergences appear as of third order in the (say, pair-) interaction perturbation, no matter how well-behaved the interaction potential. Infinite partial summations, as well as completely different procedures, can be carried out to renormalize away these infinities and there finally results[2] the clearly low-density expansions for the ground state energy per particle

$$E/N \simeq \frac{2\pi\hbar^2}{m}\rho a\Big[1 + C_1(\rho a^3)^{1/2} + C_2\,\rho a^3 \ln \rho a^3 + + C_3\,\rho a^3 + O\{(\rho a^3)^{3/2}\ln \rho a^3\}\Big], \tag{1}$$

where a is the S-wave scattering length of the assumed two-body (central) potential acting between the N bosons of mass m, $\rho \equiv N/V$ with V the volume and $C_1 = 128/\pi \cdot 15$, $C_2 = 8(\frac{4\pi}{3} - \sqrt{3})$ while C_3 and higher coefficients are not only unknown but also potential shape-dependent, unlike C_1 and C_2. These higher-order coefficients have not been calculated in part due to lack of motivation since a) equation (1) has been thought to be good (if at all!) only at very low densities and, furthermore, b) as it stands, the expression leads to complex quantities for $a<0$, a very common fact with realistic two-body potentials.

The first difficulty might in principle be surmounted with the help of modern extrapolation techniques[3] and the second one by considering a very simple expansion, namely,

$$a \underset{\lambda \ll 1}{=} a_0 + a_1 \lambda + a_2 \lambda^2 + O(\lambda^3), \qquad (2)$$

where $\lambda \geqslant 0$, in the van der Waals spirit, is related to the attractive interaction strength. Substituting (2) into (1) leads to the perturbation expansion

$$E/N \simeq \sum_{i=0}^{\infty} \epsilon_i(x) \lambda^i; \qquad x \equiv \sqrt{\rho a_0^3} \qquad (3)$$

which is now clearly real even if λ is large enough to make a negative. The coefficients $\epsilon_i(x)$ $(i=0,1,2,\dots)$ are themselves low-density but non-analytic expressions in x.

If the pair-interaction is a pure hard sphere potential then $\lambda \equiv 0$ and $a = a_0 \equiv c$, the hard sphere diameter. For this system four Green Function Monte Carlo (GFMC) data points are available[4], for N = 256 bosons. Also, we expect on physical grounds that the true (non-relativistic, Schrödinger) ground state energy, will diverge at some finite value of x corresponding to the random close packing (RCP) density of hard spheres. This divergence, being an uncertainty principle phenomenon, must be a second order pole, and should occur at a density lower than the well-known[5] classical (Bernal) RCP density $\rho_B = 0.86\sqrt{2}/c^3$ since quantum effects make the sphere diameter appear larger. In fact, since the high-energy total scattering cross section for two hard spheres is $2\pi c^2$, instead of the classical (geometric) value πc^2 , we can estimate that the quantum Bernal density will be $0.86\sqrt{2}/(2^{1/2}c)^3 \simeq 0.304\sqrt{2}/c^3$. Here, $\sqrt{2}/c^3$ is the density value corresponding to face-centered-cubic (fcc) close packing, presumably the maximum possible density for hard spheres. A satisfactory approximant to the hard sphere energy per particle $\epsilon_0(x)$, i.e., eq. (1) with $a = c$, was found[6] to be

$$\epsilon_0(x) \hat{=} \frac{2\pi\hbar^2}{mc^2} \frac{x^2}{\left[1 - \frac{\frac{1}{2}C_1 x}{1 - \frac{2C_2}{C_1}x\left(\ln x + \frac{[C_3 - \frac{3}{4}C_1^2]}{2C_2}\right)}\right]^2} . \quad (4)$$

This form achieves an excellent global fit to the GFMC data and allows the extraction of a value of 26.2 for C_3 (hard spheres), which is to be compared with $C_1 \simeq 4.81$ and $C_2 \simeq 19.6$. The second-order pole predicted by (4) then occurs at $x = 0.7082$ or $\rho_B \simeq 0.35\sqrt{2}/c^3$, i.e., slightly above the simple estimate made above.

4. FERMIONS

The Rayleigh-Schrödinger perturbation series for the many-fermion ground state energy, on the other hand, has divergences only if the (again, say, pair-) potential is infinite--- as with hard cores or as in coulombic systems like the electron gas problem. Again, infinite partial summation can remove these divergences. For short-ranged interactions one is left with the low-density expansion for the ground state energy[7]

$$\begin{aligned} \frac{10mE}{3\hbar^2 k_F^2 N} \equiv \epsilon = & \; 1 + C_1 k_F a + C_2 (k_F a)^2 + \\ & + \left[\tfrac{1}{2} C_3 \frac{r_0}{a} + C_4 \frac{A_1(0)}{a^3} + C_5\right](k_F a)^3 + \\ & + C_6 (k_F a)^4 \ln|k_F a| + \left[\tfrac{1}{2} C_7 \frac{r_0}{a} + C_8 \frac{A_0''(0)}{a^3} + C_9\right](k_F a)^4 + \\ & + \ldots ; \qquad \rho \equiv N/V \equiv \nu k_F^3 / 6\pi^2 . \end{aligned} \quad (5)$$

Here $\hbar k_F$ is the Fermi momentum, ν is the number of intrinsic degrees of freedom of each fermion and the pure numbers $C_1, C_2, \ldots$ are ν-dependent. The two-body dynamics appear in the S-wave scattering length a and effective range r_0, the P-wave $A_1(0)$, and $A_0''(0)$. The first three parameters are shape-independent since they can be determined from the low-energy phase shifts alone through the well-known formulas

$$k \cot \delta_0(k) \underset{k\to 0}{\simeq} -\frac{1}{a} + \frac{1}{2} r_0 k^2 + O(k^4),$$

$$k^3 \cot \delta_1(k) \underset{k\to 0}{\simeq} -\frac{1}{A_1(0)} + O(k^2). \tag{6}$$

The fourth parameter $A_0''(0)$, to our knowledge, is shape-dependent and defined through the integral

$$A_0''(0) = -\frac{m}{3\hbar^2}\int_0^\infty dr\, r^4\, v(r)\, \psi_0(0,r) \tag{7}$$

with $v(r)$ the pair-potential and $\psi_0(0,r)$ the zero-scattering-energy wave function for $\ell=0$. The shape-independent parameters a, r_0 and $A_1(0)$ are also expressible as integrals, namely,

$$a = \frac{m}{\hbar^2}\int_0^\infty dr\, r^2\, v(r)\, \psi_0(0,r), \quad r_0 = \frac{2}{a^2}\int_0^\infty dr[(r-a)^2 - r^2\psi_0^2(0,r)],$$

$$A_1(0) = \frac{m}{3\hbar^2}\int_0^\infty dr\, r^3 v(r)\, \psi_1(0,r), \tag{8}$$

with $\psi_1(0,r)$ the $\ell=1$, zero-energy wave function. Equations (7) and (8) provide a very accurate means of determining[8)] numerical values for any $v(r)$ (e.g., Lennard-Jones, Aziz et. al, Kolos-Wolniewicz, etc. potentials), and is far superior to the indirect method of determining first $\delta_\ell(k)$ $(\ell=0,1)$ for small k and then extracting the constants by fitting equations (6). High accuracy is required if one is to have the first few derivatives of these four scattering parameters in the attractive well depth λ, required so as to substitute into (5) to get the double series in $x \equiv k_F a_0$ and λ analogous to (3).

For the hard core square well (HCSW) potential to be studied below, however, the four parameters (7) and (8) can be determined analytically, and expansion in powers of λ thus carried out explicitly. For a HCSW of hard core diameter c, depth V_0 and square well range R, defining the dimensionless parameters

$$\alpha \equiv \frac{R-c}{c}, \quad \lambda \equiv \frac{m v_0}{\hbar^2}(R-c)^2, \tag{9}$$

we have[9], for example, that

$$a = c\left[1+\alpha\left(1-\frac{\tan\sqrt{\lambda}}{\sqrt{\lambda}}\right)\right] \underset{\lambda\to 0}{\simeq} c\left[1+\alpha\left(-\frac{1}{3}\lambda-\frac{2}{15}\lambda^2-\dots\right)\right] \tag{10}$$

and similarly for r_0, $A_1(0)$ and $A_0''(0)$. If $\lambda \equiv 0$ one has the pure hard sphere problem for which $a = c$, $r_0 = \frac{2}{3}c$, $A_1(0) = \frac{1}{3}c^3$ and $A_0''(0) = -\frac{1}{3}c^3$. The energy expression (5) then becomes a fourth-order polynomial in $x \equiv k_F c$ for $\nu=2$ fermions (since in this case $C_6 \equiv 0$), and a third-order polynomial plus the log term for $\nu=4$ fermions (since in this case C_7 and higher-order coefficients are unknown).

All simple Padé approximants to the $\nu=2$ fourth-order polynomial of $\epsilon_0(x)^{-1/2}$ were constructed. A possible zero of these would denote, of course, the desired second-order (Bernal) divergence in the hard spheres energy. Figure 1 summarizes[10] our findings: clearly all extra-

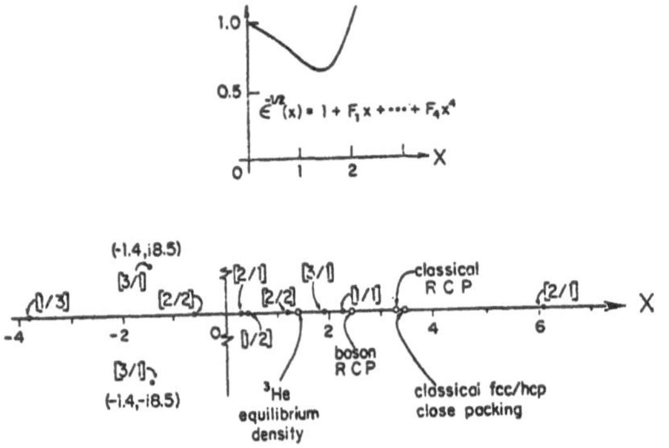

Fig. 1. Random close packing energy divergence for $\nu=2$ fermion hard spheres

polants except the [3/1](x) are to be discarded. Firstly, because we expect (since we now deal with fermions) that the Bernal density must lie below the corresponding boson value found above. Secondly, the [1/1] is eliminated since it represents correctly only the first two coefficients of the energy series, and is thus indistinguishable from the corresponding approximant to the ladder approximation which for the energy is known not to diverge for any finite k_F. Thirdly, all approximants falling below the empirical ^{3}He equilibrium density are dismissed for obvious reasons. The same holds, of course, for negative or complex χ-valued cases. The predicted Bernal density for $\nu=2$ fermion hard spheres is 0.174 $\sqrt{2}/c^3$. A similar though more laborious (because of the log term) search for $\nu=4$ fermions leads[10] to a Bernal density of 0.173 $\sqrt{2}/c^3$. Figure 2 summarizes the equation-of-state

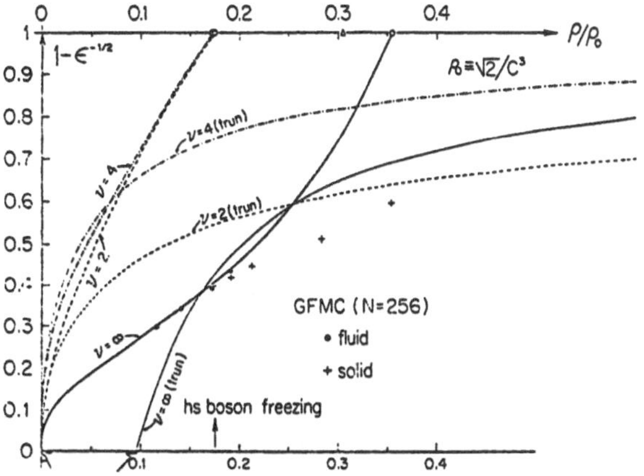

Fig. 2. Fermion and Boson equations of state for hard spheres

results for fermions (ν = 2 and 4) as well as bosons ($\nu = \infty$). We note that all the truncated expressions unphysically predict no divergence for finite densities. On the other hand, the various approximants predict, as would appear to be expected: i) a Bernal density above the simple quantum estimate of 0.304 $\sqrt{2}/c^3$ (triangle in figure) for bosons and ii) Bernal densities very close to each other for ν = 2 and 4 fermion hard spheres and below the quantum estimate. Needless to say, GFMC results for fermion hard sphere systems are certainly needed at this stage.

For homework neutrons (v_0 interaction) numerical integration of (7) and (8) was required and Figure 3 displays the results for $(E/N)^{-1/2}$

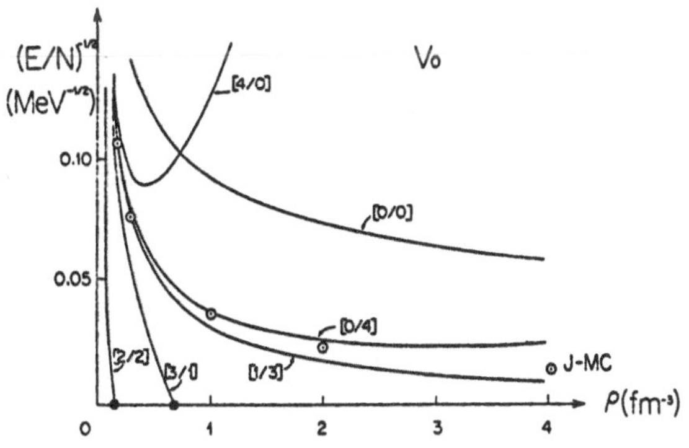

Fig. 3. $(E/N)^{-1/2}$ vs. density ρ for homework neutrons. J-MC points refer to Jastrow-Monte Carlo results

vs density ρ, where [0/0] denotes the ideal gas limit. The truncated expression [4/0] is obviously useless except at very low densities, if we accept the claim that the Jastrow-Monte Carlo (J-MC) energies[11] are about a percent higher (lower, on our graph) than the "true" values. Both the [2/2] and [3/1] approximants are now to be discarded since they predict a finite-density random close packing divergence which for soft repulsions can only occur at infinite density. Of the only remaining candidates [0/4] and [1/3], only the former is consistent with the lower density J-MC results, which are presumably upper bounds (lower, on our graph). However, the [0/4] approximant predicts a crystallization (at about 20 times nuclear matter density of 0.17fm^{-3}), contrary to the J-MC findings in ref.[11] (We find entirely analogous results for Lennard-Jones, r^{-12}, as well as Aziz interaction soft spheres.) This is because the coefficient of the x^4 is negative in both the original polynomial $(E/N)^{-1/2}$, and consequently also the analogous coefficient in the [0/4] Padé approximant. Hence, a zero in the denominator eventually develops, making E have a zero and thus $\partial P/\partial \rho$ also, where the pressure $P \equiv \rho^2 \frac{\partial E/N}{\partial \rho}$. The energies predicted by the [0/4] extrapolant are everywhere below the J-MC energies, as well as below the best variational calculations (labelled FHNC/4 with JF in ref.[12]) yet carried out on this system.

5. HELIUM LIQUIDS

We finally mention briefly some results[13] for systems with attractions. The predicted equation of state for liquid ^{3}He using the HCSW potential, as parametrized by Burkhardt[14], greatly overbound[15] (by a factor of four), as was also the case in the calculations of ref.[14]. We thus arbitrarily varied the HCSW parameters c, R and v_0 so that our best equilibrium energy value had zero error with respect to experiment, simply to observe the interrelations between different Padé's in λ and hence study the convergence of the present perturbative scheme. However, two restrictions were imposed on the potential parameters: that i) the empirical bound on the scattering length a_4 for ^{4}He (helium four) which is[16] $|a_4| > 20\,\text{Å}$ and ii) the conjecture $a_3 < 0$, both be obeyed. The results are shown in Figure 4, where both energy per particle and density are expressed in units of the empirical equilibrium values. The indicated Padé's are in λ; thus $[L/0](\lambda)$ is just straight L^{th} order perturbation theory about the hard sphere fluid $[0/0](\lambda)$, which in turn was represented in x by the $[3/1](x)$ approximant to $\epsilon_0^{-1/2}(x)$ as discussed above. Figure 5 is an enlargement; we observe fast convergence, particularly from 3rd to 4th order, but note that the $[1/1](\lambda)$ approximant clearly splits off from the rest.

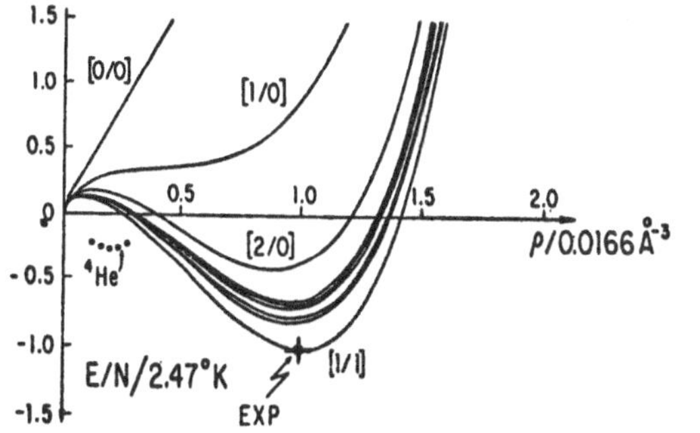

Fig. 4. Energy per particle vs. density (in units of the empirical values for liquid ^3He) for new HCSW parameters

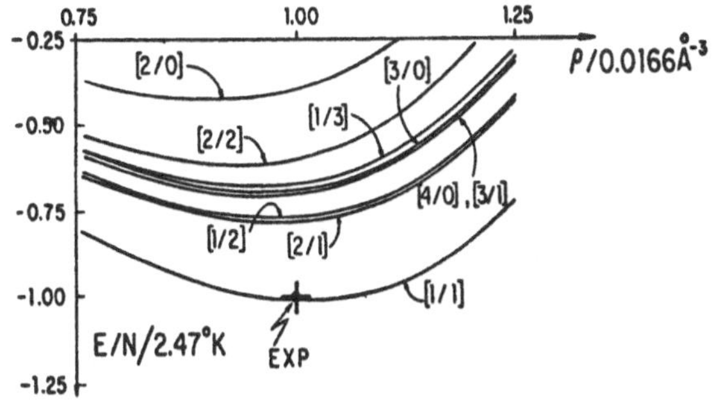

Fig. 5. Enlargement of Fig. 4

We now test for global consistency by employing the same HCSW just found for ^3He in the ^4He boson problem (3) using, for lack of a better choice, the value of the coefficient C_3 deduced above for pure hard spheres. The equilibrium minimum thus found is shown as the dotted curve in Fig. 4; the binding is just over 25%, and the equilibrium density about 35%, of the empirical ^4He values of $-7.14K$ and 0.022 Å^{-3}, respectively. The discrepancy is probably attributable to a strong HCSW shape-dependence in C_3, which is unknown in general for bosons. (Compare the coefficients of the ρ^2 term in the energy of (1) and of (5)).

6. CONCLUSIONS

On the basis of known low-density expansions it has been possible to construct, with the help of Padé approximants and generalizations thereof, a good equation of state for the boson hard spheres system, and probably also for fermion systems, for all physical densities up to and including the random close packing density. The values of these ultimate densities turn out to be relatively small ---0.174 ($\gamma = 2$), 0.173($\gamma = 4$) and 0.35 ($\gamma = \infty$)--- compared with the classical value of 0.86, not to speak of the regular (fcc) close packing value of 1. Since they are the values against which equilibrium densities of attractive systems are gauged our findings suggest that these liquids are not at so low a density after all. The extrapolants dealt with go beyond previous perturbative as well as variational schemes in at least one sense: by definition, they reproduce twice as many low-density coefficients.

The ultimate objective of investigating how close one can come of the Schrödinger ground state energy of the interacting many-particle system, starting only from the low-density expansions, will of course be within grasp once one is able to handle pair interactions (e.g., LJ and Aziz) for which GFMC results for both bosons and fermions may become available soon.

ACKNOWLEDGEMENTS

The authors appreciate discussions and collaborations with V.C. Aguilera-Navarro, L.P. Benofy, G. Gutiérrez, S.M. Peltier, A. Plastino, O. Rojo and W.C. Stwalley. One of us (M. de Ll.) acknowledges support from CONACyT and ININ (México), as well as a DAAD fellowship.

REFERENCES

1. J.A. Barker & D. Henderson, Rev. Mod. Phys. 48 (1976) 587.
2. N.M. Hugenholtz & D. Pines, Phys. Rev. 116 (1959) 489.
3. G.A. Baker, Jr. & P. Graves-Morris, Padé Approximants, in Encyl. of Math. and its Appl., ed. by G.-C. Rota (Addison-Wesley, 1981) vols. 13 & 14.
4. M.H. Kalos, D. Levesque & L. Verlet, Phys. Rev. A9 (1974) 2178.
5. G.D. Scott & D.M. Kilgour, J. Phys. D2 (1969) 863.
6. G.A. Baker, Jr., M. de Llano & J. Pineda, Phys. Rev. 24 (1981) 6304.
7. G.A. Baker, Jr., Rev. Mod. Phys. 43 (1971) 479.
8. G. Gutiérrez, M. de Llano & W.C. Stwalley, submitted to Phys. Rev. A, A.
9. G.A. Baker, Jr., L.P. Benofy, M. Fortes, M. de Llano, S.M.Peltier & A. Plastino, Phys. Rev. A 26 (1982) 3575.
10. G.A. Baker, Jr., G. Gutiérrez & M. de Llano, submitted to Ann. Phys. (N.Y.).
11. D. Ceperley, G.V. Chester & M.H. Kalos, Phys. Rev. B 16 (1977) 3081.
12. J.G. Zabolitzky, Phys. Rev. A 16 (1977) 1258.
13. G.A. Baker, Jr., M. Fortes & M. de Llano, to be published.
14. T.W. Burkhardt, Ann. Phys. (N.Y.) 47 (1968) 516.
15. G. Gutiérrez, unpublished.
16. J.P. Toennies & K. Winckelmann, J. Chem. Phys. 66 (1977) 3965.

A DIRECT ACCESS TO MANY-BODY PERTURBATION THEORY

W. Kutzelnigg
Lehrstuhl für Theoretische Chemie
Ruhr-Universität Bochum, Bochum FRG

1. Introduction

Brueckner's pioneering work[1] on many-body perturbation theory (MBPT) was strictly in the frame of standard RS theory for bound states. Since Brueckner had difficulties in proving the cancellation of 'unlinked clusters' in a simple and general way, and since Goldstone[2] was able to derive a completely linked expansion in an elegant manner, using concepts from quantum field theory, MBPT has become a branch of applied quantum field theory rather than a subdomain of the theory of stationary states. If one wants to use MBPT for bound state problems a few questions are in order:

1) Why should one use a very complicated (and not yet fully consolidated) theory such as quantum field theory as the basis for a much simpler and essentially better understood matter, namely for bound states?

2) Why use a time-dependent theory for a purely time-independent problem and why introduce a static perturbation as a tricky limit of an adiabatically switched perturbation, a limit in which the wave function does not exist?

3) Why is the wave operator of Goldstone's expansion obtained in intermediate normalization although it had been derived from the time-evolution operator which is certainly unitary (at least isometric)?

4) What is the basic physical reason for the linked diagram theorem?

5) Why is Goldstone's derivation, in spite of its shortcomings, superior to all other known approaches to MBPT. Is this due to

a) the time-dependent formulation?

b) the use of diagrams?

c) the particle-hole formalism?

d) the Fock space formulation (of the time-evolution operator)?

It will be a message of this paper that the points a) and b) are rather irrelevant, that c) is quite useful, but that d) is essential. We shall give a very simple derivation of the Goldstone expansion by taking advantage of the aspects c) and d) only. The formalism[3] that we use has been developed mainly for the quasidegenerate case and is more powerful than is apparent from this paper.

2. The Fock space Hamiltonian

We start by choosing a finite orthonormal one-electron basis $\{\psi_p\} = \{\varphi_{P\alpha}, \varphi_{P\beta}\}$ of dimension 2m. We define the matrix elements in this basis

$$h_q^p = \langle\psi_q|h|\psi_p\rangle \; ; \; v_{rs}^{pq} = \langle\psi_r(1)\psi_s(2)\,|\frac{1}{r_{12}}|\,\psi_p(1)\psi_q(2)\rangle \qquad (2.1)$$

and the excitation (transition) operators

$$a_p^q = a^q a_p = a_q^+ a_p \; ; \; a_{pq}^{rs} = a^r a^s a_q a_p \text{ etc.} \qquad (2.2)$$

as normal products of (the same number of) creation ($a^p = a_p^+$) and annihilation operators (a_p). The Hamiltonian [3]

$$H = h_q^p a_p^q + \frac{1}{2} v_{rs}^{pq} a_{pq}^{rs} \qquad (2.3)$$

with the Einstein summation convention and the summation indices going from 1 to 2m, is invariant under U(2m), the group of unitary transformations of the basis functions [4].

The eigenstates of H transform as irr. rep. of U(2m). They can be classified by Young diagrams, but only those allowed for fermions must be taken.

If H is spin-independent, it can be expressed via the spinfree excitation operators

$$E_Q^P = a_{Q\alpha}^{P\alpha} + a_{Q\beta}^{P\beta} \; ; \; E_{RS}^{PQ} = \sum_{\eta,\zeta=\alpha}^{\beta} a_{R\eta S\zeta}^{P\eta Q\zeta} \qquad (2.4)$$

and spinfree matrix elements as [3]

$$H = h_Q^P E_P^Q + \frac{1}{2} V_{RS}^{PQ} E_{PQ}^{RS} \qquad (2.5)$$

where the sums now go over (spinfree) orbitals (cap. letters) rather than spinorbitals (l.c. letters) as in (2.3). The eigenstates of H transform as irr. rep. of U(m), for which one can choose a Gelfand-Tsetline basis [4]. The irr. rep. that are allowed for electrons can be classified by the particle number n, and total spin quantum number S. They can be represented by Young diagrams with at most two columns. Our philosophy is to specify n and S as late as possible [3].

3. Diagonalization of the Fock space Hamiltonian

The usual way to get the eigenvalues and eigenfunctions of H as given by (2.3) or (2.5) consists in projecting H first to an n-particle Hamiltonian H_n, in a matrix representation in terms of the $\binom{2m}{n}$ possible Slater determinants and to diagonalize the $\underline{\underline{H}}_n$ matrix. In quantum chemistry this procedure (by which one 'exhausts' the one electron basis) is called 'full CI' (This can be improved successively by enlarging the basis, at least so in principle; in practice one reaches soon astronomic dimensions of the secular equations).

Since H_n is more complicated than H the question arises: 'Why not diagonalize H directly before one specifies n?' We shall hence search for a non-singular Fock space operator W which transforms H to a diagonal operator L [3]

$$W^{-1}HW = L; \; L = L_D; \; HW = WL \tag{3.1}$$

where $L = L_D$ means that the operator L is equal to its 'diagonal part'. There are various possibilities to define the diagonal part B_D of a Fock space operator B.

One possibility, that is the generalization [3] of a definition originally given by Primas [5] is as follows. Let

$$B = B_V + B^P_Q \, E^Q_P + \frac{1}{2} \, B^{PQ}_{RS} \, E^{RS}_{PQ} + \ldots \tag{3.2}$$

and let a one-particle reference (unperturbed) Hamiltonian H_o be given

$$H_o = e_P \, E^P_P \tag{3.3}$$

which is diagonal in the given basis (or the eigenstates of which are chosen as the φ_R). Then we define

$$B_D = B_V + \delta(e_P, e_Q) \, B^P_Q \, E^Q_P + \frac{1}{2} \, \delta(e_P + e_Q, \, e_R + e_S) \, B^{PQ}_{RS} E^{RS}_{PQ} + \ldots \tag{3.4}$$

while the non-diagonal part B_N of B is

$$B_N = B - B_D \tag{3.5}$$

Another definition appropriate for the theory of effective Hamiltonians in Fock space is based on the division of the one-particle Hilbert space into two subspaces, of 'active' and 'inactive orbitals' respectively [6].

Active orbitals will be labeled as $\varphi_X, \varphi_Y, \varphi_Z$, inactive orbitals as $\varphi_A, \varphi_B, \varphi_C \ldots$.

We divide accordingly basis operators into four categories.

C (closed):	upper and lower indices active, e.g. E^{XY}_{ZU}
A (closed from above):	upper indices active, at least one of the lower indices inactive, e.g. A^{XY}_{AZ}
B (closed from below):	lower indices active, at least one of the upper indices inactive, e.g. E^{AB}_{XY}
O (open):	at least one lower and one upper index inactive, e.g. E^{YB}_{AX} .

We further define

$$B_D = B_C + B_O \; ; \; B_N = B_A + B_B \tag{3.6}$$

or alternatively (if we use the intermediate normalization)

$$B_D = B_C + B_O + B_A \; ; \; B_N = B_B \qquad (3.7)$$

Eqn. (3.1) can with

$$H = H_o + \lambda V \; ; \; \Delta L = L - H_o \qquad (3.8)$$

be rewritten as

$$\left[H_o, W\right] = W\Delta L - \lambda V W \qquad (3.9)$$

This equation can be solved for W either iteratively [3,6] or by perturbation theory. In either case must one be able to 'invert a commutator'. Let

$$\left[H_o, B\right] = C \qquad (3.10)$$

with B and C expanded as (3.2). One gets explicitly

$$(e_P - e_Q)\, B^P_Q = C^P_Q ; \; B^P_Q = (e_P - e_Q)^{-1}\, C^P_Q$$

$$(e_P + e_Q - e_R - e_S)\, B^{PQ}_{RS} = C^{PQ}_{RS} \; ; \; B^{PQ}_{RS} = (e_P + e_Q - e_R - e_S)^{-1}\, C^{PQ}_{RS} \qquad (3.11)$$

A solution B of (3.10) exists if the diagonal part C_D of C vanishes i.e. if $C = C_N$ (only such definitions of the diagonal part of an operator are acceptable for which this holds). The solution B of (3.10) is not unique, since one can always add to the B given by (3.11) an operator D which commutes with H_o.

We introduce a shorthand notation for the solution B of (3.10)

$$B = - C_H = B_N \qquad (3.12)$$

3. Perturbation theory

One tries to solve (3.9) by a power series expansion in λ of W and L with $W^{(o)} = 1$, $L^{(o)} = H_o$

$$\left[H_o, W^{(k)}\right] = \sum_{l=1}^{k} W^{(k-l)}\, L^{(l)} - V\, W^{(k-1)} \qquad (4.1)$$

The solution is not unique and one must impose a 'normalization condition' for W. We only consider three of many more possibilities [3,6]

(a) 'intermediate' $W_D = 1$

(b) canonical unitary $W^+W = WW^+ = 1$; $W^+_D = W_D$

(c) separabel unitary $W = e^{\sigma}$, $\sigma = -\sigma^+$, $\sigma_D = 0$ (4.2)

The intermediate normalization (a) leads to the simplest expression, but L is not hermitean; variant (b) leads to 3rd order just to hermitized terms. Variant (c) is the only one that guarantees inconditionally[3] a connected diagram expansion of both L and $\sigma = \ln W$. We only indicate the lowest order in variant (a). More detailed formulae can be found elsewhere [3,6]

$$L^{(1)} = V_D \; ; \quad W^{(1)} = V_H ; \quad L^{(2)} = (VV_H)_D$$

$$L^{(3)} = \{V(VV_H)_H\}_D - \{V(V_H V_D)_H\}_D \tag{4.3}$$

The first term in $L^{(3)}$ is referred to as a 'direct term', the second as 'renormalization term'.

5. Generalized Wick theorem and diagram notation

The expressions (4.3) are products of Fock space operators, i.e. not yet in normal product form. A generalization [3] of Wick's theorem allows us easily to express products of basis operators as sums of basis operators. The following examples illustrate the recipe [5]

$$E^P_Q E^R_S = E^{PR}_{QS} + \delta^R_Q E^P_S \tag{5.1a}$$

$$E^{PQ}_{RS} E^T_U = E^{PQT}_{RSU} + \delta^T_R E^{PQ}_{US} + \delta^T_S E^{PQ}_{RU} \tag{5.1b}$$

A product of operators is the sum of the normal product plus all possible contractions. Indices contract from lower left to upper right. Unlike in the original Wick theorem there is no sign rule, but the pairings of upper and lower indices must be kept.

MBPT is hardly imaginable without diagrams. We use vertices as matrix elements with ingoing and outgoing lines as in Goldstone's diagrams, but our diagrams [3] are strictly time-independent. Energy denominators are not the result of a time integration, but must be marked explicitly. We do this by enclosing the vertex or the part of the diagram in a rectangular box. For the 2nd definition of the diagonal part (sec. 3) we mark active lines by double arrows and inactive ones by single arrows. The diagrams that contribute to $L^{(1)}$, $W^{(1)}$ and $L^{(2)}$ as given by (4.3) are shown on fig. 1. More diagrams can be found elsewhere [3,6]

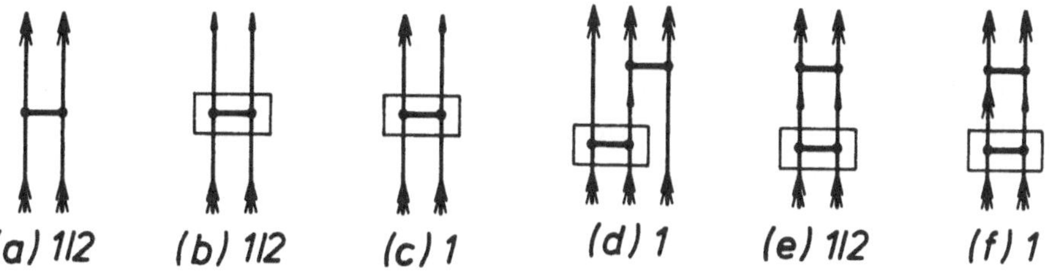

fig. 1 Diagrams to $L^{(1)}$:a, $W^{(1)}$:b,c and $L^{(2)}$:d,e,f

Our expansion of L contains both 'direct diagrams' with successive energy denominators and renormalization terms, where the order of denominators disagrees with that of the vertices. Such diagrams are not present in the traditional MBPT for quasidegenerate states [7], but instead so called folded diagrams appear. The relation between these two types of diagrams is shown for a special case on fig. 2

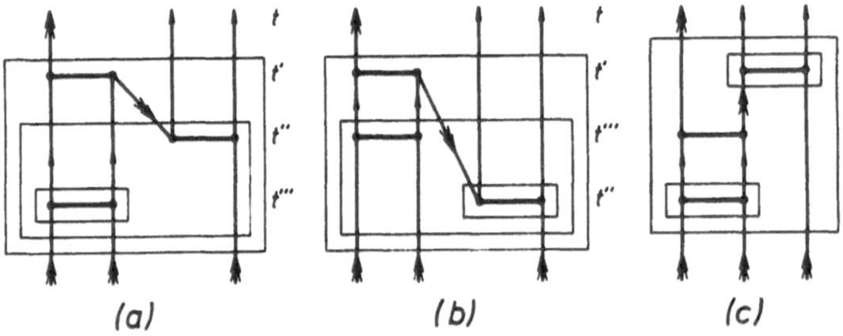

fig. 2 Illustration of the equivalence of the sum of two 'folded diagrams' (a) and (b) to a renormalization diagram (c)

6. Particle hole formalism

Traditionally one defines a closed-shell Slater determinant Φ as 'physical vacuum'. Let the spin-orbitals ψ_i, ψ_j be occupied in Φ, the ψ_a, ψ_b unoccupied. Then one introduces quasiparticle creation and annihilation operators b_p^+ and b_p as

$$b_i^+ = a_i;\ b_i = a_i^+;\ b_a = a_a;\ b_a^+ = a_a^+ \qquad (6.1)$$

To choose $b^i = a_i$ would be inconsistent with the transformation properties under U(2m). Also spin-pairing would get lost. One most abandon the b-operators and only remember that in a 'normal product in particle hole sense' the a^a and a_i must be left of the a_a and a^i. We define $\tilde{a}^{pq}_{rs}$ as a normal product operator in the particle hole sense with a factor +1 or -1 depending on the parity of the permutation that leads from one normal product to the other. So one gets e.g.

$$\tilde{a}^{ia}_{jb} = a^{ia}_{jb} - \delta^i_j\, a^a_b \qquad (6.2)$$

or after summation over spin

$$\tilde{E}^{IA}_{JB} = E^{IA}_{JB} - \delta^{I}_{J} E^{A}_{B} \tag{6.3}$$

The generalized Wick theorem for the $\tilde{E}$ operators is e.g. [6]

$$\tilde{E}^{A}_{B}\tilde{E}^{C}_{D} = \tilde{E}^{AC}_{BD} + \delta^{C}_{B}\tilde{E}^{A}_{D}$$

$$\tilde{E}^{I}_{J}\tilde{E}^{K}_{L} = \tilde{E}^{IK}_{JL} - \delta^{I}_{L}\tilde{E}^{K}_{J} \tag{6.4}$$

Hole lines contract from upper left to lower right. Every hole line contraction introduces a factor -1, every closed loop an additional factor -2.

7. Separability and connected diagram theorem

Assume that the one-particle Hilbert space consists of two non-interacting subspaces, such that matrix elements h^{P}_{Q} or V^{PQ}_{RS} vanish unless all labels refer to the same subspace. Let

$$H = H_1 + H_2;\ H_1W_1 = W_1L_1;\ H_2W_2 = W_2L_2 \tag{7.1}$$

where the subscripts 1 and 2 refer to the subspaces. Then since operators in different subspaces commute, it follows from (7.1) that [3]

$$HW = WL,\ W = W_1W_2,\ L = L_1 + L_2 \tag{7.2}$$

Both H and L are 'additively separable' while W is 'multiplicatively separable'. If one writes W as e^{σ}, σ is additively separable.

For additively separable quantities a connected diagram theorem holds (provided that the normalization condition for W is separable as well) [3].

We discuss first a related 'joint-cluster theorem' [8]. We call a term disjoint if it can be written as a product such that there are no common labels between the two factors. An additively separable operator cannot contain disjoint terms. One can namely divide the one-particle space into two subspaces such that one factor of the disjoint term belongs to one subspace, the other to the other subspace. If one switches off the interaction between the two subspaces, the disjoint term does not disappear, but it must be absent for non-interacting subspaces, so it must be absent even for interacting subspaces.

While the notions 'joint' and 'disjoint' are defined in terms of orbital labels, the terms 'connected' and disconnected' refer to the topology of diagrams and are independent of labels. What Brueckner conjectured was, in our nomenclature, rather a 'joint cluster theorem'.

The argument just given does not exclude the presence of disconnected but joint diagrams in additively separable operators. Such diagrams are only excluded if one willingly pays no attention to orbital labels, i.e. if one sums over all orbital labels independently with no special treatment for coinciding labels (in a current but somewhat misleading language one must 'ignore the exclusion principle').

There is an alternative more formal proof of the connected-diagram theorem via the Lie-algebraic structure of the theory in terms of $W = e^{\partial}$. (A criticism of this proof by Brandow [9] is due to a misunderstanding. There must be a Lie-algebraic structure in Fock space, not in n-particle Hilbert space [3,6]).

8. A simple derivation of Goldstone's expansion

Goldstone's linked cluster expansion can simply be derived in the following way. We choose a closed-shell Slater determinant as physical vacuum and define particle $(\psi_i, \psi_j, \ldots)$ and hole states $(\psi_a, \psi_b \ldots)$ in terms of it, but make no further distinction between active and inactive states. We define operators as (a special case of the definition in sec. 3)

B_C (closed) if there are no external lines

B_B (closed from below): external lines only above the diagram

B_A (closed from above): external lines only below the diagram

B_O (open): external lines both above and below

We further choose (3.7) and solve

$$HW = WL; \; L = L_C = E \tag{8.1}$$

i.e. we are only interested in the closed part (the 'vacuum expectation value) of L. This is achieved by a W that only consists of W_B and W_C. We choose the intermediate normalization such that

$$W = 1 + W_B \tag{8.2}$$

and get for the perturbation expansion of W and L

$$W^{(1)} = V_{BH}; \; L_C^{(1)} = V_C; \; W^{(2)} = (VV_{BH})_{BH} - (V_H V_C)_{BH}$$

$$L_C^{(2)} = (VV_{BH})_C; \; L_C^{(3)} = (V(VV_{BH})_{BH})_C - (V(V_{BH}V_C))_C \tag{8.3}$$

etc.

The expansion for L contains both direct and renormalization terms. However one sees easily that the renormalization terms as the 2nd term of $L_C^{(3)}$ are necessarily disconnected (see fig. 3). They can hence not contribute. In using the connected diagram theorem and limiting the expansion of L to connected terms, one is immediately lead to an expansion of L as the sum of all connected <u>direct</u> terms, i.e. Goldstone's linked-cluster expansion. One sees that for this derivation only the consequent Fock space formulation and the particle-hole formalism are necessary.

An alternative, but slightly more complicated derivation of the Goldstone expansion can be based on the $W = e^S$-ansatz for the intermediately normalized W of Coester and Kümmel [10].

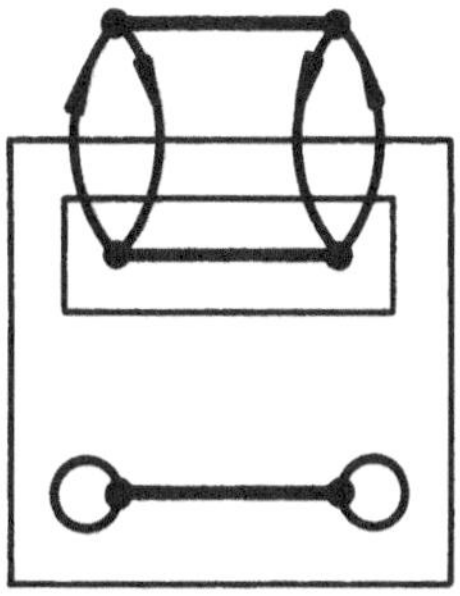

fig. 3 Illustration that renormalization terms to the vacuum energy such as the 2nd term to $L_C^{(3)}$ are disconnected since they contain a closed factor.

References

1 K.A. Brueckner, Phys. Rev. 97, 1353 (1955), 100, 76 (1955)

2 J. Goldstone, Proc. Roy. Soc. A 239, 267 (1957)

3 W. Kutzelnigg, Chem. Phys. Letters 83, 156 (1981)
W. Kutzelnigg, J. Chem. Phys. 77, 3081 (1982)

4 see e.g. 'The unitary group, Lecture Notes in chemistry 22', edited by J. Hinze (Springer, Berlin 1981)

5 H. Primas, Helv. Phys. Acta 34, 331 (1961)

6 W. Kutzelnigg, J. Chem. Phys. accepted

7 C. Bloch, Nucl. Phys. 6, 329 (1958)
T. Morita, Progr. Theor. Phys. 29, 351 (1963)
B.H. Brandow, Rev. Mod. Phys. 39, 771 (1967)
P.G.H. Sandars, Adv. Chem. Phys. 14, 365 (1969)
I. Lindgren, J. Phys. B 7, 2441 (1974)

8 W. Kutzelnigg, unpublished

9 B.H. Brandow in 'Effective Interactions and Operators in Nuclei', edited by B.R. Barret (Springer, Berlin 1977)

10 F. Coester and H. Kümmel, Nucl. Phys. 17, 477 (1960)

BEYOND THE THOMAS-FERMI-WEIZSÄCKER-DIRAC THEORY OF ELECTRONIC STRUCTURE

Eduardo V. Ludeña
Centro de Química
Instituto Venezolano de Investigaciones Científicas, I.V.I.C.
Apartado 1827, Caracas 1010-A, Venezuela

ABSTRACT

A universal energy functional which depends only on the charge density operator ρ and which approximates the Hartree-Fock energy expression, is advanced. This functional, whose construction is based on the use of Harriman's orbitals, goes beyond the usual plane-wave approximations. Closed analytic expressions for the kinetic and exchange energies, are presented

I. INTRODUCTION

Ever since the beginnings of quantum mechanics, the quest for a formulation which would be based on the particle density instead of on the more abstract and quite often less attainable wavefunction, has provided the motivation for the development of what is now known as density functional theory.

Some of the ground work for this theory was laid by the early and important efforts of Thomas[(1)] and Fermi[(2)] in their statistical description of atoms, as well as by Dirac[(3)], in his local treatment of the exchange interaction in many-electron systems.

The Thomas-Fermi-Dirac approximation was based on a plane-wave description of a many particle system and lacked, therefore, elements for an adequate treatment of non-constant densities. An attempt to remedy this shortcoming was made by Weizsacker [(4)], who introduced kinetic energy terms which depended explicitly upon the gradient of the density. A very penetrating analysis of the mathematical subtleties of these approximations has been recently given by Lieb[(5)].

An important milestone in the development of density functional theory was provided by the work of Hohenberg and Kohn[(6)] and of Kohn and Sham[(7)], who showed that the energy is a unique functional of the density and also that for a trial density, the ground state energy is an upper bound to the exact energy of the system.

The implicit expression for the energy as a functional of ρ is

$$E[\rho] = T[\rho] + V_{ne}[\rho] + J[\rho] + V_{XC}[\rho], \quad (1)$$

where $T[\rho]$ stands for the kinetic energy functional; $V_{ne}[\rho]$, for the nuclei-electron interaction:

$$V_{ne}[\rho] = -\sum_{\alpha} Z\alpha \int d1\, \rho(1)/r_{1\alpha} \;; \quad (2)$$

$J[\rho]$, for the Coulomb interaction:

$$J[\rho] = (1/2)\int d1 \int d2\, \rho(1)\rho(2)/r_{12} \;; \quad (3)$$

and finally, where $V_{XC}[\rho]$ stands for the exchange correlation functional.

There are two important points in relation to $E[\rho]$. The first is that the nuclei intervene only through $V_{ne}[\rho]$ and since it is their interaction that particularizes any given n-electron system, it follows that the sum of $T[\rho]$, $J[\rho]$ and $V_{XC}[\rho]$ should give a universal functional of the charge density. Clearly, this functional must depend on n, the number of particles, as well as on their spins. The second point is that although the exact expression for this universal functional remains unknown, several approximations have been advanced. In particular, in the present paper, we advance a novel expression for this universal functional, calculated within a framework (of Harriman's functions) which goes beyond the traditional planewave approximation.

The usual representation of the kinetic energy functional is given by the gradient expansion[(7-9)]

$$T[\rho] = T_0[\rho] + T_2[\rho] + T_4[\rho] + \ldots \quad (4)$$

where

$$T_0[\rho] = (3/10)(3\pi^2)^{2/3} \int d1\, \rho(1)^{5/3} \quad (5)$$

is the Thomas-Fermi term. $T_2[\rho]$ is $(1/9)T_w[\rho]$ where the Weizsacker term is

$$T_W[\rho] = (1/8) \int d1\, (\nabla\rho\,(1)\, \nabla\rho\,(1) \,/\, \rho\,(1)\,)\;. \quad (6)$$

$T_4[\rho]$ is the well known Hodge's correction[(9)]. The basic difficulty with Eq(4) is that it does not lead to a satisfactory functional. In fact, $T_6[\rho]$ diverges, and also $\delta T|\rho|/\delta\rho$ is incorrect.

The corresponding gradient expansion for the exchange-correlation energy term $V_{XC}[\rho]$ is also besieged with difficulties.

From general considerations on the virial theorem and the long range and cusp behavior of the various terms appearing in the kinetic energy and potential energy gradient expansions, Tal and Bader[(10)] were able to show that the only terms which remained in the gradient expansion were $T_o[\rho]$ and $T_W[\rho]$ for the kinetic energy, and the Dirac correction for the potential energy. This result clearly indicated the inadequacy of the gradient expansion and emphasized the need for a new atrategy for getting increasingly accurate approximations to the universal energy functional $E[\rho]$.

A reformulation of this problem was provided in the work of Sears, Parr and Dinur[11], who, starting from considerations based on information theory, were able to show that the Weizsacker term is the local contribution to the kinetic energy. A similar result, based on simple quantum mechanical considerations was also obtained by Gázquez and Ludeña[12]. As a consequence of these developments, the status assigned to the Weizsacker term, as a correction to the Thomas-Fermi term, was altered. Since $T_W[\rho]$ is the local term, the question emerges about the nature of the corrections to the Wizsacker term. We have dealt with this question in a previous paper[13] where it has been shown that these corrections arise from non-local components of the reduced first order density operator γ^1 (1,2). Also, we have shown recently[14] that the Thomas-Fermi term is the simplest correction to the Weizsacker term. It is, in fact, obtained when the non-local part of γ^1 (1,2) is expanded in plane-waves and when the summation over the plane-wave momenta is changed into an integral. Similarly, it has been shown[14] that in this approximation Dirac's local exchange correction also ensues from this non-local part of γ^1 (1,2).

It is the purpose of the present work to construct an approximation that goes beyond the Thomas-Fermi-Weizsacker-Dirac theory. We review in Section II the representation of γ^1 (1,2) in terms of its local and non-local components and discuss the relationship of the latter with the statistical correlation factor. In Section III we use Harriman's orbitals in order to get an approximate closed analytic expression for γ^1 (1,2). Finally, in Section IV, we obtain an approximate expression for the total energy as a functional of ρ and discuss its possible applications.

II. Statistical Correlation and the Reduced First Order Density Operator γ^1 (1,2)

In terms of an orthonormal set of single particle functions $\{\phi_1;\ i \in \{1 \ldots m\},\ m \geq n\}$ where n is the number of electrons, γ^1 (1,2) becomes

$$\gamma^1 (1,2) = \phi^+ (1)\ \Gamma^1\ \phi (2) \tag{7}$$

In this expression, ϕ (2) is the column vector

$$\phi (2) = \begin{bmatrix} \phi_1 (2) \\ \vdots \\ \phi_m (2) \end{bmatrix} \tag{8}$$

ϕ^+ is its conjugate transpose and Γ^1 is the reduced first order density matrix. The charge density operator is obtained from Eq(7) when both coordinates are the same:

$$\rho (1) = \gamma^1 (1,1) \tag{9}$$

The density operator ρ (1) is, according to the statistical interpretation[15], the average number of particle density at point (1). The conditional average number of particle density at point (2), different from point (1), when at point (1) there is exactly one particle, is

$$\rho_C (;1) = 2\,\gamma^2 (1,2;\ 1,2)\ /\ \rho(1) \tag{10}$$

where γ^2 (1,2 ; 1,2) is the reduced second order density operator. The statistical correlation factor is given by

$$f(1,2) = \rho_C(2;1)/\ \rho(2) = 2\,\gamma^2\ (1,2;1,2)/\ (\ \rho(1)\ \rho(2)\ ) \tag{11}$$

The usual definition of correlation, however, is not the statistical one discussed above; rather, by correlation it is understood all the corrections that go beyond the restricted Hartree-Fock approximation. In order to establish the connection between these two different types of correlations, it is convenient to express γ^2 (1,2;1,2) as follows

$$\gamma^2\ (1,2;1,2) = \gamma^2_{HF}(1,2;1,2)(\ 1 + f_{CORR}(1,2)\ ) \tag{12}$$

where the subscripts HF and CORR stand for Hartree-Fock and correlation, respectively. Clearly, $f_{CORR}(1,2)$ is the factor that includes the usual type of correlation which is missing in $\gamma^2_{HF}(1,2;1,2)$. It is well known, however, that

$$\gamma^2_{HF}(1,2;1,2) = (1/2)(\rho_{HF}(1)\,\rho_{HF}(2) - \gamma^1_{HF}(1,2)\ \gamma^1_{HF}(2,1)\ ) \tag{13}$$

Introducing the correlation factor $f_F(1,2)$ in order to describe the Fermi type correlation among electrons, we may rewrite

$$\gamma^2_{HF}(1,2;1,2) = (1/2)\,\rho_{HF}(1)\ \rho_{HF}(2)(1 + f_F(1,2)\ ) \tag{14}$$

where the Fermi or exchange correlation factor is defined as

$$f_F(1,2) = -\gamma^1_{HF}(1,2)\ \gamma^1_{HF}(2,1)/(\rho_{HF}(1)\ \rho_{HF}(2)\ ) \tag{15}$$

We may now ask what the average number of particles density at point (2) is, when at point (1) there is exactly one particle <u>described by the Hartree-Fock density</u> $\rho_{HF}(1)$. By analogy with Eq(10) we have

$$\rho_C(2;1/HF) = 2\gamma^2(1,2;1,2)/\,\rho_{HF}(1) \tag{16}$$

Similarly, the statistical correlation factor with respect to Hartree-Fock densities is

$$f(1,2\ /HF) = \rho_C(2;1/HF)/\ \rho_{HF}(2) = 2\gamma^2(1,2;1,2)/(\rho_{HF}(1)\,\rho_{HF}(2)\ ) \tag{17}$$

The relationship between this new statistical correlation factor and the usual correlation factors $f_F(1,2)$ and $f_{CORR}(1,2)$ may now be easily obtained using Eqs(12-17):

$$f(1,2/HF) = (1 + f_F\ (1,2))(1 + f_{CORR}(1,2)\ )\ . \tag{18}$$

In a previous work[13] we have discussed the representation of $\gamma^1(1,2)$ in terms of its local and non-local components. Restricting ourselves to the HF part of $\gamma^1(1,2)$, we may write

$$\gamma^1_{HF}(1,2) = (\rho_{HF}(1))^{1/2}\ (\rho_{HF}(2))^{1/2}\ G_{HF}(1,2) \tag{19}$$

where the non-local component is

$$G_{HF}(1,2) = (\phi^+\ (1)\Gamma^1_{HF}\ \phi(2)\phi^+(1)\Gamma^1_{HF}\phi(2)/(\rho_{HF}(1)\,\rho_{HF}(2)))^{1/2} \tag{20}$$

Using Eqs(19) and (15), we obtain a very useful expression for the Fermi correlation factor:

$$f_F(1,2) = G_{HF}(1,2)\ G_{HF}(2,1) \tag{21}$$

Clearly then, the possibility of obtaining a closed analytical functional for the energy depends on whether we can write $f_F(1,2)$ and $f_{CORR}(1,2)$ as functionals of ρ . We show in what follows that is indeed possible to obtain an approximation to $G_{HF}(1,2)$ and hence to $f_F(1,2)$.

III. An Approximate Analytical Expression for $G_{HF}(1,2)$

Harriman[16] has introduced a set of single particle functions $\{ u_k(1);$ $k \in \{1,\dots m\}\ ;\ m \geq n\}$ defined by

$$u_k(1) = (\rho(1))^{1/2} \exp(ikf(a)) \tag{22}$$

Similar functions had been previously advanced by Macke[17], Gilbert[18], and Lieb[19]. In a three dimensional space, a in Eq(22) can be any one of the coordinates. This freedom in choosing a leads in fact to a non-unique definition of Harriman's orbitals. For simplicity we take a as the radial coordinate in a spherical polar system. The phase factor in Eq(22) is defined by

$$f(r_1) = 2\int_0^{r_1} dr\ r^2\ \bar{\rho}(r) \tag{23}$$

where $\bar{\rho}(r)$ is the charge density when the angular coordinates have already been integrated.

The approximation which we now introduce in the evaluation of $G_{HF}(1,2)$ is the following. We assume that the orbitals forming the vectors in Eq(20) (see Eq(8)), are in fact Harriman's orbitals. We also take Γ^1_{HF} to be an n by n unit matrix in this representation. This amounts to calculating $\gamma^1(1,2)$ from a single Slater determinant constructed from the first n Harriman orbitals. To differentiate $G_{HF}(1,2)$ from its approximation, we put a bar over the latter in what follows.

Introducing Eq(22) into Eq(20) we obtain

$$\bar{G}_{HF}(1,2) = ((1/n^2) \sum_{k=1}^{n} \sum_{\ell=1}^{n} \exp(i(k+\ell)F(r_1,r_2))^{1/2} \tag{24}$$

We have defined in Eq(24)

$$F(r_1,\ r_2) = f(r_2) - f(r_1) \tag{25}$$

and we have assumed all spins to be the same. Noticing that

$$\bar{G}_{HF}(1,2) = (1/n) \sum_{k=1}^{n} \exp(ikF) \tag{26}$$

it can be easily shown[20] that

$$\bar{G}_{HF}(1,2) = (1/n)\exp((i/2)(n+1)F)\sin((n/2)F)/\sin(F/2) \tag{27}$$

Using Eqs(27) and (21), we can now calculate the approximate Fermi correlation factor

$$\bar{f}_F(1,2) = (1/n^2)\ \sin^2((n/2)F)/\sin^2(F/2) \tag{28}$$

IV. An Approximate Universal Energy Functional

In terms of the exact reduced first and second order density operators, the exact energy is given by

$$E[\gamma^1,\gamma^2] = (1/2)\int d1\nabla_1\nabla_2\gamma^1(1,2)_{2\to 1} + V_{ne}[\rho] + \int d1\int d2\ \gamma^2(1,2;1,2)/r_{12} \qquad (29)$$

In view of Eq(18) and noticing that we may write $\gamma^1(1,2) = \gamma^1_{HF}(1,2) + \gamma^1_{CORR}(1,2)$ (with a similar equation for ρ, we can decompose the exact energy as follows:

$$E[\gamma^1,\gamma^2] = E[\gamma^1_{HF},\gamma^2_{HF}] + E_{CORR} \qquad (30)$$

where we have defined

$$E[\gamma^1_{HF},\gamma^2_{HF}] = (1/2)\int d1\ \nabla_1\nabla_2\gamma^1_{HF}(1,2)_{2\to 1} + V_{ne}[\rho_{HF}]$$

$$+\int d1\int d2\ \gamma^2_{HF}(1,2;1,2)/r_{12} \qquad (31)$$

and where E_{CORR} is the difference between the exact energy and $E[\gamma^1_{HF},\gamma^2_{HF}]$. Let us now consider the approximation to $\gamma^1_{HF}(1,2)$ and to $f_F(1,2)$ (and hence to $\gamma^2_{HF}(1,2;1,2)$) discussed in the previous section. It is clear, because of its functional form, that $\bar{G}_{HF}(1,2)$ depends on the density and also on the phase factor $f(r)$. However, as it is seen in Eq(23), $f(r)$ is in turn a functional of ρ. Hence, in Harriman's orbital representation the Hartree-Fock energy functional becomes just a functional of the change density. This may be succintly put as

$$E[\bar{\gamma}^1_{HF},\bar{\gamma}^2_{HF}] \Rightarrow E[\bar{\rho}_{HF}] \qquad (32)$$

An explicit evaluation of the kinetic energy term in Eq(31), using Eq(19), yields

$$T[\bar{\rho}_{HF}] = T_W[\bar{\rho}_{HF}] + (1/2)\int d1\ \bar{\rho}_{HF}(1)\nabla_1\nabla_2\ \bar{G}_{HF}(1,2)_{2\to 1} \qquad (33)$$

Also, in view of Eqs(14) and (21), the electron-electron energy becomes

$$V_{ee}[\bar{\rho}_{HF}] = J[\bar{\rho}_{HF}] - (1/2)\int d1\int d2\bar{\rho}_{HF}(1)\bar{\rho}_{HF}(2)\bar{G}_{HF}(1,2)\bar{G}_{HF}(2,1)/r_{12} \qquad (34)$$

Introducing Eq(27) into Eqs(33) and (34), we are finally led to the general expression for the total energy functional

$$E[\bar{\rho}_{HF}] = T_W[\bar{\rho}_{HF}] + ((2n^2+3n+1)/12)\int d1\bar{\rho}_{HF}(\nabla_1 f(r_1))^2 + V_{ne}[\bar{\rho}_{HF}] + J[\bar{\rho}_{HF}]$$

$$-(1/2)\int d1\int d2\ (\bar{\rho}_{HF}(1)\bar{\rho}_{HF}(2)/r_{12})(\sin^2(n/2)F/\sin^2(F/2)) \qquad (35)$$

Elsewhere[20] we have shown that for spherically symmetric densities, Eq(35) can be greatly simplified so that both the Coulomb and exchange interactions can be rigorously written in terms of local potentials.

As a final remark let us mention that one of the important characteristics of the present functional is that Fermi correlation is introduced through a function which depends only indirectly upon the position coordinates, as the argument in the sine functions is $f(r_2)-f(r_1)$. Hence, the term for the exchange correlation takes into account the variations of the density at different points in space. This is so because f(r) is in itself a functional of the charge density. Another important aspect of the present functional is that it depends on both the number of particles as well as on their spins[20].

The present work is closely related to that of Nyden and Parr[21]. The presence of closed analytic expressions for the energy functional generalizes, in a sense, these previous results and also lays the groundwork for a very simple treatment of correlation effects in many electron theory.

REFERENCES

1. L.H. Thomas, Proc. Cambridge Philos. Soc. 23, 542 (1927)
2. E. Fermi, Z. Phys. 48, 73 (1928)
3. P.A.M. Dirac, Proc. Cambridge Philos. Soc. 26, 376 (1930)
4. C.F.V. Weizsacker, Z. Phys. 96, 431 (1935)
5. E.H. Lieb, Rev. Mod. Phys. 53, 603 (1981)
6. P. Hohenberg and W. Kohn, Phys. Rev. B 136, 864 (1964)
7. W. Kohn and L. Sham, Phys. Rev. A 140, 1133 (1965)
8. D.A. Kirzhnits, Zhur. Eksp. Teor. Fiz. 32, 115 (1957) [Sov. Phys. JETP 5, 64 (1957)]
9. C.H. Hodges, Can. J. Phys. 51, 1428 (1973)
10. Y. Tal and R.F.W. Bader, Int. J. Quantum Chem. Symp. 12, 153 (1978); see also L. Szasz and I. Berrios, Z. Naturforsch. A30, 1516 (1975)
11. S.B. Sears, R.G. Parr and V. Dinur, Israel J. Chem. 19, 165 (1980)
12. J.L. Gázquez and E.V. Ludeña, Chem. Phys. Lett. 83, 145 (1981)
13. E.V. Ludeña, J. Chem. Phys. 76, 3157 (1982)
14. E.V. Ludeña, Int. J. Quantum Chem. 23, 127 (1983)
15. G. Sperber, Int. J. Quantum Chem. 5, 177 (1971)
16. J.E. Harriman, Phys. Rev. A24, 680 (1981)
17. W. Macke, Phys. Rev. 100, 992 (1955); Ann. Physik. 17, 1 (1955)
18. T.L. Gilbert, Phys. Rev. B12, 2111 (1975)
19. E.H. Lieb, in Physics as Natural Philosophy: Essays in Honor of Lazlo Titza on his 75th Birthday, A. Shimony and H. Feshback, Eds., (M.I.T. Press, Cambridge, (1982) p. 141
20. E.V. Ludeña, "An Approximate Universal Functional in Density Functional Theory", J. Chem. Phys. (to be published)
21. M.R. Nyden and R.G. Parr, J. Chem. Phys. 78, 4044 (1983); M.R. Nyden, J. Chem. Phys. 78, 4048 (1983)

THE CLOSED TIME-PATH GREEN'S FUNCTION FORMALISM IN MANY-BODY THEORY

Guang-zhao ZHOU, Zhao-bin SU, Bai-lin HAO, and Lu YU
Institute of Theoretical Physics,
Academia Sinica, Beijing, CHINA.

I. INTRODUCTION

The application of the field-theoretical technique in many-body theory has proved to be highly successful.[1] However, only limited progress has been made in studying the nonequilibrium phenomena beyond the linear response using this technique. The closed time-path Green's function (CTPGF) formalism, first suggested by Schwinger,[2] and further elaborated by Keldysh and others,[3] can be applied to equilibrium as well as nonequilibrium systems. It seems to us that the potential advantages of the CTPGF formalism have not yet been fully exploited.

For the last few years we have combined the generating functional technique and the path integral representation with the CTPGF approach and have developed a unified framework to describe both equilibrium and nonequilibrium systems with symmetry breaking and dynamic coupling between order parameter and elementary excitations.[4-11] We will give here a brief description of the formalism itself and some of its applications.[4,6,9-16]

In Sec.II we define the generating functional (GF) for the CTPGF. In parallel with the closed time-path representation we introduce also the "physical" one in terms of retarded, advanced and correlation functions. We indicate the special normalization condition for the CTPGF GF and the causality naturally embodied in the theory. We discuss there also the potential condition and the fluctuation-dissipation theorem (FDT) valid for thermoequilibrium and nonequilibrium stationary state (NESS). In Sec.III we describe a practical calculation scheme for CTPGF. A system of coupled, self-consistent equations for the order parameter, the fermion quasiparticles and the collective excitations, is derived from the equation for the vertex functional. A systematic loop expansion for the self-energies of these excitations is developed. In Sec.IV we briefly describe some results of applying the CTPGF formalism to different physical systems. The presentation here is rather schematic, while a more detailed explanation can be found in Refs. 4,6,9 and 11.

II. THE GENERATING FUNCTIONAL

For simplicity let us consider the real boson field $\hat{\varphi}_\alpha(x)$, $\alpha = 1,2 \cdots N$. The GF for the CTPGF is defined as

$$Z[J(x)] = tr\{T_p[\exp(i\int_p \hat{\varphi}(x)J(x))]\hat{\rho}\}, \tag{2.1}$$

where $\hat{\rho}$ is the density matrix and T_p is the time ordering operator along the closed time-path consisting of positive $(-\infty, +\infty)$ and negative $(+\infty, -\infty)$ branches, while the external sources $J(x_+), J(x_-)$ on different branches are assumed to be different. Expanding $Z[J(x)]$ in $J(x)$ we find the n-point CTPGF

$$G_p(1\cdots n) \equiv (-i)^n tr\{T_p(\hat{\varphi}(1)\cdots\hat{\varphi}(n))\} = i(-1)^{n-1}\frac{\delta^n Z[J]}{\delta J(1)\cdots\delta J(n)}. \tag{2.2}$$

The GF for the connected Green's functions

$$W[J(x)] \equiv -i\ \ln Z[J(x)] \tag{2.3}$$

and the vertex functional

$$\Gamma[\bar{\varphi}(x)] \equiv W[J(x)] - \int_p J(x)\bar{\varphi}(x) \tag{2.4}$$

are defined in the usual way with

$$\bar{\varphi}(x) = \frac{\delta W[J(x)]}{\delta J(x)}. \tag{2.5}$$

It can be shown also that

$$\frac{\delta\Gamma[\bar{\varphi}]}{\delta\bar{\varphi}} = \left.\frac{\delta I[\varphi]}{\delta\varphi(x)}\right|_{\varphi(x)=\bar{\varphi}(x)+i\frac{\delta}{\delta J(x)}} = -J(x), \tag{2.6}$$

where $I[\varphi]$ is the action of the system. This is the basic equation of the CTPGF formalism.

By differentiating Eqs. (2.5) and (2.6) we obtain

$$\int_p G_p(x,z)\Gamma_p(z,y) = \int_p \Gamma_p(x,z)G_p(z,y) = \delta_p^d(x-y), \tag{2.7}$$

where $\delta_p^d(x)$ is the δ - function defined on the closed time-path and

$$G_p(x,y) = -i\begin{pmatrix} \langle T(\varphi(x)\varphi(y))\rangle & \langle\varphi(y)\varphi(x)\rangle \\ \langle\varphi(x)\varphi(y)\rangle & \langle\tilde{T}(\varphi(x)\varphi(y))\rangle \end{pmatrix} \tag{2.8}$$

with T and $\tilde{T}$ as time-ordering and antiordering operators respectively. This matrix function is connected with the ordinary retarded (G_r), advanced (G_a) and correlation (G_c) functions in such a way that

$$G_{\alpha\beta}(x,y) = \tfrac{1}{2}\xi_\alpha\eta_\beta G_r(x,y) + \tfrac{1}{2}\eta_\alpha\xi_\beta G_a(x,y) + \tfrac{1}{2}\xi_\alpha\xi_\beta G_c(x,y), \tag{2.9}$$

where

$$\xi_\pm = 1, \qquad \eta_\pm = \pm 1. \tag{2.10}$$

In terms of these functions the Dyson equation (2.7) can be rewritten as

$$G_r \cdot \Gamma_r = \Gamma_r \cdot G_r = 1 , \qquad (2.11)$$

$$G_a \cdot \Gamma_a = \Gamma_a \cdot G_a = 1 , \qquad (2.12)$$

$$G_c = - G_r \cdot \Gamma_c \cdot G_a , \qquad (2.13)$$

where the retarded (Γ_r), advanced (Γ_a) and correlation (Γ_c) two-point vertex functions are connected to Γ_p by a relation similar to (2.9).

It is useful to introduce new pairs of source and field variables

$$J_c(x) = \frac{1}{2} \xi_\alpha J_\alpha(x) , \qquad J_\Delta(x) = \eta_\alpha J_\alpha(x) , \qquad (2.14)$$

$$\varphi_c(x) = \frac{1}{2} \xi_\alpha \varphi_\alpha(x) , \qquad \varphi_\Delta(x) = \eta_\alpha \varphi_\alpha(x) . \qquad (2.15)$$

The GFs (2.1), (2.3) and (2.4) can be then expressed in these variables with (J_Δ, φ_c) and (J_c, φ_Δ) as conjugate pairs. It can be shown from the normalization of the density matrix $\hat{\rho}$ that

$$Z[J_\Delta(x), J_c(x)]\Big|_{J_\Delta = 0} = 1 , \qquad W[J_\Delta(x), J_c(x)]\Big|_{J_\Delta = 0} = 0 . \qquad (2.16)$$

It is worthwhile pointing out that here we require only the equality of the external source at the two time branches but not its vanishing. As consequences of Eq. (2.16) one finds also

$$\Gamma[\varphi_\Delta(x), \varphi_c(x)]\Big|_{\varphi_\Delta = 0} = 0 \qquad (2.17)$$

and the causality relations

$$\frac{\delta^n W[J_\Delta(x), J_c(x)]}{\delta J_\Delta(1) \cdots \delta J_\Delta(m) \delta J_c(m+1) \cdots \delta J_c(n)}\Bigg|_{J_\Delta = J_c = 0} = 0 , \qquad (2.18)$$

$$\frac{\delta^n \Gamma[\varphi_\Delta(x), \varphi_c(x)]}{\delta \varphi_\Delta(1) \cdots \delta \varphi_\Delta(m) \delta \varphi_c(m+1) \cdots \delta \varphi_c(n)}\Bigg|_{\varphi_\Delta = 0,\ \varphi = \varphi_c} = 0 \qquad (2.19)$$

provided any of t_j , $m+1 \le j \le n$, is greater than all of t_i, $1 \le i \le n$. Under the physical condition $J_\Delta = 0$, $\varphi_c(x)$ is the symmetry breaking in the external field, while φ is that in its absence.

We have studied[8] the time reversal symmetry of NESS from a microscopic point of view and have shown that if

$$Im\, G_r(\omega = 0, \vec{x}, \vec{y}) = 0, \qquad (2.20)$$

there exists a potential $\mathcal{F}[J(\vec{x})]$ such that

$$\frac{1}{2} \frac{\delta W[J(x)]}{\delta J(x_\sigma)}\Bigg|_{J_+ = J_-,\ t = 0} = - \frac{\delta \mathcal{F}[J(\vec{x})]}{\delta J(\vec{x})} \qquad (2.21)$$

where the functional argument on the r.h.s. depends only on $\vec{r}$, but not on t . A similar potential $\mathcal{G}[\varphi(\vec{x})]$ can be introduced as a counterpart of $\Gamma[\varphi(x)]$.

Furthermore, we have shown[8] that the FDT for equilibrium system

$$G_c(\omega) = \coth\frac{\omega}{2T}\,(G_r(\omega) - G_a(\omega)) \tag{2.22}$$

can be generalized to NESS as

$$i\,\frac{\partial G_c}{\partial t} = 2\,T^{eff}\,(G_r - G_a), \tag{2.23}$$

where T^{eff} is the effective temperature.

III. COUPLED EQUATIONS OF ORDER PARAMETER AND ELEMENTARY EXCITATIONS

Consider a system of fermions interacting via a boson field $Q(x)$ which may be nonpropagating at the tree level like the Coulomb field. The order parameter itself may be a constituent field or a composite operator. The radiative correction will in general make $Q(x)$ a dynamic variable and the fluctuations around the mean field value will form the collective excitation. The system is thus characterized by $Q_c(x)$ and the energy spectrum, dissipation and particle distribution for fermion and collective excitation.

The GF for this system can be written as[11]

$$Z_p[h, M, K] = \int_p [d\psi^+][d\psi][dQ]\exp\{i(I_{eff} + hQ + \tfrac{1}{2}QMQ + \psi^+K\psi)\}, \tag{3.1}$$

where

$$I_{eff} = I_o[\psi^+, \psi] + I_o[Q] + I_{int}[\psi^+, \psi, Q] + W_p^N[\psi^+, \psi, Q], \tag{3.2}$$

$$QMQ \equiv \iint_p d^4x\,d^4y\,Q(x)M(x,y)Q(y) \qquad \text{etc.} \tag{3.3}$$

with

$$I_o[\psi^+, \psi] = \int d^4x\,d^4y\,\psi^+(x)\,S_o^{-1}(x,y)\,\psi(y), \tag{3.4}$$

$$I_o[Q] = \frac{1}{2}\int d^4x\,d^4y\,Q(x)\,\Delta_o^{-1}(x,y)\,Q(y) \tag{3.5}$$

and W_p^N taking care of the contribution from the density matrix. It follows from the GF for the connected part $W[h, M, K]$ that[11]

$$\frac{\delta W_p}{\delta h(x)} = Q_c(x), \tag{3.6}$$

$$\frac{\delta W_p}{\delta M(y,x)} = \frac{1}{2}\,(Q_c(x)\,Q_c(y) + i\Delta(x,y)), \tag{3.7}$$

$$\frac{\delta W_p}{\delta K(y,x)} = -i\,G(x,y), \tag{3.8}$$

where Δ, G are the second order connected CTPGF for boson and fermion fields respectively. It is easy to deduce from the vertex GF defined as

$$\Gamma_p[Q_c, \Delta, G] = W[h, M, K] - hQ_c - \frac{1}{2}tr(M(Q_cQ_c + i\Delta)) - i\,tr(KG), \tag{3.9}$$

that

$$\frac{\delta \Gamma_p}{\delta Q_c(x)} = -h(x) - \int_p d^4y\, M(x,y)\, Q_c(y), \tag{3.10}$$

$$\frac{\delta \Gamma_p}{\delta \Delta(x,y)} = \frac{1}{2i}\, M(y,x), \tag{3.11}$$

$$\frac{\delta \Gamma_p}{\delta G(x,y)} = i\, K(y,x). \tag{3.12}$$

The system of equations (3.10) — (3.12) determine self-consistently the order parameter and the second order CTPGF provided Γ_p is known as a functional of Q_c, Δ and G.

We have extended the loop expansion technique proposed by Cornwall, Jackiw and and Tomboulis[17] in the quantum field theory, to the CTPGF case to get[11]

$$\Gamma_p[Q_c, \Delta, G] = I[Q_c] - \frac{i\hbar}{2} tr[\ln(\Delta_0^{-1}\Delta) - \Delta_0^{-1}\Delta + 1] + i\hbar\, tr[\ln(S_0^{-1}G) - G_0^{-1}G + 1] + \Gamma_{2p}[Q_c, \Delta, G], \tag{3.13}$$

where

$$I[Q_c] = I_{eff}[\psi^\dagger, \psi, Q]\Big|_{\substack{\psi=\psi^\dagger=0\\ Q=Q_c}}, \tag{3.14}$$

$$\Delta_0^{-1} = \frac{\delta^2 I_{eff}}{\delta Q(x)\, \delta Q(y)}\Big|_{\psi=\psi^\dagger=Q=0}, \tag{3.15}$$

$$S_0^{-1} = -\frac{\delta^2 I_{eff}}{\delta \psi^\dagger(x)\, \delta \psi(y)}\Big|_{\psi=\psi^\dagger=Q=0}, \tag{3.16}$$

$$G_0^{-1} = -\frac{\delta^2 I_{eff}}{\delta \psi^\dagger(x)\, \delta \psi(y)}\Big|_{\psi=\psi^\dagger=0,\, Q=Q_c} \tag{3.17}$$

To compute Γ_{2p} first shift the field $Q(x)$ in I_{eff} by Q_c and keep only terms cubic and higher in $\psi, \psi^\dagger$ and Q as interaction vertices. The Γ_{2p} then is calculated as a sum of all two-particle irreducible (2PI) vacuum diagrams constructed by these vertices with full Δ, G as propagators. Switching off the external sources from Eqs. (3.10)—(3.12), we obtain the following equations:

$$\frac{\delta \Gamma_p}{\delta Q_c(x)} = \frac{\delta I[Q_c]}{\delta Q_c(x)} + i\hbar\, tr\left\{\frac{\delta G_0^{-1}}{\delta Q_c(x)} G\right\} + \frac{\delta \Gamma_{2p}}{\delta Q_c(x)} = 0, \tag{3.18}$$

$$\frac{2i}{\hbar} \frac{\delta \Gamma_p}{\delta \Delta(y,x)} = \Delta^{-1}(x,y) - \Delta_0^{-1}(x,y) + \Pi(x,y) = 0, \tag{3.19}$$

$$\frac{i}{\hbar} \frac{\delta \Gamma_p}{\delta G(y,x)} = G^{-1}(x,y) - G_0^{-1}(x,y) + \Sigma(x,y) = 0, \tag{3.20}$$

where the self-energy parts

$$\Pi(x,y) = \frac{2i}{\hbar} \frac{\delta \Gamma_{2p}}{\delta \Delta(y,x)}, \qquad \Sigma(x,y) = -\frac{i}{\hbar} \frac{\delta \Gamma_{2p}}{\delta G(y,x)}. \tag{3.21}$$

Rewritten in the ordinary time variables, Eq. (3.18) is the generalized Ginzburg-Landau equation for the order parameter, while Eqs. (3.19) and (3.20) are the Dyson equations for the retarded, advanced and correlation functions. The loop expansion of Γ_{2p} is a series in the Planck constant $\hbar$. The mean field theory is recovered if the Γ_{2p} is neglected altogether. Usually we can limit ourselves to the first few terms of expansion or a partial summation for certain class of important diagrams.

IV. SOME APPLICATIONS

(i) Critical Phenomena At the critical point the long wave-length thermal fluctuation dominates. The CTPGF formalism is quantum mechanical in nature, but the classical limit can be easily taken. We have derived[6,10] from the vertex equation the generalized Ginzburg-Landau equations for the macrovariables including the order parameter and the conserved quantities. By comparing these equations with the Ward-Takahashi (WT) identities following from the symmetry properties we have separated the matrix of kinetic coefficients into reversible and irreversible parts. The reversible terms describe the mode coupling which is usually put into the Langevin equation by hand. Moreover, we have shown[6,10] that the Martin-Siggia-Rose (MSR)[18] field theory is nothing but the classical limit of CTPGF expressed in retarded, advanced and correlation functions. It turns out that the current Lagrangian field theory of critical dynamics is retrieved if the one loop approximation in the random source and the second cumulant of φ_Δ are kept in the CTPGF path integration.

(ii) Quenched Random Systems It is known that in this case one should average the free energy instead of partition function. A special replica trick has been suggested[19] to handle this difficult problem. The CTPGF formalism provided us with a dynamic approach to it.[15] Using the normalization of GF (2.16) one can show that the magnetic moment is obtained by taking functional derivative of the averaged GF, while analog of the Edwards-Anderson order parameter $q(t,t')$ appears as an integral part of the second order connected CTPGF

$$G_{pij}(t,t') = \langle G_{pij}(t,t',J_{ij})\rangle_J + i\, q_{ij}(t,t') \qquad (4.1)$$

with

$$q_{ij}(t,t') = \langle \sigma_i(t,J_{ij})\, \sigma_j(t',J_{ij})\rangle_J - \langle \sigma_i\rangle_J \langle \sigma_j\rangle_J . \qquad (4.2)$$

The steady state solution and the time evolution of q can be deduced from an equation derived from the Dyson equation for G. We have shown[15] that a physical boundary is found for the stable region in the plane q-$|\chi|$, where χ is the susceptibility. Above T_c, the Fischer line[20] lies entirely in the stable region and q tends to its fixed point q_0 exponentially in time. Below T_c, the Fischer line intersects the stability boundary at q_1 which in general in not a fixed point. After reaching the order parameter will further decay as a power law to its fixed point q_c . The physical boundary is shown to be temperature independent, so the magnetization does not depend on the temperature, while the entropy is independent of the magnetic field. This is just the basic assumption of the projection hypothesis[21] which appears naturally in our formalism.

(iii) Nonlinear Response Theory The linear response theory near thermoequilibrium has been well established. The CTPGF formalism including the causality as its integral part furnishes a unambiguous definition for all multipoint functions. As a simple application of the CTPGF GF the nonlinear response to arbitrary order of external disturbance has been derived.[12] The formal relations following from the KMS[22] condition and the time reversal symmetry as well as the possible generalization of the FDT in nonlinear case have also been worked out.[12]

(iv) Superconductivity Umezawa et al.[23] have derived a coupled system of equations for the order parameter and the weak electromagnetic field in the ground state superconductor. We have found out[13] that these equations follow directly from the vertex functional and the WT identity and, therefore, can be easily generalized to finite temperature, even to some nonequilibrium situation. This allows us to avoid the ambiguity associated with the "dynamical mapping" and the "boson transformation" used in the earlier work.[23] Our derivation also clarifies to further extent the physical meaning of the procedure involved.

(v) Laser System As example of far-from-equilibrium phenomena, the CTPGF formalism has been applied to rederive the semiclassical Lamb equation[24] for the unimode laser coupled with the two-energy-level electrons. We have also used the WT identity to derive a generalized Goldstone theorem in a slowly varying in time system.[14] As its consequence, the pole in the Green's function splits into two, each one with the same energy but different dissipation. Combined with the order parameter, these two kinds of quanta provided us with a complete description of the order-disorder transition of the phase symmetry in the saturation state of laser.

(vi) Quasi-One-Dimensional System The soliton model proposed by Su, Schrieffer and Heeger and others[25] for systems displaying Peierls instability, has stimulated great interest. The order parameter, the staggered displacement of the linear conjugate molecule, is considered as a classical quantity in this model. However, the recent Monte-Carlo calculation shows[26] that the quantum fluctuations of the ground state are significant. We have applied the CTPGF formalism to analytically study the consequences of these fluctuations. A coupled system of equations for the order parameter and the Green's function has been derived and solved numerically.[16] It turns out that the order parameter becomes a complex quantity and the energy gap is reduced compared with the adiabatic case.

V. CONCLUDING REMARKS

To summarize, we have outlined in this paper some characteristic features and the calculation scheme of the CTPGF formalism as well as some of its applications. As seen from the presentation, the theoretical framework is flexible and powerful enough to handle both equilibrium and nonequilibrium phenomena on a unified basis. Up to now we have mainly tested this formalism on systems which in principle can be also treated by other methods. But, in most cases we do find some new results, or new insight into the problem, or some significant simplification. It seems to us that the benefit we gain from using this formalism is much greater than the price paid for the apparent complexity. We believe that its potential advantages will be exploited to fuller extent if it is applied to attack wider range of problems in condensed matter, nuclear and plasma physics as well as in particle physics and cosmology.

REFERENCES

1. See, e.g., G.D. Mahan, Many-Particle Physics (Plenum, N.Y.,1981).
2. J. Schwinger, J. Math. Phys. 2,407 (1961).
3. L.V. Keldysh, Sov. Phys. -JETP 20, 1018 (1965); D. Langreth in " Linear and Nonlinear Electronic Transport in Solids ", eds. J. Devrees and V. Van Doren (Plenum, N.Y., 1976).
4. ZHOU Guang-zhao and SU Zhao-bin, Ch. 5 in " Progress in Statistical Physics " (Science Press, Beijing, in Chinese, 1981).
5. ZHOU Guang-zhao and SU Zhao-bin, Physica Energiae Fortis et Physica Nuclearis (Beijing), 3, 304,314 (1979).
6. G.Z. ZHOU, Z.B. SU, B.L. HAO, and L. YU, Phys. Rev. B22,3385 (1980).
7. ZHOU Guang-zhao, YU Lu, and HAO Bai-lin, Acta Physica Sinica, 29, 878 (1980).
8. ZHOU Guang-zhao and SU Zhao-bin, Acta physica Sinica, 30, 164,401 (1980).
9. ZHOU Guang-zhao, SU Zhao-bin, HAO Bai-lin, and YU Lu, Commun. Theor. Phys. (Beijing) 1, 295,307,389 (1982).
10. ZHOU Guang-zhao, SU Zhao-bin, HAO Bai-lin, and YU Lu, Acta Physica Sinica, 29, 961,969 (1980).
11. SU Zhao-bin, YU Lu, and ZHOU Guang-zhao, Preprints AS-ITP-83-019,025.
12. B.L. HAO, Physica,109A, 221 (1981); WANG Wei-yong, LIN Zhong-heng, SU Zhao-bin, and HAO Bai-lin, Acta Physica Sinica, 31, 1483, 1493 (1982).
13. SU Zhao-bin and ZHOU Guang-zhao, Commun. Theor. Phys. (Beijing), 1,669 (1982).
14. ZHOU Guang-zhao and SU Zhao-bin, Acta Physica Sinica, 29,618 (1980).
15. SU Zhao-bin, YU Lu, and ZHOU Guang-zhao, Preprints AS-ITP-83-010, 012.
16. SU Zhao-bin, WANG Ya-xin, and YU Lu, to be published.
17. J.M. Cornwall, R. Jackiw, and E. Tomboulis, Phys. Rev. D10, 2428 (1974).
18. P.C. Martin, E. Siggia, and H.A. Rose, Phys. Rev. A8, 423 (1973).
19. S.F. Edwards and P.W. Anderson, J. Phys. F5, 965 (1975).
20. K.H. Fischer, Phys. Rev. Lett. 34, 1438 (1975).
21. G. Parisi and G. Toulouse, J. Physique Lett. 41, 361 (1980).
22. R.Kubo, J.Phys. Soc. Japan, 12, 570 (1957); P.C. Martin and J. Schwinger, Phys. Rev. 115, 1342 (1959).
23. H. Matsumoto and H. Umezawa, Fort. der Phys. 24, 357 (1976).
24. See, e.g., H. Haken in " Laser Handbook ",ed.F.Arecchi et al.(N.H.,Amsterdam,1972)
25. W.P.Su, J.R.Schrieffer, and A.J.Heeger, Phys. Rev. Lett. 42,1698 (1979); M.J. Rice, Phys. Lett. 71A, 152 (1979).
26. W.P.Su, Solid State Commun. 42,497 (1982);J.E.Hirsch et al. Phys. Rev. Lett. 49,402 (1982).

MONTE CARLO EVALUATIONS IN FINITE FERMIONIC SYSTEMS

Y. Avishai
Department of Physics
Ben Gurion University of the Negev
Beer Sheva, Israel

1. Introduction

The connection between classical statistical-mechanics and quantum mechanics is well known ever since Feynman's path integral formulation of quantum physics. It has recently been employed by theoreticians in using Monte-Carlo simulation to study gauge field models[1,2]. A pedagogical approach to the use of Monte Carlo methods in nonrelativistic particle quantum mechanics has appeared recently[3]. The technique in both cases is similar; one starts with an imaginary time Feynman path integral formulation (which is mathematically equivalent to partition function) and defines the integration measure by path discretization. This then leads to a multidimensional integral for the partition function. In some cases, logarithmic derivatives of the partition function (with respect to $1/\hbar$) are then expressible as statistical average

$$< A > \quad \int D\underline{x}\, A(\underline{x})\exp[-S(\underline{x})]/ \int D\underline{x}\, \exp[-S(\underline{x})] \qquad (1)$$

where $\underline{x} = (x_1,x_2,\ldots,x_n)$, $D\underline{x} = \pi dx_i$ and the action $S(\underline{x})$ is a known real function. Although the normalization integral is not generally known, it is sufficient to have a normalized distribution

$$P(\underline{x})D\underline{x} = \exp[-S(\underline{x})]D\underline{x}/ \int \exp[-S(\underline{x})]D\underline{x} \qquad (2)$$

at hand for the application of Metropolis type Monte Carlo evaluation of $< A >$[4].

For systems with infinite number of fermionic degrees of freedom, the representation (1) is not valid in general. First the anticommutation rules for fermion operators lead to the occurrence of Grassman numbers. This result is well known in field theory[5] and is less known in the theory of many fermion systems[6]. Second, the postulates of wave function antisymmetrization result in positive and negative contributions to the normalization integral and there is no real function $S(\underline{x})$ such that $\exp[-S(\underline{x})]$ can serve as a probability function. In field theory, this difficulty is termed as the sign of the fermionic determinant.

For finite fermion systems the first difficulty (namely, the appearance of Grassman numbers) can be eliminated in several ways. The one closest in spirit to field theory is based on the path integral representation for the partition function of fixed number of fermions. To achieve such a representation, one can start for example from the σ field approach[7], or, as we do here, from real coherent states of Slater determinants (RCSSD). The application of Monte Carlo method to evaluate partition function of A fermions starting from its path integral representation in terms of RCSSD is the subject of section 2.

It should be stated here that the path integral representation does not cure the second pathology related to fermions, namely, the existence of negative contributions to the normalization integral. In the RCSSD approach one is sometimes lucky in the sense that the contribution of the negative terms is small as explained in section 2. However, this should be checked in any specific case and should not be taken always as safely true. Clearly, a more rigorous algorithm should be employed. The author tried to transform into the boson representation[8] but so far without success.

For this reason we introduce in section 3 another method which does not rely on states (e.g., Slater determinants) but evaluates the partition function $Z = tr(e^{-\beta H})$ directly from the matrix elements H_{ij} themselves. This algorithm (which is not based on path integral representation) is applicable for bosonic and fermionic systems as well. It is based on the use of stochastic methods in matrix algebra, originated in an unpublished work of Ulam and Von-Newmann[9].

We close the introduction section by noticing the growing interest in the pertinent theme as is exemplified by other contributions to this conference[10].

2. The Method of RCSSD

Consider a system of n_t identical fermions, which in the absence of interaction fill the "reference" state (a Slater determinant) $|\phi_o\rangle = (n_t!)^{-\frac{1}{2}} \prod_{i=1}^{n_t} a_i^\dagger |0\rangle$ where $|0\rangle$ is the vacuum ($a_i|0\rangle = 0$ for all i) and $a_i^\dagger$, a_i are particle creation and annihilation operators obeying the standard anticommutation rules. Besides the lowest n_t single particle states, there are also n_p different higher states so that the single particle space contains $B = n_p + n_t$ states. In most cases of interest $n_p \geq n_t$.

The RCSSD are defined by the following excitations

$$|X\rangle = \exp(\sum_{pt} x_{pt}\, a_p^\dagger\, a_t)|\phi_o\rangle \tag{3}$$

where x_{pt} ($p = 1,\ldots n_p$, $t = 1,\ldots n_t$) are elements of a real matrix $\underline{X}$ with n_p rows and n_t columns. The important properties of RCSSD are: a) scalar product $\langle X_1|X_2\rangle = \det[1 + \underline{X}_1^T\, \underline{X}_2]$; b) resolution of unity $\int dXf(X)|X\rangle\langle X| = 1$ (unit operator) with $f(X) = 1/[\det(1 + \underline{X}^T\underline{X})]^{B/2+1} > 0$ and $\int dX = \prod_{Pt} \int_{-\infty}^{\infty} dx_{pt}$;

c) matrix elements of any one and two-body operators between two RCSSD (and in particular those of the Hamiltonian H) can be computed. Thus, $\langle X_1|H|X_2\rangle = H(\underline{X}_1,\underline{X}_2)$ is a known rational function of $2xn_pxn_t$ real variables; d) trace of operators, $tr(A) = \int dXf(X)\langle X|A|X\rangle$.

Consider now the partition function $Z = tr(\exp(-\beta H))$ where β is a positive parameter. It is well known that all the important information on the pertinent system is contained in Z. Here we shall limit ourselves to the evaluation of the ground state energy $E_o = -\lim_{\beta\to\infty} \partial(\ell n\, Z)/\partial\beta$.

The properties of RCSSD discussed above now allow us to get a finite N approximation for Z namely $\lim_{N\to\infty} Z_N = Z$ with

$$Z_N = \int \prod_{i=1}^{N} dX_i\, f_i\, P_i(1-\beta Y_i/N) \tag{4}$$

where $f_i \equiv f(X_i)$, $P_i \equiv \langle X_i|X_{i+1}\rangle$, $Y_i \equiv \langle X_i|H|X_{i+1}\rangle/P_i$ and $\underline{X}_{N+1} = \underline{X}_1$.

The Monte Carlo calculations are meaningful only if the integrand has a definite sign, say positive. Hence, we restrict ourselves to domains where

$$P_i = \langle X_i|X_{i+1}\rangle > 0 \qquad (i = 1,2,\ldots,N) \tag{5}$$

This is precisely the positivity problem of the normalization integral discussed at the introduction. We then find

$$Z_N = \int \prod_{i=1}^{N} dX_i \exp\{-\sum_{i=1}^{N} [\beta Y_i/N - \log(P_i f_i)]\} \equiv \int \prod_{i=1}^{N} dX_i \exp(-S) \tag{6a}$$

$$E_o = \lim_{\beta\to\infty} \int \prod_{i=1}^{N} dX_i (N^{-1} \sum_{j=1}^{N} Y_j) \exp(-S)/Z_N \tag{6b}$$

The question to what extent eq. (6a) is a discretized form of a path integral representation of Z will be discussed at the end. Presently, we are concerned with numerical evaluations which require a discretized form anyhow.

Equation (6b) is our starting point for a Metropolis type Monte Carlo evaluation. A few comments on the numerical procedure are in order, since the present scheme exhibits some peculiar features which are not encountered in simple models;
1) It is very helpful (though not crucial) that in eq. (6b) one has

$$Y_i = \langle X_i|H|X_{i+1}\rangle/\langle X_i|X_{i+1}\rangle > 0 \ . \tag{7}$$

A necessary condition (but not sufficient) is that the eigenvalues of H are all positive, which can be achieved by properly shifting the reference energy.
2) An essential step is the re-exponentiation procedure. The error in passing from eq. (4) to eq. (6a) comes from the replacement $(1-\beta Y_i/N) \approx \exp(-\beta Y_i/N)$ in each factor, with N sufficiently large. The total error θ is then approximately given by

$$\theta = \sum_{i=1}^{N} (\beta Y_i/N)^2/2(1-\beta Y_i/N) \tag{8}$$

Thus, for a fixed θ (e.g. 10%) and a given N, one cannot increase β freely, since this would bring about an error greater than θ. It is therefore necessary to evaluate the maximal β which is allowed by (8) right at the zeroth iteration (for which $Y_i^{(0)} \equiv Y$) and reject any configuration (in the course of Monte Carlo iterations) for which the r.h.s. of (8) is greater than θ. Evidently

$$\beta_{max} = (1/Y)[(2N\theta)^{\frac{1}{2}}-\theta] \tag{9}$$

Consequently, the limit of large β can be reached only at the expense of having large $\sqrt{N}$. This is a heavy price to pay, since the number of integration variables is $N \times n_p \times n_t$. 3) A single term in the exponent on the r.h.s. of eq. (6a) is

$$S_i \equiv (\beta/N) <X_i|H|X_{i+1}>/P_i - \log(P_i f_i) \tag{10}$$

According to point 1), one starts with $P_i^{(0)} = P^{(0)} > 0$. If now a Monte Carlo iteration suggests a configuration for which P_i is very small it will eventually be rejected by the condition $e^{-S} > r$ ($0 < r < 1$ a random number). Hence, loosely speaking, P_i is bounded from below by a positive number, and the first term on the r.h.s. of eq. (10) can be made small for large N.

We have performed numerical calculations pertaining to a simple two fermion system with $n_p = n_t = 2$. Our results could be summarized as follows: 1) The ground state energy has been evaluated to within a few percents. 2) The Monte Carlo algorithm which is most suitable for the present case is the one in which the new states are close to the old ones. This algorithm is detailed in Ref. 3 at the bottom half of p. 439. Indeed, an algorithm which is based on the standard Metropolis method has been tried by Troudet and Koonin [11] and failed to work. 3) The choice of the initial configuration is essential not only to reduce the number of iterations, but, as turns out, also to remain always within the domain of positive normalization integral. In our calculations we have in fact never stepped on configurations for which $P_i < 0$.

Before concluding, we mention that eq. (6a) can be considered as a discretized form of path integral representation for Z only if every function is expressed in terms of $\underline{X}_i$ and the differences $\underline{X}_{i+1} - \underline{X}_i$ (which become derivatives). However the exponentiation procedure is crucially based on the assumption (which is not always justified) $\lim_{N\to\infty} < \underline{X}_i(|\underline{X}_{i+1}> - |\underline{X}_i>) = 0$. In fact, a careless use of this relation can easily lead to an absurdity. Alternatively, one can interprete S as the free energy of our lattice with nearest neighbor interaction Y_i. The first term is then the internal energy while the second one is the entropy.

We have recently learned about an interesting approach to the problem by Klauder[12] based on Langevin equations.

3. New Stochastic Method

We shall now describe an algorithm (believed to be very effective) for the stochastic evaluation of matrix operations (specifically, $tr(e^{-\beta H})$) over matrices of enormous size. Hence it is clear that this method is relevant to quantum field theory, condensed matter physics and nuclear physics. So far it has been employed in calculating the inverse of a matrix[1]. In order to evaluate $\mathrm{tr}[e^{-\beta H}]$ we need to modify it slightly.

For fixed large integer N, we denote by T the transfer matrix $T = 1-\beta H/N$ (so that $e^{-\beta H} \approx T^N$) and decompose the matrix element T_{ij} $(i,j = 1,2,\ldots,M)$ into $T_{ij} = P_{ij}R_{ij}$ such that $1 > P_{ij} > 0$ and $\sum_{j=1}^{M} P_{ij} = 1$ for all i. The matrix $P = \{P_{ij}\}$ is a stochastic matrix. Consider now an N step random walk on the domain of integers 1,2,...,M. The walk starts at some point i (selected in advance) and proceeds from point to point with probabilities P_{jk}. It then stops after exactly N steps at some point k_N. The probability to reach k_N along the chain $\gamma_i \equiv i \to k_1 \to k_2 \ldots \to k_N$ is then $P_{\gamma_i} \equiv P_{ik_1} P_{k_1k_2} \ldots . P_{k_{N-1}k_N}$. (Notice the Markov property; the probability to arrive at j from i in n+1 steps, $P_{ij}^{(n+1)}$ is equal to $\sum_k P_{ik}^{(n)} P_{kj}$.)

When the walk stops at k_N, a score S_{γ_i} (depending on the chain γ_i) is recorded, defined by

$$S_{\gamma_i} \equiv R_{ik_1} R_{k_1k_2} \ldots . R_{k_{N-1}k_N} \delta_{k_N i}$$

If we now compute the expectation value of the random variable S_{γ_i} we find (summing over chains of N steps initiating at i)

$$\langle S_{\gamma_i} \rangle = \sum_{k_1,k_2,..k_N} P_{\gamma_i} S_{\gamma_i} = (T^N)_{ii}$$

Hence, the problem is reduced to the statistical evaluation of expectation value with the probability P_{γ_i} (clearly, $\sum P_{\gamma_i} = 1$) which is feasible by a Metropolis type Monte Carlo evaluation. It is not difficult to evaluate the variance σ_i of S_{γ_i} and estimate the statistical error in $(T^N)_{ii}$. It is also possible to evaluate $\mathrm{tr}(T^N)$ in one step simply by renormalizing the probabilities and performing the average also with respect to i. A test of the above method is now in study. Alternatively, one can calculate $(e^{-\beta H})_{ij}$ in a way almost indentical with that of Ref. (1) starting from the Taylor expansion. The only difference is that each score S_{ij} recorded there after k steps should be divided by k!. Evidently the problem of convergence does not exist in the present case.

References

1) J. Kuti, Phys. Rev. Lett. 49 (1982) 183.

2) M. Creutz, L. Jacobs and C. Rebbi, Phys. Rev. Lett. 42 (1979) 1390.

3) M. Creutz and B. Freedman, Ann. Phys. 132 (1981) 427.

4) J.M. Hamersley and D.C. Handscomb, "Monte Carlo Methods", (John Willey, N.Y. 1964).

5) J.E. Hirsch, D.J. Scalapino, R.L. Sugar and R. Blankenbechler, Phys. Rev. Lett. 47 (1981) 1628.

6) H. Orland, "Developpement des Theories de Champ Moyen en Physique Nucleaire et dans les Milieux Desordones" - Thèse, Université Paris-Sud 1981.

7) S. Levit, J.W. Negele and Z. Paltiel, Phys. Rev. C21 (1980) 1603.

8) P. Garbaczewski, Phys. Rep. 36 (1978) 35.

9) G.E. Forsythe and R.A. Leibler, MTAC 4 (1950) 127.

10) See contributions by J.W. Negele and J.G. Zabolitzky.

11) Y. Avishai and J. Richert, Phys. Rev. Lett. 50 (1983) 1175. A comment by S. Koonin and T. Troudet and a reply by the authors have been submitted recently. S.E. Koonin, In Proceeding of Nuclear Theory Summer Workshop at SB, August 1981. Y. Avishai and J. Richert, contributions to Hirschegg X (1982) and Hirschegg XI (1983).

12) J.R. Klauder, J. Phys. A (1983) L317.

APPLICATION OF GREEN'S FUNCTION MONTE CARLO TO ONE-DIMENSIONAL LATTICE FERMIONS

Michael A. Lee and Kazi A. Motakabbir
Kent State University
Kent, Ohio 44242

K.E. Schmidt
Los Alamos National Laboratory
Los Alamos, New Mexico 87545

A Monte-Carlo procedure for obtaining the ground state of a class of one-dimensional lattice Fermion Hamiltonians is presented. Variational calculations and Green's Function Monte Carlo (GFMC) results are presented for a simple Hubbard model.

One-dimensional models of quantum many-body systems have received a great deal of attention over the years,[1] not only because they are more mathematically tractable,[2] but also because they are experimentally accessible. Among the most popular theoretical models of quasi-one-dimensional are the Hubbard[3] Hamiltonians. These have served as models for the study of organic conductors as well as the investigation of solitons in materials like polyacetylene. In its simplest form the Hamiltonian

$$H = \sum_{i=1}^{L}\left[- C_i^+C_{i+1} - C_i^+C_{i-1} + V_{nn}C_i^+C_iC_{i+1}^+C_{i+1}\right] \tag{1}$$

consists of a "hopping" term and an interaction, V_{nn}, between nearest neighbor particles. Here C_i^+ and C_i are the usual Fermion creation and annihilation operators for the ith site of an L site lattice with periodic boundary conditions. This Hamiltonian is much simplified since the interaction could depend on spin and interparticle separations as well as position. Since the method we wish to describe is easily generalized to treat such complications, we will consider only this simple Hamiltonian and thus avoid the myriad of subscripts and notational inconvenience of the more complex interactions.

Recently Hirsch, Scalapino and co-workers[4] have published results of a Monte Carlo procedure for treating systems with this type of Hamiltonian at finite temperature. Their method is similar to that introduced by Barker[5] which maps the finite temperature problem onto a space time manifold and then discretizes the time (inverse temperature). These methods suffer at low temperatures because the number of discrete reciprocal temperature increments become excessive. This is necessary if the approximation inherent in discretization is to remain small. This low temperature inefficiency was noted by Whitlock and Kalos[6] in their work on quantum hard spheres, even though their method did not contain a discretization approximation.

We wish to describe here a simple and efficient Monte Carlo simulation method of obtaining the zero temperature (ground state) solution to this class of lattice Hamiltonians. The end result is in essence the discrete analog of the Green's Function Monte Carlo (GFMC) method developed by Kalos[7] and co-workers for quantum fluids. Such methods are exact (except for statistical uncertainties) when applied to a boson ground state. With one exception,[8] it has not been possible to generalize this method exactly to many-Fermion problems; however, several accurate approximate procedures exist.[9] We will see that in one dimension, this problem can be circumvented.

The Green's Function G, central to GFMC methods, is defined as the inverse of the Hamiltonian

$$HG = I \quad . \tag{2}$$

The procedure to be carried out is the iterative process

$$|\phi_{n+1}\rangle = E_T G|\phi_n\rangle \tag{3}$$

of repeatedly applying the Green's Function to some initial trial state $|\phi_0\rangle$. This process geometrically converges to the ground state and the correct energy is the value of E_T which maintains the normalization, $\langle\phi_{n+1}|\phi_{n+1}\rangle = \langle\phi_n|\phi_n\rangle$. The only practical way of carrying out this procedure for a complicated many-body problem is through Monte Carlo methods.[7]

Brevity prevents us from presenting any more than a terse outline of the derivation of the method. Further results will be published elsewhere. We begin by considering a Hamiltonian equivalent to Eq. (1), with its eigenvalue spectrum shifted by an amount $2N+V_0$ so that they are now positive definite. The number of particles is N.

$$H = V_0 + \sum_i 2C_i^+C_i - C_i^+C_{i-1} - C_i^+C_i + V_{nn}C_i^+C_iC_{i+1}^+C_{i+1} \tag{4}$$

In order to solve this problem stochastically, we need to deal with wave functions and lattice sites instead of operators and state vectors. To this end, we will establish a representation using the basis states

$$|J\rangle \equiv \prod_{i=1}^{N} C_{j_i}^+ |0\rangle \equiv |j_1 j_2 \ldots j_N\rangle \quad .$$

The value of j_i is the site of the ith particle. We will use uppercase letters as indicating a particular set of particle locations. These basis states together with Eq. (2) can be used to obtain a matrix equation for G. Substituting into

$$\sum_I \langle M|H|I\rangle\langle I|G|J\rangle = \delta_{MJ}$$

we, after some algebra, obtain

$$A_M\{V_M G_{M,J} + \sum_{i=1}^{N} 2G_{M,J} - G_{M(i+1),J} - G_{M(i-1),J}\} = \delta_{MJ} \tag{5}$$

Here, V_M is the potential energy of configuration M, M(i+1) is configuration M with the ith particle moved one site to the right, A_M is an antisymmetrization operator acting on the first index of G only, and $\delta_{MJ} = \langle M|J\rangle$, not a product of Kronacker deltas.

It is perhaps not immediately apparent that this equation has the structure of a diffusion equation (random walk) on an N-dimensional lattice. The term in the summation is, in fact, a discrete laplacian. If we were treating a boson problem, the simulation of a diffusion process is well-known,[7] even on a lattice,[10] and could be treated by methods not unlike those employed by Metropolis[11] over two decades ago. The condition of antisymmetry results in the Green's Function (and the eigenfunction) not being positive definite. It is a unique property of one-dimensional Fermions that we know the location of the nodes (zeros) of the Green's Function and the ground state, half of the time. We will state without proof, that for odd N, the only nodes present are those which occur when two particles occupy the same site. The difference between even and odd N can be seen by explicitly solving the noninteracting Hamiltonian for N=2 and N=3. The two particle ground state has additional nodes.

It has been stated[9] that if the nodes of the Fermion ground state were known, then GFMC techniques could be modified to obtain an exact solution. This amounts to imposing the boundary condition on the boson Green's Function that it be zero at the nodes, i.e., when two particles overlap in this case. This "fixed node" Green's Function will satisfy the same defining relationship, Eq. (5), as the Fermion Green's function, except that the antisymmetrization operation is removed and the right-hand side becomes a genuine Kronacker delta.

It can be shown[10] that the random walk algorithm which iterates Eq. (3), amounts to performing a random walk to produce configurations at M from initial configurations at J (in other words, sampling G_{MJ}) according to the equation

$$G_{M,J} = \frac{\delta_{MJ}}{U + 2N} - \left[\frac{U - V_J}{U + 2N}\right] G_{M,J} + \frac{2N}{U + 2N} \sum_i \left(\frac{G_{M(i+1),J} + G_{M(i-1),J}}{2N}\right) \tag{6}$$

where U is chosen to be greater than V_J.

The procedure to be carried out involves the following steps:

1. Choose an initial population ($\sim$ 1000) of configurations randomly.
2. For each configuration in the current population, call it J, sample a new position(s) M from $E_T G_{MJ}$. The result is a new population and Eq. (3) has been iterated once.
3. Repeat step 2 with the new population.

The sampling of M from $E_T G_{MJ}$ can occur in Eq. (6) three ways: a) First term: Set $M = J$ with probability $E_T/(U + 2N)$. b) Second term: With probability $(U - V_J)/(U + 2N)$ sample M from $E_T G_{MJ}$, i.e., add configuration J to the current population. c) With probability $2N/(U + 2N)$, move a random particle one position right or left and if no particles overlap return the new configuration to the current population (otherwise, discard it).

The procedure described above is not the most efficient possible, but it is quite generally applicable and has proven effective in all cases tested. For the simple cases, we are considering step b could be eliminated by choosing $U = V_J$, but we have retained it for the sake of generality.

It is instructive to test our assertion that the procedure works correctly for odd N by solving the non-interacting Fermion problem both numerically and analytically. In Fig. 1 we have plotted the energy of a half-filled lattice ($N/L = 1/2$) for various even and odd values of N. The analytic results oscillate as each shell (pair of +k and -k states) is filled while the simulation results smoothly approach the correct infinite system limit. Clearly the differences become negligible for $N \gtrsim 20$.

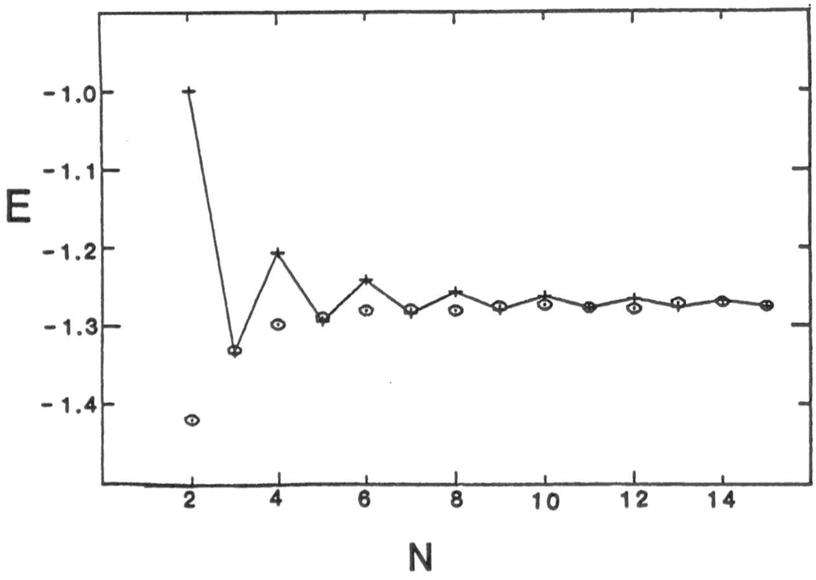

FIG. 1. Energy per particle for various size systems. Crosses are analytic results.

Although there are some analytic solutions for some simple Hubbard models, there appear to be none when V_{nn} is non-zero. We have obtained results of high statistical accuracy using the current method for a wide range of interaction strengths. Results obtained for the energy per particle for V_{nn} = -2 to +2 are presented in

Fig. 2 as a function of density, N/L. The statistical uncertainty of these results is generally smaller than the plotted symbol. For values of V_{nn} outside this range, the method works, but the behavior of the system becomes uninteresting. For $V_{nn} > 2$, the particles rarely occupy adjacent sites, so the structure is like a charge density wave; and for $V_{nn} < -2$, the entire group of particles form a single chain most of the time. In both extremes, the pair correlations are slowly decaying and the finite size of their simulation becomes relevant. There is no particular reason to restrict our attention to small N, since the expense of the simulation grows only linearly with N instead of quadratically or, at worst, like N! for some other Fermion calculations.

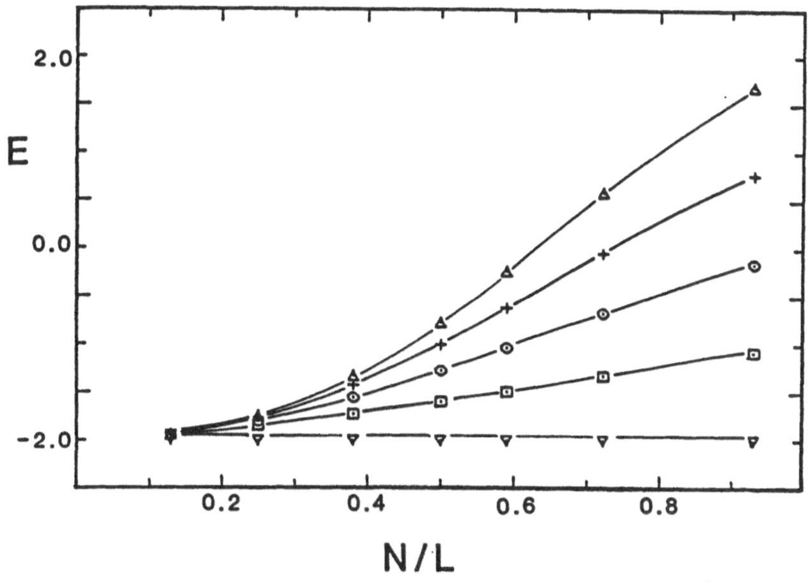

FIG. 2. GFMC results for energy per particle versus density with N=13. V_{nn} = -2, -1, 0, 1, 2 going from bottom to top.

One of the most surprising results of this work occurred when we tried to compare the GFMC results to a Variational Monte Carlo (VMC) calculation. It was decided to do a simple Jastrow-Slater VMC calculation. The most naive and simple wavefunction possible is a determinant of plane waves multiplied by a symmetric two-body correlation factor. To be even cruder, the two-body factor was chosen to have a range of only one lattice site, hence only one variation parameter. The results cannot be displayed graphically, because over the entire range of density and interaction strength, the difference between VMC and GFMC is almost indistinguishable. Table 1 displays these two sets of results for the energy per particle for the same range of densities and nearest neighbor interactions, as shown in Fig. 2. The results are so close that, because of statistical fluctuations, the VMC energy is occasionally below the GFMC energy. In any case, the differences are hardly statistically significant. The uncertainty is generally less than 0.01, but some of the variational results could possibly be optimized further.

TABLE 1. Comparison of GFMC and VMC results for energy per particle. The upper number is from GFMC.

N/L \ V_{NN}	-2	-1	0	1	2
$\frac{13}{100}$	-2.000 -1.968	-1.946 -1.955	-1.944 -1.945	-1.940 -1.941	-1.936 -1.936
$\frac{13}{52}$	-1.996 -1.953	-1.843 -1.864	-1.800 -1.802	-1.764 -1.765	-1.739 -1.738
$\frac{13}{34}$	-2.000 -1.955	-1.726 -1.702	-1.554 -1.555	-1.425 -1.413	-1.331 -1.319
$\frac{13}{26}$	-2.000 -1.965	-1.593 -1.602	-1.271 -1.276	-1.004 -1.000	-0.773 -0.765
$\frac{13}{22}$	-1.999 -1.974	-1.489 -1.485	-1.031 -1.037	-0.617 -0.623	-0.235 -0.226
$\frac{13}{18}$	-2.000 -1.979	-1.323 -1.318	-0.667 -0.679	-0.045 -0.041	0.589 0.584
$\frac{13}{14}$	-1.996 -2.000	-1.075 -1.077	-0.153 -0.154	0.770 0.769	1.691 1.690

In summary, we have presented an efficient GFMC method for obtaining stochastic solutions for the ground state of a class of one-dimensional lattice Fermion problems. The method cannot be readily extended beyond one dimension for essentially the same reason that the finite temperature methods of Ref. 4 are restricted to one-dimensional problems. We have also found that the most simplistic Slater-Jastrow variational wavefunctions are nearly in quantitative agreement with GFMC calculations. These wavefunctions can, of course, be employed in three-dimensional problems and we feel that the success in one dimension may suggest that the accuracy for higher dimensional lattice models may be even better than past experience with quantum fluids might indicate.

REFERENCES

1. *Highly Conducting One-Dimensional Solids*, edited by J.T. Devreese, R.P. Evrard, and V.E. Van Doren (Plenum Press, New York, 1979).
2. *Mathematical Physics in One Dimension*, edited by E.H. Lieb and D.C. Mattis (Academic Press, New York, 1966).
3. J. Hubbard, Proc. Roy. Soc. London A276, 238 (1963).
4. J.E. Hirsch and D.J. Scalipino, Phys. Rev. B27, 7169 (1983) and references therein.
5. J.A. Barker, J. Chem. Phys. 70, 2914 (1979).
6. P.A. Whitlock and M.H. Kalos, J. Comp. Phys. 30, 361 (1979).
7. M.H. Kalos, D. Levesque, and L. Verlet, Phys. Rev. A9, 2178 (1974); M.H. Kalos, M.A. Lee, P.A. Whitlock, and G.V. Chester, Phys. Rev. B24, 115 (1981); and D.M. Ceperley and M.H. Kalos in *Monte Carlo Methods in Statistical Physics*, edited by K. Binder (Springer-Verlag, New York, 1979).
8. D.M. Arnow, M.H. Kalos, M.A. Lee, and K.E. Schmidt, J. Chem. Phys. 77, 5562 (1982).
9. M.A. Lee, K.E. Schmidt, M.H. Kalos, and G.V. Chester, Phys. Rev. Lett. 46, 728 (1981); J.W. Moskowitz, K.E. Schmidt, M.A. Lee, and M.H. Kalos, J. Chem. Phys. 77, 349 (1982); D. Ceperley and B. Alder, Phys. Rev. Lett. 45, 566 (1980).
10. G.H. Weiss, Amer. Sci. 71, 65 (1983).
11. N. Metropolis, in *Symposium on Monte Carlo Methods*, edited by H.A. Mayer (Wiley, New York, 1956).

ON THE INVERSE PROBLEM IN MANY BODY SYSTEMS: FROM CORRELATIONS TO DISTRIBUTION FUNCTION

L. Reatto
Dipartimento di Fisica, G.N.S.M., Università di Parma, Italy
and
G.L. Masserini
Dipartimento di Fisica, G.N.S.M., Università di Milano, Italy

Abstract. - In a classical fluid the inverse problem consists in deducing the interatomic interaction from structural data. In a quantum fluid a similar inverse problem is established starting from a maximum overlap criterion between the exact ground state and a model wave function. This criterion is an alternative and in many respects a better method than the energy one for determining the best wavefunction within a certain approximation. We present the result of a successful computation of the maximum overlap Jastrow function for the Lennard Jones Bose fluid and for liquid ^{4}He. The computation is based on a predictor corrector iterative scheme. Preliminary results for the inverse problem for a classical Lennard Jones fluid are also discussed.

Introduction

A basic problem in classical statistical mechanics is the computation of correlation functions, in particular the radial distribution function (rdf) g(r), starting from the probability distribution $P(r_1..r_N)$ in configurational space, i.e. $\exp(-V(r_1..r_N)/K_BT)$ where $V(r_1..r_N)$ is the potential energy of the system and T the absolute temperature. Perturbation expansions, the method of integral equations and simulation methods have been developed for this purpose.[1] Viceversa with the inverse problem one assumes known g(r) or other correlation functions and wants to deduce $V(r_1..r_N)$. In general this inverse problem is not well defined unless one puts some restriction on the form of $V(r_1..r_N)$, for instance that it contains only pair terms

$$V(r_1..r_N) = \frac{1}{2} \sum_{i\neq j} v(r_{ij}) \qquad (1)$$

It is commonly believed, but we are not aware of any general formal proof, that given g(r) there is a unique v(r). This inverse problem has been attacked with the method of integral equations [2,3] but the approximations inherent to these equations do not allow in general a computation of v(r) free of unctrolled errors and the problem has still to be considered as open. On the other hand it is clear that the problem is of fundamental interest because its solution would allow the determination of the interatomic interaction in condensed phase starting from scattering measurements that give the structure factor S(k). The recent and the foreseen progress in these measurements due to the availability of higher flux sources and of a larger range of momentum transfer calls, in our opinion, for an adequate theoretical effort to solve this inverse problem.

In a quantum fluid at zero temperature the probability P(R) of finding the particles in the configuration $R \equiv (r_1..r_N)$ equals the modulus square of the wave function $\psi_o(R)$ and the direct problem is the determination of g(r) starting from P(R). Methods similar to those of the classical case, i.e. integral equations and simulation methods, have been successfully applied to this problem. The interest in the inverse problem has been spurred by the existence of "exact" simulations of quantum systems like the ones provided by the Green Function Monte Carlo (GFMC) method for Bose fluid. [4] With this method a sequence of configurations drawn from the exact ψ_o is generated but the wave function is not produced by the computation. In this way one can compute averages and correlations but it is clear that one would also like to extract some information on the structure of ψ_o. One of us [5] has shown that such information can be obtained from a maximum overlap criterion: chosen a subspace of the Hilbert space of the system spanned by a model wave function ψ_M the maximum of the overlap integral $\langle\psi_M|\psi_o\rangle$ with respect to ψ_M leads to certain equalities for correlation functions. For instance in the case of a Bose fluid if we choose the subspace of Jastrow functions:

$$\psi_J(r_1..r_N|u) = \Pi_{i<j} \exp(-\tfrac{1}{2}u(r_{ij})), \tag{2}$$

the maximum overlap obtains for that pseudopotential $\bar{u}(r)$ for which

$$g_J(r|\bar{u}) = g_{mxd}(r|\bar{u}). \tag{3}$$

Here g_J and g_{mxd} are the r d f corresponding, respectively, to the probabilities ψ_J^2 and $\psi_o\psi_J$. From GFMC computations it is known [4] that ψ_J is close enough to ψ_o that the difference $\psi_o-\psi_J$ can be treated as a perturbation to first order. In this case the exact r d f is given by $g_o(r) = 2g_{mxd}(r)-g_J(r)$ and the maximum overlap condition (3) becomes

$$g_J(r|\bar{u}) = g_o(r). \tag{4}$$

This is an equation equivalent to that found in the classical inverse problem: one must find that pseudopotential $\bar{u}$ that reproduces a given r d f . In this case $\bar{u}$ has the role of $v(r)/K_BT$ in the classical system. In the next Section we show how we succeeded in solving this problem by a novel approach.

Best Jastrow function for a Bose fluid

The method we use to solve the inverse problem (4) is based on a predictor-corrector algorithm. Given a non optimum pseudopotential $u(r)$ from the difference $g_o(r)-g_J(r|u)$ we obtain a first estimate of $\bar{u}$ that we call $\bar{u}^{(1)}(r)$. Due to the approximate nature of the predictor $\bar{u}^{(1)}$ is not yet $\bar{u}(r)$ but by a Monte Carlo run we compute the corresponding $g_J(r|\bar{u}^{(1)})$. We use now $g_o(r)-g_J(r|\bar{u}^{(1)})$ as a new starting point to get a new estimate $\bar{u}^{(2)}$. The procedure is repeated until $g_o(r)$ is reproduce to within the required accuracy.

As predictor we have used two different approximations. The first is the random phase approximation (RPA) and this gives the following estimate [5] for $\delta^{(i)}_{RPA}(r) = \bar{u}^{(i)}(r) - \bar{u}^{(i-1)}(r)$:

$$\delta^{(i)}_{RPA}(r) = \frac{1}{(2\pi)^3\rho}\int d^3k\, e^{i\vec{k}\cdot\vec{r}}(S_o^{-1}(k)-S_J^{-1}(k|\bar{u}^{(i-1)})). \tag{5}$$

Here $S_o(k)$ is the exact structure factor corresponding to $g_o(r)$. This predictor is not appropriate if $u(r)$ and $\bar{u}(r)$ differ also at short distances in the core region because it is known that RPA does not handle correctly short range correlations. In this case we have used as predictor an algorithm [5] based on the modified hypernetted chain equation [6] (MHNC). This equation is known in the field of classical fluids to be very accurate in the core region. The basic approximation of the MHNC equation consists in the assumption that the bridge functions [1] (tne sum of the elementary diagrams) corresponding to the unknown $\bar{u}$ and to u are equal. Then the estimate for $\delta^{(i)}$ is [5]

$$\delta^{(i)}_{MHNC}(r) = \frac{1}{(2\pi)^3\rho}\int d^3k\, e^{i\vec{k}\cdot\vec{r}}(S_o^{-1}(k)-S_J^{-1}(k|\bar{u}^{(i-1)})) +$$

$$+ g_o(r)-g_J(r|\bar{u}^{(i-1)})+\ln\left[g_J(r|\bar{u}^{(i-1)})/g_o(r)\right] . \tag{6}$$

The integral term coincides with the RPA estimate and is due to the difference of the Ornstein-Zernike direct correlation functions corresponding to g_o and to g_J. The remaining terms of (6) are specific to the MHNC equation.

In our computations we have used both predictors. Usually we start the iterations with $\delta^{(i)}_{RPA}$ in order to get the intermediate range part of $\bar{u}$, then we perform one or more iterations with δ_{MHNC} if after the first stage g_o and g_J differ in the core region. We prefer to use as a small number of MHNC iterations as possible because this predictor is more sensitive to the statistical noise present in the r d f given by simulation in particular in the core region.

Lennard Jones Bose fluid

We have implemented the previous scheme in the case of a Bose fluid interacting with the Lennard Jones (LJ) potential $v(r)=4\epsilon[(\sigma/r)^{12}-(\sigma/r)^{6}]$. As exact $g_o(r)$ we take the result [7] of a GFMC computation for 64 particles and we use this same number of particles in our MC computations. Both the RPA and the MHNC predictor make use of the structure factor so that one has to extend both $g_o(r)$ and $g_J(r|\bar{u}^{(i)})$ at distances beyond L/2, half of the size of the simulation box, before making the Fourier transform. We have employed the same method used in ref.(7): the r d f beyond the first maximum of g(r) is fitted by the damped oscillating function $1+Ar^{-1}\exp(-Br)\cos(Cr+D)$ with respect to the coefficients A, B, C and D and this gives a tail that continues the simulation g(r). It is clear that any specific effect due to the presence of long range correlations in ψ_o will not be taken into account by this extension. This is the case, for instance, of the effect of long wawelength phonons [8] so that the structure factor obtained with this extension algorithm does not have the correct limit S(0)=0. However, this limit can be recovered in a second time by including the long range tail of u(r) by the RPA approximation, for instance.

Evidence for the basic soundness of the extension procedure comes from the following observation . Due to the finite size of the simulation cube there is a cut off on phonons at $k_c=\pi/L$. In this case S(0) is not zero but we can estimate its value as:

$$S(0) \simeq \hbar k_c/2mc \qquad (7)$$

If we use the sound velocity c of the LJ fluid [7] we find from (7) for a 64 particle system $S_o(0)=0.074$ at the equilibrium density $\rho_{eq}\sigma^3=0.3648$ and

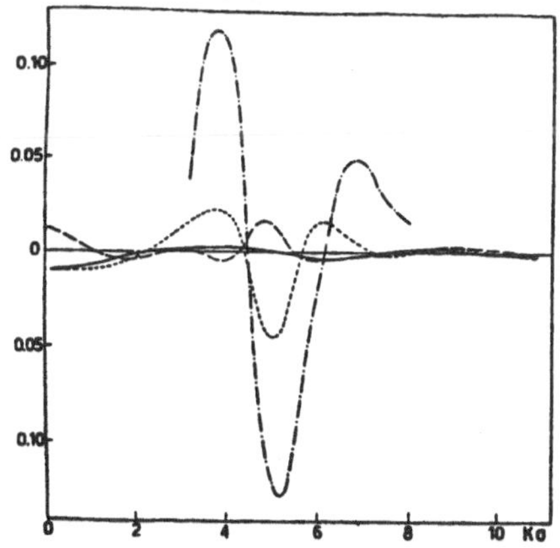

Fig.1-Deviations $S_J(k|\bar{u}^{(n)})-S_o(k)$ for the LJ bose fluid at ρ_{eq} for n=0 (---), for n=2(- - -) and for n=6(——) and $S_J(k|u)$-$S_o(k)$ for the McMillan's u(r) (— · —).

$S_o(0)=0.056$ at $\rho=1.2x\rho_{eq}$. With the extension procedure we find, respectively $S_o(0)=0.083$ and $S_o(0)=0.057$.

The first computation of $\bar{u}$ has been performed at the equilibrium density. Many variational computations of the energy have been performed [4] and the short range behaviour of u(r) can be considered as well established so that we use the RPA predictor. We find that six iterations give convergence at the level of the statistical noise (of order of 1% when $k\sigma \tilde{>} 1$) as it can be verified in Fig.1 where some of the computed $S_J(k,\bar{u}^{(i)})-S_o(k)$ are plotted. At the sixth iteration also $|g_J(r|\bar{u})-g_o(r)|$ is below the noise level. [9] The best pseudopotential $\bar{u}$ is plotted in Fig.2 together with some of the intermediate $\bar{u}^{(i)}$ as well as u(r) used as starting point. This was taken from a variational computation of the energy. [10] We have tested [11] the convergence property of the method by starting also from a different pseudopotential: within the statistical noise the two computations gave the same $\bar{u}(r)$. For comparison we have plotted in Fig.1 and 2 also the result for the widely used McMillan's pseudopotential: $u(r)=(b/\sigma r)^5$.

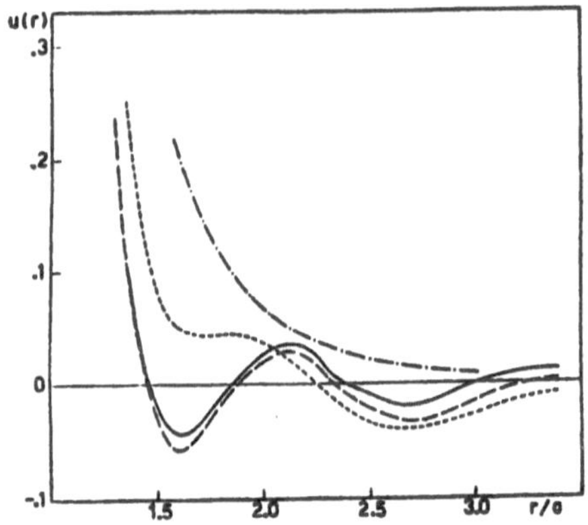

Fig.2 - Pseudopotentials $\bar{u}^{(n)}$ as in Fig.1

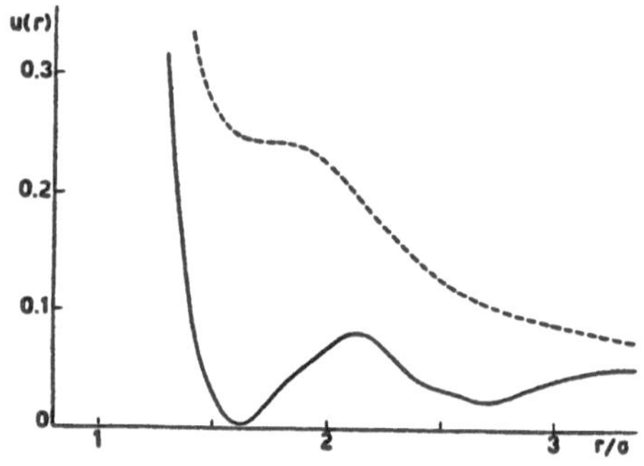

Fig.3 Pseudopotential $\bar{u}^{(6)}+\delta_{Ph}$(—) and u(r)(---) from ref.12.

From Fig.1 it can be noticed that S_J-S_o converges very quickly for large k but at k=0 the deviation is not substantially reduced. This is due to the extension method of $g_J(r)$ because any statistical fluctuation present in $g_J(r)$ for $r\sim L/2$ tends to be amplified in the tail. This effect, however, can introduce only a slowly varying component in $\bar{u}(r)$. As a consequence the absolute values of $\bar{u}$ at its extreme are not determined with great accuracy. On the contrary the very existence of these extreme in $\bar{u}$ and their positions are not affected by this small k problem and must be considered as well established.

The most noticeable aspect of $\bar{u}(r)$ is the presence of a well defined structure at intermediate distance. Also some variational computations of the energy have produced a u(r) with some structure [10] or shoulder [12,13] at intermediate distance. One of these results is, in fact, the initial u(r) of the present computation and plotted in Fig.1. Another one, plotted in Fig.3, was obtained [12] by a functional minimization of the energy in the paired phonon analysis. Since this last result includes the effect of long wavelength phonons, in this figure we have added to our result for $\bar{u}(r)$ the contribution $\delta_{Ph}(r)$ of the long wavelength phonons. $\delta_{Ph}(r)$ is obtained from

$$\delta_{Ph}(r) = \frac{1}{(2\pi)^3\rho}\int d^3k\, e^{i\vec{k}.\vec{r}}(S^{-1}_{J,extr.}(k|\bar{u})-S^{-1}_{J}(k|\bar{u})) \tag{8}$$

where $S_{J,extr.}$ is $S_J(k|\bar{u})$ modified at small k so that it extrapolates to zero at k=0 with the known sound velocity.

The energy criterion and the maximum overlap one do not give, in general, the same best pseudopotential. It is expected, however, that they should be close if the true ψ_o is close to a Jastrow state. It is gratifying to find that the $\bar{u}(r)$ that we find gives an expectation value of the energy rather close to that found with the energy criterion. For instance at ρ_{eq} we find E= =-5.76 K/particle in comparison with the best energy [10] E_o=-6.10 corresponding to $u^{(o)}$ of Fig.2.

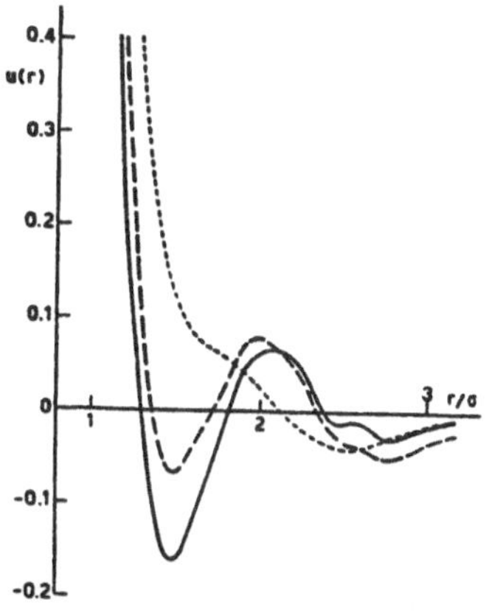

Fig.4 - Pseudopotentials for the L.J. Bose fluid at $\rho=1.2x\rho_{eq}$ for n=0(---), for n=3(- $\frac{eq}{}$) and for n=9 (—).

The convergence at $\rho=1.2x\rho_{eq}$, a density close to solidification, is slower than at ρ_{eq} (again the starting u(r) was taken from ref.10).However, after 8 iterations with the RPA algorithm and one with the MHNC one the convergence was as good as at ρ_{eq}: $:S_J(k,\bar{u})$ differs from $S_o(k)$ for less than $3x10^{-3}$ when $k\sigma>1$. Also in this case $\bar{u}$ has a structure at intermediate distance and this is more pronounced than at ρ_{eq} (see Fig.4).

The origin of the structure of $\bar{u}(r)$ at intermediate distance is not well understood. It has been suggested [10] that such structure is due to the zero point motion of rotons. Our result is compatible with this origin since the position of the structure of $\bar{u}(r)$ roughly scales with $\rho^{-1/3}$ as expected for a collective effect: the position of the first minimum of $\bar{u}$ decreases by 8% when the density increases from ρ_{eq} to 1.2 ρ_{eq} and this can be compared with $(\rho/\rho_{eq})^{1/3}$ = 1.06.

Liquid ^{4}He

The experimental determination of the structure factor S(k) of liquid ^{4}He at low temperature gives the possibility of computing the best Jastrow function for real ^{4}He. In fact one can solve the equation (4) where $g_o(r)$ is determined from S(k). We have obtained [9] $\bar{u}(r)$ using the method described above with the RPA predictor. $\bar{u}(r)$ is rather similar to that of the LJ potential, but it has a more pronounced structure at intermediate distance and it is softer in the core region. This is the expected behaviour at the light of the difference between the best He-He potential compared with the L.J. one. [14]

Classical fluids

The inverse problem for a classical fluid can be attacked exactly with the same methods that we have discussed in the previous sections. In this case $v(r)/K_BT$ takes the place of the pseudopotential $\bar{u}(r)$. As a test of the method we have taken g(r) computed [15] by simulation for the L.J. potential and we try to reproduce this potential. In Fig.5 we show the result of a preliminary computation [16] for the state $\rho\sigma^3=0.65$ and $T/\varepsilon=0.9$. In this computation the MHNC predictor has been used only for $r>\sigma$ whereas for $r<\sigma$ we have used the correct L.J. potential. The initial estimate of v(r) is obtained assuming for the bridge function [6] the Percus Yevick form for hard spheres of diameter $d=\sigma$. For the extension of g(r) for $r>L/2$ we have used in this case the Verlet's algorithm. [15] These preliminary results indicate that our method is useful also for classical fluids.

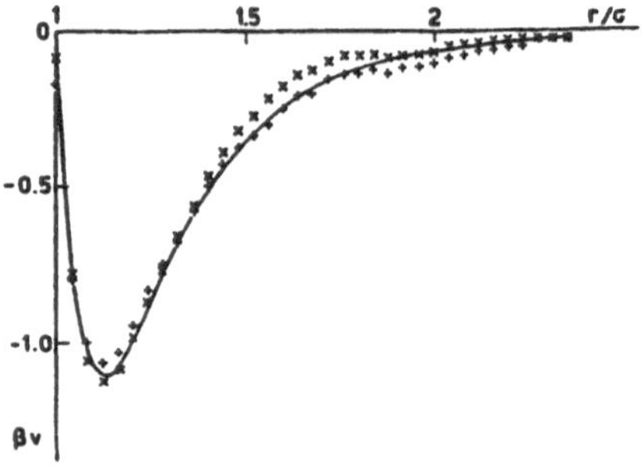

Fig.5 - $\beta v(r)$ at the first (xxx) and at the third (+++) iteration for classical L.J. fluid and the true βv (——).

Discussion

We have solved an inverse problem for a Bose quantum fluid. We believe that this is the first time that an inverse problem in condensed matter has been solved to within the statistical accuracy of the input data. The method is flexible and our preliminary results indicate that it is useful for classical fluids.

The pseudopotentials that give the maximum overlap with the ground state of the L.J. Bose fluid and of liquid ^{4}He have a structure at intermediate distance similar, but more pronounced, to that found on the basis of energy computations. This structure roughly scales as $\rho^{-1/3}$ thus indicating its origin as a collective effect.

Several extensions of the present computation can be indicated like the computation of three body terms in the wave function, the study of the solid phase, the study of ^3He and the use of the maximum superposition $\bar{u}(r)$ for ^4He in order to obtain information on the pair interaction in the condensed phase.

References

1. See, for instance, J.P. Hansen and I.R. McDonald, Theory of simple liquids (Academic Press, New York, 1976).
2. M.D. Johnson, P. Hutchinson and N.H. March, Proc. R. Soc. A 282, 283 (1964).
3. M. Brennan, P. Hutchinson, M.J.L. Sangster and P. Schofield, J. Phys. C 7, L411 (1974).
4. For a review, see D.M. Ceperley and M.H. Kalos in Monte Carlo method in statistical physics, edited by K. Binder (Springer, Berlin, 1979).
5. L. Reatto, Phys. Rev. B 26, 130 (1982).
6. F. Lado, Phys. Rev. A 8, 2548 (1973).; Y. Rosenfeld and N.W. Ashcroft Phys. Rev. A 20, 1208 (1979).
7. P.A. Whitlock, D.M. Ceperley, G.V. Chester and M.H. Kalos, Phys. Rev. B 19, 5598 (1979).
8. L. Reatto and G.V. Chester, Phys. Rev. 155, 88 (1967).
9. G.L. Masserini and L. Reatto, Proceedings of the Conference on ^4He, St. Andrews (Scotland, Aug. 1983).
10. G. Gaglione, G.L. Masserini and L. Reatto, Phys. Rev. B 22, 1237 (1980).
11. G.L. Masserini and L. Reatto in "Monte Carlo Methods in Quantum Problems" edited by M.H. Kalos (to be published).
12. C.E. Campbell and F. Pinski, Phys. Lett. B 79, 23 (1978).
13. R. Smith, A. Kallio, M. Puoskari and P. Toropainen, Nucl. Phys. A 328, 186 (1979).
14. M.H. Kalos, M.A. Lee, P.A. Whitlock and G.V. Chester, Phys. Rev. B 24, 115 (1981).
15. L. Verlet, Phys. Rev. 165, 201 (1968).
16. This computation has been performed by G. Diana and A. Scotti, C.C.R. Euratom, Ispra (Italy).

THE INTERPOLATING EQUATIONS METHOD IN QUANTUM FLUIDS

S.Rosati and A.Fabrocini

Istituto Nazionale di Fisica Nucleare,sect.of Pisa and
Dipartimento di Fisica dell'Università, Pisa, Italy

and M.Viviani

Dipartimento di Fisica dell'Università, Pisa, Italy.

1) Introduction and outline of the method.

The knowledge of the radial distribution function (r.d.f.) g(r) of a classical gas is an obliged step in evaluating the thermodynamical properties of the system. Several methods have been employed in solving this problem: "exact" (Monte Carlo and Molecular Dynamics tecnhniques) or approximated ones (BBGKY, Hyper Netted Chain and Percus Yevick equations). These last approaches have been widely applied for their simplicity. The interest in this problem has been renewed in the past years for the close analogy between the calculations of the classical r.d.f. and of the ground state two-body distribution function of a Bose quantum fluid described by a Jastrow correlated wave function

$$\Psi_J(1,..,A)=\prod_{i<j}^{A} f(r_{ij}) \quad . \tag{1}$$

A is the number of the particles enclosed in a volume Ω with constant ρ density in the thermodynamical limit ($A, \Omega\to\infty$). $f(r_{ij})$ is the correlation factor between the i and the j particles: for simplicity, we will take it depending only on the interparticle distance . The evaluation of the quantum g(r) is equivalent to the one of a classical gas interacting via the effective potential

$$u(r)= - KT \ln f^2(r) \quad . \tag{2}$$

The approximations in the aforementioned methods are reflected in an uncorrect description of some properties of the classical system. As an example, the HNC approach gives the Helmholtz free energy as an integrable function[1] but the equivalence between the isothermal compressiblity as computed by the classical pressure derivative and by the compressibility integral[2] (Compressibility

Consistent Condition, CCC) is no longer preserved. For these reasons, interpolated equations between HNC and PY have been often proposed(3). These interpolations are obtained by a parameter chosen to satisfy the CCC. Two of us have recently proposed (4) a method which employs an r-dependent interpolating function and consists in solving the following set of equations (HNC/α):

$$g(r)= f^2(r) \exp(N(r)+E(r)) , \quad (3)$$

$$E(r)= \ln ((\exp(\alpha(r)N(r))-1)/\alpha(r)) +1) - N(r) , \quad (4)$$

$$\alpha(r)= 1 + \alpha_o E_{PY}(r)/N_{PY}(r) . \quad (5)$$

The nodal function N(r) is evaluated by the Ornstein-Zernike relation, E(r) is the sum of the elementary diagrams contributions and the subscript PY indicates that the function has been derived in that approximation. The free parameter α_o is fixed by the CCC.

2) Results for short-tailed correlations.

The reliability of the HNC/α equations may be tested by the results obtained for liquid ^{4}He, at T=0 K, with a Jastrow factor of the short ranged and commonly used form

$$f^2(r)= \exp(- (b\sigma/r)^5) , \quad (6)$$

with b=1.17 and σ=2.556 Å. The expectation value of the Hamiltonian

$$H= -\hbar^2/(2m) \sum_{i=1,A} \nabla_i^2 + \sum_{i<j}^{A} V(r_{ij}) \quad (7)$$

has been computed in conjunction with the 6-12 Lennard-Jones potential (ϵ=10.22 K) and the Jackson-Feenberg form of the kinetic energy. The energies per particle are shown in fig.(1) and they are compared with the MC results(5) and with the results obtained using a constant interpolating parameter. For all the considered densities, a very satisfactory agreement is obtained. In fig.(2), the r.d.f. is depicted.

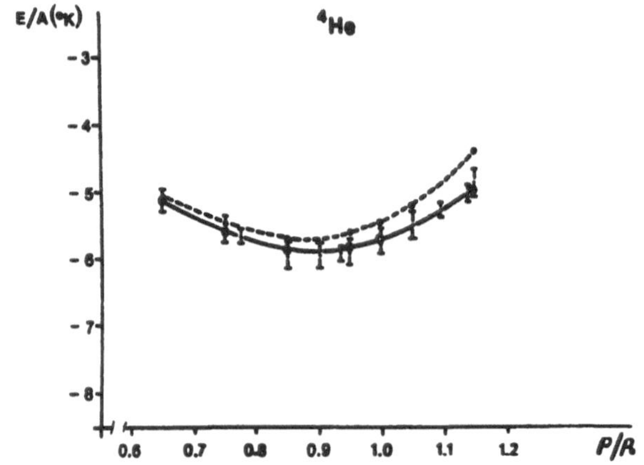

Fig.(1) ^{4}He energies per particle with the L.J. potential. The solid line gives the results in HNC/α, the dashed line the ones obtained by a constant parameter. The points with error bars are M.C. results. ρ_0 is the ^{4}He equilibrium density.

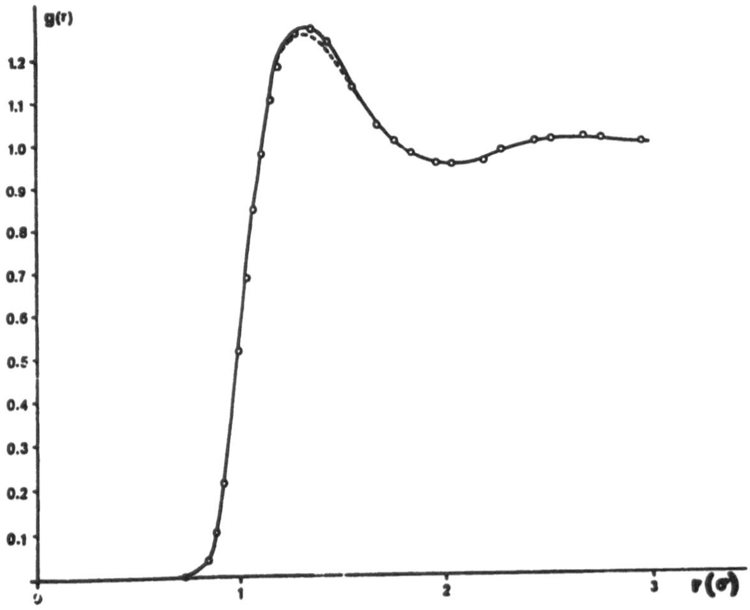

Fig.(2) The ^{4}He g(r) in HNC/α at the equilibrium density and the M.C. results (white circles). For the dashed line see fig.(1).

Another interesting quantity in ^{4}He is the one-body density matrix $\rho(r)$

$$\rho(r_{11'}) = \rho \int \Psi^*(1,..,A)\, \Psi(1',..,A)\, d\underline{r}_2..d\underline{r}_A \Big/ \int |\Psi|^2\, d\underline{r}_1..d\underline{r}_A\ , \tag{8}$$

and its Fourier transform, the momentum distribution n(k). The results obtained for $\rho(r)$, at the experimental ^{4}He equilibrium density ρ_0=0.02185 $\mathring{A}^{-3}$, are reported in fig.(3). In table (1), some properties of $\rho(r)$ are given, namely the zero momentum condensate fraction n_0 (n_0^{MC}=0.105±0.005), the density condition ($\rho(0)/\rho_0$=1) and the kinetic energy condition (K.E.= $1/(8\pi^3\rho_0)\ \hbar^2/(2m)\int k^2 n(k)\ d\underline{k}$). It must be noticed that, in the HNC/ α equations for $\rho(r)$, the fuction $\alpha(r)$ as derived in the energy calculation has been used.

	n_0	$\rho(0)/\rho_0$	K.E.(n)	K.E.(J.F.)
HNC	0.113	1.18	44.42	14.66
HNC/α	0.111	1.06	17.03	14.10

Table (1) Some properties of $\rho(r)$ for ^{4}He at the experimental equilibrium density (see text). K.E.(n) as computed via n(k), K.E.(J.F.) via the J.F. identity, both in K.

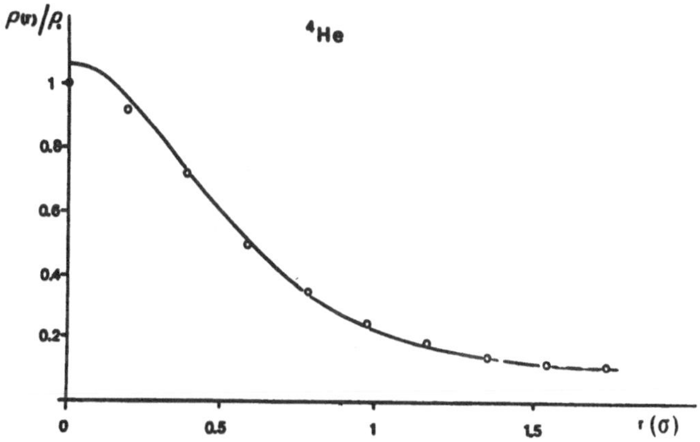

Fig.(3) Comparison of $\rho(r)/\rho_0$ for ^{4}He at the equilibrium density with the M.C. results.

The interpolating technique may be easily extended to Jastrow-Slater correlated Fermi systems(4). In fig.(4), the energies per particle obtained for liquid ^{3}He are shown. The correlation factors adopted have the form (6) with the parameter b chosen as in the MC calculations. Results are also given for the constant parameter case and for a fictitious mass three Bose model employed to derive the $\alpha(r)$

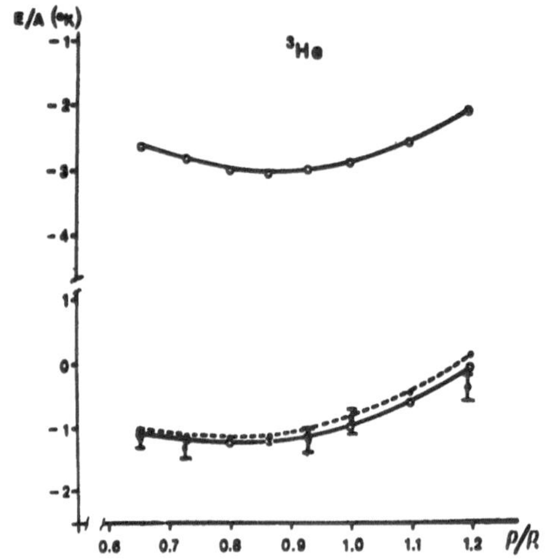

Fig.(4) ^{3}He energies per particle with the L.J. potential in HNC/α (lower solid line), constant interpolating parameter (dashed line) and Bose mass three model (upper solid line). Also shown is a M.C. estimate. $\rho_0(^3\text{He})=0.0164\ \text{A}^{-3}$.

function. Table (2) shows the energies for a system of spin aligned deuterium with two and three equally populated nuclear spin states ($D^{\downarrow}_{2,3}$). The potential of Kolos and Wolniewicz[6] and a correlation factor following the prescriptions

$\rho\sigma^3$	$D^{\downarrow}_2$	$D^{\downarrow}_3$	$D^{\downarrow}_3$(MC)
0.13	0.146	-0.018	-0.019 ± 0.010
0.14	0.137	-0.028	-0.040 ± 0.010
0.15	0.133	-0.033	$-0.039+0.010$
0.16	0.133	-0.034	-0.040 ± 0.012
0.17	0.139	-0.030	-0.035 ± 0.015
0.18	0.148	-0.021	

Table (2) Energies per particle (in K) for $D^{\downarrow}_{2,3}$. The MC results are from reference (8). σ=3.69 A.

of Pandharipande and Bethe[7] ($d=2.2\ r_o$ with $r_o=(4\pi\rho/3)^{-1/3}$) have been used.

3) Preliminary results for long-tailed correlations.

The results given for ^{4}He in section 2) are based on a short tailed correlation factor. This choice gives satisfactory results for the ground state energy but fails to reproduce some short wave-lenghts properties. It is known, from the Feynman theory[9], that the ^{4}He static structure factor $S(k)$ must be linear for small k values ($S(k)\cong\gamma k$ when $k\to 0$). The choice (6) does not reproduce this behavior. A Jastrow factor giving the best variational energy in the Jastrow correlated wave function space is consistent with that statement. To derive this factor, we have to solve the Euler-Lagrange (EL) equation

$$\delta(\langle H\rangle/A)/\ \delta g(r) = 0\ . \tag{9}$$

With the JF kinetic energy form, from eq.(9) we obtain the Schroedinger like equation[10,11]

$$-\hbar^2/m\ \ \nabla^2 g^{\frac{1}{2}}(r) + g^{\frac{1}{2}}(r)[\ V(r)+w_o(r)+w_E(r)\ ] = 0\ , \tag{10}$$

where $w_o(r)$ is the HNC induced potential and the elementary induced potential $w_E(r)$ is

$$w_E(r)= \hbar^2/(4m)\ \nabla^2 E(r) + \hbar^2/(4m)\int \nabla^2 g(s)\ \delta E(s)/\delta g(r)\ d\underline{s}\ . \tag{11}$$

When the LJ potential is adopted, the HNC solution ($w_E=0$) has the following behavior

$$S(k)\cong \gamma k + \beta k^3 + O(k^4) \tag{12}$$

if k goes to zero, and it satisfies the Feynmann condition. Because of the short range of the elementary diagrams, we expect that the full $g(r)$ also accomplishes this property.

The method here presented gives $E(r)$ as $E\{N(r)\}$, then we must examine the EL equation in this particular case. We must remember that the PY approximation also has this form, in fact $E_{PY}(r)=\ln[1+N(r)]-N(r)$.

We can rewrite w_E in the form

$$w_E(r)= \hbar^2/(4m)\ \nabla^2E(r) + \hbar^2/(4m)\int \nabla^2 g(s)\ \partial E\{N(s)\}/\partial N(s)\cdot F(t)\ d\underline{s} \qquad (13)$$

with $t=|\underline{r}-\underline{s}|$ and

$$F(x)= 1/(8\pi^3\rho\ )\int e^{-i\ \underline{k}\cdot\underline{r}}\ (S^2(k)-1)/S^2(k)\ d\underline{k}\ . \qquad (14)$$

In examining the PY approximation, the first term in the r.h.s. of eq.(13) has an r^{-6} behavior for large r-values. This feature is consistent with a solution as eq.(12). The second term in eq.(13) has an r^{-1} tail. This long range function is not compatible with the requested solution. The conclusion is that the PY approximation cannot give the correct S(k).

Unfortunately, eq.(4) has the same behaviors, therefore the interpolated form also does not treat in a correct way the terms arising from the variation of the elementary diagrams.

In table (3) some results obtained by the EL equation are given. In solving it, we have adopted two approximations: a) the elementary function has been derived by the short range solution and no more reevaluated, b) the elementary induced potential has been approximated by $w_E(r)=\hbar^2/(4m)\ \nabla^2E(r)$. In table (4) the values of S(k) for small k's are shown.

An attempt to overcome the previously described difficulties of the method may be done by noticing that the second term in the r.h.s. of eq.(11) corresponds to the sum of all the elementary diagrams containing once the $\nabla^2 g(r)$ function. It is possible to construct all the HNC and PY diagrams of that type and then interpolate to obtain an estimate of this new elementary function. By this method, if $S(k)=\gamma k$, then the second term of eq.(13) has an r^{-8} tail, consistently with the EL solution. This term may be source of rearrangement in the evaluated energy. Futhermore, we think that our results may have some little variation by evaluating E(r) in a consistent way with the EL solution. Work is in progress in estimating these two effects.

ρ/ρ_o	SR	EL
0.85	-5.84	-6.07
1.00	-5.67	-5.94
1.15	-4.97	-5.37

Table (3) ^{4}He energies per particle, in K, for the short range correlation factor (6) and for the optimal one with the LJ potential.

	S(k)		
$k(A^{-1})$	HNC	HNC/α	ref.(11)
0.05	0.020	0.018	0.018
0.10	0.038	0.035	0.036
0.15	0.056	0.050	0.054
0.20	0.072	0.063	0.071
0.25	0.087	0.077	0.088
0.30	0.101	0.089	0.103

Table (4) The structure factor of ^{4}He at $\rho=\rho_o$ from the HNC and HNC/α EL equations, also shown are results obtained by a parametrized g(r) with an estimate of the elementary diagrams of ref.(11).

REFERENCES.

1) K. Hiroike, J.of Phys.Soc.of Japan 12,326(1957).
2) A.Münster, "Statistical Thermodynamics", vol.I, N.Y.1974.
3) J.S.Rowlinson, Mol.Phys., 9,217(1965); D.D.Carley and F.Lado, Phys.Rev. A137,42(1965); T.Morita, Progr.Theor.Phys. 41,339(1969).
4) A.Fabrocini and S.Rosati, Nuovo Cimento D1,567(1982) and D1,615(1982).
5) D.Schiff and S.Verlet, Phys.Rev. 160,208(1967); M.H.Kalos, D.Levesque and S.Verlet, Phys.Rev. A9,2178(1974).
6) W.Kolos and L.Wolniewicz, J.Chem.Phys. 43,2429(1965); Chem.Phys.Lett. 24,457(1974).
7) V.R.Pandharipande and H.A.Bethe, Phys.Rev. C7,1312(1973).
8) R.M.Panoff, J.W.Clark, M.A.lee, K.E.Schmidt, M.H.Kalos and G.V.Chester, preprint 1982.
9) R.P.Feynman , Phys.Rev. 94,262(1954).
10) L.J.Lannto, A.D.Jackson and P.J.Siemens, Phys.Lett. 68B,311(1977).
11) R.A.Smith, A.Kallio, M.Puoskari and P.Toropainen, Nucl.Phys. A328,186(1979).

Third International Conference on Recent Progress in Many-Body Theories

Summary Talk *)

L. H. Nosanow
Division of Materials Research,
National Science Foundation,
Washington, D.C. 2o55o, U.S.A.

In the decade of the 195o's, several breakthroughs occured in the theory of systems composed of many particles. Among the foremost of these were: (1) the development of systematic perturbation theory, which received a major impetus from Brueckner's conjecture of the linked-cluster theorem and his introduction of the t-matrix, (2) the treatment of the ground states of the hard-sphere Bose and Fermi gases by Lee, Yang, and Huang, (3) the Bardeen-Cooper-Schrieffer theory of superconductivity, and (4) the Green's function approach introduced by Martin and Schwinger. Since that time, there have been many notable advances, along with a steady development of new concepts and improved techniques. Thus, Many-Body Theory has now become a very highly developed subject with an impressive body of knowledge. This conference has been devoted to an important part of the forefront of research in Many-Body Theory.

The subject matter covered by this conference can be classified under the following four major headings:

(1) systems of nucleons,
(2) quantum liquids,
(3) systems of electrons,
(4) techniques.

Each of these can, in turn, be classified into sub-headings, along with the last names of the speakers, whose talk fell into each category.
These are given in Tables I to IV. Although this classification scheme is clearly not unique, e.g., some authors talks fit under

*) The opinions expressed herein are those of the author and do not necessarily reflect the views of the National Science Foundation.

more than one sub-heading, it, nevertheless, serves as a useful outline of the activity of this conference.

Table I - Systems of Nucleons

A. Fundamental Approaches
   1. Quarks and QCD (Negele, Vary)
   2. Mesons (Coester, Muether, Zabolitzky)
   3. Three-body forces (Lejeune, Wiringa)

B. Nuclear Matter (Broad Range of Conditions)
   1. Phase diagram (Pethick)
   2. Heavy-ion collisions (Friman)
   3. Pi-zero condensation (Benhar, Takatsuka)
   4. Fermi liquid theory with tensor forces (Haensel)

C. Conventional Questions
   1. Effective mass (Mahaux)
   2. Correlations and deformations (Guardiola)
   3. Correlated pairs near the Fermi surface (Piechocki)

Table II - Quantum Liquids

A. "Old" Quantum Liquids
   1. Helium-3 (Clark and Krotscheck, Pines)
   2. Helium-4 (Lande, Pines)
   3. Helium surface (Miller)
   4. Monolayers (Bruch)
   5. Impurities (Kuerten, Szprynger)
   6. Droplets (Pieper)

B. "New" Quantum Liquids
   1. Liquid metallic hydrogen and deuterium (Ashcroft, Lantto)
   2. Spin-polarized helium-3 (Lhuillier, Bedell)
   3. Spin-polarized atomic hydrogen (Goldman)

Table III - Electron Systems

A. Dynamical Properties (Clark and Krotscheck, Green, Pines, Singwi)

B. Surfaces
   1. Metal surfaces (Woo)
   2. Phase transitions on surfaces (Vashishta)
   3. Two-dimensional electron-hole liquid (Chakraborty)

C. Quantum Fluids
   1. Liquid metallic hydrogen and deuterium (Ashcroft, Lantto)
   2. Superconductors (Rainer)
   3. Quantized Hall effect (extensive private discussions)

<u>Table IV - Techniques</u>

A. Correlated Basis Functions (Clark and Krotscheck, Fantoni, Sandler)
B. Coupled Cluster Method (Arponen, Bishop, Emrich, Monkhorst)
C. Perturbation Theory (de Llano, Lande, Kutzelnigg, Smith)
D. Green's Functions (Lande, Smith, Yu)
E. Density Functional Method (Ludena)
F. Monte Carlo Methods (Avishai, Lee)
G. Other (Fabrocini, Kallio, Reatto)

Perhaps, the area which received the greatest emphasis at this conference was that of technique. Indeed, the technical developments reported were impressive: For example, the correlated-basis-function approach has now been developed to the point that studies of excited states and dynamical properties can be undertaken. The coupled-cluster method, which has been so successfully applied to nuclear systems is now being applied to electron systems. A major factor in this technical development has been the concurrent development in computers. Techniques such as the Green's-Function-Monte-Carlo method have produced important results that can not be obtained by other methods. For example, exact results on the ground-state energy of the electron gas provided a standard against which many of the approximate calculations discussed at the conference could be judged. Finally, Green's-function-perturbation theory through the summation of parquet graphs has produced reasonable numerical results for the ground state of helium-four.

A number of promising directions emerged during the conference. These are classified in Table V.

<u>Table V - Trends and Opportunities</u>

A. Fundamental Approaches to Nucleon Systems Starting from Mesons or Quarks
B. Study of Nucleon Matter under Extreme Conditions (Astrophysics; Heavy-Ion Physics)
C. "New" Quantum Liquids
D. Surfaces and Interfaces

E. Technique Development
   1. Elementary excitation
   2. Dynamical properties
   3. Temperature greater than zero
F. Use of Advanced Computers.

One of the most exciting was the discussion of "new" quantum liquids. These systems can excist only under very special conditions and are currently being sought after or studied by many experimental groups. This development is part of a much broader trend in condensed matter research toward the preparation and study of new materials. An example of special interest is the two-dimensional electron gas which, because of the discovery of the fractional quantum Hall effect, is believed to be a new quantum fluid. The techniques discussed at this conference, in particular correlated basis functions, should be well suited for future studies of this problem.

Another area which has a bright future is that of surfaces and interfaces. These systems are complex theoretically for many reasons not the least of which are the loss of translational invariance and the diffuse nature of real surfaces. The density functional approach has yielded many interesting results on such systems; however, it is not known how to construct this functional exactly. Thus, the study of surfaces and interfaces using some of the other techniques discussed at this conference will undoubtedly yield most interesting results.

Although, perturbation theory yields a well-defined scheme to evaluate the properties of many-body systems at positive temperatures, the scheme is so complicated that numerical results have not been obtained. Further, attempts to extend the Green's-Function Monte-Carlo approach to finite temperature have not yet borne fruit. Therefore, the initial results on finite temperature properties of liquid helium reported at this conference are most welcome. This is an important problem worth of attention.

Finally, it is clear that much of the future work on many-body systems will involve large scale computational efforts. In particular, as many-body theorists learn to exploit the coming generation of computers with vector or parallel processing architectures, heretofore infeasible calculations will become feasible. In addition the combination of the various techniques discussed at this conference with this computer power will certainly yield much deeper insight into many-body theory.

List of Participants

Y. Akaishi, Department of Physics, Hokkaido University, Kita-10, Nishi-8, Sapporo 060, Japan

M. Alexanian, Centro de Investigacion, y Estudios Avanzados, J.P.N. Apto. Postal 14-740, 0700 Mexico, D.F., Mexico

J. Arponen, Research Institute for Theoretical Physics, University Helsinki, Siltavuorenpenger 2oc, oo17o Helsinki 17, Finland

N.W. Ashcroft, Laboratory of Atomic and Solid, State Physics, Clark Hall, Cornell University, Ithaca, New York 14853, U.S.A.

Y. Avishai, Physics Department, Ben-Gurion University, P.O.Box 653, Beer-Sheva 841o5, Israel

Cheng-guang Bao, Institute of High Energy Physics, P.O.Box 918, Beijing, China

K.S. Bedell, Department of Physics, State University of New York, Stony Brook, New York 11794, U.S.A., and NORDITA, Blegdamsvej 17, DK-21oo Copenhagen Ø, Denmark

O. Benhar, I.N.F.N., Sezione Sanita, Viale Regina Elena 299, I-oo161 Roma, Italy

W. Biem, Institut für Theoretische Physik, Universität Giessen, Heinrich-Buff-Ring 16, D-63oo Giessen, W.-Germany

R.F. Bishop, UMIST, Dept. of Mathematics, University of Manchester, P.O.Box 88, GB-Manchester M 6o 1QD, England, U.K.

W. Borrmann, Max-Planck-Institut für Festkörperforschung, Heisenbergstrasse 1, Postfach 8oo665, D-7ooo Stuttgart 8o, W.-Germany

Y. Brana, Prirodno-Matematicki Fakultet, V. Putnika 43, YU-71oo Sarajevo, Yugoslavia

L.W. Bruch, Department of Physics, University of Wisconsin, 115o University Ave., Madison, Wi. 537o6 U.S.A.

C.E. Campbell, School of Physics and Astronomy, University of Minnesota, 116 Church Street S.E., Minneapolis, Minn. 55 455, U.S.A.

T. Chakraborty, School of Physics and Astronomy, University of Minnesota, 116 Church Street S.E., Minneapolis, Minn. 55 455, U.S.A.

J.W. Clark, Department of Physics, Washington University, St. Louis, Mo. 6313o, U.S.A.

F. Coester, Physics Division, Bldg. 2o3, Argonne National Laboratory, 97oo South Cass Ave., Argonne, Ill. 6o 439, U.S.A.

K. Emrich, Institut für Theoretische Physik II, Ruhr-Universität Bochum, Postfach 1o2148, D-463o Bochum, W.-Germany

A.R. Engelmann, Uppsala University, Department of Quantum Chemistry, Box 518, S-751 2o Uppsala, Sweden

P. Entel, Theoretische Physik FB 1o, Universität-Gesamthochschule Duisburg, Postfach 1o1619, D-41oo Duisburg 1, W.-Germany

A. Fabrocini, Istituto di Fisica, Universita di Pisa, Piazza Torricelli 2, I-561oo Pisa, Italy

S. Fantoni, Istituto di Fisica, Universita di Pisa, Piazza Torricelli 2, I-561oo Pisa, Italy

P. Fazekas, Institut für Theoretische Physik, Universität zu Köln, Zülpicherstraße 77, D-5ooo Köln 41, W.-Germany

M.F. Flynn, Department of Physics, Washington University, St. Louis, Missouri 6313o, U.S.A.

B. Friman, Dept. of Physics, Abo Akademi, Porthansgatan 3-5, SF-2o 5oo Abo 5o, Finland

W. Gasser, Theoretische Physik T 3o, Physik-Department, TU München, James-Franck-Straße, D-8o46 Garching, W.-Germany

V. Goldman, Natuurkundig Laboratorium, Universiteit van Amsterdam, Valckenierstr. 65, NL-1o18 XE Amsterdam, The Netherlands

S. Goulart Rosa, JR. Universidade de Sao Paulo, Instituto de Fisica e Quimica de Sao Carlos, Caixa Postal 369, 1356o Sao Carlos S.P., Brazil

F. Green, School of Physics, The University of New South Wales, P.O.Box 1, Kensington, N.S.W. 2o33, Australia

M. Grypeos, Department of Theoretical Physics, University of Thessaloniki, Thessaloniki, Greece

J. Gspann, Institut für Kernverfahrenstechnik, Universität Karlsruhe, Postfach 364o, D-75oo Karlsruhe, W.-Germany

R. Guardiola, Departamento Fisica Nuclear, Facultad de Sciencias, Universidad de Granada, E-Granada, Spain

P. Haensel, Polish Academy of Sciences, N. Copernicus Astronomical Center, ul. Bartycka 18, PL-00-716 Warszawa, Poland

M.J. Haftel, Naval Research Laboratory, Washington, D.C. 2o375, U.S.A.

E.F. Hefter, Institut für Theoretische Physik, Universität Hannover, Appelstraße 2, 3ooo Hannover 1, W.-Germany

O. Hipolito, Universidade de Sao Paulo, Departamento de Fisica, Sao Carlos, 1356o Sao Carlos S.P., Brazil

P. Horsch, Max-Planck-Institut für Festkörperforschung, Heisenbergstr. 1, Postfach 8o o6 65, D-7ooo Stuttgart 8o, W.-Germany

Chen-Shiung Hsue, Institute of Physics, National Tsing Hua University, 855 Kuang Fu Road, Hsinchu, Taiwan, China

A. Huber, Institut für Theoretische Physik, Universität Kiel, Physikzentrum, Leibnizstraße, D-23oo Kiel 1, W.-Germany

S. Huber, Beaver College, Department of Physics, Glenside, Pa.19o38, U.S.A.

P. Huguenin, Institut de Physique, Rue Breguet 1, CH-2ooo Neuchatel, Switzerland

A. Kallio, University of Oulu, Dept. of Theoretical Physics, Linnanmaa, SF-9o 57o, Oulu 57, Finland

T.A. Kaplan, Department of Physics and Astronomy, Michigan State University , East Lansing, Mi. 48 842, U.S.A.

T. Katayama, Institut für Theoretische Physik, Universität Hannover, Appelstraße 2, D-3ooo Hannover 1, W.-Germany

I.M. Khalatnikov, Academy of Sciences, Landau Institute for Theoretical Physics, 11794o Moscow B-334, USSR

S. Klic, Fakultet Gradevinski Znanosti, 58oo Split, Yugoslavia

R. Kishore, Instituto de Pesquisas Espaciais-INPE, S.P. 515, 12 2oo S.J. Campos, S.P. Brazil

E. Krotscheck, Max-Planck-Institut für Kernphysik, Postfach 1o398o, D-69oo Heidelberg, W.-Germany

H. Kümmel, Institut für Theoretische Physik II, Ruhr-Universität Bochum, Postfach 1o2148, 463o Bochum, W.-Germany

K.E. Kürten, Courant Institute of Mathematical Sciences, New York University, 251 Mercer Street, New York, N.Y. 1oo21, U.S.A.

W. Kutzelnigg, Institut für Theoretische Chemie, Ruhr-Universität Bochum, Postfach 1o2148, D-463o Bochum 1, W.-Germany

Nai-Hang Kwong, Max Planck-Institut für Kernphysik, Postfach 1o398o, D-69oo Heidelberg, W.-Germany

A. Lande, Institute for Theoretical Physics, University of Groningen, P.O.Box 8oo W.S.N., NL-Groningen, The Netherlands

L. Lantto, University of Oulu, Dept. of Theoretical Physics, Linnanmaa, SF-9o 57o Oulu 57, Finland

L.F. Lathouwers, University of Antwerp (RUCA), Groenenborgerlaan 171, B-2o2o Antwerp, Belgium

M.A. Lee, Department of Physics, Kent State University, Kent, Ohio 44242, U.S.A.

A. Lejeune, Université de Liège, Physique nucleaire theorique B5, Sart Tilman, B-4ooo Liège 1, Belgium

C. Lhuillier, Laboratoire de Spectroscopie, Hertzienne de l'E.N.S., 24 rue Lhomond, F-75 231 Paris Cedex o5, France

T.K. Lim, Department of Physics, Drexel University, Philadelphia, Pa 191o4, U.S.A.

W. von der Linden, Max-Planck-Institut für Festkörperforschung, Heisenbergstraße 1, Postfach 8oo665, 7ooo Stuttgart, W.-Germany

M. de Llano, Instituto de Fisica, Universidad Nacional, Autonoma de Mexico, Apto. Postal 2o/364, Mexico 2o, D.F., Mexico

E.V. Ludena, Instituto Venezolano de Investigaciones cientificas Quimica, Apto. 1827, Caracas, 1o1o-A, Venezuela

W.D. Lukas, Max-Planck-Institut für Festkörperforschung, Heisenbergstr. 1, Postfach 8oo665, 7ooo Stuttgart 8o, W.-Germany

W. Macke, Institut für Theoretische Physik, Johannes Kepler-Universität, Bachlbergweg 77, A-4o4o Linz, Austria

C. Mahaux, Institut de Physique B5, Université de Liège, Sart-Tilman, B-4ooo Liège 1, Belgium

K. Maschke, Institut de Physique Appliquée, Ecole Politechnique Federale, PHB-Ecublens, CH-1o15 Lausanne, Switzerland

E. Mavrommatis, Tandem Accelerator Laboratory, Nuclear Research Center Democritos, Aghia Paraskevi, Attiki, Greece

H. Miesenböck, Institut für Theoretische Physik, Johannes Kepler-Universität Linz, Bachlbergweg 77, A-4o4o Linz, Austria

M. Miller, Department of Physics, Washington State University, Pullman, WA 99164, U.S.A.

H. Mimura, Chijoda Institute of Technology, 5-3o Shitaya - 1, Taito-ku, Tokyo 11o, Japan

V.K. Mishra, Nordita, Blegdamsvej 17, DK-21oo Copenhagen-Ø, Denmark

R. Mittet, Fysisk Institutt, NLHT, Universitetet i Trondheim, N-7o55 Dragvoll, Norway

H.J. Monkhorst, Quantum Theory Project, University of Florida, Williamson Hall 36, Gainesville, Florida 32 611, U.S.A.

H. Moraal, Institut für Theoretische Physik, Universität zu Köln, Zülpicher Straße 77, D-5ooo Köln 41, W.-Germany

S.A. Moszkowski, Physics Department, UCLA, Los Angeles, Ca.9oo24, U.S.A.

H. Müther, Institut für Theoretische Physik, Universität Tübingen, Auf der Morgenstelle 14, D-74oo Tübingen 1, W.-Germany

V. Mujica, Quantum Chemistry Group, University of Uppsala, Box 518, S-751 2o Uppsala, Sweden

S. Nakaichi-Maeda, Institut für Theoretische Physik, Universität Tübingen, Auf der Morgenstelle 14, D-74oo Tübingen, W.-Germany

J.W. Negele, 6-3o8 Massachusetts Institute of Technology, Cambridge, Ma. o2 139, U.S.A.

J. Nitsch, Institut für Theoretische Physik, Universität zu Köln, Zülpicher Straße 77, D-5ooo Köln 41, W.-Germany

L. Nosanow, Division of Materials Research, National Science Foundation, 18oo G.Str. N.W., Washington D.C. 2o55o, U.S.A.

E. Pajanne, Research Institute for Theoretical Physics, Siltavuorenpenger 2oc, SF-oo17o Helsinki 17, Finland

J.K. Percus, Courant Institute of Mathematical Sciences, New York University, 251 Mercer Str., New York, N.Y. 1oo12, U.S.A.

C.J. Pethick, Nordita, Blegdamsvej 17, DK-21oo Copenhagen Ø, Denmark

M. Pfitzner, Physik Department, TU München, James-Franck-Str., D-8o46 Garching, W.-Germany

W. Piechocki, Institute of Nuclear Research, ul. Hoza 69, PL-OO-681, Warszawa, Poland

S.C. Pieper, Argonne National Laboratory, Physics Building 2o3, Argonne, Ill. 6o439, U.S.A.

D. Pines, Department of Physics, University of Illinois, 111o W.Green Str., Urbana, Ill. 618o1, U.S.A.

A. Polls, Departamento Fisica Nuclear, Universidad de Granada, E-Granada, Spain

D. Rainer, Physikalisches Institut, Universität Bayreuth, Postfach 3oo8, 858o Bayreuth, W.-Germany

L. Reatto, Istituto di Fisica, Universita di Parma, Via M. D'azeglio 85, I-Parma, Italy

M.L. Ristig, Institut für Theoretische Physik, Universität zu Köln, Zülpicher Str. 77, D-5ooo Köln 41, W.-Germany

M. Roger, CEA, DPh SPSRM, Orme des Merisiers, CEA-CENS, F-91191 Gif sur Yvette Cedex, France

S. Rosati, Istituto di Fisica, Universita di Pisa, Piazza Torricelli 2, I-561oo Pisa, Italy

M. Saarela, University of Oulu, Department of Theoretical Physics, Linnanmaa, SF-9o57o Oulu 57, Finland

D.G. Sandler, Department of Physics, University of Illinois, Urbana-Champaign, 111o W.Green Street, Urbana, Ill. 618o1, U.S.A.

R. Sartor, Institut de Physique B5, Université de Liège, Sart Tilman, B-4ooo Liège 1, Belgium

D. Schütte, Institut für Theoretische Kernphysik, Universität Bonn, Nussallee 14-16, D-53oo Bonn, W.-Germany

N. Schulz, Institut für Theoretische Physik Universität zu Köln, Zülpicher Str. 77, D-5ooo Köln 41, W.-Germany

V.F. Sears, Atomic Energy of Canada Limited, Chalk River, Ontario, Canada, KOJ 1JO, Canada

G. Senger, Institut für Theoretische Physik, Universität zu Köln, Zülpicher Straße 77, D-5000 Köln 41, W.-Germany

K.S. Singwi, Department of Physics, Northwestern University, Evanston, Ill. 60201, U.S.A.

R.A. Smith, Department of Physics, Texas A and M University, College Station, Texas 77843, U.S.A.

S. Sunaric, Univerzitet Dzemal Bijedic u Mostaru, Masinski Fakultet Mostar, YU-7900 Mostar, Yugoslavia

A. Szprynger, Institute of Low Temperature and Structure Research, Polish Academy of Sciences, P.O.Box 937, PL-50-950 Wroclaw, Poland

L. Szybisz, Institut für Theoretische Physik, Universität zu Köln, Zülpicher Str. 77, D-5000 Köln 41, W.-Germany

T. Takatsuka, College of Humanities and Social Sciences, Iwate University, Morioka 020, Japan

H. Tanaka, Department of Physics, Faculty of Science, Hokkaido University, Kita 10, Nishi 8 Kitaku, Sapporo 060, Japan

Z. Tesanovic, School of Physics and Astronomy, University of Minnesota, 116 Church Street S.E., Minneapolis, Minnesota 55 455, U.S.A.

J.P. Vary, Physics Department, Ames Laboratory, Iowa State University, Ames, Iowa 50011, U.S.A.

P. Vashishta, Solid State Sciences Division, Argonne National Laboratory, Argonne, Ill. 60 439, U.S.A.

P. Vasilopoulos, Max-Planck-Institut für Festkörperforschung, Heisenbergstraße 1, Postfach 80 06 65, D-7000 Stuttgart 80, W.-Germany

M. Viviani, Istituto di Fisica, Universita di Pisa, Piazza Torricelli 8, I-56100 Pisa, Italy

I.T.M. Walraven, Natuurkundig Laboratorium der Universiteit Amsterdam, Valkenierstraat 65, NL-1018 XE Amsterdam, The Netherlands

L. Wiesenfeld, C.N.R.S. - Ecole Normale Superieure, Lab. Spectrs. Hertzienne, 24 rue Lhomond, F-75 231 Paris Cedex 05, France

R.B. Wiringa, Physics Building 203, Argonne National Laboratory, Argonne, Ill. 60439, U.S.A.

W. Wolff, Institut für Theoretische Physik, Universität zu Köln, Zülpicher Str. 77, D-5000 Köln 41, W.-Germany

R.G. Wolff, Max-Planck-Institut für Physik und Astronomy, Karl-Schwarzschild-Str. 1, D-8046 Garching, W.-Germany

Chia-Wei Woo, University of California, San Diego and San Francisco State University, Office of the President, 1600 Holloway Avenue, San Francisco, CA 94132, U.S.A.

Chung-en Wu, Institute of High Energy Physics, P.O.Box 918, Beijing, China

Lu Yu, Institute of Theoretical Physics, Academia Sinica, P.O.Box 2735, Beijing, China

J.G. Zabolitzky, Institut für Theoretische Physik, Universität zu Köln, Zülpicher Str. 77, D-5000 Köln 41, W.-Germany

J.S. Zmuidzinas Jet Propulsion Laboratory, California Institute of Technology, 4800 Oak Grove Drive, Pasadena, CA 91109, U.S.A.

A. Zuker, Lab.Phys.Theorique, CNRS,Strasbourg, B.P.20, F-67037 Strasbourg Cedex, France

Texts and Monographs in Physics

J. Kessler: **Polarized Electrons.** 1976.

W. Rindler: **Essential Relativity: Special, General, and Cosmological.** Revised Second Edition. 1977.

K. Chadan, P. C. Sabatier: **Inverse Problems in Quantum Scattering Theory.** 1977.

C. Truesdell, S. Bharatha: **The Concepts and Logic of Classical Thermodynamics as a Theory of Heat Engines: Rigourously Constructed upon the Foundation Laid by S. Carnot and F. Reech.** 1977.

R. D. Richtmyer: **Principles of Advanced Mathematical Physics.** Volume I. 1978. Volume II. 1981.

R. M. Santilli: **Foundations of Theoretical Mechanics.** Volume I: The Inverse Problem in Newtonian Mechanics. 1978. Volume II: Birkhoffian Generalization of Hamiltonian Mechanics. 1983.

A. Böhm: **Quantum Mechanics.** 1979.

H. Pilkuhn: **Relativistic Particle Physics.** 1979.

M. D. Scadron: **Advanced Quantum Theory and Its Applications Through Feynman Diagrams.** 1979.

O. Bratteli, D. W. Robinson: **Operator Algebras and Quantum Statistical Mechanics.** Volume I: C*-and W*-Algebras. Symmetry Groups. Decomposition of States. 1979. Volume II: Equilibrium States. Models in Quantum Statistical Mechanics. 1981.

J. M. Jauch, F. Rohrlich: **The Theory of Photons and Electrons: The Relativistic Quantum Field Theory of Charged Particles with Spin One-half.** Second Expanded Edition. 1980.

P. Ring, P. Schuck: **The Nuclear Many-Body Problem.** 1980.

R. Bass: **Nuclear Reactions with Heavy Ions.** 1980.

R. G. Newton: **Scattering Theory of Waves and Particles.** Second Edition. 1982.

G. Ludwig: **Foundations of Quantum Mechanics I.** 1983.

G. Gallavotti: **The Elements of Mechanics.** 1983.

F. J. Yndurain: **Quantum Chromodynamics: An Introduction to the Theory of Quarks and Gluons.** 1983.

Springer-Verlag
Berlin
Heidelberg
New York
Tokyo

Lecture Notes in Physics

Vol. 173: Stochastic Processes in Quantum Theory and Statistical Physics. Proceedings, 1981. Edited by S. Albeverio, Ph. Combe, and M. Sirugue-Collin. VIII, 337 pages. 1982.

Vol. 174: A. Kadić, D.G.B. Edelen, A Gauge Theory of Dislocations and Disclinations. VII, 290 pages. 1983.

Vol. 175: Defect Complexes in Semiconductor Structures. Proceedings, 1982. Edited by J. Giber, F. Beleznay, J. C. Szép, and J. László. VI, 308 pages. 1983.

Vol. 176: Gauge Theory and Gravitation. Proceedings, 1982. Edited by K. Kikkawa, N. Nakanishi, and H. Nariai. X, 316 pages. 1983.

Vol. 177: Application of High Magnetic Fields in Semiconductor Physics. Proceedings, 1982. Edited by G. Landwehr. XII, 552 pages. 1983.

Vol. 178: Detectors in Heavy-Ion Reactions. Proceedings, 1982. Edited by W. von Oertzen. VIII, 258 pages. 1983.

Vol. 179: Dynamical Systems and Chaos. Proceedings, 1982. Edited by L. Garrido. XIV, 298 pages. 1983.

Vol. 180: Group Theoretical Methods in Physics. Proceedings, 1982. Edited by M. Serdaroğlu and E. İnönü. XI, 569 pages. 1983.

Vol. 181: Gauge Theories of the Eighties. Proceedings, 1982. Edited by R. Raitio and J. Lindfors. V, 644 pages. 1983.

Vol. 182: Laser Physics. Proceedings, 1983. Edited by J. D. Harvey and D. F. Walls. V, 263 pages. 1983.

Vol. 183: J. D. Gunton, M. Droz, Introduction to the Theory of Metastable and Unstable States. VI, 140 pages. 1983.

Vol. 184: Stochastic Processes – Formalism and Applications. Proceedings, 1982. Edited by G. S. Agarwal and S. Dattagupta. VI, 324 pages. 1983.

Vol. 185: H. N. Shirer, R. Wells, Mathematical Structure of the Singularities at the Transitions between Steady States in Hydrodynamic Systems. XI, 276 pages. 1983.

Vol. 186: Critical Phenomena. Proceedings, 1982. Edited by F.J.W. Hahne. VII, 353 pages. 1983.

Vol. 187: Density Functional Theory. Edited by J. Keller and J.L. Gázquez. V, 301 pages. 1983.

Vol. 188: A. P. Balachandran, G. Marmo, B.-S. Skagerstam, A. Stern, Gauge Symmetries and Fibre Bundles. IV, 140 pages. 1983.

Vol. 189: Nonlinear Phenomena. Proceedings, 1982. Edited by K. B. Wolf. XII, 453 pages. 1983.

Vol. 190: K. Kraus, States, Effects, and Operations. Edited by A. Böhm, J. W. Dollard and W. H. Wootters. IX, 151 pages. 1983.

Vol. 191: Photon Photon Collisions. Proceedings, 1983. Edited by Ch. Berger. V, 417 pages. 1983.

Vol. 192: Heidelberg Colloquium on Spin Glasses. Proceedings, 1983. Edited by J. L. van Hemmen and I. Morgenstern. VII, 356 pages. 1983.

Vol. 193: Cool Stars, Stellar Systems, and the Sun. Proceedings, 1983. Edited by S. L. Balliunas and L. Hartmann. VII, 364 pages. 1984.

Vol. 194: P. Pascual, R. Tarrach, QCD: Renormalization for the Practitioner. V, 277 pages. 1984.

Vol. 195: Trends and Applications of Pure Mathematics to Mechanics. Proceedings, 1983. Edited by P. G. Ciarlet and M. Roseau. V, 422 pages. 1984.

Vol. 196: WOPPLOT 83. Parallel Processing: Logic, Organization and Technology. Proceedings, 1983. Edited by J. Becker and I. Eisele. V, 189 pages. 1984.

Vol. 197: Quarks and Nuclear Structure. Proceedings, 1983. Edited by K. Bleuler. VIII, 414 pages. 1984.

Vol. 198: Recent Progress in Many-Body Theories. Proceedings, 1983. Edited by H. Kümmel and M. L. Ristig. IX, 422 pages. 1984.

Selected Issues from

Lecture Notes in Mathematics

Vol. 836: Differential Geometrical Methods in Mathematical Physics. Proceedings, 1979. Edited by P. L. García, A. Pérez-Rendón, and J. M. Souriau. XII, 538 pages. 1980.

Vol. 837: J. Meixner, F. W. Schäfke and G. Wolf, Mathieu Functions and Spheroidal Functions and their Mathematical Foundations Further Studies. VII, 126 pages. 1980.

Vol. 838: Global Differential Geometry and Global Analysis. Proceedings 1979. Edited by D. Ferus et al. XI, 299 pages. 1981.

Vol. 840: D. Henry, Geometric Theory of Semilinear Parabolic Equations. IV, 348 pages. 1981.

Vol. 841: A. Haraux, Nonlinear Evolution Equations- Global Behaviour of Solutions. XII, 313 pages. 1981.

Vol. 842: Séminaire Bourbaki vol. 1979/80. Exposés 543-560. IV, 317 pages. 1981.

Vol. 843: Functional Analysis, Holomorphy, and Approximation Theory. Proceedings. Edited by S. Machado. VI, 636 pages. 1981.

Vol. 845: A. Tannenbaum, Invariance and System Theory: Algebraic and Geometric Aspects. X, 161 pages. 1981.

Vol. 849: P. Major, Multiple Wiener-Itô Integrals. VII, 127 pages. 1981.

Vol. 851: Stochastic Integrals. Proceedings, 1980. Edited by D. Williams. IX, 540 pages. 1981.

Vol. 852: L. Schwartz, Geometry and Probability in Banach Spaces. X, 101 pages. 1981.

Vol. 856: R. Lascar, Propagation des Singularités des Solutions d'Equations Pseudo-Différentielles à Caractéristiques de Multiplicités Variables. VIII, 237 pages. 1981.

Vol. 858: E. A. Coddington, H. S. V. de Snoo: Regular Boundary Value Problems Associated with Pairs of Ordinary Differential Expressions. V, 225 pages. 1981.

Vol. 861: Analytical Methods in Probability Theory. Proceedings 1980. Edited by D. Dugué, E. Lukacs, V. K. Rohatgi. X, 183 pages. 1981.

Vol. 866: J.-M. Bismut, Mécanique Aléatoire. XVI, 563 pages. 1981.

Vol. 878: Numerical Solution of Nonlinear Equations. Proceedings, 1980. Edited by E. L. Allgower, K. Glashoff, and H.-O. Peitgen. XIV, 440 pages. 1981.

Vol. 881: R. Lutz, M. Goze, Nonstandard Analysis. XIV, 261 pages. 1981.

Vol. 888: Padé Approximation and its Applications. Proceedings, 1980. Edited by M. G. de Bruin and H. van Rossum. VI, 383 pages. 1981.

Vol. 898: Dynamical Systems and Turbulence, Warwick, 1980. Proceedings. Edited by D. Rand and L.-S. Young. VI, 390 pages. 1981.

Vol. 901: Séminaire Bourbaki vol. 1980/81 Exposés 561–578. III, 299 pages. 1981.

Vol. 904: K. Donner, Extension of Positive Operators and Korovkin Theorems. XII, 182 pages. 1982.

Vol. 905: Differential Geometric Methods in Mathematical Physics. Proceedings, 1980. Edited by S.J. Andersson, H.-D. Doebner, and. H.R. Petry. VI, 309 pages. 1982.

Vol. 909: Numerical Analysis. Proceedings, 1981. Edited by J.P. Hennart. VII, 247 pages. 1982.

Vol. 912: Numerical Analysis. Proceedings, 1981. Edited by G. A. Watson. XIII, 245 pages. 1982.

Vol. 920: Séminaire de Probabilités XVI, 1980/81. Proceedings. Edité par J. Azéma et M. Yor. V, 622 pages. 1982.

Vol. 921: Séminaire de Probabilités XVI, 1980–81 Supplément: Géométrie Différentielle Stochastique. Proceedings. Edité par J. Azéma et M. Yor. III, 285 pages. 1982.

Vol. 922: B. Dacorogna, Weak Continuity and Weak Lower Semicontinuity of Non-Linear Functionals. V, 120 pages. 1982.

Vol. 923: Functional Analysis in Markov Processes. Proceedings, 1981. Edited by M. Fukushima. V, 307 pages. 1982.

Vol. 926: Geometric Techniques in Gauge Theories. Proceedings, 1981. Edited by R. Martini and E.M.de Jager. IX 219 pages. 1982.

Vol. 927: Y. Z. Flicker, The Trace Formula and Base Change for GL (3). XII, 204 pages. 1982.

Vol. 928: Probability Measures on Groups. Proceedings 1981. Edited by H. Heyer. X, 477 pages. 1982.

Vol. 929: Ecole d'Eté de Probabilités de Saint-Flour X - 1980. Proceedings, 1980. Edited by P.L. Hennequin. X, 313 pages. 1982.

Vol. 930: P. Berthelot, L. Breen, et W. Messing, Théorie de Dieudonné Cristalline II. XI, 261 pages. 1982.

Vol. 931: D.M. Arnold, Finite Rank Torsion Free Abelian Groups and Rings. VII, 191 pages. 1982.

Vol. 932: Analytic Theory of Continued Fractions. Proceedings, 1981. Edited by W.B. Jones, W.J. Thron, and H. Waadeland. VI, 240 pages. 1982.

Vol. 934: M. Sakai, Quadrature Domains. IV, 133 pages. 1982.

Vol. 935: R. Sot, Simple Morphisms in Algebraic Geometry. IV, 146 pages. 1982.

Vol. 936: S.M. Khaleelulla, Counterexamples in Topological Vector Spaces. XXI, 179 pages. 1982.

Vol. 937: E. Combet, Intégrales Exponentielles. VIII, 114 pages. 1982.

Vol. 938: Number Theory. Proceedings, 1981. Edited by K. Alladi. IX, 177 pages. 1982.

Vol. 942: Theory and Applications of Singular Perturbations. Proceedings, 1981. Edited by W. Eckhaus and E.M. de Jager. V, 363 pages. 1982.

Vol. 953: Iterative Solution of Nonlinear Systems of Equations. Proceedings, 1982. Edited by R. Ansorge, Th. Meis, and W. Törnig. VII, 202 pages. 1982.

Vol. 956: Group Actions and Vector Fields. Proceedings, 1981. Edited by J.B. Carrell. V, 144 pages. 1982.

Vol. 957: Differential Equations. Proceedings, 1981. Edited by D.G. de Figueiredo. VIII, 301 pages. 1982.

Vol. 963: R. Nottrot, Optimal Processes on Manifolds. VI, 124 pages. 1982.

Vol. 964: Ordinary and Partial Differential Equations. Proceedings, 1982. Edited by W.N. Everitt and B.D. Sleeman. XVIII, 726 pages. 1982.

Vol. 968: Numerical Integration of Differential Equations and Large Linear Systems. Proceedings, 1980. Edited by J. Hinze. VI, 412 pages. 1982.

Vol. 970: Twistor Geometry and Non-Linear Systems. Proceedings, 1980. Edited by H.-D. Doebner and T.D. Palev. V, 216 pages. 1982.

Vol. 972: Nonlinear Filtering and Stochastic Control. Proceedings, 1981. Edited by S.K. Mitter and A. Moro. VIII, 297 pages. 1983.

Vol. 978: J. Ławrynowicz, J. Krzyż, Quasiconformal Mappings in the Plane. VI, 177 pages. 1983.

Vol. 979: Mathematical Theories of Optimization. Proceedings, 1981. Edited by J.P. Cecconi and T. Zolezzi. V, 268 pages. 1983.